Fundamental Theories of Physics

Volume 198

The international monograph series "Fundamental Theories of Physics" aims to stretch the boundaries of mainstream physics by clarifying and developing the theoretical and conceptual framework of physics and by applying it to a wide range of interdisciplinary scientific fields. Original contributions in well-established fields such as Quantum Physics, Relativity Theory, Cosmology, Quantum Field Theory, Statistical Mechanics and Nonlinear Dynamics are welcome. The series also provides a forum for non-conventional approaches to these fields. Publications should present new and promising ideas, with prospects for their further development, and carefully show how they connect to conventional views of the topic. Although the aim of this series is to go beyond established mainstream physics, a high profile and open-minded Editorial Board will evaluate all contributions carefully to ensure a high scientific standard.

More information about this series at http://www.springer.com/series/6001

Valia Allori · Angelo Bassi · Detlef Dürr ·
Nino Zanghì
Editors

Do Wave Functions Jump?

Perspectives of the Work of GianCarlo Ghirardi

Springer

Editors
Valia Allori
Department of Philosophy
Northern Illinois University
Naperville, IL, USA

Angelo Bassi
Department of Physics
University of Trieste
Trieste, Italy

Detlef Dürr
Mathematisches Institut
LMU
Munich, Germany

Nino Zanghi
Dipartimento di Fisica INFN-Sezion
Universitá di Genova
Genoa, Italy

ISSN 0168-1222 ISSN 2365-6425 (electronic)
Fundamental Theories of Physics
ISBN 978-3-030-46776-0 ISBN 978-3-030-46777-7 (eBook)
https://doi.org/10.1007/978-3-030-46777-7

This Springer imprint is published by the registered company Springer Nature Switzerland AG
The registered company address is: Gewerbestrasse 11, 6330 Cham, Switzerland

Preface

On 1 June 2018, GianCarlo Ghirardi died unexpectedly in his summer resort home in Grado, Italy. He was born in Milan on 28 October 1935. He had maintained lifelong loyalty to Trieste where he became professor in 1959. Trieste was also home to the internationally renowned ICTP—a centre of science and education which allowed him to host many scientists in his field of research. Many of the contributors to this volume came to visit him and the beautiful coastal landscape which he treasured so much.

His contributions to the foundations of quantum physics gave him a worldwide reputation of excellence. He maintained a close and mutually enriching friendship with John Stewart Bell which ended with Bell's death in 1990. It was Bell who coined the acronym "GRW theory" for the theory which Ghirardi had invented with his long time collaborators Alberto Rimini and Tullio Weber: a theory of physically induced wave function collapse—of wave function jumps—and John Bell was the first to catch on.

This book in honour of the philosopher-physicist GianCarlo Ghirardi is not a "that was your life" review. We, the editors, who were among GianCarlo's closest friends and collaborators, felt that GianCarlo would have much preferred a collection of contributions by leading scientists and "youngsters" whose research has been influenced or in some way triggered by his own writings. Since we mention "youngsters", GianCarlo spent much of his time teaching young researchers what quantum theory ought not to be about, namely "observers". GRW theory is a "quantum theory without observers", a notion also coined by John Bell. The conferences and workshops organized with GianCarlo were always structured as platforms for young researchers to meet, learn about, and discuss the modern foundations of quantum physics. It is therefore no surprise that the contributors to this volume range from young researchers to established senior scientists.

When we learned of GianCarlo's death, we suggested to Springer editor Angela Lahee, who was well acquainted with GianCarlo, that it would be interesting to publish an honorary volume with hand-selected original contributions, spanning a wide range of topics, including philosophy, physics, and mathematical physics, and of a rigorous, speculative, or prospective nature. She agreed at once. The overwhelming response to our invitations for contributions was, however, unexpected. For easier access, we decided on a rough categorization of the contributions, although we were well aware that no categorization would ever be perfect: Part I *History and Honour*, Part II *Philosophy*, Part III *Mathematical Physics*, Part IV *Theoretical Physics*, and Part V *Experimental Physics*.

Let us say a few words about each of the chapters in each part. The order is always alphabetical.

Part I: History and Honour

In "GianCarlo Ghirardi: Passing the Torch on Collapse Models", **Stephen L. Adler** (IAS Princeton) briefly reviews his acquaintance with GianCarlo, which led to a fruitful collaboration with Angelo Bassi, a former Ph.D. student of GianCarlo.

The nonlocality of nature, discovered by John Bell (it's not so easy to get that discovery across to people), was always one of GianCarlo's concerns, because of the tension it creates with relativity. In "EPR–Bell–Schrödinger Proof of Nonlocality Using Position and Momentum", **Jean Bricmont** (Louvain la Neuve) and **Sheldon Goldstein** (Rutgers, NJ.) deliver a new and surprising proof of nonlocality using the original variables considered by Einstein, Podolsky, and Rosen in their seminal paper.

GianCarlo had an excellent understanding of the nature of statistical reasoning in physics, based on Ludwig Boltzmann's insights. Actually on my (D.D.'s) first visit to the ICTP, GianCarlo informed me that, Boltzmann had committed suicide nearby, in Duino, in 1906. Even now, Boltzmann's statistical reasoning is sometimes questioned. In "Typicality in the Foundations of Statistical Physics and Born's rule", **Detlef Dürr** (LMU) and **Ward Struyve** (KU-Leuven) repeat once again the essential insights of that reasoning, which go hand in hand with the role of chance in physics and in particular in quantum physics.

GRW theory is but one of many possible collapse theories which were invented in parallel, for example by Lajos Diosi. While GRW theory localizes the wave function in a discrete time jump process, other models, including Diosi's, involve continuous localization. In "Presentation of Collapse Models", **Luca Ferialdi** (University of Trieste) gives an expert up-to-date overview of collapse models.

GRW theory is a quantum theory without observers. The collapse of the wave function just happens by dictate of the theory. That's easily said, but was it understood? In "Appreciating What He Did", **Tim Maudlin** (NYU) describes the true physical depth of GRW theory in his inimitable and precise way.

Part II: Philosophy

In his chapter "The GRW Theory and the Foundations of Statistical Mechanics", **David Albert** (Columbia University) wishes to explore how the GRW theory contributed to the foundations of statistical mechanics. In Albert's reconstruction of the Boltzmann explanatory schema of statistical mechanics, in order to derive the macroscopic laws which govern thermodynamic phenomena from the microscopic Newtonian dynamics, one would also need a statistical postulate, to ground the meaning of the probabilities arising from the statistical derivation of the laws of thermodynamics (in addition to the assumption that the universe started out with a low entropy to guarantee that entropy will likely increase in the future but not in the past). Albert argues that for the GRW theory there would be no need for such an additional postulate because this theory can provide a *dynamical* explanation of probabilities. If so, the statistical mechanical probabilities would just be the quantum mechanical probabilities. The idea is that entropy-decreasing (abnormal) microstates are very unstable, since they are almost always surrounded by normal entropy-increasing (normal) microstates. So, a wave function collapse in the GRW theory, with overwhelming likelihood, will make an abnormal microstate "jump" into a normal one. As a consequence, Albert argues, one can account for the second law of thermodynamics entirely in terms of the microscopic constituents and the dynamics, without postulating anything else.

Aside from the chapters by Albert and Laudisa (see below), the other chapters in this part discuss the implications of the GRW theory for the foundations of quantum theory. In particular, they engage with the issue of whether or not the wave function can be associated with a suitable material ontology, and if this is not the case, which of the alternative possibilities is to be preferred. GianCarlo never thought of quantum theory in instrumentalist terms, merely as a tool for predicting measurement results. It is clear from the title of the 1986 paper, in which he (with his colleagues and friends Alberto Rimini and Tullio Weber) introduced his theory of spontaneous collapse, that he wanted to provide a dynamics which could unify the macroscopic and the microscopic world, without ever resorting to the notion of measurement in the formulation of the theory. Later, however, he became sceptical of the idea that the wave function could describe physical objects, and he accordingly postulated that matter is described by a continuous matter field in three-dimensional space, defined in terms of the wave function. The GRW theory with GianCarlo's matter density ontology, in the literature dubbed GRWm, is an example of a theory with a primitive ontology, in which matter is not described by the wave function but rather by some other entity in three-dimensional space or four-dimensional space-time. GRWm is compared with other proposals for the ontology of matter. One now famous theory called GRWf was put forward by John S. Bell, who suggested that the world might be a collection of instantaneous "flashes". Another alternative is a particle ontology for the GRW theory, dubbed GRWp.

The idea of a particle GRW theory is taken up by **Valia Allori** (Northern Illinois University). In her "Spontaneous Localization Theories with a Particle Ontology", she first argues that a primitive ontology is needed because a satisfactory realist theory should provide a microscopic, dynamical explanation of the phenomena (rather than "closing the circle" from experience and back), and that can only be accomplished in terms of a three-dimensional ontology. She then explores the possibility of having a particle GRW theory, starting by pointing out the advantages of such a theory in terms of super-empirical virtues (simplicity, explanatory power, compatibility with scientific realism). In addition, she uses the same criteria to rank the various particle GRW theories proposed in the literature, after having analysed their tenability on the basis of criteria such as empirical adequacy, equivariance, and empirical coherence. Finally, she compares the best particle GRW theory with GRWm and GRWf, pointing out that considerations of relativistic invariance will determine the final ranking, and sketching ways in which a relativistic particle GRW could be constructed.

In his "From the Measurement Problem to the Primitive Ontology Programme", **Michael Esfeld** (University of Lausanne) also argues that one needs some primitive ontology and discusses the advantages and disadvantages of GRWf over GRWm and GRWp. He argues that the flash ontology is the most desirable for GRW because the problems of GRWm and GRWp do not appear in GRWf, while objections to GRWf are generally less severe than the problems with GRWm.

On a different front, in his "Might Laws of Nature 'Ground' Phenomena?", **Federico Laudisa** (University of Trento) takes up the issue of the nature of laws in the GRW framework. In more detail, the GRW theory allows for a realist interpretation of quantum theory, and this raises two broad sets of questions: what is matter according to the theory, and what laws are there to govern the behaviour of such matter. While the first question leads to discussions about the nature of the wave function, the second question requires us to investigate the nature of these laws. Laudisa argues against the Humean view, according to which laws of nature are mere regularities, and defends a primitivist account, in which laws are unanalysed primitives. Then he uses the metaphysical notion of grounding to provide a new characterization of the idea of a governing law, proposing that a law grounds its instances.

In his "On Closing the Circle", **Peter J. Lewis** (Dartmouth College) goes back to the issue of the ontology of matter, and disputes the necessity of a primitive ontology for the GRW theory. He argues that the theory can satisfactorily "close the circle" from human experiences and back, provided that one thinks of the wave function structurally. However, Lewis concludes that there is another problem, namely that the GRW theory cannot account for the concept of particle we use in microscopic quantum mechanical explanations. One could bypass this problem with a particle GRW theory, but Lewis concludes that would give rise to other, more serious, problems.

Part III: Mathematical Physics

Completely positive master equations guarantee full consistency of quantum dynamics. After showing that the master equation associated with the GRW model is Markovian and completely positive, **F. Benatti** and **F. Gebbia** (University of Trieste) consider a generalization of the GRW model that leads to a non-Markovian master equation. Since general considerations on complete positivity of such extended GRW models are extremely difficult, the authors focus on specific examples to shed light on the complexity of the problem.

One of GianCarlo's early studies concerned the decay of unstable systems. This inspired young researchers to delve into the mathematical physics of the exponential decay of nuclei. A short discussion of the modern state of the art is given by **Robert Grummt** and **Nicola Vona** (LMU) in "Energy–Lifetime Relations".

According to the original GRW theory, wave functions jump. In other collapse models the jumps are smoothed out and the change is continuous. In "On the Continuum Limit of the GRW Model", **Günter Hinrichs** (University of Augsburg) explains a strategy for making the transition from jumps to continuous change of the wave function.

For collapse models to be of physical relevance, in any measurement situation they must provide a transition of an initial wave function to a final one according to Born's rule. When the transition is examined with mathematical rigour, the dynamical behaviour in continuous collapse processes is found to be surprisingly complex, and one must ask whether signals of that complexity are experimentally verifiable. In "Continuous Collapse Models on Finite Dimensional Hilbert Spaces", **Antoine Tilloy** (MPI for quantum optics Munich) explains in a didactic manner how complexity is accompanied in his models by the appearance of "spikes" (very sharp transition peaks).

Part IV: Theoretical Physics

The collapse mechanism (narrowing the wave function) results in an energy increase which in the case of relativistic models leads to yet another divergence. In "Collapse Models, Relativity, and Discrete Spacetime", **Daniel J. Bedingham** (University of London) shows how the divergent behaviour can be regularized by a strategy discretising relativistic space-time.

Continuous spontaneous collapse models offer predictions that are experimentally distinguishable from standard quantum mechanics. In "Opto-Mechanical Test of Collapse Models", **Matteo Carlesso** (University of Trieste) and **Mauro Paternostro** (Queen's University Belfast) discuss current and proposed experimental tests of spontaneous collapse models, focusing on non-interferometric tests. An opto-mechanical set-up for testing collapse models is reviewed in detail and it is shown how the sensitivity of the experiment is affected by system size, properties of the mechanical system, and the surrounding environment.

Recent experimental tests of the Continuous Spontaneous Localisation (CSL) model involve the use of opto-mechanical systems, such as cantilevers. This class of experiments exploits the dependence of extra CSL-induced noise on the specific geometry of the test particle, by using probes with cavities or a layered structure. **Lajos Diósi** (Winger Center) provides a more robust mathematical description of the phenomenon in his "Two Invariant Surface-Tensors Determine CSL of Massive Body Wave Function". In particular, the author shows that, under certain regimes of approximation, the effect of CSL on opto-mechanical probes is a surface effect, and can be fully characterized in terms of the body's surface tensors and the mass density.

One of the characteristic traits of collapse models is radiation emission from any charged particle, induced by the noise causing the collapse of the wave function. It is important to analyse this process to analyse because it sets strong bounds on the collapse parameters. **Sandro Donadi** (IAS Frankfurt) gives a historical overview of the problem, summarizing the main theoretical results for the CSL model, as well as for the non-Markovian QMUPL model.

A different approach, yet close to GRW ideas, in that the collapse of the state is part of the theory, is presented in a very accessible manner by **Jürg Fröhlich** (ETH). In his chapter "Relativistic Quantum Theory", reality is to be read off from a Tree of Histories of Events (this is one interpretation of the acronym "ETH theory" which Jürg gave to his theory). This presentation could also have been categorized under *Mathematical Physics*, as it draws substantially from results on operator algebras.

The intersection between quantum mechanics and gravity is still not well understood by physicists. In the chapter "Classically Gravitating Quantum Systems" **André Großardt** (University of Jena) reviews experiments that are designed to see whether gravity is a fundamentally classical force. This chapter discusses how spontaneous collapse models coupled to gravity predict that gravity is classical and how they might give a solution to the problem of causality with classical gravity.

The modified Schrödinger equation of collapse models does not rely on the presence of an observer, and the predictions of these theories are falsifiable. Motivated by these facts, in "Collapse Models and Cosmology", **Jérôme Martin** and **Vincent Vennin** (IAP, Paris), inspired by the CSL model, implement a dynamical collapse model in a cosmological setting. They calculate the associated inflationary power spectrum, and compare their results with those coming from CMB measurements, constraining the two parameters of the model. They stress the importance of considering a cosmological setting as an arena to test collapse models.

In cosmology, the emergence of the primordial seeds of structure in the Universe as a result of quantum fluctuations during the inflationary epoch has become the accepted idea within the community. However, in "Spontaneous Collapse Theories and Cosmology", **Daniel Sudarsky** (ICN-UNAM) points out some conceptual difficulties in the standard approach. In the framework of semiclassical general relativity, he incorporates spontaneous collapse theories as a possible solution to these difficulties. In particular, his model can account for the lack of detection of primordial tensor modes at the level predicted by the standard approach.

The tension between Lorentz invariance and wave function collapse (or more generally between nonlocality and relativity) was always GianCarlo's main concern. His hope was to find a gratifying solution to this problem, following John Bell's remarks about the GRW invention: *It takes away the ground of my fear that any exact formulation of quantum mechanics must conflict with fundamental Lorentz invariance.* In 2004, **Roderich Tumulka** (University of Tübingen) devised a much acclaimed Lorentz invariant collapse model—albeit a non-interacting one. In this model, reality is imagined to be constituted by the formation of collapse centres called "flashes". In "A Relativistic GRW Flash Process with Interaction", he presents the generalization to an interacting collapse theory.

Dynamical reduction models and the theory of open quantum systems share the same mathematical structure in their description of the dynamics of the density matrix. **Bassano Vacchini** (University of Milan) reviews this link in "Non-Markov Processes in Quantum Theory". The author pays particular attention to non-Markovian processes because they are the most general class of non-unitary dynamics, and can, therefore, describe a broader spectrum of collapse models.

Part V: Experimental Physics

Technological advancement is bringing experimental tests of alternative quantum theories closer and closer. In this context, **N. Ares**, **A. N. Pearson**, and **G. A. D. Briggs** (University of Oxford) provide an account of the future prospects for addressing fundamental questions in quantum theory. They do so by proposing a manifesto of eight questions that addresses the interplay between theories and experiment, as well as the various open questions in quantum mechanics, such as the role of gravity.

The creation of a superposition and the measurement of its interference fringes in a matter-wave experiment is the most natural way to test spontaneous collapse models. In "Interferometric Tests of Wave-Function Collapse", **Stefan Gerlich**, **Yaakov Y. Fein**, and **Markus Arndt** (University of Vienna) present a comprehensive overview of these experiments, with special reference to collapse model testing.

Space offers a unique environment for testing spontaneous collapse models and the limits of the quantum theory. In "Tests in Space", **Rainer Kaltenbaek** (University of Ljubljana) describes the state of the art of Earth-based experiments and the advances that could be made by bringing them into space.

Collapse models predict that charged particles should radiate when interacting with collapse noise. In "Sneaking a Look at Ghirardi's Cards: Collapse Models Mapped with the Spontaneous Radiation", **K. Piscicchia** (Centro Fermi & INFN), **R. Del Grande**, **M. Laubenstein**, and **C. Curceanu** (INFN) analyse X-ray spectra measured in high precision low-background experiments and compare their results with the predictions of the CSL and DP models.

A plethora of different interferometric and non-interferometric experiments are potentially able to test spontaneous collapse models. In "New Avenues for Testing Collapse Models", **Andrea Vinante** and **Hendrik Ulbricht** (University of Southampton) provide a survey of such experimental efforts and offer a panorama for possible future research.

Naperville, USA	Valia Allori
Trieste, Italy	Angelo Bassi
Landsberg, Germany	Detlef Dürr
Genoa, Italy	Nino Zanghì
March 2020	

Contents

History and Honour

GianCarlo Ghirardi: Passing the Torch on Collapse Models

Stephen L. Adler

Abstract I describe how I became interested in the quantum measurement problem, and how I learned about the important work of Ghirardi, Rimini, and Weber and of Pearle. I first met Ghirardi in Trieste in 1999, and through him met Angelo Bassi, with whom I have had a long and fruitful friendship and collaboration focusing on the phenomenology and experimental tests of collapse models.

My interest in collapse models began in the late 1990s. For several years I had been working on ideas stemming from the paper "Generalized quantum dynamics as pre-quantum mechanics" which I wrote with my graduate student Andrew Millard in 1996 [1]. This work, which suggested a pre-quantum theory based on a non-commutative extension of classical mechanics, later became the basis for the "trace dynamics" on which I wrote my 2004 book "Quantum Theory as an Emergent Phenomenon" [2]. Ever since an anonymous reviewer's report for my earlier book on "Quaternionic Quantum Mechanics and Quantum Fields" [3] asked whether quaternionic quantum theory solves the quantum measurement problem, I had started to read the quantum measurement literature. I soon became convinced that the old advice "shut up and calculate" wasn't good enough; there really *is* a problem, and any linear quantum dynamics, whether complex linear or quaternion linear, cannot resolve it. So along with thinking about technicalities of the trace dynamics program, I also started to think about how a pre-quantum theory could lead to a phenomenology for explaining how measurements take place in standard quantum theory.

I first became aware of the work of Ghirardi, Rimini, and Weber, and of Pearle, through coming across papers by Fivel [4] on dynamical reduction and wave function collapse. Reading the GRW and Pearle papers, I was struck with how the idea of a stochastic reduction mechanism fits neatly into my basic idea of the trace dynamics program, that quantum theory is the thermodynamics of an underlying pre-quantum dynamics. Classical thermodynamics has an underlying statistical mechanics, which leads to fluctuation or Brownian motion effects. Similarly, a thermodynamics giv-

S. L. Adler (✉)
Institute for Advanced Study, Einstein Drive, Princeton, NJ 08540, USA
e-mail: adler@ias.edu

V. Allori et al. (eds.), *Do Wave Functions Jump?*, Fundamental Theories of Physics 198,
https://doi.org/10.1007/978-3-030-46777-7_1

ing rise to quantum theory should also have stochastic corrections, and the objective reduction model ideas fit this framework: thermodynamics of trace dynamics gives linear quantum dynamics; fluctuation corrections lead to the phenomenology of reduction models, and an explanation of quantum measurements.

Around the same time I came across a paper by Hughston [5] giving a geometric reformulation of stochastic state vector reduction, and the sketch of a proof, missing one key identity in the general case, of reduction leading to the Born rule. In collaboration with Horwitz [6], I derived additional needed identities and wrote a paper giving a general proof of state vector collapse with probabilities obeying the Born rule. We also showed that the proof could be simply cast in terms of the density matrix evolution, avoiding the complicated geometric structure on which Hughston's approach is based.

My paper with Horwitz came to the attention of Ghirardi, with the effect that when I came to Trieste in June, 1999 to give my Dirac Medal lecture, Ghirardi came to talk to me. He pointed out that in the original Continuous Spontaneous Localization paper of Ghirardi et al. [7], there is a section which I had not appreciated proving that CSL implies reduction to the Born Rule. Ghirardi also made a point of introducing me to his then graduate student Angelo Bassi. This led to what is now a 20 year friendship, and collaboration, with Angelo on phenomenology of reduction models. I have been particularly interested in turning reduction models from a subject of philosophical discussion to a subject of active experimentation, with the ultimate aim of verifying, or falsifying, these models as an explanation of how quantum measurements occur. Angelo has enthusiastically embraced this approach, and has a large group in Trieste working on aspects of the theory and experimental tests of models for wave function collapse.

Looking back, I see my first meeting with GianCarlo Ghirardi and his introduction of me to Angelo Bassi as a symbolic "passing of the torch". As the years have gone by the main work on collapse models in Trieste has passed to the Bassi group, which is now at the forefront of efforts to assess whether collapse models, which I find to be theoretically beautiful and compelling as a phenomenology of measurement, are in fact Nature's choice.

References

1. S. L. Adler and A. C. Millard, Nucl. Phys. B **473**, 199 (1996).
2. S. L. Adler, *Quantum Theory as an Emergent Phenomenon*, Cambridge University Press, Cambridge (2004).
3. S. L. Adler, *Quaternionic Quantum Mechanics and Quantum Fields*, Oxford University Press, New York (1995).
4. D. I. Fivel, Phys. Lett. A **248**, 139 (1998); see also D. I. Fivel, arXiv:quant-ph/9611016 and quant-ph/9710042.
5. L. P. Hughston, Proc. Roy. Soc. London, Ser. A **452**, 953 (1996).
6. S. L. Adler and L. P. Horwitz, J. Math. Phys. **41**, 2485 (2000).
7. G. C. Ghirardi, P. Pearle, and A. Rimini, Phys. Rev. A **42**, 78 (1990), Section II B.

EPR-Bell-Schrödinger Proof of Nonlocality Using Position and Momentum

Jean Bricmont, Sheldon Goldstein, and Douglas Hemmick

We dedicate this paper to the memory of Giancarlo Ghirardi, who devoted his life to understanding quantum mechanics. He was a friend of John Bell, who was inspired by Giancarlo's work. He was also a friend of two of us (J.B. and S.G.)

Abstract Based on his extension of the classical argument of Einstein, Podolsky and Rosen, Schrödinger observed that, in certain quantum states associated with pairs of particles that can be far away from one another, the result of the measurement of an observable associated with one particle is perfectly correlated with the result of the measurement of another observable associated with the other particle. Combining this with the assumption of locality and some "no hidden variables" theorems, we showed in a previous paper [11] that this yields a contradiction. This means that the assumption of locality is false, and thus provides us with another demonstration of quantum nonlocality that does not involve Bell's (or any other) inequalities. In [11] we introduced only "spin-like" observables acting on finite dimensional Hilbert

To appear in: Do wave functions jump? Perspectives on the work of G. C. Ghirardi. Editors: V. Allori, A. Bassi, D. Dürr, N. Zanghì.

J. Bricmont
IRMP, Université catholique de Louvain, chemin du Cyclotron 2, 1348 Louvain-la-Neuve, Belgium
e-mail: jean.bricmont@uclouvain.be

S. Goldstein (✉)
Department of Mathematics, Rutgers University, Hill Center, 110 Frelinghuysen Road, Piscataway, NJ 08854-8019, USA
e-mail: oldstein@math.rutgers.edu

D. Hemmick
66 Boston Drive, Berlin, MD 21811, USA
e-mail: jsbell.ontarget@gmail.com

V. Allori et al. (eds.), *Do Wave Functions Jump?*, Fundamental Theories of Physics 198,
https://doi.org/10.1007/978-3-030-46777-7_2

spaces. Here we will give a similar argument using the variables originally used by Einstein, Podolsky and Rosen, namely position and momentum.

1 Introduction

In 1935, Einstein, Podolsky and Rosen (EPR) argued that quantum mechanics is incomplete by considering two particles in one dimension moving in opposite directions and whose joint wave function (see (3.2.1) below) was such that the measurement of the position of one of the particles immediately determined the position of the other particle and, similarly, the measurement of the momentum of one of the particles immediately determined the momentum of the other one.

Since, said EPR, a measurement made on one particle obviously could not possibly influence the physical state of the other particle, situated far away from the first particle, and since the wave function of both particles specifies neither the position nor the momentum of those particles, this quantum mechanical description of the state of both particles provided by this wave function must be *incomplete* in the sense that other variables, such as the values of the positions and momenta of both particles, must be included in a complete description of that physical system.

EPR's argument had been widely misunderstood and misrepresented or ignored by almost everybody at that time. But not by Schrödinger, who, in his "cat paper," originally published in German [34], as well as in the papers [35, 36], understood the "paradox" raised by EPR and deepened the perplexity that it causes.

Schrödinger showed that for certain states, called now maximally entangled (see Sect. 2.1), it is not just that the positions and the momenta of the particles are perfectly correlated. He showed that, for every observable associated with the first particle, there is another observable associated with the second particle such that the results of the measurements of both observables are perfectly correlated.

In [11], following [24, 25], we explained that, if one assumes locality, meaning that there is no effect whatsoever on the state of the second particle due to a measurement carried out on the first particle (when both particles are sufficiently spatially separated), there must exist what we call a "non-contextual value map" v which assigns to each observable A a value $v(A)$ that pre-exists its measurement and is simply revealed by it. The word "non-contextual" refers to the fact that, since it pre-exists the measurement, the value $v(A)$ does not depend on the procedure used to measure A.

However several theorems, originally due to Bell [3] and to Kochen and Specker [27], preclude the possibility of a non-contextual value map.[1] Since the existence of this map is a logical consequence of the assumption of locality and of the perfect correlations, the assumption of locality is false.

[1] In the literature on quantum mechanics, these theorems are often called "no hidden variables" theorems. But we prefer the expression "inexistence of a non-contextual value map" because, as we will discuss in Sect. 5, the expression "hidden variables" is really a misnomer.

In this paper, we first summarize the arguments of our previous paper [11] (Sect. 2). We then turn to the EPR paper as well as related work by Einstein alone and Schrödinger (Sect. 3). Next, we provide a proof of nonlocality similar to the one of Sect. 2, but using only functions of the EPR variables, namely positions and momenta (Sect. 4). This argument relies on a theorem of Clifton [14].

We then consider what happens in Bohmian mechanics (Sect. 5): in that theory, particles have, at all times, both a position and a momentum and one might therefore think that this would imply the existence of a non-contextual value-map for functions of those variables. We explain however, through an analysis of what a measurement of momentum means in that theory, that this is not the case. Finally we briefly discuss how nonlocality manifests itself in Bohmian mechanics.

For a discussion of the relationship between this work and previous ones, including [1, 12, 13, 20, 26, 37], see Sect. 7 of [11].

2 Proof of Nonlocality Based on Perfect Correlations

We will first discuss special quantum states, called maximally entangled, for pairs of physical systems that can possibly be located far apart, and having the property that, for each quantum observable of one of the systems, there is an associated observable of the other one such that the result of the measurement of that observable is perfectly correlated with the result of the measurement on the first one.

2.1 Maximally Entangled States

Consider a finite dimensional (complex) Hilbert space $\mathcal{H}$, of dimension N, and orthonormal bases ψ_n and ϕ_n in $\mathcal{H}$ (we will assume below that all bases are orthonormal). A unit vector Ψ in $\mathcal{H} \otimes \mathcal{H}$ is *maximally entangled* if it is of the form

$$\Psi = \frac{1}{\sqrt{N}} \sum_{n=1}^{N} \psi_n \otimes \phi_n. \tag{2.1.1}$$

Since we are interested in quantum mechanics, we will refer to those vectors as *maximally entangled states* and we will associate, by convention, each space in the tensor product with a "physical system," namely we will consider the set $\{\phi_n\}_{n=1}^N$ as a basis of states for physical system 1 and the set $\{\psi_n\}_{n=1}^N$ as a basis of states for physical system 2.

Now, given a maximally entangled state, one can associate to each operator of the form $\mathbb{1} \otimes O$ (meaning that it acts non-trivially only on particle 1) an operator of the form $\tilde{O} \otimes \mathbb{1}$ (meaning that it acts non-trivially only on particle 2). Here $\mathbb{1}$ denotes the identity operator on $\mathcal{H}$.

Define the operator U mapping $\mathcal{H}$ to $\mathcal{H}$ by setting

$$U\phi_n = \psi_n, \tag{2.1.2}$$

$\forall n = 1, \ldots, N$, and extending U to an anti-linear operator on all of $\mathcal{H}$:

$$U\left(\sum_{n=1}^{N} c_n\phi_n\right) = \sum_{n=1}^{N} c_n^* U\phi_n = \sum_{n=1}^{N} c_n^*\psi_n \tag{2.1.3}$$

where * denotes the complex conjugate.

Using the operator U, the state Ψ in (2.1.1) can be written as:

$$\Psi = \frac{1}{\sqrt{N}}\sum_{n=1}^{N} U\phi_n \otimes \phi_n. \tag{2.1.4}$$

It is easy to check that this formula is the same for any basis, see [11, Eq. 3.1.8].

U thus determines, and is uniquely determined by, a maximally entangled state Ψ.

Given such a state Ψ, and hence U, we may associate to every operator of the form $\mathbb{1} \otimes O$ an operator of the form $\tilde{O} \otimes \mathbb{1}$ by setting

$$\tilde{O} = UOU^{-1}. \tag{2.1.5}$$

Suppose ϕ_n are eigenstates of O, with eigenvalues λ_n,

$$O\phi_n = \lambda_n\phi_n. \tag{2.1.6}$$

Then, the states $\psi_n = U\phi_n$ are eigenstates of $\tilde{O}$, also with eigenvalues λ_n:

$$\tilde{O}\psi_n = \lambda_n\psi_n. \tag{2.1.7}$$

This implies and is in fact equivalent to the following relationship between the operators O and $\tilde{O}$:

$$(O \otimes \mathbb{1} - \mathbb{1} \otimes \tilde{O})\Psi = 0, \tag{2.1.8}$$

directly expressing the fact that, in the state Ψ, $O \otimes \mathbb{1}$ and $\mathbb{1} \otimes \tilde{O}$ are perfectly correlated.

We may summarize this as follows:

Theorem 2.1 *Consider a finite dimensional Hilbert space $\mathcal{H}$, of dimension N, and a maximally entangled state $\Psi \in \mathcal{H} \otimes \mathcal{H}$. Then, for any self-adjoint operator O acting on $\mathcal{H}$, there exists a self-adjoint operator $\tilde{O}$ acting on $\mathcal{H}$ such that (2.1.8) holds.*

Remarks

1. A simple example of a maximally entangled state is:

$$|\Psi\rangle = \frac{1}{\sqrt{2}}(|\uparrow\rangle|\downarrow\rangle - |\downarrow\rangle|\uparrow\rangle), \tag{2.1.9}$$

where the right factors refer to system 1 and left ones to system 2. That state, according to ordinary quantum mechanics, means that the spin measured on system 1 will have equal probability to be up or down, but is perfectly anti-correlated with the spin measured on system 2.
In the notation of (2.1.1), one has:

$$\phi_1 = |\uparrow>, \quad \phi_2 = |\downarrow>, \quad \psi_1 = -|\downarrow>, \quad \psi_2 = |\uparrow>,$$

and therefore,

$$U|\uparrow> = -|\downarrow>,$$
$$U|\downarrow> = |\uparrow> .$$

If one takes

$$O = \begin{pmatrix} 1 & 0 \\ 0 & -1 \end{pmatrix} \tag{2.1.10}$$

which corresponds to the spin operator for system 1 and has eigenvectors ϕ_1 with eigenvalue 1 and ϕ_2 with eigenvalue -1, one computes that

$$\tilde{O} = UOU^{-1} = \begin{pmatrix} -1 & 0 \\ 0 & 1 \end{pmatrix} = -O, \tag{2.1.11}$$

which means that the spin operator for systems 1 and 2 are perfectly anti-correlated, since $\tilde{O}$ is minus the spin operator for system 2.
We will use later the following:

2. *Products of maximally entangled states are maximally entangled states*: If one has two Hilbert spaces $\mathcal{H}_1$, $\mathcal{H}_2$, and two maximally entangled states $\Psi_i \in \mathcal{H}_i \otimes \mathcal{H}_i$, $i = 1, 2$, then it is easy to check that the state $\Psi = \Psi_1 \otimes \Psi_2$ is maximally entangled in $\mathcal{H} \otimes \mathcal{H}$, where $\mathcal{H} = \mathcal{H}_1 \otimes \mathcal{H}_2$ (under the canonical identification of $(\mathcal{H}_1 \otimes \mathcal{H}_1) \otimes (\mathcal{H}_2 \otimes \mathcal{H}_2)$ with $\mathcal{H} \otimes \mathcal{H}$).

Let us now see what this notion of maximally entangled state implies for quantum measurements.

Suppose that we have a pair of physical systems, whose states belong to the same finite dimensional Hilbert space $\mathcal{H}$. And suppose that the quantum state Ψ of the pair is maximally entangled, i.e. of the form (2.1.1).

Any observable acting on system 1 is represented by a self-adjoint operator O, which has therefore a basis of eigenvectors. Since the representation (2.1.4) of the state Ψ is valid in any basis, we may choose, without loss of generality, as the set

$\{\phi_n\}_{n=1}^N$ in (2.1.1) the eigenstates of O. Let λ_n be the corresponding eigenvalues, see (2.1.6).

If one measures that observable O, the result will be one of the eigenvalues λ_n, each having equal probability $\frac{1}{N}$. If the result is λ_k, the (collapsed) state of the system after the measurement, will be $\psi_k \otimes \phi_k$. Then, the measurement of observable $\tilde{O}$, defined by (2.1.5), (2.1.2), on system 2, will necessarily yield the value λ_k.

Reciprocally, if one measures an observable $\tilde{O}$ on system 2 and the result is λ_l, the (collapsed) state of the system after the measurement, will be $\psi_l \otimes \psi_l$, and the measurement of observable O on system 1 will necessarily yield the value λ_l.

To summarize, we have derived the following consequence of the quantum formalism:

Principle of Perfect Correlations. *In any maximally entangled quantum state, of the form* (2.1.1)*, there is, for each operator O acting on system 1, an operator $\tilde{O}$ acting on system 2 (defined by* (2.1.5), (2.1.2)*), such that, if one measures the physical quantity represented by operator $\tilde{O}$ on system 2 and the result is the eigenvalue λ_l of $\tilde{O}$, then, measuring the physical quantity represented by operator O on system 1 will yield with certainty the same eigenvalue λ_l, and vice-versa.*[2]

2.2 Schrödinger's "Theorem"

The following property will be crucial in the rest of the paper.

Locality. *If systems 1 and 2 are spatially separated from each other, then measuring an observable on system 1 has no instantaneous effect whatsoever on system 2 and measuring an observable on system 2 has no instantaneous effect whatsoever on system 1.*

Finally, we must also define:

Non-contextual value-maps. Let $\mathcal{H}$ be a finite dimensional Hilbert space and let $\mathcal{A}$ be the set of self-adjoint operators on $\mathcal{H}$. Suppose $\mathcal{H}$ is the quantum state space for a physical system and $\mathcal{A}$ is the set of quantum observables. Suppose there are situations in which there are observables A for which the result of measuring A is determined already, before the measurement. Suppose, that is, that A has, in these situations, a pre-existing value $v(A)$ revealed by measurement and not merely created by measurement. Of course, this implies that for every experiment $\mathcal{E}_A$ measuring A, the result $v(\mathcal{E}_A)$ of that experiment, in the situation under consideration, must be $v(A)$. And suppose finally that the situation is such that we have a pre-exiting value $v(A)$ for every $A \in \mathcal{A}$.

We would then have a *non-contextual value-map*, namely a map $v : \mathcal{A} \to \mathbb{R}$ that assigns the value $v(A)$ to any experiment associated with what is called in quantum mechanics a measurement of an observable A. There can be different ways to measure the same observable. The value-map is called non-contextual because all such

[2] The correlations mentioned here are often called anti-correlations, for example when $\tilde{O} = -O$, as in the example of the spin in remark 1 above.

experiments, associated with the same quantum observable A, are assigned the same value.

This notion of value-map is not a purely mathematical one, since it involves the notion of an experiment that measures a quantum observable A, which we have not mathematically formalized. However, we shall need only the following obvious purely mathematical consequence of non-contextuality.

A non-contextual value-map has the fundamental property that, if $A_i, i = 1, \ldots, n$, are mutually commuting self-adjoint operators on $\mathcal{H}$, $[A_i, A_j] = 0, \forall i, j = 1, \ldots, n$, then, if f is a function of n variables and $B = f(A_1, \ldots, A_n)$, then

$$v(B) = f(v(A_1), \ldots, v(A_n)). \tag{2.2.1}$$

It is a well-known property of quantum mechanics that, since all the operators $A_1, \ldots, A_n, B$ commute, they are simultaneously measurable and the result of those measurements must satisfy (2.2.1).

But, and this is what we emphasized in [11], (2.2.1) follows trivially from the non-contextuality of the value-map. Indeed, a valid quantum mechanical way to measure the operator $B = f(A_1, \ldots, A_n)$ is to measure $A_1, \ldots, A_n$ and, denoting the results $\lambda_1, \ldots, \lambda_n$, to regard $\lambda_B = f(\lambda_1, \ldots, \lambda_n)$ as the result of a measurement of B. Since, by the non-contextuality of the map v, all the possible measurements of B must yield the same results, (2.2.1) holds.

Thus, once one has a non-contextual value-map, *one does not even need to check* (2.2.1).

Now we will use the perfect correlations and locality to establish the existence of a non-contextual value-map v, for a maximally entangled quantum state of the form (2.1.1) or, equivalently, (2.1.4). By the principle of perfect correlations, or any operator O on system 1, there is an operator $\tilde{O}$ on system 2, defined by (2.1.5), (2.1.2), which is perfectly correlated with O through (2.1.8).

Thus, if we were to measure $\tilde{O}$, obtaining λ_l, we would know that

$$v(O) = \lambda_l \tag{2.2.2}$$

concerning the result of then measuring O. Therefore, $v(O)$ would pre-exist the measurement of O. But, by the assumption of locality, the measurement of $\tilde{O}$, associated with the second system, could not have had any effect on the first system, and thus, this value $v(O)$ would pre-exist also the measurement of $\tilde{O}$ and this would not depend upon whether $\tilde{O}$ had been measured. Letting O range over all operators on system 1, we see that there must be a non-contextual value-map $O \to v(O)$.

To summarize, we have shown:

Schrödinger's "Theorem". Let $\mathcal{A}$ be the set of self-adjoint operators on the component Hilbert space $\mathcal{H}$ of a physical system in a maximally entangled state (2.1.1). Then, assuming locality and the principle of perfect correlations, there exists a non-contextual value-map $v : \mathcal{A} \to \mathbb{R}$.

Remark

We put "Theorem" in quotation marks because the statement concerns physics and not just mathematics. Its conclusions are nevertheless inescapable assuming the hypothesis of locality and the empirical validity of the principle of perfect correlations, a principle which is, as we showed, a consequence of the quantum formalism.

2.3 The Non-existence of Non-contextual Value-Maps

The problem posed by the non-contextual value-map v whose existence is implied by Schrödinger's "theorem" is that such maps simply do not exist (and that is a purely mathematical result). Indeed, one has the:

"Theorem": Non-existence of non-contextual value-maps. Let $\mathcal{A}$ be the set of self-adjoint operators on the Hilbert space $\mathcal{H}$ of a physical system. Then there exists no non-contextual value-map $v : \mathcal{A} \to \mathbb{R}$.

This "theorem" is an immediate consequence of the following theorem, since (2.3.1), (2.3.2) are consequences of (2.2.1).[3]

Theorem 2.2 *Let $\mathcal{H}$ be a finite dimensional Hilbert space of dimension at least three, and let $\mathcal{A}$ be the set of self-adjoint operators on $\mathcal{H}$. There does not exist a map $v : \mathcal{A} \to \mathbb{R}$ such that:*

(1) $\forall O \in \mathcal{A}$,

$$v(O) \text{ is an eigenvalue of } O \tag{2.3.1}$$

(2) $\forall O, O' \in \mathcal{A}$ *with* $[O, O'] = OO' - O'O = 0$, *and for any real valued function f of two real variables,*

$$v(f(O, O')) = f(v(O), v(O')). \tag{2.3.2}$$

See [11] for a discussion of the proof of the theorem, which is a consequence of stronger theorems, originally due to Bell [3] and to Kochen and Specker [27], with simplified proofs of Theorem 2.2 due to Mermin [28], and to Peres [31, 32].

2.4 Nonlocality

The conclusion of Schrödinger's "theorem" and of the "Theorem" on the non-existence of non-contextual value-maps plainly contradict each other. So, the assump-

[3]This is obvious for (2.3.2), a special case of (2.2.1). For (2.3.1) we observe that, since O is self-adjoint, we can write $O = \sum_i \lambda_i P_{\lambda_i}$ where P_{λ_i} is the projector on the subspace of eigenvectors of eigenvalue λ_i of O and thus we have that $f(O) = \sum_i f(\lambda_i) P_{\lambda_i}$. If we choose any f whose range is the set of eigenvalues of O and is such that $f(\lambda_i) = \lambda_i \ \forall i$, we have that $O = f(O)$ and, by (2.2.1), we obtain that $v(O) = v(f(O)) = f(v(O))$ and thus $v(O)$ is an eigenvalue of O.

tions of at least one of them must be false. Moreover, the stronger Theorem 2.2 is a purely mathematical result. To derive Schrödinger's "theorem," we assume only the perfect correlations and locality. The perfect correlations are an immediate consequence of quantum mechanics. The only remaining assumption is locality. Hence we can deduce:

Nonlocality "Theorem". The locality assumption is false.

See [11, Sects. 5, 7] for a discussion of the relation between this proof and other proofs of nonlocality.

3 The Original EPR Argument

Let us now turn to the original EPR argument [18] and explain its connection to the notion of locality. EPR gave both a general argument and a specific example.

3.1 EPR's General Setup

For their general argument, EPR considered a system of two particles, 1 and 2, in one dimension, that may be far apart and a physical quantity represented by a self-adjoint operator O that acts on system 1. We shall assume that O has an orthonormal basis of eigenvectors $\phi_n(x_1)$ with eigenvalues λ_n.

One can then write the joint state of both particles as:

$$\Psi(x_1, x_2) = \sum_{n=1}^{\infty} \psi_n(x_2)\phi_n(x_1), \tag{3.1.1}$$

where $\psi_n(x_2)$ are the (x_2 dependent) coefficients of that expansion.[4]

After a measurement of O on system 1, if the result is λ_l, then the state collapses to $\psi_l(x_2)\phi_l(x_1)$, i.e. $\phi_l(x_1)$ for the first particle and $\psi_l(x_2)$ for the second.

If, on the other hand, one considers a physical quantity represented by an operator O' that acts on system 1, and one assumes that O' has eigenvectors $\phi'_s(x_1)$ and eigenvalues μ_s, one can write the joint state as:

$$\Psi(x_1, x_2) = \sum_{s=1}^{\infty} \psi'_s(x_2)\phi'_s(x_1) \tag{3.1.2}$$

[4]This resembles a maximally entangled state, like (2.1.1), but it is not one because the sum in (3.1.1) extends to infinity and, for (3.1.1) to be a maximally entangled state, the set $\{\psi_n\}_{n=1}^{\infty}$ should be orthonormal. But then the norm of (3.1.1) would be infinite and thus (3.1.1) would not belong the Hilbert space.

After a measurement of O' on system 1, if the result is μ_k, then the state collapses to $\psi'_k(x_2)\phi'_k(x_1)$, i.e. $\phi'_k(x_1)$ for the first particle and $\psi'_k(x_2)$ for the second.

We will discuss the implications of that observation after giving the concrete examples of the operators considered by EPR.

3.2 The Example of Position and Momentum

For their specific example, EPR introduced a two particle wave function[5]:

$$\Psi_{EPR}(x_1, x_2) = \int_{-\infty}^{\infty} \exp(i(x_1 - x_2 + x_0)p)dp \tag{3.2.1}$$

(putting $\hbar = 1$). This can be written, by analogy with (3.1.1), i.e. with sums replaced by integrals, as:

$$\Psi_{EPR}(x_1, x_2) = \int_{-\infty}^{\infty} \psi_p(x_2)\phi_p(x_1)dp \tag{3.2.2}$$

with: $\phi_p(x_1) = \exp(ix_1 p)$, and $\psi_p(x_2) = \exp(-i(x_2 - x_0)p)$.

It will be useful to introduce the Fourier transform of a wave function Ψ:

$$\widehat{\Psi}(p_1, p_2) = \frac{1}{2\pi} \int \exp(-i(p_1 x_1 + p_2 x_2))\Psi(x_1, x_2)dx_1 dx_2, \tag{3.2.3}$$

whose inverse is:

$$\Psi(x_1, x_2) = \frac{1}{2\pi} \int \exp(i(p_1 x_1 + p_2 x_2))\widehat{\Psi}(p_1, p_2)dp_1 dp_2. \tag{3.2.4}$$

EPR took the operator O to be the momentum operator

$$P_1 = -i\frac{d}{dx_1}$$

acting on the first particle and on a suitable set of functions (see [33, Chap. VIII] for precise definitions).

[5]This is a generalized wave function, which means that it is not an element of the Hilbert space $L^2(\mathbb{R}^2)$, but rather a distribution, namely a linear function acting on a space of smooth functions that decay rapidly at infinity (see [33, Sect. 5.3] for a short introduction to distributions). We will not try to be rigorous about these generalized functions here, but we will give a regularized version of the same wave function in Sect. 3.6.

We know that $\phi_p(x_1) = \exp(ix_1 p)$ is a (generalized) eigenstate of P_1 of eigenvalue p, and $\psi_p(x_2) = \exp(-i(x_2 - x_0)p)$ is a (generalized) eigenstate of eigenvalue $-p$ of the momentum operator

$$P_2 = -i\frac{d}{dx_2}$$

acting on the second particle.

Alternatively, $P_j, j = 1, 2$, can be defined by its action on $\widehat{\Psi}(p_1, p_2)$:

$$P_j\Psi(x_1, x_2) = \frac{1}{2\pi}\int \exp(i(p_1x_1 + p_2x_2))p_j\widehat{\Psi}(p_1, p_2)dp_1dp_2\ , \quad j = 1, 2\ . \tag{3.2.5}$$

EPR took the operator O' to be the position operator $Q_1 = x_1$ acting on the first particle.

Using a standard identity for distributions ($\int_{-\infty}^{\infty}\exp(ixp)dp = 2\pi\delta(x)$) one can write the state (3.2.1), as:

$$\begin{aligned}\Psi_{EPR}(x_1, x_2) &= 2\pi\delta(x_1 - x_2 + x_0)\\ &= 2\pi\int_{-\infty}^{\infty}\delta(x - x_2 + x_0)\delta(x_1 - x)dx\\ &= \int_{-\infty}^{\infty}\psi'_x(x_2)\phi'_x(x_1)dx,\end{aligned} \tag{3.2.6}$$

with $\psi'_x(x_2) = \sqrt{2\pi}\delta(x - x_2 + x_0)$ and $\phi'_x(x_1) = \sqrt{2\pi}\delta(x_1 - x)$. The last formula is analogous to (3.1.2).

The (generalized) eigenfunctions of the operator $Q_1 = x_1$ are $\phi'_x(x_1) = \sqrt{2\pi}\delta(x_1 - x)$, with eigenvalue x, and $\psi'_x(x_2) = \sqrt{2\pi}\delta(x - x_2 + x_0)$ is a (generalized) eigenvector of the operator $Q_2 = x_2$, with eigenvalue $x + x_0$.

Therefore, depending on whether we choose to measure the operator O or O' on the first particle, one can produce two different states, $\psi_p(x_2) = \exp(-i(x_2 - x_0)p)$ and $\psi'_x(x_2) = \sqrt{2\pi}\delta(x - x_2 + x_0)$, for the second particle, which can be, in principle, as far as one wants from the first one.

Moreover, the states $\psi_p(x_2) = \exp(-i(x_2 - x_0)p)$ and $\psi'_x(x_2) = \sqrt{2\pi}\delta(x - x_2 + x_0)$ are (generalized) eigenfunctions of two non-commuting operators, P_2 and Q_2.

3.3 *The Conclusions of the EPR Paper by EPR*

Since EPR assumed no actions at a distance, they concluded that the values of two non commuting observables, like P_2 and Q_2, for the second particle, far away from where the measurements on the first particle take place, must have "simultaneous reality" when the system is in the quantum state (3.2.1). Thus, say EPR, quantum mechanics, i.e., the description provided by the state (3.2.1), is an incomplete description of physical reality.

But they could have made a simpler argument: *considering only one variable is enough to show that quantum mechanics is incomplete*. Indeed, I can *know* the position of the second particle by measuring the position of the first one. If that measurement, being made far away from the second particle, does not affect the state of the second particle, then the position of that second particle (which is left undetermined by the state (3.2.1)) must exist independently of any measurement on the first particle.

And, since one can reason by exchanging the two particles, one can also know the position of the first particle by measuring the one of the second particle, so that the position of the first particle must also exist independently of any measurements.

Of course, they could have made the same argument about the momentum of either particle, but there was no need to bring in both quantities.

3.4 The Conclusions of the EPR Paper by Einstein

In a June 19, 1935 letter to Schrödinger, Einstein complained that the EPR paper had been written by Podolsky "for reasons of language" and that the main point "was buried, so to speak, by erudition" [19].

Then Einstein explains what is, for him, the main point: in the notation used here, see (3.1.1), if one measures quantity O on system 1, the state collapses to some state $\psi_l(x_2)$ for the second particle. Similarly, if one measures a quantity O' on system 1, see (3.1.2), the state collapses to some *different* state $\psi'_k(x_2)$ for the second particle.

For the state Ψ_{EPR}, (3.2.2), (3.2.6) one obtains either a state of the form $\psi_p(x_2) = \exp(-i(x_2 - x_0)p)$, if one measures the momentum of the first particle or a state of the form $\psi'_x(x_2) = \sqrt{2\pi}\delta(x - x_2 + x_0)$, if one measures the position of the first particle.

The fact that one can obtain *two different states* for the second particle by acting on the first particle, far away from the second one, proves that the wave function description in quantum mechanics is incomplete (assuming of course locality) since a more complete description would be provided by both states together.

Einstein said that "he could not care less" [21, p. 38] about the fact that those states, $\psi_p(x_2) = \exp(-i(x_2 - x_0)p)$ and $\psi'_x(x_2) = \sqrt{2\pi}\delta(x - x_2 + x_0)$, are or are not eigenstates of some observable (related to the second particle).

This is indeed different, and simpler, than the conclusion of the EPR paper, but it is still more complicated than the argument that we gave in Sect. 3.3.

3.5 Schrödinger's Extension of EPR

What Schrödinger did in his 1935 paper[6] [34] and in [35, 36], was to reflect on the EPR paper [18]. He introduced what we call here maximally entangled states and concluded that the value of every observable O for the first system can be determined by the measurement of the corresponding observable $\tilde{O}$ on the second system, distant from the first one. That puzzled him a lot. Of course, like EPR, Schrödinger always assumed locality.

To illustrate his puzzlement, Schrödinger used the following example. Let O be the energy of the harmonic oscillator, $O = \frac{1}{2}(p^2 + \omega^2 x^2)$ with $p = -i\frac{d}{dx}$. It is well known that the eigenvalues of the operator O are of the form $\omega(n + \frac{1}{2})$, $n = 0, 1, 2, \ldots$. But, argued Schrödinger, if those values can be determined by measuring a similar operator $\tilde{O}$ acting on a distant system, they must pre-exist the measurement of O, and that should hold true *for every value of* ω. But, by the EPR reasoning, the values of the position x and the momentum p of the first system can also be determined by measuring either the operator $\tilde{x}$ or the operator $\tilde{p}$ on the second system, so the values of x and p must also pre-exist their measurements. But it is impossible for the quantity $\frac{1}{2}(p^2 + \omega^2 x^2)$ to belong to the set $\{\omega(n + \frac{1}{2}) | n = 0, 1, 2, \ldots\}$, for arbitrary values of ω and any given values of x and p.

It is interesting to compare Schrödinger's attitude to that of von Neumann a little before 1935 [39] (von Neumann's book was published in German in 1932 but translated into English only in 1955); von Neumann proved a "no hidden variable theorem" similar in its conclusion to our Theorem 2.2, but by making the much stronger assumption that (2.3.2) holds even for non-commuting operators O and O', at least for the function $f(x, y) = x + y$, and he concluded that the "value-map" cannot exist. If one assumes that (2.3.2) holds for non-commuting operators, then it is very simple to prove the non-existence of a value-map. Take $O = \frac{1}{\sqrt{2}}\sigma_x$, $O' = \frac{1}{\sqrt{2}}\sigma_y$, where σ_x and σ_y are the Pauli matrices corresponding to the spin along the x and y axes. Then $O + O' = \frac{\sigma_x + \sigma_y}{\sqrt{2}}$ corresponds to the spin at an angle of 45° between the x and y axes. All the Pauli matrices have eigenvalues equal to ± 1 and so does $O + O'$. Thus $v(O) = v(O') = \pm\frac{1}{\sqrt{2}}$, and we have $v(O) + v(O') = \pm\sqrt{2}$ or 0. But we also have $v(O + O') = v\big((\sigma_x + \sigma_y)/\sqrt{2}\big) = \pm 1$. Thus (2.3.2) cannot hold for this choice of O and O' and $f(x, y) = x + y$.

If Schrödinger had reasoned like von Neumann he would also have derived a "no hidden variable theorem," using his example of the harmonic oscillator: Indeed, if $O = \frac{1}{2}(p^2 + \omega^2 x^2)$, and one applies (2.3.2) even to non-commuting operators, one gets $v(O) = \frac{1}{2}(v(p)^2 + \omega^2 v(x)^2) = \omega(n + \frac{1}{2})$ for some $n = 0, 1, 2, \ldots$, which, as Schrödinger observed, would be impossible for arbitrary $v(p)$, $v(x)$ and ω. But Schrödinger's goal was *not* to prove that a value-map was impossible, since the point of his "theorem" was to show that it existed (assuming locality of course). He was just baffled by the situation: recognizing that this relationship between values

[6]This paper remained famous for his example of the cat that is "both dead and alive", but that example will not concerned us here.

suggested by the form of $O = \frac{1}{2}(p^2 + \omega^2 x^2)$ could not always hold, he wondered what relationship, if any, might exist among the relevant values. Of course, had Schrödinger made the (unwarranted) assumption of von Neumann and applied (2.3.2) to non-commuting operators, he would have been even more baffled, since he would probably have been led to question the locality assumption.

Finally, note that in 1966, much later than 1935, John Bell constructed in [3] an explicit counter-example to von Neumann's conclusions, by giving a simple example of a "hidden variables theory" that reproduces the quantum mechanical results *for a single spin operator* (but, of course, without satisfying (2.3.2) for non-commuting operators). Bohmian mechanics (see Sect. 5) also provides a counter-example to von Neumann's conclusions, but a more comprehensive one.

3.6 A Regularized EPR State

A way to avoid dealing with generalized functions or distributions such as (3.2.1), (3.2.6) is to put a cutoff both in the spatial and the momentum variables, x and p. A convenient way to do that is to require that x take values in a finite (but arbitrarily large) box on a lattice of (arbitrarily small) spacing a, which amounts to putting a cutoff in the momentum variable p.

So, let $x \in \Lambda_a = [-L, L] \cap a\mathbb{Z}$, or $x = na$, $n \in \mathbb{Z}$, $|n| \leq M$, with $M = [\frac{L}{a}]$, and $[\cdot]$ denoting the integer part.

Let $\widehat{\Lambda}_a$ be the dual of Λ_a:

$$\widehat{\Lambda}_a = \left\{ p = \frac{2\pi k}{a(2M+1)}, k \in \mathbb{Z}, |k| \leq M \right\}.$$

Then, one has the orthogonality relation: $\forall x \in \Lambda_a$

$$\sum_{p \in \widehat{\Lambda}_a} \exp(\pm ixp) = \sqrt{2M+1}\delta_{a,L}(x) \equiv (2M+1)\delta_{x,0}, \tag{3.6.1}$$

where $\delta_{x,0}$ is the Kronecker delta.

And, $\forall p \in \widehat{\Lambda}_a$,

$$\sum_{x \in \Lambda_a} \exp(\pm ixp) = \sqrt{2M+1}\delta_{a,L}(p) \equiv (2M+1)\delta_{p,0}. \tag{3.6.2}$$

Let, $\forall x_1, x_2, x_0 \in \Lambda_a$,

$$\Psi^{a,L}_{EPR}(x_1, x_2) = \sum_{p \in \widehat{\Lambda}_a} \exp(i(x_1 - x_2 + x_0)p), \tag{3.6.3}$$

where the sum $x_1 - x_2 + x_0$ is modulo $2aM$.

Using (3.6.1),

$$\Psi^{a,L}_{EPR}(x_1, x_2) = \sqrt{2M+1}\delta_{a,L}(x_1 - x_2 + x_0) \tag{3.6.4}$$

can be written as:

$$\Psi^{a,L}_{EPR}(x_1, x_2) = \sum_{x \in \Lambda_a} \delta_{a,L}(x - x_2 + x_0)\delta_{a,L}(x_1 - x). \tag{3.6.5}$$

One can also introduce the finite Fourier transform:

$$\widehat{\Psi}(p_1, p_2) = \frac{1}{2M+1} \sum_{x_1, x_2 \in \Lambda_a} \exp(-i(x_1 p_1 + x_2 p_2))\Psi(x_1, x_2) \tag{3.6.6}$$

whose inverse is:

$$\Psi(x_1, x_2) = \frac{1}{2M+1} \sum_{p_1, p_2 \in \widehat{\Lambda}_a} \exp(i(x_1 p_1 + x_2 p_2))\widehat{\Psi}(p_1, p_2). \tag{3.6.7}$$

The analogues of the operators P_1, P_2, Q_1, Q_2 of Sect. 3.2 are:

$$P_j\Psi(x_1, x_2) = \sum_{p_1, p_2 \in \widehat{\Lambda}_a} \exp(i(x_1 p_1 + x_2 p_2)) p_j \widehat{\Psi}(p_1, p_2), \quad j = 1, 2, \tag{3.6.8}$$

and

$$Q_j\Psi(x_1, x_2) = x_j\Psi(x_1, x_2), \quad j = 1, 2. \tag{3.6.9}$$

These operators have proper (not generalized) eigenvectors:

$$P_j \exp(-i(x_1 p_1 + x_2 p_2)) = p_j \exp(-i(x_1 p_1 + x_2 p_2)) \tag{3.6.10}$$

and

$$Q_j\delta_{a,L}(x_1 - x_{0,1})\delta_{a,L}(x_2 - x_{0,2}) = x_{0,j}\delta_{a,L}(x_1 - x_{0,1})\delta_{a,L}(x_2 - x_{0,2}). \tag{3.6.11}$$

Thus, if one applies the collapse rule for the measurement of the observable P_1 to $\Psi^{a,L}_{EPR}(x_1, x_2)$, when the observed value is p, the resulting state will be proportional to $\exp(i(x_1 - x_2 + x_0)p)$, meaning that the state of the second particle will be proportional to $\exp(-i(x_2 - x_0)p)$. And, if one applies the collapse rule for the measurement of the observable Q_1 to $\Psi^{a,L}_{EPR}(x_1, x_2)$, when the observed value is x, the resulting state will be proportional to $\delta_{a,L}(x - x_2 + x_0)\delta_{a,L}(x_1 - x)$, meaning that the state of the second particle will be proportional to $\delta_{a,L}(x - x_2 + x_0)$.

4 Proof of Nonlocality Using the EPR Variables

Given a state like (3.2.1), (3.2.6), we can almost repeat the arguments of Sect. 2 in order to prove nonlocality. First observe that one has an analogue of a Schrödinger theorem. Consider a generalized state for four particles in one dimension:

$$\delta(x_1 - x_3 + x_0)\delta(x_2 - x_4 + x_0), \tag{4.1}$$

which is just the product of two copies of the EPR state (up to a $4\pi^2$ factor, see (3.2.6)), one for the pair of particles $(1, 3)$, the other for the pair of particles $(2, 4)$. Alternatively, one may regard this as a state of two particles in two dimensions, with coordinates (x_1, x_2) and (x_3, x_4). In our previous notation, system 1 will consist of particles 1 and 2 and system 2 will consist of particles 3 and 4.[7]

One may also replace that state by its regularized version, see (3.6.4):

$$\delta_{a,L}(x_1 - x_3 + x_0)\delta_{a,L}(x_2 - x_4 + x_0). \tag{4.2}$$

By Remark 2 in Sect. 2.1, the state (4.2) is maximally entangled and so the state (4.1) is also (formally) maximally entangled.[8]

We need to introduce standard operators Q_1, Q_2, Q_3, Q_4, that act as multiplication on a suitable set of functions in $L^2(\mathbb{R}^4)$:

$$Q_j\Psi(x_1, x_2, x_3, x_4) = x_j\Psi(x_1, x_2, x_3, x_4)\ , \quad j = 1, 2, 3, 4\ , \tag{4.3}$$

and operators P_1, P_2, P_3, P_4 that act by differentiation on a suitable set of functions in $L^2(\mathbb{R}^4)$:

$$P_j\Psi(x_1, x_2, x_3, x_4) = -i\frac{\partial}{\partial x_j}\Psi(x_1, x_2, x_3, x_4)\ , \quad j = 1, 2, 3, 4\ . \tag{4.4}$$

Or, using the Fourier transform (3.2.3) of Ψ (for four variables):

$$P_j\Psi(x_1, x_2, x_3, x_4) =$$
$$\frac{1}{(2\pi)^2}\int \exp(i(p_1x_1 + p_2x_2 + p_3x_3 + p_4x_4))p_j\widehat{\Psi}(p_1, p_2, p_3, p_4)dp_1dp_2, dp_3dp_4, \tag{4.5}$$

for $j = 1, 2, 3, 4$.

Consider the eight operator Q_1, Q_2, Q_3, Q_4, P_1, P_2, P_3, P_4, defined by (4.3) and (4.4), (4.5).

Let $\mathcal{B}$ be the set of products of analytic functions of one of the operators Q_1, Q_2, P_1, P_2 defining a self-adjoint operator, and let $\tilde{\mathcal{B}}$ be the set of sums of products

[7] We need two copies of the EPR state only in order to prove Theorem 4.1 below.

[8] Formally, since the state itself is not a vector in a finite dimensional Hilbert space. But, since we are not concerned here with mathematical rigor, we will put aside that issue.

of analytic functions of one of the operators Q_3, Q_4, P_3, P_4 defining a self-adjoint operator.

Given the maximally entangled state (4.1), for every operator $\tilde{O} \in \tilde{\mathcal{B}}$, there is a corresponding (in the sense of the Principle of Perfect Correlations) operator $O \in \mathcal{B}$, and vice-versa. (For $x_0 = 0$, O is obtained by changing in $\tilde{O}$ the index 3 to 1 and the index 4 to 2). And, by Schrödinger's theorem, assuming locality, there is a non-contextual value-map $v : \mathcal{B} \to \mathbb{R}$ that satisfies (2.2.1) and therefore also the property (2.3.2).

However this is contradicted by a theorem of Clifton [14], proven in the appendix.

Theorem 4.1 Non-existence of pre-existing values for positions and momenta.

Consider the set of analytic functions of one of the operators Q_1, Q_2, P_1, P_2. And let $\mathcal{B}$ be the set of products of such functions defining a self-adjoint operator. Then, there does not exist a map

$$v : \mathcal{B} \to \mathbb{R} \tag{4.6}$$

such that:

(1)

$$v(f(O)) = f(v(O)), \tag{4.7}$$

for any real valued function f of a real variable.

(2) $\forall O, O' \in \mathcal{B}$ *with* $[O, O'] = OO' - O'O = 0$, *(2.3.2) for* $f(x, y) = xy$ *holds:*

$$v(OO') = v(O)v(O'). \tag{4.8}$$

In particular, there cannot exist a non-contextual value-map.

So, combining the EPR argument with the previous theorem, we again establish nonlocality, without using Bell's inequalities.

The logic is the same as in Sect. 2:

1. EPR show that the perfect correlations plus locality imply that the values of some physical quantities (the values $v(O)$ of the operators $O \in \mathcal{A}$ in Sect. 2.3 or the operators $O \in \mathcal{B}$ here), must exist independently of whether one measures them or not, and that defines a non-contextual value-map.
2. Theorems 2.2 or 4.1 show that merely assuming the existence of such a map leads to a contradiction.

Therefore the locality assumption is false!

5 What Happens in Bohmian Mechanics?

In Bohmian mechanics, or pilot-wave theory, the complete state of a closed physical system composed of N particles is a pair (|quantum state>, $\mathbf{X}$), where |quantum state> is the usual quantum state (given by the tensor product of wave functions with some possible internal states), and $\mathbf{X} = (X_1, \ldots, X_N)$ is the configuration representing the positions of the particles (that exist, independently of whether one "looks" at them or one measures them; each $X_i \in \mathbb{R}^3$).[9]

These positions are the "hidden variables" of the theory, in the sense that they are not included in the purely quantum description |quantum state>, but they are not at all hidden: it is only the particles' positions that one detects directly, in any experiment (think, for example, of the impacts on the screen in the two-slit experiment). So the expression "hidden variables" is really a misnomer, at least in the context of Bohmian mechanics.

Both objects, the quantum state and the particles' positions, evolve according to deterministic laws, the quantum state guiding the motion of the particles. Indeed, the time evolution of the complete physical state is composed of two laws (we consider, for simplicity, spinless particles):

1. The wave function evolves according to the usual Schrödinger's equation.
2. The particle positions $\mathbf{X} = \mathbf{X}(t)$ evolve in time according to a guiding equation determined by the quantum state: their velocity is a function of the wave function. If one writes[10]:

$$\Psi(x_1, \ldots, x_N) = R(x_1, \ldots, x_N) e^{iS(x_1, \ldots, x_N)},$$

then:

$$\frac{dX_k(t)}{dt} = \nabla_k S(X_1(t), \ldots, X_N(t)), \tag{5.1}$$

where ∇_k is the gradient with respect the coordinates of the kth particle.

In order to understand why Bohmian mechanics reproduces the usual quantum predictions, one must use a fundamental consequence of that dynamics, *equivariance*: If the probability density $\rho_{t_0}(\mathbf{x})$ for the initial configuration $\mathbf{X}_{t_0}$ is given by $\rho_{t_0}(\mathbf{x}) = |\Psi(\mathbf{x}, t_0)|^2$, then the probability density for the configuration $\mathbf{X}_t$ at any time t is given by

$$\rho_t(\mathbf{x}) = |\Psi(\mathbf{x}, t)|^2, \tag{5.2}$$

[9]For elementary introductions to this theory, see [10, 38] and for more advanced ones, see [5, 7–9, 15–17, 23, 30]. There are also pedagogical videos made by students in Munich, available at: https://cast.itunes.uni-muenchen.de/vod/playlists/URqb5J7RBr.html.

[10]We use lower case letters for the generic arguments of the wave function and upper case ones for the actual positions of the particles.

where $\Psi(\mathbf{x}, t)$ is a solution to Schrödinger's equation. This follows easily from equation (5.1).

Because of equivariance, the quantum predictions for the results of measurements of any quantum observable are obtained if one assumes that the initial density satisfies $\rho_{t_0}(\mathbf{x}) = |\Psi(\mathbf{x}, t_0)|^2$. The assertion that configurational probabilities at any time t_0 are given by this "Born rule" is called the *quantum equilibrium hypothesis*. The justification of the quantum equilibrium hypothesis—and, indeed, a clear statement of what it actually means—is a long story, too long to be discussed here (see [15]).

In Bohmian mechanics, particles have a velocity at all times and therefore they have what we would be inclined to call a momentum (mass $\times$ velocity). So one might ask, what sort of probability does Bohmian mechanics supply for the latter: will it agree with the quantum mechanical probability for momentum? The answer, as we will see in the next subsection, is no!

One may also ask: isn't having both a position and a velocity at the same time contradicted by Heisenberg's inequalities? Moreover, since Bohmian mechanics is deterministic, the result of any quantum experiment must be pre-determined by the initial conditions of the system being measured and of the measuring device. But why doesn't that provide a non-contextual value-map whose existence is precluded by Theorem 4.1? We will discuss these issues in the following subsections and this will also provide an example of how nonlocality manifests itself in Bohmian mechanics.

5.1 The Measurement of Momentum in Bohmian Mechanics

To understand what is going on, we should analyze "momentum measurements," i.e., what are called momentum measurements in standard quantum mechanics. Consider a simple example, namely a particle in one space dimension with initial wave function $\Psi(x, 0) = \pi^{-1/4} \exp(-x^2/2)$. Since this function is real, its phase $S = 0$ and the particle is at rest (by equation (5.1): $\frac{dX(t)}{dt} = \frac{\partial S(X(t),t)}{\partial x}$). Nevertheless, the measurement of momentum p must have, according to the usual quantum predictions, a probability distribution whose density is given by the square of the Fourier transform of $\Psi(x, 0)$, i.e. by $|\hat{\Psi}(p)|^2 = \pi^{-1/2} \exp(-p^2)$. Isn't there a contradiction here? Isn't there a clear disagreement with the quantum predictions?

In order to answer this question, one must focus on the quantum mechanical *measurement* of momentum. One way to do this is to let the particle move freely and to detect its asymptotic position $X(t)$ as $t \to \infty$. Then, one sets $p = \lim\limits_{t\to\infty} \dfrac{X(t)}{t}$ (putting the mass $m = 1$).

Consider the free evolution of the initial wave function at $t_0 = 0$, $\Psi(x, 0) = \pi^{-1/4} \exp(-x^2/2)$. The solution of Schrödinger's equation with that initial condition is:

$$\Psi(x, t) = \frac{1}{(1 + it)^{1/2}} \frac{1}{\pi^{1/4}} \exp\left[-\frac{x^2}{2(1 + it)}\right], \tag{5.1.1}$$

and thus

$$|\Psi(x,t)|^2 = \frac{1}{\sqrt{\pi\left[1+t^2\right]}} \exp\left[-\frac{x^2}{1+t^2}\right]. \tag{5.1.2}$$

If one writes $\Psi(x,t) = R(x,t)\exp\left[iS(x,t)\right]$, one gets (up to a t-dependent constant):

$$S(x,t) = \frac{tx^2}{2(1+t^2)}, \tag{5.1.3}$$

and the guiding equation (5.1) becomes:

$$\frac{d}{dt}X(t) = \frac{tX(t)}{1+t^2}, \tag{5.1.4}$$

whose solution is:

$$X(t) = X(0)\sqrt{1+t^2}. \tag{5.1.5}$$

This gives the explicit dependence of the position of the particle as a function of time. If the particle is initially at $X(0) = 0$, it does not move; otherwise, it moves asymptotically, when $t \to \infty$, as $X(t) \sim X(0)t$. Thus, $p = \lim_{t\to\infty} X(t)/t = X(0)$.

Now, assume that we start with the quantum equilibrium distribution:

$$\rho_0(x) = |\Psi(x,0)|^2 = \pi^{-1/2}\exp(-x^2).$$

This is the distribution of $X(0)$. Thus, the distribution of $p = \lim_{t\to\infty} X(t)/t = X(0)$ will be $\pi^{-1/2}\exp(-p^2) = |\hat{\Psi}(p,0)|^2$. This is the quantum prediction! But the detection procedure (measurement of $X(t)$ for large t) does *not* measure the initial velocity (which is zero with probability 1).

Remarks

1. Although the particles do have, at all times, *a position and a velocity*, there is no contradiction between Bohmian mechanics and the quantum predictions and, in particular, with Heisenberg's uncertainty principle. The latter is simply a relation between variances of results of measurements. It implies nothing whatsoever about what exists or does not exist outside of measurements, since those relations are simply mathematical consequences of the quantum formalism which, strictly speaking, dictates only what takes place during a measurement.
2. Bohmian mechanics shows that what are called measurements of quantum observables other than positions are typically merely *interactions* between a microscopic physical system and a macroscopic measuring device whose statistical results coincide with the quantum predictions.

   To use a fashionable expression, one might say for both Bohmian mechanics and standard quantum mechanics, values of most observables are *emergent*. But it is only in Bohmian mechanics that one can understand how that emergence comes about.

5.2 *The Contextuality of the Momentum Measurements in Bohmian Mechanics*

The reader might nevertheless worry that there *is* in fact an intrinsic property of the particle that is revealed in a momentum measurement, for example its original position, since, as we showed in the previous subsection, $p = \lim_{t\to\infty} X(t)/t = X(0)$ in the simple case considered there. Of course, if one were to measure the position one would also find an intrinsic property of the particle (namely its position!).[11] But doesn't that contradict our Theorem 4.1 (our example could of course be formulated in two dimensions by taking a product of wave functions of the form (5.1.1))? After all, the latter theorem asserts that *there does not exist* a value-map that assigns to a quantum system pre-existing values that are revealed by quantum measurements and here we seem to have just defined such a map.

However, as we shall explicitly show, the map provided by Bohmian mechanics would be contextual (see the Appendix for the concrete operators that we use in the proof of Theorem 4.1). In particular the value $v(O)$ will depend on which other operators O', O'', ..., one measures together with O. Hence relations like (4.8) that are needed to prove Theorem 4.1 will not be valid: for example, if one writes $v(OO') = v(O)v(O')$ and $v(OO'') = v(O)v(O'')$, the value $v(O)$ will in general be different in the two relations.

We will now show in particular that the measurement of momentum is contextual, using a modified version of the example given by (5.1.1).

Take that quantum state (5.1.1) and write $\Psi_0(x)$ for $\Psi(x, 0)$. Consider the corresponding Gaussian wave functions:

$$\Psi_{+k}(x) = \Psi_0(x)e^{ikx} \tag{5.2.1}$$

and

$$\Psi_{-k}(x) = \Psi_0(x)e^{-ikx} \tag{5.2.2}$$

where $k > 0$. We will assume below that k is large.

Consider first the initial wave function $\Psi_{+k}(x) = \Psi_0(x)e^{ikx}$. This is a right-moving Gaussian wave packet moving with speed k. Thus at time t it will be centered at kt. Explicitly, the solution of Schrödinger's equation is:

[11] The fact that the measurements of both the momentum and the position reveal the same intrinsic property may sound strange but that is just a peculiarity of the example considered here.

$$\Psi_{+k}(x,t) = \frac{1}{(1+it)^{1/2}}\frac{1}{\pi^{1/4}}\exp\left(ikx - \frac{ik^2t}{2} - \frac{(x-kt)^2}{2(1+it)}\right), \tag{5.2.3}$$

which can also been seen immediately from (5.1.1) using Galilean invariance. For this wave packet we have that $p = \lim_{t\to\infty} \frac{X(t)}{t} \approx k$ for $k \gg 1$.

Now form an $N = 2$ entangled state Ψ from the wave functions (5.2.1), (5.2.2)[12]:

$$\Psi(x, y) = A[\Psi_{+k}(x)\Psi_{+k}(y) + \Psi_{-k}(x)\Psi_{-k}(y)], \tag{5.2.4}$$

with A the normalization constant. Let $O = P_x$. Consider two different experiments that measure O:

Experiment$_1(O)$: measure O alone by the procedure described in Sect. 5.1, with result corresponding to the solution to the guiding equation (5.1) associated with the solution of Schrödinger's equation.

Experiment$_2(O)$: first measure at time 0 the position Q_y of the second particle, then measure O by the above procedure.

For Experiment$_1(O)$, we claim that the result is

$$v(\text{Experiment}_1(O)) \approx \text{sgn}\,(X(0) + Y(0))k \tag{5.2.5}$$

for k large.

To prove (5.2.5), introduce the variables:

$$\begin{aligned} w &= \frac{x+y}{\sqrt{2}}, \\ z &= \frac{x-y}{\sqrt{2}}. \end{aligned} \tag{5.2.6}$$

In terms of these variables, we can rewrite (5.2.4) as

$$\Psi(w, z) = A(\Psi_{+k'}(w) + \Psi_{-k'}(w))\Psi_0(z). \tag{5.2.7}$$

with $k' = \sqrt{2}k$.

So the solution of Schrödinger's equation factorizes into a function $\Psi(w,t)$ of (w,t) and a function $\tilde{\Psi}(z,t)$ of (z,t). We have that $\tilde{\Psi}(z,t)$ is given by (5.1.1) with x replaced by z, while for $\Psi(w,t)$ we get a sum of two wave functions like (5.2.3), one with k replaced by k', the other with k replaced by $-k'$:

$$\Psi(w, t) = A(\Psi_{+k'}(w, t) + \Psi_{-k'}(w, t)) \tag{5.2.8}$$

with $\Psi_{\pm k'}(w, t)$ of the form (5.2.3).

---

[12]This state ressembles a maximally entangled one, but it does not fit the definition of maximally entangled, since the Hilbert space here in infinite dimensional.

For large t, $|\Psi(w,t)|^2$ is a sum of two more or less non-overlapping terms, one corresponding to the part of the wave function with k' (whose support is around $k't$), the other one corresponding to the part of the wave function with $-k'$ (whose support is around $-k't$):

$$|\Psi(w,t)|^2 \approx A^2(|\Psi_{+k'}(w,t)|^2 + |\Psi_{-k'}(w,t)|^2). \tag{5.2.9}$$

Since the solution of Schrödinger's equation factorizes into a function of (w,t) and one of (z,t), the guiding equations (5.1) for $W(t)$ and $Z(t)$ are decoupled. For $Z(t)$ we obtain a solution like (5.1.5) ($Z(t) \approx Z(0)t$ as $t \to \infty$).

To analyze $W(t)$, note that one property of the dynamics (5.1) is that, in one dimension, trajectories cannot cross.[13] Since there is a symmetry between the two parts of the wave function (5.2.8) (upon reflection, $\Psi_{+k'}$ becomes $\Psi_{-k'}$), if the initial condition $W(0) > 0$, the particle must stay on the right, while if $W(0) < 0$, the particle must stay on the left. Moreover, by equivariance, the particle evolves so as to be in the support of $|\Psi(w,t)|^2$, which, by (5.2.9), consists of two non-overlapping terms supported around $\pm k't$ for large t. So, for large k and large times, we get that $W(t) \approx \text{sgn}\, W(0)k't = \text{sgn}\, W(0)\sqrt{2}kt$.

Rewriting what we've found in terms of the $X(t)$ and $Y(t)$ variables, we get that $X(t) = \frac{W(t)+Z(t)}{\sqrt{2}} \approx \frac{1}{\sqrt{2}}(\text{sgn}\, W(0)\sqrt{2}kt + Z(0)t)$ and thus, $v(\text{Experiment}_1(O)) = \lim_{t\to\infty} \frac{X(t)}{t} \approx \text{sgn}\,(X(0) + Y(0))k$, for k large, which is (5.2.5).

For Experiment$_2(O)$, if Y is the result of the measurement of Q_y, the wave function (5.2.4) collapses, yielding for the wave function of the x system[14]:

$$\Psi(x) = A(Y)(c_+(Y)\Psi_{+k}(x) + c_-(Y)\Psi_{-k}(x)). \tag{5.2.10}$$

with $c_\pm(Y) = \Psi_{\pm k}(Y)$ and $A(Y)$ the normalization coefficient.

The solution of Schrödinger's equation with this initial condition is again a sum of two wave functions like (5.2.3), one with $+k$, the other with $-k$, multiplied by coefficients $c_\pm(Y)$:

$$\Psi(x,t) = A(Y)(c_+(Y)\Psi_{+k}(x,t) + c_-(Y)\Psi_{-k}(x,t)), \tag{5.2.11}$$

where $\Psi_{\pm k}(x,t)$ of the form (5.2.3).

We can now more or less reason as we just did for the $\Psi(w,t)$ given by (5.2.8), except that because of the coefficients $c_\pm(Y)$ there is no symmetry here between the two parts of the wave function—unless the complex exponentials in $c_\pm(Y)$ are real (i.e. $e^{ikY} = \pm 1$). Nonetheless, the effect of the coefficients in (5.2.11) is merely to replace the $\cos kx$, which would arise there if $c_\pm(Y) > 0$ (i.e. $e^{ikY} = 1$), by its translate $\cos(kx + kY)$. Thus the $|\Psi|^2$ probability of the interval $[X_m, \infty)$ will be 1/2

---

[13]That is because there is a unique solution of the first order equation (5.1) if the position is fixed at a given time.

[14]In Bohmian mechanics, in fact, there is an actual collapse of the (conditional) wave function of a system upon measurement; see [8, Sect. 6.1], [4].

for some X_m with $|X_m| < \frac{\pi}{2k}$.[15] Thus, by no-crossing and equivariance, we get that for large times $X(t) \approx \text{sgn}\,(X(0))kt$ for $k \gg 1$, and thus

$$v(\text{Experiment}_2(O)) = \lim_{t\to\infty} \frac{X(t)}{t} \approx \text{sgn}\,(X(0))k. \tag{5.2.12}$$

Comparing (5.2.5) and (5.2.12), we see that the measurement of momentum is contextual, since it may depend on whether or not one measures another operator Q_y together with $O = P_x$.

5.3 An Example of Nonlocality in Bohmian Mechanics

It would go far beyond the scope of this paper to really explain how nonlocality appears in Bohmian mechanics in general, but we saw an example of nonlocality in Bohmian mechanics in the previous subsection: the particles with coordinates x and y having the entangled quantum state (5.2.4), can be (in principle) as far apart as one wants and the result of the measurement of $O = P_x$ will depend on whether or not one measures Q_y before measuring P_x, and, since the time interval between these two measurements can be arbitrarily small, we have indeed here an example of an instantaneous action at a distance. Here we should regard the measurement of P_x as taking a (large but) finite time, and x and y as referring to different (distant) origins.

The fact that Bohmian mechanics is nonlocal is obviously a merit rather than a defect, since we know that any theory accounting for the quantum phenomena must be nonlocal, as shown in Sects. 2–4 (and many other places).

6 Summary and Conclusions

Both EPR and Schrödinger argued that the quantum mechanical description of a system by its wave function is incomplete in the sense that other variables must be introduced in order to obtain a complete description. Their argument was very simple: if I can determine the result of a measurement carried at one place by doing another measurement far away from that place, then that result must pre-exist its measurement. The wave function alone does not tell us what that result is. Therefore, the quantum mechanical description of a system by its wave function is incomplete.

However, there was a crucial assumption in the reasoning of EPR and Schrödinger, which was too obvious for them to question it: that doing a measurement at one place cannot possibly affect instantaneously the physical situation far away, or what is now called the assumption of locality.

---

[15] In fact, X_m must lie between 0 and the nearest maximum of $\cos^2(kx + kY)$.

The history of the EPR-Schrödinger argument is complicated, because although their conclusion about incompleteness of quantum mechanics was right, their assumption of locality was not. The completion of quantum mechanics was found by de Broglie in 1927 and developed by Bohm in 1952. Bohm showed that one may consistently assume that particles have trajectories and explained on that basis how to understand measurements as consequences of the theory and not, as they are in ordinary quantum mechanics, as a *deus ex machina* [7].

The falsity of the locality assumption was shown by John Bell in 1964 [2] and by subsequent experiments. Bell first recalled that, if one assumes locality, then, as the EPR argument correctly showed, there must exist other variables than the quantum state to characterize a physical system. But then Bell showed that the distribution of those variables must satisfy some constraints that are violated by quantum predictions, predictions that were later verified experimentally (see [22] for a survey).

Here and in [11] we give a simpler argument, but using the maximally entangled states introduced by Schrödinger: for those states, one can, for each observable associated to one system, construct another observable associated to the second system, possibly far away from the first one, such that the results of the measurement of both observables are perfectly correlated. Then, assuming locality, those results must pre-exist their measurement. But assuming that, in general, observables have values before their measurement leads to a contradiction. Hence, the assumption of locality is false.

The difference between this paper and [11] is that here we use the position and momentum variables used by EPR, while in [11] we used spin variables, such as those in terms of which the EPR argument was reformulated by Bohm [6].

Next one might ask how Bohmian mechanics deals with this impossibility of the pre-existence of measurement results prior to measurements, since it is a deterministic theory, and in such a theory everything is pre-determined by the initial condition. In [11] we reviewed that the measurements of spin variables are contextual, in fact should not properly be called measurements at all. Here we illustrate the contextuality of momentum. In both cases, the contextuality is linked to nonlocality, as it must be, since as explained here and in [11], if locality were true, then measurements must (sometimes) be non-contextual. Bohmian mechanics is an extremely natural version of quantum mechanics, involving the obvious ontology evolving the obvious way. A proper appreciation of the role of contextuality in Bohmian mechanics can help dispel the widespread uneasy feeling that somehow there must be something amiss in that theory.

Appendix 1: Proof of Clifton's Theorem 4.1

The proof we give here is taken from a paper by Myrvold [29], which is a simplified version of the result of Clifton [14] and is similar to proofs of Mermin [28], and to Peres [31, 32] in the case of spins. We note that the same proof would apply to the regularized EPR state of Sect. 3.6.

Proof of Theorem 4.1
We will need the operators $U_j(b) = \exp(-ibQ_j)$, $V_j(c) = \exp(-icP_j)$, $j = 1, 2$, with Q_j, P_j defined by formulas (4.3), (4.4), but acting in $L^2(\mathbb{R}^2)$ instead of $L^2(\mathbb{R}^4)$, and $b, c \in \mathbb{R}$. They act as

$$U_j(b)\Psi(x_1, x_2) = \exp(-ibx_j)\Psi(x_1, x_2) \ , \quad j = 1, 2 \ , \tag{A.1}$$

which follows trivially from (4.3), and

$$V_1(c)\Psi(x_1, x_2) = \Psi(x_1 - c, x_2) \ , \tag{A.2}$$

and similarly for $V_2(c)$. Equation (A.2) follows from (4.4) by expanding both sides in a Taylor series, for functions Ψ such that the series converges, and by extending the unitary operator $V_2(b)$ to more general functions Ψ (see, e.g., [33, Chap. 8] for an explanation of that extension).

We choose now the following functions of the operators Q_i, P_i:

$$A_1 = \cos(aQ_1) \ , \quad A_2 = \cos(aQ_2) \ , \quad B_1 = \cos\frac{\pi P_1}{a} \ , \quad B_2 = \cos\frac{\pi P_2}{a} \ , \tag{A.3}$$

where a is an arbitrary constant, and the functions are defined by (A.1), (A.2), and the Euler relations:

$$\begin{aligned} \cos(aQ_j) &= \frac{\exp(iaQ_j) + \exp(-iaQ_j)}{2} \ , \\ \cos\frac{\pi P_j}{a} &= \frac{\exp(i\pi P_j/a) + \exp(-i\pi P_j/a)}{2} \ , \end{aligned} \tag{A.4}$$

for $j = 1, 2$. Note that A_1, A_2, B_1, B_2 are self-adjoint. By applying (4.8) several times to pairs of commuting operators made of products of such operators, we will derive a contradiction.

We have the relations

$$[A_1, A_2] = [B_1, B_2] = [A_1, B_2] = [A_2, B_1] = 0 \ , \tag{A.5}$$

since the relevant operators act on different variables.

We can also prove:

$$A_1B_1 = -B_1A_1 \ , \qquad A_2B_2 = -B_2A_2 \ . \tag{A.6}$$

To show (A.6), note that, from (A.1) and (A.2), one gets

$$U_j(b)V_j(c) = \exp(-ibc)V_j(c)U_j(b) \ , \tag{A.7}$$

for $j = 1, 2$, which, for $bc = \pm\pi$, means

$$U_j(b)V_j(c) = -V_j(c)U_j(b)\ . \tag{A.8}$$

Now use (A.4) to expand the product $\cos(aQ_j)\cos(\pi P_j/a)$, for $j = 1, 2$, into a sum of four terms; each term will have the form of the left-hand side of (A.7) with $b = \pm a$, $c = \pm\pi/a$, whence $bc = \pm\pi$. Then applying (A.8) to each term proves (A.6).

The relations (A.5) and (A.6) imply that

$$[A_1A_2, B_2B_1] = 0 \tag{A.9}$$

since two anticommutations (A.6) suffice to move the B's to the left of the A's. Similarly we have that

$$[A_1B_2, A_2B_1] = 0. \tag{A.10}$$

We also have, using (A.6) once, that

$$A_1A_2B_2B_1 = -A_1B_2A_2B_1. \tag{A.11}$$

Thus, with $C = (A_1A_2)(B_2B_1)$ and $D = (A_1B_2)(A_2B_1)$, we have that

$$C = -D. \tag{A.12}$$

Now suppose there is a value map v as described in Theorem 4.1. Then, from (4.7) with $f(x) = -x$, we have that

$$v(C) = -v(D). \tag{A.13}$$

But by (A.5), (A.9) and (A.10), we also have, by (4.8), that

$$v(C) = v(A_1A_2)v(B_2B_1) = v(A_1)v(A_2)v(B_2)v(B_1) \tag{A.14}$$

and

$$v(D) = v(A_1B_2)v(A_2B_1) = v(A_1)v(B_2)v(A_2)v(B_1). \tag{A.15}$$

Thus $v(C) = v(D)$. This is a contradiction unless $v(C) = 0$, i.e. unless at least one of $v(A_i)$, $v(B_i)$, $i = 1, 2$ vanishes. But, by (4.7),

$$v(A_i) = \cos(av(Q_i))$$

and

$$v(B_i) = \cos\left(\frac{\pi}{a}v(P_i)\right),$$

and thus a can be so chosen that $v(A_i)$ and $v(B_i)$ are all nonvanishing. ■

References

1. P.K. Aravind, Bell's theorem without inequalities and only two distant observers, *Foundations of Physics Letters* **15**, 399–405 (2002)
2. J.S. Bell: On the Einstein–Podolsky–Rosen paradox, *Physics* **1**, 195–200 (1964). Reprinted as Chap. 2 in [5]
3. J.S. Bell: On the problem of hidden variables in quantum mechanics, *Reviews of Modern Physics* **38**, 447–452 (1966). Reprinted as Chap. 1 in [5]
4. J.S. Bell: de Broglie–Bohm, delayed-choice double-slit experiment, and density matrix, *International Journal of Quantum Chemistry*: Quantum Chemistry Symposium **14**, 155–159 (1980). Reprinted as Chap. 14 in [5]
5. J.S. Bell: *Speakable and Unspeakable in Quantum Mechanics. Collected Papers on Quantum Philosophy*, 2nd edn, with an introduction by Alain Aspect, Cambridge University Press, Cambridge, 2004; 1st edn 1987
6. Bohm D., *Quantum Theory*, Dover Publications, New York, 1989.
7. D. Bohm: A suggested interpretation of the quantum theory in terms of "hidden variables", Parts 1 and 2, *Physical Review* **89**, 166–193 (1952). Reprinted in [40] pp. 369–390
8. D. Bohm and B.J. Hiley: *The Undivided Universe*, Routledge, London, 1993
9. J. Bricmont, *Making Sense of Quantum Mechanics*, Springer, Berlin, 2016
10. J. Bricmont, *Quantum Sense and Nonsense*, Springer International Publishing, Switzerland, 2017
11. J. Bricmont, S. Goldstein, D. L. Hemmick, Schrödinger's paradox and proofs of nonlocality using only perfect correlations, preprint; arXiv:1808.01648
12. H.R. Brown, G. Svetlichny, Nonlocality and Gleason's lemma. Part I:. Deterministic theories, *Foundations of Physics* **20**, 1379–1386 (1990)
13. A. Cabello, Bell's theorem without inequalities and without probabilities for two observers, *Physical Review Letters* **86**, 1911–1914 (2001)
14. R. Clifton: Complementarity between position and momentum as a consequence of Kochen–Specker arguments, *Physics Letters A* **271**, 1–7 (2000)
15. D. Dürr, S. Goldstein and N. Zanghì: Quantum equilibrium and the origin of absolute uncertainty, *Journal of Statistical Physics* **67**, 843–907 (1992)
16. D. Dürr, S. Teufel, *Bohmian Mechanics. The Physics and Mathematics of Quantum Theory*, Springer, Berlin-Heidelberg, 2009.
17. D. Dürr, S. Goldstein and N. Zanghì: *Quantum Physics Without Quantum Philosophy*, Springer, Berlin, Heidelberg, 2012
18. A. Einstein, B. Podolsky and N. Rosen: Can quantum mechanical description of physical reality be considered complete?, *Physical Review* **47**, 777–780 (1935)
19. A. Einstein: Letter to Erwin Schrödinger, 19 June 1935, in [21, p. 35]
20. A. Elby, Nonlocality and Gleason's Lemma. Part 2. *Foundations of Physics* **20**, 1389–1397 (1990)
21. A. Fine: *The Shaky Game: Einstein Realism and the Quantum Theory*, University of Chicago Press, Chicago, 1986
22. S. Goldstein, T. Norsen, D.V. Tausk and N. Zanghì: Bell's theorem, *Scholarpedia* 6(10): 8378 (2011)
23. S. Goldstein: Bohmian mechanics, *The Stanford Encyclopedia of Philosophy*, Edward N. Zalta (ed.), https://plato.stanford.edu/entries/qm-bohm
24. D. L. Hemmick, Hidden variables and nonlocality in quantum mechanics, Doctoral thesis, Rutgers University, October, 1996, available on https://arxiv.org/abs/quant-ph/0412011v1
25. D. L. Hemmick, A. M. Shakur, *Bell's theorem and quantum realism. Reassessment in light of the Schrödinger paradox*, Springer, Heidelberg, 2012
26. P. Heywood, M. L. G. Redhead, Nonlocality and the Kochen-Specker Paradox, *Foundations of Physics* **13**, 481–499 (1983)
27. S. Kochen and E. P. Specker: The problem of hidden variables in quantum mechanics, *Journal of Mathematics and Mechanics* **17**, 59–87 (1967)

28. D. Mermin: Hidden variables and the two theorems of John Bell, *Reviews of Modern Physics* **65**, 803–815 (1993)
29. W.C. Myrvold: On some early objections to Bohm's theory, *International Studies in the Philosophy of Science* **17**, 7–24 (2003)
30. T. Norsen, *Foundations of Quantum Mechanics: An Exploration of the Physical Meaning of Quantum Theory*, Springer International Publishing, Switzerland, 2017
31. A. Peres: Incompatible results of quantum measurements, *Physics Letters A* **151**, 107–108 (1990)
32. A. Peres: Two simple proofs of the Kochen–Specker theorem, *Journal of Physics A: Math. Gen.* **24**, L175–L178 (1991)
33. M. Reed and B. Simon: *Methods of Modern Mathematical Physics I: Functional Analysis*, Academic Press, New York, 1972
34. E. Schrödinger: Die gegenwärtige Situation in der Quantenmechanik, *Naturwissenschaften* **23**, 807–812; 823–828; 844–849 (1935). English translation: The present situation in quantum mechanics, translated by J.- D. Trimmer, *Proceedings of the American Philosophical Society* **124**, 323–338 (1980). Reprinted in [40] pp. 152–167
35. E. Schrödinger: Discussion of probability relations between separated systems, *Mathematical Proceedings of the Cambridge Philosophical Society* **31**, 555–563 (1935)
36. E. Schrödinger: Probability relations between separated systems, *Mathematical Proceedings of the Cambridge Philosophical Society* **32**, 446–452 (1936)
37. A. Stairs: Quantum Logic, Realism and Value-Definiteness, *Philosophy of Science* **50**, 578–602 (1983)
38. R. Tumulka: Understanding Bohmian mechanics – A dialogue, *American Journal of Physics* **72**, 1220–1226 (2004)
39. J. von Neumann, *Mathematical Foundations of Quantum Mechanics*, Princeton University Press, Princeton, 1955. First edition in German, *Mathematische Grundlagen der Quantenmechanik*, 1932
40. J.A. Wheeler and W.H. Zurek (eds): *Quantum Theory and Measurement*, Princeton University Press, Princeton, 1983

Typicality in the Foundations of Statistical Physics and Born's Rule

Detlef Dürr and Ward Struyve

Abstract Typicality has always been in the minds of the founding fathers of probability theory when probabilistic reasoning is applied to the real world. However, the role of typicality is not always appreciated. An example is the article "Foundations of statistical mechanics and the status of Born's rule in de Broglie-Bohm pilot-wave theory" by Antony Valentini (Valentini in Statistical Mechanics and Scientific Explanation. World Scientific, [1]), where he presents typicality and relaxation to equilibrium as *distinct* approaches to the proof of Born's rule, while typicality is in fact an overriding necessity. Moreover the "typicality approach" to Born's rule of "the Bohmian mechanics school" is claimed to be inherently circular. We wish to explain once more in very simple terms why the accusation is off target and why "relaxation to equilibrium" is neither necessary nor sufficient to justify Born's rule.

Nino Zanghì and D.D. remember vividly the discussions with GianCarlo Ghirardi on Boltzmann's insights into statistical physics and its relation to the random theory he himself had proposed (with his coworkers) and had worked on for many decades until his untimely death. Not only was GianCarlo an admirer of Boltzmann, he also had a full grasp of Boltzmann's ideas and on the role of typicality. The GRW theory is intrinsically random and the $|\psi|^2$-distribution arises from the collapse mechanism built into the theory and he understood that the appeal to typicality, for empirical assertions, cannot be avoided. We miss GianCarlo Ghirardi, our invaluable friend, coworker and colleague and we dedicate this work in memoriam to him.

D. Dürr
Mathematisches Institut, Ludwig-Maximilians-Universitat München,
Theresienstr. 39, 80333 Munich, Germany

W. Struyve (✉)
Instituut voor Theoretische Fysica & Centrum voor Logica en Filosofie van de Wetenschappen,
KU Leuven, Leuven, Belgium
e-mail: ward.struyve@fys.kuleuven.be

V. Allori et al. (eds.), *Do Wave Functions Jump?*, Fundamental Theories of Physics 198,
https://doi.org/10.1007/978-3-030-46777-7_3

1 Why "Most" Cannot Be Avoided

Typicality has always been in the minds of the founding fathers of probability theory when probabilistic reasoning is applied to the real world. Nevertheless, still it's role is often not understood. An example is [1], where Valentini presents typicality and relaxation to equilibrium as *distinct* approaches to the proof of Born's rule, while typicality is in fact an overriding necessity. Valentini writes in the abstract of his article:

> *We compare and contrast two distinct approaches to understanding the Born rule in de Broglie-Bohm pilot-wave theory, one based on dynamical relaxation over time (advocated by this author and collaborators) and the other based on typicality of initial conditions (advocated by the 'Bohmian mechanics' school). It is argued that the latter approach is inherently circular and physically misguided.*

The accusation of circularity concerns the proof of Born's rule in de Broglie-Bohm pilot-wave theory, or "Bohmian mechanics" for short, given in [3]. It is an important proof, as it explains the observed regularity concerning the outcomes of measurements on ensembles of identically prepared systems. As such, Valentini's accusation is at the same time an onslaught to the ideas underlying statistical physics. We wish to explain once more in very simple terms why the accusation is off target and why "relaxation to equilibrium" is neither necessary nor sufficient to justify Born's rule.

In the history of mathematics pointing out circularities in important proofs have been sometimes pathbreaking. An example is provided by what we would call now the "PhD thesis" of Georg Simon Klügel (1739–1812), who showed that all existing proofs (about 27 of them) of the 11th Postulate of Euclid on the uniqueness of parallels were circular in that they used in the proofs equivalents of the postulate as (hidden) assumptions. That thesis has led to the discovery of non-Euclidean geometry!

The accusation of circularity in the derivation of Born's rule is however less breathtaking; it is simply off target. The criticism misses the point of statistical physics entirely. That may be partly due to the loose manner of speaking about probability and distributions which is common in statistical physics and which clouds the meaning of these objects. Instead, the notion of typicality is necessary to understand what the statistical predictions of a physical theory really mean. In fact, typicality (though the word may not have been directly used) has always been in the minds of probabilists and physicists (from Jacob Bernoulli $\sim$ 1700 over Ludwig Boltzmann $\sim$ 1850 to Kolmogorov's axiomatics of probability $\sim$ 1930) when probabilistic reasoning is applied to the real world [2]. We try to explain once more in simplest terms why this need be.

Let's start with a simple example: A (fair) coin is thrown, say, a thousand times (the number thousand is chosen only because it is kind of large). We obtain a head-tail sequence of length 1000 and ask prior to inspection of the sequence: Roughly how many heads are there in the sequence? Some would perhaps prefer to be agnostic about the answer but most would say—perhaps after some time of reflection—that the number will be roughly 500. Actually they will find out by counting the heads that they were right. Why roughly 500? Well, the relative frequency of heads (or tails) is then $1/2$, the probability[1] which we naturally assign for the sides of a (fair) coin. What matters here is that there is obviously some relation between the factual occurrence of the relative frequency of heads and the number $1/2$. That needs to be explained. Why? Because other sequences are possible as well, for example sequences which show less than 300 heads. The question which needs to be answered is: Why don't they show up in practice? The (mathematical) way that the regularity of roughly 500 heads is explained is by the *law of large numbers (LLN)*, which establishes the closeness of the empirical distribution of heads, i.e., the distribution which counts the relative frequency of heads in the sequence of length 1000 (which is the large number in the LLN) and the number $1/2$.

How does the LLN explain that? By counting sequences! Here are some telling numbers: There are about 10^{300} sequences with about 500 heads. There are about 10^{260} sequences with about 300 heads. So the proportion of sequences with 500 heads versus 300 heads is about 10^{40}. For sharpening our intuition about the power of such numbers note that the age of the universe in seconds is about 10^{17}. Thus most[2] sequences show a law-like regularity, namely that the relative number of heads is roughly $1/2$. Wouldn't that suggest that it is most likely that the observed sequence has roughly equal numbers of heads and tails? Well, most likely is just another way of saying "with high probability". But then, what does probability mean here? It is better and simpler to say that the *typical* sequences will have roughly 500 heads. The LLN says nothing more than that. It is a typicality statement. We remark for later that in introductory courses to probability theory the counting is introduced as Laplace probability which is then a normalized quantity by dividing the numbers of sequences of interest by the total number of sequences and which we better refer to by the role it plays in our example as Laplace-typicality. The point of this example is that mere counting of head-tail sequences (or $0 - 1$ sequences if one wants) gives two insights (where the second one we take it as being intuitively clear without going into details):

1. Most sequences show the law-like regularity that the empirical distribution of heads (the relative frequency of heads) is near $1/2$.
2. The succession of heads and tails (or $1's$ and $0's$) in a typical sequence looks random, unpredictable, while randomness never entered. It's just the way typical sequences look like.

Agnostics may still complain that explains nothing: What needs to be explained is why we only see typical sequences! That's actually the deep question underlying the meaning of probability theory from its very beginning and Antoine-Augustin Cournot (1801–1877) coined once what became known as Cournot's principle, which in our own rough words just says that we should only be concerned with typical events. The point we wish to make with the simple example is that appeal to typicality cannot be avoided. Sequences with drastically unequal number of heads and tails are physically

[1]Luckily, for what we have to say here, we can ignore the issues related to the question what the notion of "probability 1/2" *really* means. It does not matter.

[2]"Most" should be understood as "overwhelmingly many".

possible. The reason they do not appear in practice is because there are much too few of them, they are atypical. There is no way around that. That is what the founders of probability theory understood and had to swallow. Note that typicality, through Cournot's principle, only tells us what to expect or not. It does not allow to associate a probability to, say, the sequence with 300 heads: In terms of the Laplace-typicality only values near zero or one matter—atypical or typical. The notion of typicality is distinct from the notion of probability.

Let's go a step further and consider the coin tossing as a physical process, because that is what it is after all: There is a hand which tosses the coin, thereby providing the coin with an initial momentum and position which determine its flight through the air. The trajectory of the coin is determined, given the initial conditions, by the laws of physics (here Newtonian) and hence it defines a function which maps initial conditions to head or tail (0 or 1). But the hand is just a physical system itself—a coin tossing machine so to say. The machine picks up the coin, throws it, and after the landing the machine notes down head or tail, picks the coin up again, throws it and so on and so forth. Thus the resulting sequence of heads and tails depends only on the initial conditions, i.e., on the phase space point which determines the whole process of the coin tossing machine. The physical description and analysis may not be that easy, but at least the principle is clear[3]: It shows that the head-tail-sequences are the images of a function $F = (F_1, \ldots, F_N)$ of the high dimensional phase space[4] variables q—the initial conditions. Here the component F_k maps to the outcome $\delta \in \{0, 1\}$ of the kth tossing of the machine. Such a function F (a coarse-graining function by the way) is usually called a random variable. The point is that in this description where phase space variables play the decisive role, counting is not anymore possible, as classical phase space is a continuum. What then replaces the counting? That is a measure—a *typicality measure*. In classical physics, which would be appropriate for studying coin tossing as a physical process, the measure commonly used is the "Liouville-measure"—the volume measure in phase space. It recommends itself by the property of being stationary,[5] an observation which was promoted in the works of Ludwig Boltzmann. It is a measure which is suggested by the physical law itself and not by an arbitrary human choice. The fact that, with this measure, typicality is a timeless notion is of great help for proving the LLN.

The role of the Liouville-measure is to define the notion of "most" for the phase space of classical mechanics. In mathematical terms, the above mentioned Laplace-typicality emerges then ideally as the *image measure* of the more fundamental Liouville-measure defined by the function F. To express the LLN in this more fundamental setting, it is useful to introduce the empirical distribution $\rho^N_{\text{emp}}(q, \delta)$, the function which counts the relative number of heads and tails and which is a function of the phase space variables q and the image variables $\delta \in \{0, 1\}$.

---

[3]For more on this, see the chapter "Chance in Physics" in [4].

[4]Dimension of the size of Avogadro's number perhaps.

[5]In classical mechanics, there are many more measures which share this property, but that does not matter for our concerns here.

$$\rho^N_{\text{emp}}(q,\delta) = \frac{1}{N}\sum_{k=1}^{N} \mathbf{1}_{\{\delta\}}(F_k(q))\,.$$

Here $\mathbf{1}_{\{\delta\}}(F_k(q)) = 1$ if $F_k(q) = \delta$ and 0 otherwise. The LLN (if it would be proven for a physically realistic coin tossing machine) would then say something like[6]: *For most phase points q (w.r.t. the Liouville-measure) and for large enough N the empirical distribution $\rho^N_{\text{emp}} \approx 1/2$, or, Liouville-measure typically $\rho^N_{\text{emp}} \approx 1/2$.*

The reference to typicality cannot be avoided, as there are phase points which are mapped to sequences with, say, 300 heads, i.e., "most" cannot be replaced by "all".

One further point should be noted which is often used to actually justify the use of statistical methods in physics: It is almost impossible to know in a realistic physical system exactly which initial conditions lead to which outcomes (as for example in the case of the coin tossing machine). The power of typicality is that exact details are not needed. It suffices that for most initial conditions the observed statistical regularities obtain.

Coin tossing is not a process which happens only here and now but which happens at arbitrary locations and times. To explain the statistical regularities in such generality, we still need to lift the whole discussion to a universal level. The universally relevant LLN would then have to say (very) roughly something like:

For most universes in which coin tossing experiments are done, i.e., for Liouville measure-typical such universes, it is the case that the empirical distribution of heads in long enough sequences in coin tossing experiments is approximately 1/2.

The typicality assertion concerning Born's law is very analogous to this and has been proven in [3]. Before we turn to that we shortly look at another rather simple classical system. Everything that will be said for this example can be carried over to the case of Bohmian mechanics.

Consider an ideal gas of point particles in a rectangular box, lets say with elastic collisions of the particles at the walls. The gas is in equilibrium when the gas molecules fill the box approximately homogeneously. Most configurations (with respect to the Liouville-typicality measure) are like that, like most 0, 1-sequences have about equal numbers of heads and tails. In the course of time, there will be fluctuations of the number of molecules in a given region in the box, but those will escape our gross senses. Most configurations will stay in equilibrium over time. Now suppose we start with a gas that is occupying only one half of the box, the other half being empty. This would count as a non-equilibrium configuration. What will happen in the course of time? Well, eventually the gas molecules will fill the box approximately homogeneously. Will this relaxation happen for any possible configuration of gas that starts in one half of the box? The answer is no. For our simple example one

[6] A technical remark on the side: To model the coin tossing experiment in which the coin is thrown a great number of times in a physically realistic way is not so easy and to prove the LLN may turn out hard: The stochastic independence of the different tosses of the coin is easily said, but to prove that in a physically realistic model is far from being easy (see [4], chapter "Chance in Physics" for an elaboration on that).

can easily construct configurations which will never look like the equilibrium ones. There will be configurations for which it takes an enormous amount of time to evolve into ones which look like equilibrium. And some never will. Why is that important to observe? Because if one wants to make predictions about the possible behavior non-equilibrium configurations one needs to invoke typically. Namely, the idea (which is Boltzmann's insight) is that typical, i.e., *most*, non-equilibrium configurations will evolve to configurations which macroscopically look like equilibrium ones. ("Most" is again with respect to the Liouville-typicality measure, concentrated initially on the very small subset of configurations which are such that the box is only half filled.) Why? Because the equilibrium set in phase space, is so overwhelmingly larger then the tiny non-equilibrium set, so that typically trajectories will wander around and will end up in the overwhelmingly large set and stay there for a very large time. And, as we said, there exist also atypical configurations which will not at all behave like that. That is, without typicality, we have no explanation why to expect equilibration. Having said this, we should warn the reader that this just is the physical idea behind the equilibration. To turn this into a rigorous argument is famously hard, as hard as to justify the Boltzmann equation from first principles.

The warning in mind, we can think of describing the transition from non-equilibrium to equilibrium also in terms of coarse-graining densities $\rho(\mathbf{x}, t)$, which are more or less smooth functions (macro-variables) on the three-dimensional physical space with variables $\mathbf{x}$ and which should be pictured as approximations of empirical densities.[7] The uniform density i.e., $\rho_{eq}(\mathbf{x}) = \text{const.}$ would then be the equilibrium density. Hence, starting with a non-equilibrium density ρ_{neq}, it is perhaps reasonable to assume, that $\rho_{neq}(t) \to \rho_{eq}$ as t gets large. This convergence of densities is sometimes referred to as "mixing property" and we shall refer to this notion to mean just that: convergence of densities without reference to typicality. There have been attempts to show this. The mixing idea is presumably due to Willard Gibbs who had introduced the so-called ensemble view into statistical physics. An idea for a strategy for a "convergence to equilibrium proof" was suggested by Paul Ehrenfest as is recalled on page 85 by Kac in [5] and where he refers to Ehrenfest's attempt as an "amusing" theorem, since convergence to equilibrium in time does not follow at all from what Ehrenfest had shown.

But even when the mixing property, i.e., convergence of densities, were shown to be of physical relevance, the connection with the actual configuration (i.e., the empirical distribution) would still have to be established. After all, Newtonian physics is about configurations and not densities. In addition, by our arguments above, some non-equilibrium densities will never show the mixing property, for example densities which are concentrated on "bad" configurations, i.e., atypical ones.

[7]The empirical density is in this case given by $\rho^N_{\text{emp}}(q, \mathbf{x}) = \frac{1}{N}\sum_{k=1}^N \delta(\mathbf{x} - \mathbf{x}_i)$, where $\mathbf{x}_i$ is the position of the ith particle. Note the analogy with the definition in the case of the coin tossing.

2 Born's Rule

What we have said about the statistical analysis in classical physics carries over to the statistical analysis of Bohmian mechanics, where the phase space is now replaced by configuration space. Born's rule $\rho = |\psi|^2$ is a short hand for the universal LLN for the empirical distribution ρ^N_{emp} of the coordinates of the particles of subsystems in an ensemble (defined similarly as in Footnote 7). Roughly speaking, the universal LLN in the context of Born's rule says the following (for the precise formulation, see e.g. [3]):

For typical Bohmian universes hold: In an ensemble of (identical) subsystems of a universe, where each subsystem has effective wave function[8] *ψ, the empirical distribution ρ^N_{emp} of the particles coordinates of the subsystems are $|\psi|^2$ distributed.*

For this to hold sufficiently well, the number N of subsystems in the ensemble should be large. Note that in analogy with the coin tossing, the number $1/2$ is here replaced by $|\psi|^2$ and the sequence of length 1000 is here the number of subsystems in the ensemble. But instead of the Liouville-measure, the typicality measure used in [3] is the measure $\mathbb{P}^\Psi(A) = \int_A |\Psi|^2(q)\text{dq}$ (q is a generic configuration space variable), where A is a subset of the configuration space of the Bohmian universe and Ψ is the universal wave function on that space.[9] What is special about the typicality measure $\mathbb{P}^\Psi$? It is a measure which is transported equivariantly by the Bohmian flow. This means that it is a typicality measure which like the stationary Liouville measure is independent of time.[10]

The very nice property of the universal quantum equilibrium LLN is that it is empirically adequate. Up to date all tests affirm the empirical validity of Born's law.

3 Dynamical Relaxation?

Valentini dislikes the use of typicality. Instead, he proposes "dynamical relaxation" to equilibrium to explain Born's rule in the realm of Bohmian mechanics. It is however not at all clear what is meant by "dynamical relaxation" and in which way reference to typicality can be overcome. On the configurational level, i.e., on the level of empirical densities, starting in non-equilibrium our discussion of the gas in the box applies verbatim. There will always exist initial configurations of particles for which the empirical distribution will never become close to $|\psi|^2$—*the equilibrium distribution.* So why should we expect equilibrium then? Appealing to Boltzmann's idea, one could invoke typicality as in the case of the gas example. But as soon as one invokes typicality, there is no longer any need to invoke relaxation to begin with to explain

[8]Think of this wave function as the usual wave function of a system as it is used in physics courses.

[9]As an aside, note that the typicality measure which is used in the LLN is really a member of an equivalence set of measures. That is, all measures which are absolutely continuous with respect to $\mathbb{P}^\Psi$ yield a LLN for Born's law.

[10]It has been proven under very reasonable conditions in [7] that this measure is unique.

equilibrium! Namely, most configurations will be in equilibrium most of the time and hence non-equilibrium just doesn't occur–for all practical purposes—as established in [3].

Valentini also follows the Gibbs-Ehrenfest idea of mixing and provides an analytic argument for the convergence of densities. But the argument is the direct analogue of the "amusing" theorem proven by Ehrenfest, which "tells us *nothing* about the behavior of the non-equilibrium density ρ in time" [5]. Not to say that the connection to empirical densities needs to be established on top of that.

Hence the "dynamical relaxation" approach turns out be neither necessary nor sufficient.

4 Physically Misguided?

All of the quantum formalism follows from Born's rule [6].[11] There is no dispute about that. Heisenberg's uncertainty follows from Born's rule. No dispute about that either. There is actually no dispute about any of the consequences which arise from or in quantum equilibrium. So what is the dispute about then? If it is about the needed reference to typicality, then that can't be because both "approaches" need reference to typicality anyhow for physically meaningful assertions.

What then makes the use of typicality physically misguided? Because the physical law allows for atypical universes? Because a coin tossing sequence of only heads is possible by the physical law? No argument, other than denying the physical law, can make those possibilities impossible. Why then, don't we deny the law to make them go away? Because by humbly looking at the facts in our world we understand that the law-like regularities in apparently random events are in surprising harmony with the physical law: They are typical.

Acknowledgments W.S. is supported by the Research Foundation Flanders (Fonds Wetenschappelijk Onderzoek, FWO), Grant No. G066918N.

References

1. Valentini, A., Foundations of statistical mechanics and the status of the Born rule in de Broglie-Bohm pilot-wave theory. in: Statistical Mechanics and Scientific Explanation, Editor: Allori, V. Publisher: World Scientific (2020) (arXiv:1906.10761).
2. Shafer, G. & Vovk, V., The Sources of Kolmogorov's Grundbegriffe. *Statistical Science* **21**, 70–98, https://doi.org/10.1214/088342305000000467 (2006).
3. Dürr, D., Goldstein, S. & Zanghì, N. Quantum equilibrium and the origin of absolute uncertainty. J. Stat. Phys. **67**, 843–907, https://doi.org/10.1007/BF01049004 (1992).

[11]This involves in some way or another an analysis of measurement situations, which in Bohmian mechanics is straightforward, and which in let's say orthodox quantum theory needs the problematical collapse of the wave function.

4. Dürr, D. & Teufel, S. Bohmian Mechanics, The Physics and Mathematics of Quantum Theory. *Springer* (2009).
5. Kac, M., Probability and Related Topics in Physical Sciences. *Interscience Publishers Inc.* (1959).
6. Dürr, D., Goldstein, S. & Zanghì, N. Quantum equilibrium and the role of operators as observables in quantum theory. *J. Stat. Phys.* **116**, 959–1055, https://doi.org/10.1023/B:JOSS.0000 (2004).
7. Goldstein, S. & Struyve, W. On the Uniqueness of Quantum Equilibrium in Bohmian Mechanics *J. Stat. Phys.* **128**, https://doi.org/10.1007/s10955-007-9354-5 (2007).

Presentation of Collapse Models

Luca Ferialdi

The Ghirardi-Rimini-Weber (GRW) model of wave function collapse was successful in giving a solution to the measurement problem in quantum mechanics. Despite this, it lacked two desirable features—one mathematical and one physical—for a collapse model. The mathematical aspect of the GRW model that one would like to improve is its lacking of a unified description in terms of an equation for the wave function. On the physical side, the GRW model uses the first quantization language, and does not preserve the symmetry properties of indentical particles wave functions [1]. In 1990 Ghirardi, Rimini and Pearle took the desired step forward devising the Continuos Spontaneous Localisations (CSL) model [2]. They considered collapses to be driven by continuous stochastic processes (instead of having discrete collapses like in GRW), that allowed to describe the collapse dynamics by a stochastic Schrödinger equation. Moreover, the CSL model relies on the second quantization formalism, and correctly describes ensembles of identical particles. In the following years, collapse models attracted the interest of many scientists, their properties were deeply investigated and different collapse models were proposed [3–10]. The CSL model however is still the reference model in the field, and is the one against which experimental results are used to bound the collapse parameters. In this chapter we review the CSL model—focusing on its mass-dependent version—and its more recent extension to colored stochastic processes and dissipative dynamics. We also briefly review the Diósi-Penrose model, that links the wave function collapse to gravity, and the Quantum Mechanics with Universal Position Localizations (QMUPL) model, that well approximates the CSL model under certain circumstances, and has the advantage of being mathematically easier to handle.

L. Ferialdi (✉)
Department of Physics, University of Trieste, Strada Costiera 11, 34151 Trieste, Italy

Istituto Nazionale di Fisica Nucleare, Trieste Section, Via Valerio 2, 34127 Trieste, Italy
e-mail: ferialdi@ts.infn.it

V. Allori et al. (eds.), *Do Wave Functions Jump?*, Fundamental Theories of Physics 198,
https://doi.org/10.1007/978-3-030-46777-7_4

1 Modified Schrödinger Dynamics

Providing a unified description of the Schrödinger evolution and the measurement process at the wave function level, implies a modification of the Schrödinger equation. Besides being careful to the fact that one needs to recover quantum mechanical predictions for microscopic objects and classical ones for macroscopic objects, one also needs to pay attention to other technical details. The starting point is that, since one wants to kill linear superpositions of macroscopic objects, the modifications made have to be *non-linear*. Nonetheless, the density matrix dynamics has to be linear, in order to avoid faster-than-light signalling, i.e. to avoid instantaneous communication at arbitrary distances [11]. Since *non-linear and deterministic* modifications at the wave funciton level would lead to a non-linear density matrix dynamics, in order to avoid a contraddiciton with relativity, one thus has to consider *non-linear and stochastic* modifications of the Schrödinger equation. These requirements, together with the one that the norm of the wave function has to be preserved, essentially fix the structure of a consistent stochastic Schrödinger equation:

$$d|\psi_t\rangle = \left[-\frac{i}{\hbar}\hat{H}dt + \sqrt{\gamma}(\hat{A} - \langle\hat{A}\rangle_t)dW_t - \frac{\gamma}{2}(\hat{A} - \langle\hat{A}\rangle_t)^2 dt \right] |\psi_t\rangle, \tag{1}$$

where the Itô formalism is assumed (see [12] for mathematical details). $\hat{H}$ is the standard quantum Hamiltonian, $\hat{A}$ is the hermitian operator on whose basis the wave function collapses, γ is the collapse strength, and $\langle A\rangle_t$ is the expectation value of $\hat{A}$ (which makes the equation non-linear). The second term of this equation is stochastic due to the Wiener process dW_t, while the third term guarantees the average norm conservation of the state vector. It is useful to introduce the linear equation corresponding to Eq. (1):

$$\frac{d|\phi_t\rangle}{dt} = \left[-\frac{i}{\hbar}\hat{H} + \sqrt{\gamma}\,\hat{A}\,w_t - \frac{\gamma}{2}\hat{A}^2 \right] |\phi_t\rangle, \tag{2}$$

where $w_t = dW_t/dt$ is a white noise: $\mathbb{E}[w_t] = 0$, $\mathbb{E}[w_t w_s] = \delta(t-s)$. We remark that this equation does not preserve the norm of the state vectors, thus to obtain physical states one needs to normalize $|\phi_t\rangle$: $|\psi_t\rangle = |\phi_t\rangle/|||\phi_t\rangle||$.

2 The Continuos Spontaneous Localisations Model

The CSL model is still considered the collapse model par excellence, and it is currently the one most exploited for phenomenological analysis of collapse models [13]. This model has been devised is such a way that dynamics of microscopic objects only slightly differs from the Schrödinger evolution, while it guarantees that macroscopic objects are well localized in space. We introduce the creation (annihilation) operators of a particle of type i and mass m_i at the space position $\mathbf{x}$: $\hat{A}_i^\dagger(\mathbf{x})$ ($\hat{A}_i(\mathbf{x})$), and the local mass density operator

$$\hat{M}(\mathbf{x}) = \sum_i m_i \, \hat{A}_i^\dagger(\mathbf{x})\hat{A}_i(\mathbf{x}) \,, \tag{3}$$

(bold symbols denote three dimensional vectors). By replacing the collapse operator $\hat{A}$ in Eq. (1) with the local mass density operator, one obtains a model describing a collapse process that is more likely to happen where the particles mass density is higher. This is the (mass proportional) CSL model, whose stochastic Schrödinger equation reads:

$$\begin{aligned} d|\psi_t\rangle = \Bigg[& -\frac{i}{\hbar}\hat{H}\,dt + \frac{\sqrt{\gamma}}{m_0}\int d^3\mathbf{x}\left(\hat{M}(\mathbf{x}) - \langle \hat{M}(\mathbf{x})\rangle_t\right) dW_t(\mathbf{x}) \\ & -\frac{\gamma}{2m_0^2}\iint d^3\mathbf{x}\,d^3\mathbf{y}\,\mathcal{G}(\mathbf{x}-\mathbf{y})\left(\hat{M}(\mathbf{x}) - \langle \hat{M}(\mathbf{x})\rangle_t\right)\left(\hat{M}(\mathbf{y}) - \langle \hat{M}(\mathbf{y})\rangle_t\right) dt\Bigg]|\psi_t\rangle, \end{aligned} \tag{4}$$

where m_0 is a nucleon reference mass, and γ is a new parameter of the model that sets the collapse strength, whose value will be discussed later. $dW_t(\mathbf{x})$ is a family of Wiener processes such that

$$\mathbb{E}[dW_t(\mathbf{x})] = 0, \qquad \mathbb{E}[dW_t(\mathbf{x})\,dW_t(\mathbf{y})] = dt\,\mathcal{G}(\mathbf{x}-\mathbf{y})\,, \tag{5}$$

where the spatial correlation function is chosen to be

$$\mathcal{G}(\mathbf{x}-\mathbf{y}) = \frac{1}{(4\pi r_c)^{3/2}}\exp\left(-\frac{(\mathbf{x}-\mathbf{y})^2}{4r_c^2}\right), \tag{6}$$

Such a choice establishes a connection with the GRW model, both for the shape of the correlator, that recalls the GRW collapse operator, and for the correlation length r_C, the second free parameter of the model. The correlation length is set to $r_C = 10^{-7}$ m: this is a mesoscopic distance that guarantees that the structure of matter is not affected by the collapse, while macroscopic objects are. Notice that in recent years the value of r_C has been relaxed, and the range of values it can take is bounded by experiments (see contribution from Carlesso and Paternostro for details).

In order to compute expectation values of physical quantities, it is convenient to introduce the master equation for the statistical operator $\hat{\rho}(t) = \mathbb{E}[|\psi_t\rangle\langle\psi_t|]$. This can be easily calculated from Eq. (4), and reads:

$$\frac{d}{dt}\hat{\rho}(t) = -\frac{i}{\hbar}\left[\hat{H},\hat{\rho}(t)\right] + \frac{\gamma}{m_0^2}\iint d^3\mathbf{x}\,d^3\mathbf{y}\,\mathcal{G}(\mathbf{x}-\mathbf{y})\,[\hat{M}(\mathbf{x}),[\hat{M}(\mathbf{y}),\hat{\rho}(t)]]\,. \tag{7}$$

This master equation can also be rewritten in the first quantization formalism as follows [14]:

$$\frac{d}{dt}\hat{\rho}(t) = -\frac{i}{\hbar}\left[\hat{H},\hat{\rho}(t)\right] + \frac{m^2}{m_0^2}\frac{\gamma}{(2\pi\hbar)^3}\sum_{j,l=1}^{N}\int d\mathbf{Q}e^{-\frac{\mathbf{Q}^2 r_C^2}{\hbar^2}}\left(e^{\frac{i}{\hbar}\mathbf{Q}\cdot\hat{\mathbf{x}}_j}\hat{\rho}(t)e^{-\frac{i}{\hbar}\mathbf{Q}\cdot\hat{\mathbf{x}}_l} - \hat{\rho}(t)\right). \quad (8)$$

2.1 Localisation of Macroscopic Objects

One of the crucial properties of collapse models is their ability to suppress macroscopic superpositions without affecting microscopic ones. Let us see what are the CSL predictions in these regards.

We introduce the "position" states

$$|\mathbf{x_1}, \ldots \mathbf{x_N}\rangle = a^\dagger(\mathbf{x_1}), \ldots a^\dagger(\mathbf{x_N})|0\rangle\,, \quad (9)$$

describing N particles at positions $\mathbf{x_1}, \ldots \mathbf{x_N}$. We first consider a single particle, and we represent the master equation (7) on the position states obtaining:

$$\begin{aligned}\frac{d}{dt}\langle\mathbf{x}|\hat{\rho}(t)|\mathbf{y}\rangle &= -\frac{i}{\hbar}\langle\mathbf{x}|\left[\hat{H},\hat{\rho}(t)\right]|\mathbf{y}\rangle \\ &- \frac{\gamma}{\left(\sqrt{4\pi}r_C\right)^3}\left(1 - e^{-(1/4r_C^2)(\mathbf{x}-\mathbf{y})^2}\right)\langle\mathbf{x}|\hat{\rho}(t)|\mathbf{y}\rangle \end{aligned} \quad (10)$$

One finds that this equation is the same as the one for the GRW model, provided that $\gamma = (\sqrt{4\pi}r_C)^3\lambda$, where λ is the rate for the latter model. The value originally proposed by GRW for the collapse rate was $\lambda_{\text{GRW}} = 10^{-16}$ s^{-1}, that correspond to a coupling constant $\gamma \approx 10^{-30}$ cm^3 s^{-1} [1, 2]. GRW's choice is motivated by the requirement that superpositions of 6.02×10^{23} nucleons, displaced by a distance of at least r_C, be suppressed within 10^{-3} s. Later Adler proposed the value $\lambda = 10^{-8}$ s^{-1} (with an uncertainty of two orders of magnitude), motivated by the requirement that the collapse occurs already at the level of process of latent image formation in photography [15]. Accordingly, for such a choice of the CSL parameter γ, a single particle is almost never localized.

Let us now consider a macroscopic rigid body that consists of N identical particles, with position operators $\hat{\mathbf{x}}_i$, and let us call $\hat{\mathbf{X}}$ the center of mass position operator:

$$\hat{\mathbf{X}} = \frac{1}{N}\sum_{i=1}^{N}\hat{\mathbf{x}}_i\,, \quad (11)$$

Under the assumption that the center of mass Hamiltonian factorizes from the relative one, one finds that the center of mass density matrix $\hat{\rho}_{\text{CM}}$ evolves according to the following master equation:

$$\frac{d}{dt}\hat{\rho}_{\mathrm{CM}}(t) = -\frac{i}{\hbar}\left[\hat{H}_{\mathrm{CM}}, \hat{\rho}_{\mathrm{CM}}(t)\right] \\ +\lambda\left(\frac{r_C}{\sqrt{\pi}\hbar}\right)^3 \frac{m^2}{m_0^2}\int d\mathbf{Q}R(\mathbf{Q})e^{-\frac{Q^2 r_C^2}{\hbar^2}}(e^{\frac{i}{\hbar}\mathbf{Q}\cdot\hat{\mathbf{X}}}\hat{\rho}_{\mathrm{CM}}(t)e^{-\frac{i}{\hbar}\mathbf{Q}\cdot\hat{\mathbf{X}}} - \hat{\rho}_{\mathrm{CM}}(t)), \quad (12)$$

where

$$R(\mathbf{Q}) = \sum_{j,l=1}^{N} e^{\frac{i}{\hbar}\mathbf{Q}\cdot(\mathbf{r_j}-\mathbf{r_l})} \quad (13)$$

depends on the specific distribution of the particles around the center of mass of the considered object. We can however estimate the magnitude of $R(\mathbf{Q})$ as follows (see [14, 15] for details). Let us ideally divide the object in N spheres of radius r_C, and let us call n the number of particles in each sphere. Since the Gaussian factor $e^{-\frac{Q^2 r_C^2}{\hbar^2}}$ selects the values of $\mathbf{Q}$ such that $|\mathbf{Q}| < \hbar/r_C$, one finds that the particles contained in each sphere contribute as follows

$$R(\mathbf{Q}) \approx \sum_{j=1,l=1}^{n} 1 = n^2. \quad (14)$$

On the other hand, when the particles are distant more than r_C (contribution of each sphere), the phases oscillations are such that only the terms with $i = j$ survive, thus leading to the following estimate:

$$R(\mathbf{Q}) \approx \sum_{j=1}^{N} 1 = N. \quad (15)$$

By replacing these estimates in Eq. (12), one finds that the center of mass density matrix $\hat{\rho}_{\mathrm{CM}}$ evolves according to the one particle master equation (8) where λ is replaced by the enhanced factor

$$\Lambda = n^2 N\lambda\,. \quad (16)$$

This is the so called "amplification mechanism": the effective collapse rate for a macroscopic object is amplified with respect to the single particle one, by a factor proportional to the number of its constituents. Accordingly, although the collapse rate is very small, in such a way that one particle is almost never localized, the amplification mechanism guarantees that a macroscopic object collapses instantly, thus recovering a classical behaviour.

2.2 Extensions of the CSL Model

The CSL model shares with most collapse models two undesirable features: the increase of the system average energy, and the absence of a high frequency cutoff for the noise. In order to solve these issues, two extensions of the CSL model have been proposed: the dissipative CSL (dCSL) [16], and the colored CSL (cCSL) [17, 18].

dCSL: In the CSL model, the average energy of the system steadily increases, diverging asymptotically. This is a consequence of the fact that the collapse noise constantly kicks the system, acting like an infinite temperature reservoir. Since there is no dissipation mechanism in the model, such an interaction leads to a linear increase of the energy in time, that for a free particle of mass m is

$$\langle\hat{H}\rangle_t = \lambda\,\frac{3\hbar^2 m}{4r_C^2 m_0^2}\,t\,. \tag{17}$$

This issue was solved by extending the CSL model to dissipative dynamics (dCSL) [16], i.e. by modifying the collapse operator in order to account for dissipative effects. The dCSL model is described by the following stochastic Schrödinger equation

$$\begin{aligned} d|\psi_t\rangle = \Bigg[&-\frac{i}{\hbar}\hat{H}dt + \frac{\sqrt{\gamma}}{m_0}\int d^3\boldsymbol{x}\,[\hat{L}(\boldsymbol{x}) - l_t(\boldsymbol{x})]dW_t(\boldsymbol{x}) \\ &-\frac{\gamma}{2m_0^2}\int d^3\boldsymbol{x}\,[\hat{L}^\dagger(\boldsymbol{x})\hat{L}(\boldsymbol{x}) + l_t^2(\boldsymbol{x}) - 2l_t(\boldsymbol{x})\hat{L}(\boldsymbol{x})]dt\Bigg]|\psi_t\rangle, \end{aligned} \tag{18}$$

with $l_t(\boldsymbol{x}) \equiv \langle\psi_t|(\hat{L}^\dagger(\boldsymbol{x}) + \hat{L}(\boldsymbol{x}))\psi_t/2$. The collapse operator $\hat{L}$ now depends not only on position, but also on momentum, thus inducing dissipation:

$$\begin{aligned} \hat{L}(\mathbf{x}) = \sum_j \frac{m_j}{(2\pi\hbar)^3}\int d\mathbf{P}d\mathbf{Q}\,\hat{a}_j^\dagger(\mathbf{P}+\mathbf{Q})\,e^{-\frac{i}{\hbar}\mathbf{Q}\cdot\mathbf{x}} \\ \times\exp\left(-\frac{r_C^2}{2\hbar^2}\left|(1+k_j)\mathbf{Q} + 2k_j\mathbf{P}\right|^2\right)\hat{a}_j(\mathbf{P})\,, \end{aligned} \tag{19}$$

where

$$k_j \equiv \frac{\hbar}{2m_j v_\eta r_C}\,, \tag{20}$$

and v_η is a new parameter the sets the strength of the dissipative effects. One can easily check that in the limit $v_\eta \to 0$ the model recovers the CSL. The master equation associated to Eq. (18) is

$$\frac{d}{dt}\hat{\rho}(t) = -\frac{i}{\hbar}\left[\hat{H}\,,\,\hat{\rho}(t)\right] + \frac{\gamma}{2m_0^2}\int d^3\mathbf{x}\left[\hat{L}(\mathbf{x})\hat{\rho}(t)\hat{L}^\dagger(\mathbf{x}) - \frac{1}{2}\left\{\hat{L}^\dagger(\mathbf{x})\hat{L}(\mathbf{x}),\,\hat{\rho}(t)\right\}\right]. \tag{21}$$

By exploiting this master equation one can show that the dCSL model predicts the following evolution for the average energy of a free particle of mass m:

$$\langle H(t)\rangle = e^{-\chi t}\left(\langle H(0)\rangle - H_\infty\right) + H_\infty, \tag{22}$$

where the relaxation rate χ reads

$$\chi = \frac{4k\lambda m^2}{(1+k)^5 m_0^2}\,, \tag{23}$$

and H_∞ is the asymptotic value of the kinetic energy:

$$H_\infty = \frac{3\hbar^2}{16kmr_C^2}\,. \tag{24}$$

We thus see that the dissipative terms have the effect of making the system reach a finite asymptotic value. We can interpret this fact as if the system thermalizes at the noise temperature. While in the CSL model such a temperature is infinite, for the dCSL one finds that the noise temperature is

$$T = \frac{\hbar v_\eta}{4k_B r_C}, \tag{25}$$

where k_B is the Boltzmann constant. Although the system energy is not conserved, one can restore energy conservation by considering an energy exchange between the system and the noise. Furthermore, the asymptotic value is very small, and not detectable with present day technology [16].

cCSL: The collgapse noise in the CSL model is white, i.e. it has no frequency cutoff. This is an undesirable feature if one thinks about such a noise as physical field (e.g. with cosmological origin [24]), because a noise of this kind is unphysical. One thus needs to consider a colored noise, i.e. a noise with a frequency cutoff that is not delta correlated in time. The extension to the cCSL model can be obtained by generalizing equation (2) to Gaussian stochastic processes $w_t(\mathbf{x})$ such that

$$\mathbb{E}[w_t(\mathbf{x})] = 0\,, \qquad \mathbb{E}[w_t(\mathbf{x})w_s(\mathbf{y})] = \mathcal{G}(\mathbf{x}-\mathbf{y})\,D(t-s)\,, \tag{26}$$

where we have implicitly assumed that spatial and temporal correlations factorize. The linear stochastic Schrödinger equation describing the cCSL model thus reads

$$\frac{d|\phi_t\rangle}{dt} = \left[-\frac{i}{\hbar}\hat{H} + \frac{\sqrt{\gamma}}{m_0} \int d^3\mathbf{x}\, \hat{M}(\mathbf{x})\, w_t(\mathbf{x}) \right. \\ \left. -\frac{\gamma}{m_0^2} \iint d^3\mathbf{x}\, d^3\mathbf{y}\, \mathcal{G}(\mathbf{x}-\mathbf{y})\hat{M}(\mathbf{x}) \int_0^t ds\, D(t-s) \frac{\delta}{\delta w_s(\mathbf{y})} \right] |\phi_t\rangle, \tag{27}$$

where $\delta/\delta w_s(\mathbf{y})$ denotes a functional derivative. We remind that physical states are obtained by normalizing $|\phi_t\rangle$. The presence of the integral term and of the functional derivative makes it extremely difficult to investigate the features of the cCSL model. Nonetheless, a perturbative expansion allows to grasp the main properties of the model [17, 18].

3 Other Collapse Models

In this section we briefly describe two particularly relevant collapse models: the Diósi-Penrose model [3], that links the collapse of the wave function to gravity, and Quantum Mechanics with Universal Position Localization (QMUPL) [4], that is a very good compromise between mathematical simplicity and physical adequacy.

Diósi-Penrose: This model was first proposed by Diósi [3, 19–21], and has the particularly appealing feature of connecting the wave function collapse to gravity. The model has the same structure as the CSL, provided that the noise correlator is proportional to the gravitational potential:

$$\mathcal{G}(\mathbf{x}) = \frac{G}{\hbar}\frac{1}{|\mathbf{x}|}, \tag{28}$$

where $G(\mathbf{x})$ is the gravitational constant. With this definition of $\mathcal{G}$, γ is a dimensionless constant that one can set to one. One of the advantages of this choice is that the model has no free parameters (unlike CSL). The model however needs to be regularized, because the $1/|\mathbf{x}|$ potential is divergent for small distances. One thus has to introduce an effective radius R_0 below which particles are considered point-like. A natural choice for R_0 is the nucleon radius $R_0 = 10^{-15}$ m. However, also the Diósi-Penrose model is affected by a steady energy increase, and for the proposed value of cutoff, such a rate is very high (10^{-4} K/s for a proton). This issue can be avoided by choosing a larger cutoff $R_0 = 10^{-7}$ m, that gives a more reasonable rate of 10^{-28} K/s [22]. Such a cutoff is more difficult to justify on physical grounds, and the model loses its appeal of being a phenomenological model without free parameters. The introduction of dissipative effects in the model cannot be exploited because doing so one should keep the cutoff very large, or limit the validity of the model [23].

QMUPL: This model has the merit of being mathematically easier to handle compared to CSL, still providing a reliable physical description of the collapse dynamics.

This allowed to investigate this model thoroughly both under the mathematical as well as physical points of view [3, 25], and extend it to dissipative [8], or non-Markovian (colored noise) [9, 26] dynamics. Moreover, we stress that this is the only collapse model for which a dissipative and non-Markovian [10, 27] extension exists, and that exact solutions for initial Gaussian wave packets exist for all of its extensions. In this model, the stochastic field is linearly coupled to the particle position, and is described by Eq. (1) with $\hat{A}$ replaced by $\hat{x}$. The wave function dynamics of one particle is equivalently described by the following one dimensional linear equation:

$$\frac{d|\phi_t\rangle}{dt} = \left[-\frac{i}{\hbar}\hat{H} + \sqrt{\gamma}\,\hat{x}\,w_t - \frac{\gamma}{2}\hat{x}^2 \right] |\phi_t\rangle \,. \tag{29}$$

Essentially, the QMUPL model is CSL in the limit of short superposition distances with respect to r_C. This equation predicts that, for a Gaussian wave packet, the average position and momentum evolve stochastically, while the spread evolves deterministically, guaranteeing the collapse of the wave function. The non-Markovian and dissipative QMUPL model is described by the following stochastic Schrödinger equation:

$$\begin{aligned} \frac{d|\phi_t\rangle}{dt} = \Bigg[& -\frac{i}{\hbar}\left(\hat{H} + \frac{\gamma\mu}{2}\{\hat{x},\hat{p}\}\right) + \sqrt{\gamma}\left(\hat{x} + i\frac{\mu}{\hbar}\hat{p}\right) w_t \\ & -2\sqrt{\gamma}\,\hat{x}\int_0^t ds\, D(t,s)\frac{\delta}{\delta w_s} \Bigg] |\phi_t\rangle \,, \end{aligned} \tag{30}$$

where μ is a parameter that sets the strength of dissipative effects, and $D(t,s)$ is the time noise correlation function: $D(t,s) = \mathbb{E}[w_t\, w_s]$. One can easily check that by setting $\mu = 0$ (no dissipation), and $D(t,s) = \delta(t-s)$ (white noise), Eq (29) is recovered. It has been showed that, in principle, both non-Markovian and dissipative effects make the collapse process weaker. However, even at very low temperatures and with a strong frequency cutoff, the collapse remains as effective as in the non-dissipative white-noise cases [10].

While the master equation for the white-noise dissipative and non-dissipative QMUPL models is of the Lindblad type and can be readily obtained from from the respective stochastic Schrödinger equations, the master equations for the colored extensions require more elaborate calculations [28, 29].

The model simplicity allowed to find the analytical solution for all these extensions of the model (which was not possible for the CSL model), providing great insight in collapse models in general. It was shown that dissipative contributions allow to reach an asymptotic value for the energy, whose value depends on the dissipation strength. The presence of a colored noise (non-Markovian) makes the collapse less effective, depending on how much the noise correlation function departs from a Dirac delta.

References

1. G. C. Ghirardi, A. Rimini, and T. Weber, Phys. Rev. D **34**, 470 (1986).
2. G. C. Ghirardi, P. Pearle, and A. Rimini, Phys. Rev. A **42**, 78 (1990).
3. L. Diósi, Phys. Rev. A **40**, 1165 (1989).
4. L. Diósi, Phys. Rev. A **42**, 5086 (1990).
5. L. P. Hughston, Proc. Roy. Soc. London Ser. A **452**, 953 (1996).
6. F. Benatti, G. C. Ghirardi, A. Rimini, and T. Weber, Il Nuovo Cimento B **101**, 333 (1988).
7. S. L. Adler, J. Phys. A **35**, 841 (2002).
8. A. Bassi, E. Ippoliti, Phys. Rev. A **69**, 012105 (2004).
9. A. Bassi and L. Ferialdi, Phys. Rev. A **80**, 012116 (2009).
10. L. Ferialdi and A. Bassi, Phys. Rev. A **86**, 022108 (2012).
11. N. Gisin, Helv. Phys. Acta **62**, 363 (1989).
12. A. Bassi and G. Ghirardi, Phys. Rep. **379**, 257 (2003).
13. A. Bassi, K. Lochan, S. Satin, T. P. Singh, and H. Ulbricht, Rev. Mod. Phys. **85**, 471 (2013).
14. M. Toroš, A. Bassi, J. Phys. **51**, 115302 (2016).
15. S. L. Adler, J. Phys. A **40**, 2935 (2007).
16. A. Smirne and A. Bassi, Sci. Rep. **5**, 12518 (2015).
17. S.L. Adler and A. Bassi, J. Phys. A **40** 15083, (2007).
18. S.L. Adler and A. Bassi, J. Phys. A **41**, 395308 (2008).
19. R. Penrose, Gen. Relativ. Gravit. **28**, 581 (1996).
20. L. Diósi, J. Phys. A **40**, 2989 (2007).
21. L. Diósi, New J. Phys. **16**, 105006 (2014).
22. G. C. Ghirardi, R. Grassi, and A. Rimini, Phys. Rev. A **42**, 1057 (1990).
23. M. Bahrami, A. Smirne, A. Bassi, Phys. Rev. A **90**, 062105 (2014).
24. A. Bassi, D.-A. Deckert, L. Ferialdi, Europhys. Lett. **92**, 50006 (2010).
25. A. Bassi, J. Phys. A **38**, 3173 (2005).
26. A. Bassi and L. Ferialdi, Phys. Rev. Lett. **103**, 050403 (2009).
27. L. Ferialdi and A. Bassi, Phys. Rev. Lett. **108**, 170404 (2012).
28. L. Diósi, L. Ferialdi, Phys. Rev. Lett. **113**, 200403 (2014).
29. L. Ferialdi, Phys. Rev. Lett. **116**, 120402 (2016).

Appreciating What He Did

Tim Maudlin

In 1989, a NATO summer school organized by Arthur Miller was held in Erice, Sicily. By a rather unexpected set of circumstances I managed to horn my way into attending that conference, and therein lies a tale I have often told. It bears repeating here.

The highlight of the conference was John Bell's presentation of his classic paper "Against 'Measurement'". That was the first time I had heard of the GRW theory, and I was extremely puzzled by what Bell had to say. As was common at the time, I had grown up hearing about quantum theory from an exclusively Copenhagen perspective, the same approach enshrined in von Neumann's *Mathematical Foundations of Quantum Mechanics.* Apparently, the physical world obeyed two quite different dynamical laws, one which obtained when nothing was being "measured" or "observed" and quite another when a measurement happened. The former was always smooth, continuous, deterministic, predictable Schrödinger evolution, and the latter was discrete, jumpy, indeterministic and unpredictable collapse. Given this setting, the natural question cannot be avoided: What triggers the collapse? Under what conditions does the one equation take over from the other? It seems as though the trigger must be something of a very unique and profound nature, something able to switch the entire mode of physical evolution from one character to an entirely different character. Add on top of this Wigner's suggestion that the key ingredient is consciousness—the same profoundly puzzling aspect of reality that drives the mind/body problem—and the result is a heady mix of suggestion and speculation. Approached in this way, the mysteries of quantum theory become the mystery of collapse, which intimates possible connections to the biggest metaphysical enigma there is.

So when Bell turned from the conceptually inadequate accounts of quantum theory that he unearthed from physics textbooks to the possible solutions that met his criteria

T. Maudlin (✉)
New York University, New York, USA
e-mail: twm3@nyu.edu

V. Allori et al. (eds.), *Do Wave Functions Jump?*, Fundamental Theories of Physics 198,
https://doi.org/10.1007/978-3-030-46777-7_5

for professional clarity and mathematical precision I brought all my attention and concentration to bear. And when he further turned to the possibility of a precise *collapse* theory, I was riven by anticipation. Finally, a clear and exact proposal for the magical trigger of collapse would be presented. What really are the conditions that occasion such a profound alteration in the ways of nature herself?

And when the answer was revealed I was overwhelmed by a sense of profound......disappointment. You mean that the collapses just happen *at random*? There is no trigger, nothing about consciousness, no mystical key? I could hardly believe my ears.

After the lecture I went to Bell overcome by this sense of deflation. I asked him how he could be so interested in such a disenchanting resolution to the great mystery. I asked him, with evident distress, "You mean it just *happens*?".

And I can report the precise words Bell used in response. He looked at me in his usual calm and kindly way, and said "You don't appreciate what they have done".

And of course he was right.

This volume offers an opportunity to remind ourselves what GianCarlo—both in collaboration with Rimini and Weber and also on his own—achieved. There is a technical side and a conceptual side to this story, and both are important, but I will dwell on the conceptual side. One of the main contributions of the objective collapse theories that GianCarlo developed, in both discrete and continuous forms, is how they force us to directly confront the question of what the measurement problem really *is*, and what a conceptually acceptable sort of solution to it might possibly look like. Shaking the physics community from its blasé and dismissive attitude towards foundational questions requires overcoming the claim (much inculcated by Bohr) that a clear and comprehensible account of quantum phenomena *simply cannot be had*. The view that even two-slit interference (leave aside violations of Bell's inequality) cannot be *understood* but only *described* was advocated by Feynman in his *Lectures on Physics*, and Feynman's pronouncements cannot be taken lightly. So—at least at that time—one had the unmistakable sense of the scales falling from one's eyes when presented with a conceptually clear and clean physical account of what at least *might* be going on. Bell himself had had the same reaction to reading Bohm's 1952 papers on the pilot wave theory, but as a non-collapse theory that provided no obvious insight into the usual way of thinking about things. GRW did.

There are two quite different aspects of the measurement problem, and GianCarlo's work reflects his appreciation of each. Let's start with the first.

When Schrödinger presented his example of the cat in 1935, the main conceptual issue he confronted was what we might call *smeariness*. Bohr had denied the accuracy of the naïve planetary model of the atom: electrons are not, according to Bohr, little (possibly point-like) bodies that orbit the nucleus as the old quantum theory suggested. Bohr had come to insist that the wavefunction of an electron is *complete*, so the electron itself could have no physical characteristics not recoverable from the wavefunction. Hence, if the wavefunction is—in some sense—spherically symmetric and spread out in all directions around the nucleus, then the *electron itself* must be spread out in exactly the same way. The electron, on this view, is more like a cloud or a mist than a billiard ball. It is denser some places than others, but not paticularly located in any one place.

Schrödinger had no objection to electrons being smeared out in such a way in an atom. After all, we have no direct empirical access to individual electrons in atoms. If they are smeared out, so be it.

But as Schrödinger recognized, the mathematical character of his own dynamical equation for the wavefunction, and in particular its linearity, could not confine the smeariness to microscopic scale. In experimental conditions such he described the linear evolution would of necessity amplify the smeariness from microscopic to macroscopic magnitude. Absent any non-linear collapse, the dynamics would smear out the cat itself between a state in which it is alive and one in which it is dead. The smeariness would—according to Bohr and von Neumann—persist until a "measurement" or "observation" is carried out on the cat. And that, Schrödinger opined, is simply ridiculous. No one in their right mind could believe that.

Put in this way, the challenge is to get the collapse, but to get it somehow *earlier* than the moment when the experimentalist checks on the cat. That is certainly not an impossible task: one needs to identify a collapse trigger that goes off much earlier in the process than the ultimate observation. And the collapse must also be *late* enough that the observed interference effects (like two-slit interference) are not suppressed. But there is a wide range of Goldilocks locations for the collapse: not too early and not too late. One obvious way to proceed, from Schrödinger's perspective, is to find the right trigger. That done, the completeness of the wavefunction can be maintained.

The genius of the GRW theory was to cut the Gordian knot of the trigger: don't tie it to anything at all. By having the collapse rate low enough, the observed interference effects are safe. And the brilliant observation was that entanglement of positions in solid macroscopic objects (such as needles on apparatuses) ensured definite enough positions for the macroworld with no further ado. The relatively low collapse rate is overmatched by the tremendously large number of particles in familiar macroscopic objects. This is what Bell appreciated and I did not: how such a simple and small and mathematically precise change in the dynamics can get all of the collapse characteristics one wants or needs.

1 The Second Innovation

At a highly abstract level, the GRW spontaneous collapse solves one aspect of the measurement problem cleanly and precisely, with a mathematically specified dynamical law. That gives the Born rule something to be *about*: the probabilities are probabilities for the collapse to happen one way rather than another. It breaks the symmetry that creates a measurement problem in any theory according to which the wavefunction is complete. So it is tempting to say that the GRW collapses solve the measurement problem *tout court*. And for a while I certainly thought it did. But, as Bell pointed out and GianCarlo acknowledged, there is more to the story.

The illusion that the collapses alone solve all the issues arises from a misleading way of *labelling* quantum states. The aspect of the Measurement Problem that rests on the completeness of the wavefunction and the linearity of the dynamics goes like this:

In an experimental situation where, for example,

|z-spin-up > |Device-in-ready-state > $\Rightarrow$ |Cat-alive>

and

|z-spin-down > |Device-in-ready-state > $\Rightarrow$ |Cat-dead >,

(with the arrow representing the Schrödinger evolution), linearity implies that

|x-spin-up > |Device-in-ready-state > $\Rightarrow$ $\frac{1}{\sqrt{2}}$ |Cat-alive > + $\frac{1}{\sqrt{2}}$ |Cat-dead >.

If the wavefunction is complete, and|Cat-alive > represents a state in which the cat it alive while |Cat-dead > represents a state in which the cat is dead, then the final state is a problematic superposition of macroscopically different states. The collapse, of course, prevents it from forming.

But note that the argument just given *takes for granted* that |Cat-alive > represents a physical situation with a live cat and |Cat-dead > represents a situation with a dead cat. There is *no attempt at all* to explain just how that could be the case. All of the burden of making the connection between these particular wavefunctions and the conditions of real, physical cats is carried simply by the labelling "Cat-alive" and "Cat-dead". If ever Russell's quip about theft over honest toil were called for, it is here.

One might try to argue that we know that the transition

|z-spin-up > |Device-in-ready-state > $\Rightarrow$ |Cat-alive>

yields a final state with a live cat by simple empirical observation: the apparatus just *does* yield a live cat whenever a z-spin-up particle is fed in. But this observation is neither here nor there: it *makes the presupposition* that wavefunction never collapses in this case, which is part of what is at issue. And it *makes the presupposition* that the wavefunction is complete, which is also at issue. So we can't turn to the experimental records to bridge this gap in the argument.

Bohr insisted that experimental conditions and outcomes *must* be provided in "classical terms", but he never explained how they *can* be provided in classical terms. The "classical terms" have nothing to do with the details of classical dynamics as enshrined, say, in Newton's Laws of Motion. "Classical terms" just means a description in terms of the macroscopic characteristics (including position and motion) of macroscopic objects like knobs and needles and cats. And we can see why Bohr insisted on this: experimental procedures and outcomes *are* described in precisely terms like these. In particular, one positively wants to keep theoretical terms out of the description of the experiment and its outcome since the experiment is being used to test the theory, and so should not tacitly presuppose it.

In Copenhagen, the exact conceptual and physical relation between the "classical" language of the experimentalist and the mathematical formalism of the quantum theorist is left extremely hazy and puzzling. As Bell says:

"The kinematics of the world, in this orthodox [Copenhagen] picture, is given by a wavefunction (maybe more than one?) for the quantum part, and classical variables—variables which *have* values—for the classical part: $(\Psi(t, q, \ldots), X(t), \ldots)$. The Xs are somehow macroscopic. This is not spelled out very explicitly. The dynamics is not formulated very precisely either. It includes a Schrödinger equation for the quantum part, and some sort of classical mechanics for the classical part, and 'collapse' recipes for their interaction.

It seems to me that the only hope for precision with the dual (Ψ, x) kinematics is to eliminate completely the shifty split, and let both Ψ and x refer to the world as a whole. Then the xs must not be confined to some vague macroscopic scale, but must extend to all scales".[1]

In Bell's taxonomy, the xs include the local beables of the theory: they must because Ψ is not a local object. So this point of Bell's connects directly to his second main conceptual issue with standard quantum theory: the issue of local beables.

Somehow or other, a theoretical picture of physical reality must make contact with experimental data in order to be an empirical theory. And since the empirical data is ultimately taken to be recorded in the structure of localized macroscopic objects, the physical theory must somehow make contact with those. That requires the theoretical picture to have some local beables, which ought to be defined at microscopic scale. Then the "classical" world gets built up, just as we have always assumed, from large collections of microscopic entities. The theory dictates the behavior of the microscopic local beables, which in turn determine the macroscopic local objects by simple aggregation.

The programmatic necessity for local beables has been widely overlooked in foundational investigations. The duel (Ψ, X) ontology with only macroscopic Xs, Bell's presentation of Copenhagen, does not have trouble making contact with experience, but rather problems making the contact between the observable objects and the thing actually governed by quantum theory. And it suffers the severe conceptual problem of denying that macroscopic objects are nothing more than large collections of microscopic parts. It is very hard to see how to maintain the Copenhagen approach in any rigorous way, and nowadays no one even tries. But it has not generally been appreciated what sort of problems a defensible view of the relation between macro and micro must face. Bell showed us the way.

For the sake of contrast, the Many Worlds interpretation has never been provided with a clear account of either what the local beables of the theory are, or of how to understand the empirical consequences of the theory without them. There have been attempts to somehow interpret the reduced density matrix of quantum field theory as a representation of some local beable, and also attempts to have space-time itself along with some local beables in it "emerge" via functional considerations from a non-spatial-temporal foundation. This is not the place to review these efforts, but it is fair to say that there is no consensus about whether they can be made to work. With respect to GRW theory, though, Bell provided a clear ontology of local beables—the flash ontology—in his exposition. In order to appreciate Bell's point, I think it is important to note that he uses the phrase "in the theory" to mean "in the mathematical apparatus" rather than "according to the theory". For his own ontological proposal

[1]J. Bell, *Speakable and Unspeakable in Quantum Mechanics*, Cambridge University Press, Cambridge, 2004, p. 228.

here is that the world would be composed of a quantum state (directly represented by the wavefunction), a space-time, and a collection of local physical point-events in the space-time:

"There is nothing in this theory but the wavefunction. It is in the wavefunction that we must find the physical world, and in particular the arrangement of things in ordinary three-dimensional space. But the wavefunction as a whole lives in a much larger space, of 3 N-dimensions. It makes no sense to ask for the amplitude or phase or whatever of the wavefunction at a point in ordinary space. It has neither amplitude nor phase nor anything else until a multitude of points in ordinary three-space are specified. However, the GRW jumps (which are part of the wavefunction, not something else) are well localized in ordinary space. Indeed, each is centered on a particular space-time point $(\mathbf{x}, t)$. So we can propose these events as the basis of the 'local beables' of the theory. These are the mathematical counterparts in the theory to real events in definite places and times in the real world....A piece of matter is then a galaxy of such events".[2]

Although the original GRW paper did not suggest such an ontology of local beables, GianCarlo understood the point Bell was making and embraced the requirement that an acceptable theory specify some local beables. But along with the articulation of the original GRW proposal in that direction, different considerations arose from a different origin.

Adopting a flash ontology comports well with the discrete character of the GRW collapses. But there was also some desire to eliminate the jumpy two-different-evolutions dynamics in favor of a uniform continuous stochastic process. Together with Philip Pearle, that avenue of research yielded the Continuous Spontaneous Localization (CSL) model. CSL not only eliminated the discontinuous dynamics of GRW, it also seemed better situated for generalization to field theory.

Unlike a classical particle ontology, a classical field ontology postulates a spatially continuous local beable. Together with the continuous stochastic CSL dynamics, this yields a picture that is the polar opposite of the flash ontology: space-time now plays host to a continuously evolving continuous distribution of local matter-density. The formal problem is to define such a matter-density on physical space-time in a way determined by the wavefunction. This can be done in many ways, each with its own virtues and defects. (For example, seeking a fully Relativistic theory one can commit to using only the light-cone structure of the space-time in the specification of the matter-density field.[3])

---

[2]Bell 2004, pp. 204-5. Note: the points $(\mathbf{x}, t)$ are *mathematical* items. The values of the variable t, for example, are real numbers. They come to represent point of ordinary space-time via a coordinate map.

[3]For a general overview of the menu of options for collapse theories, see G. Ghirardi, "Collapse Theories" in the Stanford Encyclopedia of Philosophy, https://plato.stanford.edu/entries/qm-collapse/. For a discussion of using light-cones to specify a Lorentz-invariant theory of a matter-density field, see D. Bedingham, D. Dürr, G. Ghirardi, S. Goldstein, R. Tumulka and N. Zanghì, "Matter Density and Relativistic Models of Wave Function Collapse", Journal of Statistical Physics **154**, 623-631 (2014), https://doi.org/10.1007/s10955-013-0814-9.

So alongside Bell's flash ontology, GianCarlo and others have proposed local matter-density ontologies. And one of the upshots of all of this productive theoretical activity has been the realization that a properly and completely formulated GRW-like "collapse" theory shares a tremendous amount of abstract structure in common with the non-collapse pilot wave approach.[4] Perhaps this lesson has now been learned, but it has been a long trek of slow progress. And it is clear in retrospect why: at first glance, the collapse approach and the pilot-wave approach appear diametrically opposed. One has stochastic collapse and the other only deterministic evolution of the quantum state. In one the quantum state dynamics is linear and in the other it is not. In the GRW approach the wavefunction remains complete in Einstein's sense: given the postulates of the theory (including what the local beables are) the entire physical description of a system can be recovered from the wavefunction, while in the pilot-wave picture this is never true. Even when the need for local beables has been acknowledged, the collapse theories postulate flashes or matter densities and the pilot-wave theories generally posit particles. One could easily be forgiven for thinking that the two approaches are radically unlike each other in every respect.

But over the course of decades, with GianCarlo always leading the exploration of new possibilities, we have come to understand the various related forms that a serious, exact and defensible theory of quantum phenomena can take. As he saw from the beginning, if you want to have collapses they should be handled using precise mathematics not vague gestures. And as he came to see, along with the discussion of the wavefunction and the quantum state, there must be consideration of the local beables of the theory and the connection to empirical data. There are a plethora of ways that this can be accomplished, and they are all worthy of careful consideration and elaboration.

In sum, there are many different dimensions of appreciation for what GianCarlo Ghirardi did. Everyone knows of his seminal contribution to pioneering mathematically precise collapse dynamics. But in addition, he made equally important contributions to the development of empirically adequate theories of local beables, and played an indispensable role in the formation of a broad and unified community working in the foundations of physics, open to discussion and collaboration across a wide variety of theoretical approaches. He will always hold a place of both admiration and affection in the hearts of those of us lucky enough to have called him a friend.

[4]See V. Allori, S. Goldstein, R. Tumulka, and N. Zanghì, "On the Common Structure of Bohmian Mechanics and the Ghirardi-ARimini-Weber Theory", British Journal for the Philosophy of Science **59**, 353–389, https://doi.org/10.1093/bjps/axn012 (2008).

Philosophy

The GRW Theory and the Foundations of Statistical Mechanics

David Albert

Abstract Giancarlo Ghirardi, along with Alberto Rimini and Tullio Weber, first showed the world that there could be a scientifically respectable theory of the collapse of the quantum–mechanical wave-function. It is primarily and rightfully for that, that they will always be remembered. In this paper I mean to point to a small and somewhat neglected *side-effect* of that achievement which has perhaps received less attention than it deserves: if anything along the lines of the GRW theory should turn out to be true, then the probabilities of universal statistical mechanics are nothing other than the familiar probabilities of quantum mechanics.

It was Giancarlo Ghirardi, along with his colleagues Alberto Rimini and Tullio Weber, who first showed the world that there could be a beautiful and simple and explicit and precise and quantitative and scientifically respectable theory of the collapse of the quantum–mechanical wave-function. For something on the order of a half a century, prior to their seminal paper of 1986 [1], talk of collapses was hopelessly mired in talk of 'measurement' and of 'macroscopicness', and of 'irreversibility', and of 'indelible recordings', and of 'consciousness' and of 'the distinction between subject and object' and of 'the inescapability of classical language' and (to make a long story short) of every other imaginable variety of nonsense—nonsense that (mind you) many of us who lived through it can now only vaguely remember, and that the young physicists and philosophers of today can barely imagine—and it was Ghirardi and Rimini and Weber who, with a single brilliant stroke, swept all of that finally and decisively away, and showed us how, at long last, and for ever after, to talk about these matters *seriously*.

And it is primarily for that, and it is rightfully for that, that they will always be revered. And all I mean to do here is to point to a small and somewhat neglected *side-effect* of that achievement—something not about the foundations of quantum mechanics but about the foundations of *statistical* mechanics, and about the way that *probabilities* enter into our fundamental description of the world—which we owe

D. Albert (✉)
Columbia University, New York City, NY, USA
e-mail: da5@columbia.edu

V. Allori et al. (eds.), *Do Wave Functions Jump?*, Fundamental Theories of Physics 198,
https://doi.org/10.1007/978-3-030-46777-7_6

to that same paper of 1986, and which has perhaps received less attention than it deserves.

Think (to begin with) of two macroscopic bodies whose temperatures initially differ. And suppose that those two bodies are brought into thermal contact with each other. And suppose that they are not subsequently disturbed. Over the next ten minutes, then, the temperature difference between those two bodies will decrease. And the traditional statistical-mechanical explanation of that decrease, both in the classical and in the quantum case, runs (roughly) as follows. The initial macrocondition of this two-body system—the one in which the two bodies are in thermal contact with each other and their temperatures are different—is compatible with a continuously infinite collection (call it {C}) of that system's possible microconditions. And the microconditions in {C} come in two different varieties: the normal ones (which are the ones that happen to be sitting on trajectories which pass—ten minutes hence—through a macrocondition of the two-body system in which the temperature difference between the two bodies is lower, and lower by the right amount) and the abnormal ones (which are all the rest, the ones associated with un-thermodynamic or with anti-thermodynamic sorts of behaviors, the ones in which the temperature difference will subsequently rise, or not change at all, or oscillate, or whatever). And there happens to be a very simple and straightforward *measure* on the set of the possible microconditions of a system like this one which is preserved by the equations of motion,[1] and which our experience of the world seems to suggest is something along the lines of a measure of non-dynamical probability. And it happens that this measure counts the collection of normal points in {C} as vastly larger than the collection of abnormal points in {C}.

And that (according to the usual story) is that.

But consider at this. It happens (to begin with) that the collection of normal microconditions is vastly larger than the collection of abnormal ones—on the above-mentioned standard measure—not only over the entirety of {C}, but over every individual not-unimaginably-small microscopic *neighborhood* of {C}, and (more particularly) over every individual not-unimaginably-small microscopic neighborhood of every individual *abnormal* microcondition of {C}, as well!

And what that means (or at any rate, one of the things it means) is that the property of being a normal microcondition is extraordinarily stable under small perturbations of those two bodies, and that the property of being an *abnormal* microcondition is extraordinarily *unstable* under small perturbations of those two bodies.

And what *that* means is that if the two bodies we've been talking about here were in fact somehow being frequently and microscopically and randomly *perturbed*, then the temperatures of those two bodies would be overwhelmingly likely to approach each other *no matter which one* of the microconditions in {C} actually initially obtained.

The question, of course, is about where perturbations like that might imaginably come from. And the wonderful thing—and this is the main punch-line of this

[1]The sort of "preservation" I have in mind here is the one connected with Liouville's theorem, or (in the quantum–mechanical case) with unitarity.

paper—is that the quantum jumps in the GRW theory turn out to be just the sorts of perturbations we need. The idea (more particularly) is that it's going to turn out to be a consequence of the full stochastic dynamics of the GRW theory[2] that every single individual one of the microconditions in {C} will be overwhelmingly likely to evolve, over the subsequent ten minutes, into other microconditions in which the temperature difference between the two bodies is smaller, and (moreover) smaller by precisely the right amount.

And so if this thought is correct, and if anything along the lines of the GRW theory should turn out to be true (which will, of course, be a matter for future experiments to determine) then the probabilities of universal statistical mechanics are (as a matter of fact, when you come right down to it) nothing other than the familiar probabilities of quantum mechanics. And if this thought is correct, and if anything along the lines of the GRW theory should turn out to be true, then the tendency of the temperatures of the two bodies we've been talking about here to approach each other over time amounts to a genuine (albeit statistical) dynamical law. And if this suggestion is correct, and if anything along the lines of the GRW theory should turn out to be true, then the tendency of the temperatures of the two bodies we've been talking about here to approach each other over time can be understood *entirely* in terms of readily observable characteristics of the elementary microscopic constituents of those bodies—in precisely the same way that (say) the functioning of a mechanical clock can be understood entirely in terms of the material characteristics, and the spatial arrangements, of its parts.

And it happens that none of the other attempts to solve the quantum–mechanical measurement problem that are currently on the table—and (as a matter of fact) nothing else that has ever seriously been put forward as a fundamental dynamical theory of the world— can do anything like that.

And this will be worth going into in some detail, as it seems to have had a way (here and there) of uncannily escaping people's attention.

It has often been suggested in the literature (for example) that nothing even remotely as up-to-date as quantum mechanics is going to be required here—that (more particularly) the sorts of perturbations we were talking about above are already all over the place, if one simply stops and looks, in (say) the Newtonian picture of the world. The idea is that since none of the macroscopic two-body systems of which we have ever had any experience, and none of the macroscopic two-body systems of which we ever *will* have any experience, are genuinely isolated ones, the perturbations in question can be seen as arising simply from the interactions of the two-body system we've been talking about here with its environment. But if (as these authors always suppose) whatever constitutes the environment of these two bodies evolves in accord with precisely the same sorts of deterministic dynamical laws as the constituents of the bodies themselves do, then whatever "randomness" there is in the perturbations arising from interactions with that environment can only have gotten

[2]Which is to say, it is going to be a consequence of that dynamics *alone*; it is going to be a consequence of that dynamics without any non-dynamical addenda whatsoever.

there in virtue of precisely the same sort of probability-distribution over that environment's initial conditions that is the bread and butter of traditional formulations of statistical mechanics. And so the whole exercise gets us nowhere.

What about something like Bohm's theory? Bohm's theory has probabilities in it. The trouble is that those probabilities don't get inserted into the world in the right *place* to do the sort of job we have in mind for probabilities here. The only sorts of things that turn out to be probabilistic, according to Bohm's theory are the positions of the particles. The only sorts of fundamental probabilities there are in Bohm's theory are (more particularly) probabilities that such-and-such a collection of particles has such-and-such a spatial configuration at such-and-such a temporal instant, given that the particles' wave-functions have such-and-such an overall shape at that instant. And it happens that those parts of the laws of physics which govern the time-evolutions of the shapes of wave-functions, on Bohm's theory, are completely deterministic; and it turns out that there are wave-functions compatible with the initial macrocondition of (say) the two-body system I talked about before which (if those laws are right) will with certainty evolve, with the passage of time, into ones which determine that the temperature-difference between the two bodies will very likely have *increased*.

And there are even *collapse* theories on offer nowadays, theories on which the time-evolution of the wave-function itself is genuinely (and dynamically) probabilistic, which are nonetheless incapable of underwriting the foundations of statistical mechanics in the way that the GRW theory can. These sorts of theories (which have been defended in recent years by Roger Penrose, among others) stipulate that departures from the deterministic equations of motion require a "trigger"; that only certain particular sorts of wave-functions, the ones corresponding to superpositions of "macroscopically different states," ever undergo "collapses." And the trouble with that (insofar as the question of statistical mechanics is concerned) is that one can cook up (or at any rate one fears that one can cook up) initial wave-functions of thermodynamic systems which pick out perfectly deterministic entropy-decreasing future trajectories which entirely avoid those triggers.

And of course there are no real probabilities *at all*—or none (at any rate) of the kind we are in need of here—in *Everettian* proposals for solving the quantum–mechanical measurement problem.

And so the business of underwriting the thermodynamic regularities of the world, on *any* of the proposals for making sense of quantum mechanics I know of, with the sole exception (of course) of the GRW theory, is going to call for a story about why it is that the above-mentioned sorts of initial wave-functions—notwithstanding that they surely exist—need not worry us too much; which is to say that the business of underwriting the thermodynamic regularities of the world on any of those other theories is going to call for something along the lines of a probability-distribution over initial wave-functions, a probability-distribution which (note) is altogether unrelated and in addition to the probabilities with which those theories underwrite the statistical regularities of *quantum mechanics*.

Now, the business of deciding whether or not to take a GRW-based statistical mechanics seriously (if that turns out to be a project worth undertaking at all; if, that is, there should turn out to be experimental evidence that there are such things in the world as collapses of wave-functions) will presumably involve detailed quantitative examinations of a host of particular cases; but there are reasons for being optimistic (and the sort of thing I have in mind here, of which more in a minute, is the very same radical instability of the condition of abnormality by which all of this was first suggested) about how those examinations will ultimately come out.

Here's the idea.

Think (to begin with) of some particular individual GRW jump. And call the microcondition of the system in question just prior to that jump A, and call the microcondition of the system in question just after that jump B.

And note that the laws of jumps like that will straightforwardly entail an infinite set of probability-distributions $P_A(B)$ over all the possible destinations of any particular such jump, given the point at which that jump starts out.

And there are two particular features of the $P_A(B)$'s of the GRW theory (and of any theory more or less in the neighborhood of the GRW theory) that it will be well (for the purposes of the next paragraph or so) to bear in mind: one is that every particular one of the $P_A(B)$'s of the GRW theory turns out to be more or less centered on its own particular A, and the other is that the volume of the space of possible microconditions over which any particular one of the $P_A(B)$'s of the GRW theory has non-negligible values will typically be far smaller than the volume of any one of the macroconditions of anything that deserves the name of a thermodynamic system.

Now the sort of thing we need from these jumps—in order to get the statistical-mechanical job done—is (of course) for them to be very good at getting us from abnormal microconditions to normal ones. The sort of thing we need (that is) is for it to be the case that the scales over which the tiny individual clots of abnormal microconditions typically extend are vastly smaller than the scales over which the values of the $P_A(B)$ appreciably vary. The sort of thing we need (more particularly) is for it to be the case that the probability of abnormality that follows from every single individual one of the $P_A(B)$'s (no matter what A may happen to be) is roughly equal to the probability of abnormality that follows from the standard statistical-mechanical measure over the entirety of the macrocondition within which the A in question happens to fall.[3]

And it would seem to be an eminently plausible proposition—given the radical unimaginable submicroscopic tinyness of the clots, and given the two particular

[3]Actually, we don't need quite that, and probably can't quite have it. The trouble is that abnormal quantum states have got to be more or less orthogonal (if you think about it) to normal ones, and that no single GRW collapse can ever (in and of itself) bring about transitions between states that are (perfectly) orthogonal to one another, and that (as a matter of fact) no single GRW collapse is ever going to be able to do much of anything (in and of itself) about the abnormality of a quantum state if that state should happen to be anything along the lines of an eigenstate of the positions of the particles that make the system in question up. But none of that turns out to matter much. Let the A and B we have just now been discussing represent (instead) the before and after states of a dynamical process involving (say) two GRW collapses, or three, or twenty, with the appropriate deterministic dynamical evolutions between them (all of which is still going to be overwhelmingly likely to take place, on the GRW theory, over time-intervals which are negligibly short compared—say—with times over which the temperatures of the two bodies we were talking about before ever undergo any significant change)—and everything will come out fine.

characteristics of the jumps in the GRW theory that we took note of in the paragraph before last—that there are any number of different sorts of GRW-like perturbations that are perfectly capable of getting all that accomplished.

Nonetheless, there are hard cases, or apparently hard ones (and I am thankful to Larry Sklar and Phillip Pearle, among others, for bringing these to my attention); and there turn out to be interesting lessons in them; and it will be worth taking the trouble to think through two or three of them in some detail.

Consider (for example) an extraordinarily tiny gas, one which consists of something on the order of 10^5 molecules. Even gasses as tiny as that are known to be very likely to spread out (if space is available) over reasonable intervals of time, and yet gasses as tiny as that are very unlikely to suffer even a single GRW-type collapse over such an interval, and so an explanation of the tendencies of gasses like that to evolve like that over intervals like that in terms of GRW-type collapses of the wave-functions of their constituents is apparently out of the question.

Or consider the collection of dazzling and beautiful experiments which have actually been performed over the past forty years or so, and which are referred to in the scientific literature as "spin-echo" experiments, in which it has turned out to be possible to isolate some very large array of interacting microscopic systems from the relevant sorts of external influences—and (moreover) to replace the dynamical condition of that array, at a certain particular instant, as the array is in the midst of some entropy-increasing transformation, with its time-reverse—and (thereafter) merely to watch, in astonishment, as the array traces its previous trajectory out, dutifully, backward.

The microscopic systems in question are typically atomic nuclei. And these nuclei are typically being held at fixed spatial positions—but in such a way that the orientations of their nuclear magnetic fields are free to rotate— by intermolecular forces in a crystal. And the sort of thing that happens in these experiments is (very schematically) that the nuclei are all initially arranged with their magnetic fields pointing in the same direction—and then they're left (as it were) to their own devices, and they magnetically interact with one another, and their magnetic fields begin to pivot around, and in time the directions in which those individual fields are pointing become more and more disorganized and uncorrelated. Eventually a state of equilibrium is arrived at, in which the arrangement of the individual fields is random, in which (that is) the cumulative macroscopic magnetic field of the entire array is zero, and then (and this is the cool part) a very intense external magnetic field is turned on for a very short time, which has the effect (for reasons that need not concern us here) of turning all those tiny individual fields exactly around—and then the system is left again to its own devices, and in time, and (more particularly) in precisely the same amount of time as had elapsed between the array's first having been left to its own devices and the moment when the external field was turned on, the fields spontaneously re-align themselves!

It would seem (on the face of it) that GRW collapses can play no role whatsoever in any explanation of the initial approach to equilibrium here. The trouble is that the atomic nuclei in these experiments are very rigidly held in place—which is to say that the wave-functions of those atomic nuclei are permanently localized—which is to say that the wave-functions of those atomic nuclei are permanently frozen into that particular mathematical form which is (if you think about it) altogether impervious to the effects of GRW collapses—by the powerful intermolecular forces I mentioned above. Moreover (and this is the particularly astonishing business—and this seems powerfully confirmatory of the doubts expressed in the previous two sentences), it turns out that the approach to equilibrium can be reversed—it turns out that the original alignment of the fields can be reinstated—simply by flipping the nuclei around!

Or consider what it is, on a statistical mechanics of the sort that we have been imagining here, that guarantees that a regular-sized gas in equilibrium at t will not spontaneously explode or condense or turn into an elephant between t and t and t + ε, where ε is so short an interval that even a regular-sized gas is unlikely to suffer a GRW-type collapse in it.

Let's think through these three cases one at a time.

Take the case of a small gas. We might appeal, there, to the fact that we have no empirical experience whatsoever, that (come to think of it) we can have no empirical experience whatsoever, of a small gas which is genuinely isolated from all external influences. And so for all we now empirically know or ever will empirically know, it might not be a law of nature that gasses like that tend to spread out at all! And the behaviors of the sorts of small gasses that can actually be looked at can very plausibly be accounted for by GRW-type collapses of the wave-functions of particles in (say) their containers.

Or we could appeal to the fact that such gasses, even if they are isolated, have pasts. This will take a bit more setting up. What we will want to show, in this case, is that the GRW theory will entail that a small isolated gas which is condensed at t, and which is around for a while, is likely to be more dispersed at t + @, even if the gas in question is unlikely to undergo a single collapse in the interval between t and t + @. Good. Here's how to do it: call the average time between GRW collapses in the gas in question i, and call the gas's macrocondition at t C, and call the gas's macrocondition at (say) t – i(100,000) S. And consider the probability, on the GRW theory, given that the macrocondition at t – i(100,000) is S and that the macrocondition at t is C, that the microcondition of the gas at t will be one of the "normal" ones. And note that the instability of the property of being abnormal will entail, completely independent of what state S is, that that probability is high.

What about the case of the spin-echo experiments? Collapses in the environment will patently get us nowhere with that. The realignability of the fields, after all, amounts to a direct empirical proof that those collapses (just like the ones that hit the nuclei themselves) produce no significant short-term disruptions of the trajectories along which this system evolves. But the longer term is (of course) another matter. Given sufficient time, even in systems like this, GRW collapses will move us relentlessly away from abnormality. And so there would seem to be every reason in the world to believe that the previous history of the array of nuclei in question, whatever that history may have been, will give us just what we need.

What about large gasses over the very short term? The environment will be of no avail there either; but histories still will. And here a third strategy suggests itself.

The macroconditions of thermodynamic systems never get measured at instants. The thermodynamical regularities of our actual experience, if you stop and think about it, are relations between the physical situations of systems not at different instants but around different instants. And so maybe the right way to think of propositions like "this is a gas with such-and-such a volume and a temperature and a pressure" is to see them as asserting that certain physical properties of a certain collection of particles have persisted over a certain short interval. And if we read such propositions that way, they will entail (in conjunction with the GRW theory) that the probability that the microcondition of the gas in question is a normal one is high.

You get the idea. The crux of the matter is that the job of statistical mechanics is not (after all) to underwrite the letter of the laws of thermodynamics, but to underwrite the actual content of our thermodynamic experience. And I know of no compelling argument, at present, why a statistical mechanics based on GRW collapses should be incapable of doing that.

One can go further. If the GRW theory should turn out to be true— and this, of course, is a very big if—it may turn out (as I mentioned earlier on) that there is at bottom only a single kind of probability in nature. It may turn out (that is) that all the robust lawlike statistical regularities there are, not only in thermodynamics but (one can even imagine) in biology, and in psychology, and in sociology, and God knows where else, are at bottom nothing other than the probabilities of certain particular GRW collapses' hitting certain particular sub-atomic particles.

Reference

1. Ghirardi, G.C., Rimini, A., and Weber, T. (1986). "Unified dynamics for microscopic and macroscopic systems". *Physical Review D* 34: 470.

Spontaneous Localization Theories with a Particle Ontology

Valia Allori

Abstract Spontaneous localization theory is a quantum theory proposed by Gian-Carlo Ghirardi, together with Alberto Rimini and Tullio Weber in 1986. However, soon it became clear to Ghirardi that his work was more than just one theory: he actually developed a framework, a family of theories in which the wavefunction jumps, but where the ontology of the theory is underdetermined. After acknowledging that the wavefunction did not provide a satisfactory ontology, he assumed that matter was described by a continuous matter density field in three-dimensional space, whose evolution is governed by a stochastic wavefunction evolution. Alternatively, Bell assumed that the wavefunction would govern a spatiotemporal event ontology, dubbed 'flashes.' However, not much work has been done with the perhaps most obvious possibility, namely that physical objects are made of particles. This paper has two aims. First to explain the reason why people require spontaneous localization theory to be more than just a theory about the wavefunction. This is done by showing how the problem everyone in the foundation of quantum mechanics take to be the fundamental problem of quantum mechanics, namely the measurement problem, is a red herring. Then, the paper explores the possibility of spontaneous localization theories of particles. I argue that this discussion is not a mere exercise, as spontaneous localization theories of particles may be amenable to a relativistic extension which does not require a foliation, and because in general the peculiar type of indeterminism of spontaneous localization theories may help sheding new light on the nature of the tension between quantum theory and relativity.

Forthcoming in: V. Allori, A. Bassi, D. Dürr and N. Zanghí (eds.) *Do Wave Functions Jump? Perspectives on the Work of GianCarlo Ghirardi*. Springer.

V. Allori (✉)
Department of Philosophy, Northern Illinois University, Dekalb, IL, USA
e-mail: vallori@niu.edu
URL: https://www.valiaallori.com

V. Allori et al. (eds.), *Do Wave Functions Jump?*, Fundamental Theories of Physics 198,
https://doi.org/10.1007/978-3-030-46777-7_7

1 Introduction

In this paper I wish to discuss spontaneous localization theories of particles within the primitive ontology framework. In the first part of the paper, I argue in favor of spontaneous localization theories with a primitive ontology, while in the second part I discuss the tenability and the superiority of a particle primitive ontology for this kind of theories. First, I discuss how the origin of the interpretational problems of quantum theory is not, as commonly maintained and as explained in Sect. 2, the measurement problem. Indeed, the measurement problem is a red herring: even if one solves the measurement problem the theories so obtained are still problematical. I discuss in Sect. 3 how the real problem stems from thinking of the wavefunction as describing physical objects. In line with the primitive ontology approach (POA), in Sect. 4 I present the various proposals for spontaneous localization theories understood as theories about some microscopic ontology in three-dimensional space, in terms of a matter density field, or four-dimensional spacetime, in terms of flashes. In Sect. 5, instead I move to possible spontaneous localization particle theories, and show how only one of the alternatives is worth pursuing. In Sect. 6 I compare this theory with spontaneous localization matter density and flash theories, and finally I propose an approach that could lead to a relativistic spontaneous localization particle theory. I conclude arguing that the value of considering any of these theories does not reside in the simplicity of their ontology or law (as they are not simple when compared to alternatives such as the pilot-wave theory) but rather it lies in the lesson they may teach us about the compatibility between quantum theory and relativity.

2 The Traditional Problem of Realism and Quantum Mechanics

Since its birth quantum theory has been such an interpretative nightmare that many felt the lesson to be learned was to embrace instrumentalism. However, many others still searched for a realist interpretation of quantum mechanics, starting most famously Albert Einstein, Louis de Broglie and Erwin Schrödinger, continuing with David Bohm, High Everett, GianCarlo Ghirardi and John Stuart Bell. Most often than not, the problem of reconciling quantum theory with a realist description of the world is summarized mentioning the so-called *measurement problem*. This problem has been around since Schrödinger [1] criticized the reading of standard quantum mechanics (the one of Bohr and Heisenberg) according to which the microscopic world has a 'blurred reality,' to be contrasted with the 'definite reality' one observes macroscopically. If the microscopic world is 'blurred' but completely described by the specification of a linearly evolving wavefunction, then this 'blurriness' would immediately spread to the macroscopic scale when we couple the microscopic system (in Schrödinger's example a radioactive source) to a macroscopic one (a cat). In other words, granting that radioactive nuclei can be 'blurred', we can measure whether a nucleus has decayed or not by hooking it up to a device which would kill a cat in case of decay. Since someone's death cannot be 'blurred,' one immediately sees that this interpretation is untenable. This conclusion is often reformulated as a problem for the view that a Schrödinger evolving wavefunction provides the complete description of a physical system: if every physical system is completely described such an object, because of the linearity of the Schrödinger equation we should observe macroscopic superpositions such as a cat which is in a superposition of being both alive and dead. Since we do not observe them, this reading is empirically inadequate. Traditionally, three ways to get around the problem have been identified:[1] (1) deny that the wavefunction provides the complete description of each physical system; (2) deny that macroscopic superpositions are a problem; and (3) deny that the wavefunction evolves according to the Schrödinger equation. In the 1950s, building on some groundbreaking work done by de Broglie [3], Bohm [4] proposed a solution of this problem along route 1. In fact his theory, which many dub the pilot-wave theory or Bohmian mechanics, is often taken to be one in which there are particles and waves, and the particles' behavior is determined by the wave's behavior. Few years later, Everett [5] proposed his 'relative state formulation' of quantum theory, which goes along route 2. Everett's theory later was developed into the so-called many-worlds theory, often characterized as accepting the macroscopic superpositions as real but suitably existing in other, undetectable, worlds which do not interact with the one we are in. In the 1980s, Ghirardi et al. [6] added their solution going along the lines of route 3. In their theory, called among other names spontaneous localization theory or GRW theory, the wavefunction does not evolve according to the Schrödinger equation but it suitably collapses at random into one of the terms of the superposition, localizing in a small region of space in the case of macroscopic objects. All these theories are generally taken to be quantum theories that are amenable to a realist interpretation. The reason which is given for this is that they do not suffer from the measurement problem. In fact in the pilot-wave theory, we do not observe macroscopic superpositions because the complete description of the system is given by the specification of particles and wavefunction, and particles are always localized, just like cats, whether they are dead or alive. In the many-world theory the various terms of the superpositions 'live' in other worlds which are inaccessible to us and do not interact with us, and this explains why we do not encounter the alive counterpart of a dead cat. Finally, in the spontaneous localization theory the facts that the wavefunction localizes very fast for macroscopic objects explains why we never see macroscopic objects in superposition states, and why dead cates remain, perhaps unfortunately, dead.

[1] See e.g. Bell [2].

3 The Real Trouble for the Quantum Realist

In the previous Sect. I kept using the locutions 'traditionally,' 'often,' or 'generally.' Recently, a new approach to quantum theories, the so-called *primitive ontology approach* (POA), has been proposed,[2] and while this has not been explicitly stated anywhere, I think that the main lesson from it is that the tension between realism and quantum mechanics is not captured by the measurement problem. This implies that the pilot-wave theory is not a theory of waves and particles; the many-world theory and the spontaneous localization theories are not theories 'about' the behavior of the wavefunction. The main lesson of the POA, I argue, is that the measurement problem is a red herring. That is, even if we solve the measurement problem, the tension between realism and quantum mechanics remains open. The real issue is instead the so-called *configuration space problem*: the wavefunction, whether it provides the complete description of a system or not, is not an object which is defined in three-dimensional space. Instead it is a function whose domain is the space of the configuration of particles, if there are particles as in the pilot-wave theory, or of 'particles' in the case of the other theories.[3]

The proponents of the POA have given several argument against the tenability or desirability of a non-three dimensional ontology such as the wavefunction, but I am not going to reproduce them here.[4] Rather, let me focus on the reason why I think this approach implies that the measurement problem is not the real problem for the quantum realist. This problem is created by the existence of macroscopic superpositions, which arise from a linearly evolving entity, namely the wavefunction, which is taken to represent all physical objects. However, there is nothing intrinsically strange in superpositions, either microscopic or macroscopic: they are natural for waves, and the wave ontology has been successfully used in physics before, as in the case of light, for instance. Superpositions are a problem only if we try to describe all physical objects as wave-like, because in many occasions they show a localized, particle-like behavior. So, one obvious solution would be to deny that the theory entails that all matter is described by a wave-like ontology. However, historically, this is not what has been done. Instead, the theories presented in the previous section all maintain the wavefunction as part of the ontology.[5] This is so even in the case of the pilot-wave theory, in which it is granted that the wavefunction does not provide the complete description of physical systems. However, the problems are not over. In fact, the view that the pilot-wave theory is a theory in which the wave is material, just

[2]See Dürr et al. [7], Allori et al. [8–10], Allori [11–15].

[3]That is, regardless of whether some coordinates $r_1, r_2, \ldots, r_N$, where N is the number of particles thought to exist in the universe (roughly of the order of 10^{90}) can be interpreted as the position of real particles or not, the wavefunction is a function of a high-dimensional variable $q = (r_1, r_2, \ldots, r_N)$.

[4]Se Allori [11–16] and references therein for an exposition of these arguments.

[5]The reason for this is unclear, and I cannot fully explore this issue in this paper. Presumably however one could say that historically the theory developed and flourished after the proposal of the Schrödinger's equation, and so from that moment on it seemed unthinkable to not consider it as part of the theory.

like the particles, is a theory of N particles in three-dimensional space, and a wave in $3N$ dimensional (configuration) space. And this is still problematical: what is the cat currently on my lap made of, particles of waves? One could say that matter has a dual ontology: on the one hand, physical objects like cats are made of particles; on the other hand here is also another physical entity represented by the wavefunction, which is understood to be similar to electromagnetic fields, as another 'kind' of constituent of the world. Despite this, unlike electromagnetic fields the wavefunction is not defined in three-dimensional space, and this created another problem: how is this wave in configuration space supposed to interact with particles in three-dimensional space? Moreover, consider now the spontaneous localization theory. Before it localizes, for a brief but finite instant, the wavefunction is spread out in configuration space. That is, the cat on my lap this wavefunction describes is, for a brief but finite instant, spread out configuration space, where the emphasis is not so much on 'spread out' but on 'configuration space.' It seems only slightly counterintuitive to think that cats have infinite three-dimensional tails: we think that cats have a matter distribution that do not extend in space to infinity, but we are wrong. This is not a big problem, however, as these tails are undetectable. [6] Instead, it seems much more troublesome to think that the cat, before the wavefunction collapse, was not in three-dimensional space but rather she was in the high dimensional configuration space. That is, the cat, which we would normally think of being described by a (soft) lump of matter now localized here on my lap, before localizing here was instead in *another* space with a large number of dimensions. Now consider the many-worlds theory. In this theory physical objects are 'made of' wavefunctions, which suitably 'splits' into 'different worlds' thus avoiding the macroscopic superpositions. However, in each world the object is described by a component of the superposition, which is something in configuration space, while the physical objects we experience are not in this space. Whatever we devise to account for why we perceive what we perceive will however not remove the fact that there is another space involved before this perception happens.

If one wants to stick to the idea that the wavefunction represents physical objects, after having solved the measurement problem, one has also to answer all these question.[7] In contrast, the POA takes a completely different route. Instead of trying to explain the connection between configuration space and three-dimensional space in the various theories, I take it, the proponents of the POA deny that the wavefunction is material to start with. What represents material objects, the so-called *primitive ontology* (PO), is instead something else. It's not important exactly what it is (a field, a particle, a string, a spatiotemporal event) aside from the fact that it is in three-dimensional space (or four-dimensional spacetime). In this way, there is no configuration space problem, as everything is in the same space. Moreover, there is no measurement problem, because either the PO does not superimpose, or the macroscopic superpositions are short lived. The former situation happens when one has a PO of particles, as in the pilot-wave theory, or, as we will see in the next section,

[6] However, see Sect. 4 for more on this.

[7] See Albert [17], Ney [18, 19] and references therein for proposals to make sense of this.

a PO of flashes, as in some version of the spontaneous localization theory, while the latter happens in some version of the spontaneous localization theory.

Before moving on, let me pause for a second. To reject the wavefunction as material may seem like an outrageous proposal: abandon the very object that is taken to define quantum mechanics seems to be to deny realism about quantum mechanics itself. That may be so: perhaps just to ditch quantum mechanics as we know it is the right thing to do. However, I think this is the wrong way of thinking about this move: the POA is thinking of quantum mechanics as an effective theory, which can be understood in terms of a more fundamental theory, just like thermodynamics can be understood in terms of statistical mechanics. As in thermodynamics one understand temperature and heat in terms of molecular motion, in the POA one understands the quantum behavior in terms of motion of objects in three-dimensional world. It just so happens that in doing that we use the wavefunction, but the rest is very much in line with the practice and the spirit of physics before the advent of quantum mechanics.

Open questions in this approach are connected to the nature of the wavefunction. The wavefunction does not represent physical objects but rather is used to 'generate' the trajectories of the PO. Because of this, in the POA the wavefunction is usually taken to have a nomological role, even if it is debatable what is the best way of capturing this idea.[8]

4 Different Ontologies for Different Theories

The POA is a natural framework for the pilot-wave theory, as this theory has an obvious interpretation as a theory with a PO of particles. However, the POA generalizes to the other theories as well by specifying a PO for each of them. So, in the POA the spontaneous localization theory and the many worlds theory are theories in which matter is made of whatever the PO is, and in which the evolution of the PO is governed

[8]See Dürr et al. [7], Goldstein and Teufel [20], Goldstein and Zanghí [21], Allori [16] for a defense of the nomological approach. Since the wavefunction is part of the axioms of quantum theory, it can be naturally regarded as a Humean law (see [22–25]). There are other ways in which someone could think of the wavefunction, broadly speaking, as nomological. One can think of the wavefunction as a property which expresses some non-material aspect of the particles [26]. Similarly, one can endorse a dispositional account where laws are understood in terms of dispositions, which in turn are described by the wave-function [27, 28]. Arguably, since dispositions can be time dependent, the objection to the nomological view that laws of nature are time independent while the wavefunction evolves in time seems less compelling here. Having said that, I think these proposals are not very promising in that they rely on the notion of properties which are notoriously a rough nut to crack. As Esfeld [24] has pointed out, there are several severe problems in trying to spell out what fundamental properties are, both in the classical and the quantum domain. On a different tone, let me notice that the objection that in theories with the wavefunction only there are two spaces involved and the relationship between these spaces is a mystery closely resembles one of the objection against Cartesian dualism: if mental states are not in three-dimensional space, how are they interacting with physical states? With this analogy in mind, one can argue that the answer of the proponents of the POA will be similar to the one of the reductive physicalist, presumably a functionalist: the wavefunction is whatever function it plays to generate the empirical data (see [15] for more on this).

respectively by a spontaneously localizing wavefunction, and a Schrödinger evolving one. In contrast with the pilot-wave theory, however, it is not obvious what the PO of these theories is supposed to be. Indeed, one can chose as one wishes: particles, fields, events. In fact, the PO, like particles in classical mechanics, is postulated beforehand as the best compromise between simplicity and explanatory power. The theory is then constructed around it, adding various elements to it including the wavefunction, to successfully reproduce the empirical data, just like forces and potentials are added to Newtonian mechanics. This is clearly not what historically happened in the quantum domain, in which we got 'stuck' with the wavefunction. So, the PO for the various quantum theories have been developed 'backwards,' by keeping the wavefunction and its evolution, and then trying to figure out what the PO should be. This of course means that the choice of the PO is underdetermined. However, this is not surprising, as it is merely a restatement of the well-known fact that there are many theories that fit the data. As a consequence, the spontaneous localization theory, as well as the many-worlds theory, are better seen as families of theories, rather than one single definite theory: there is one theory for each choice of PO.

So, *what is the 'best' choice of PO for spontaneous collapse theories* and many-worlds theories? The answer is not straightforward for a variety of reasons. In this paper I will not discuss the many-world theory but only spontaneous collapse theories, even if presumably some of the considerations will also apply in that framework.[9] If one hadn't followed the literature, one might think that the natural PO for any quantum theory would be the one of particles. In fact, a particle ontology seems to be the simplest, as it only takes a point to define a particle. Instead, the two formulations of the spontaneous localization theory with a PO which were historically considered are not particle theories. The first of these theories was proposed by John Stuart Bell [2]. In this theory the PO is directly into spacetime: matter is made of those events in spacetime in which the wavefunction happens to spatiotemporally localize. Thus, matter is spatially discontinuous, like the case of particles, but also temporally so that one can say that matter is made of flashes. The flashes are divided into families, with each family corresponding, intuitively, to a single 'particle'. The wavefunction provides the conditional probability measure over the flashes, namely the probability that the next flash in a given family will be in a given spatiotemporal point. This theory is therefore known in the literature as GRWf, with an obvious notation. Bell chose this unfamiliar ontology not because it is the simples but because he noticed might help finding a relativistic invariant theory (see Sect. 6 for more on this).

The other (prominent) spontaneous localization PO theory has been proposed by Ghirardi and collaborators [29], who took matter to be described by a continuous three-dimensional matter field (which is defined in terms of the wavefunction). It is the most natural PO in the following sense: assuming that the problem with the original theory was not the wave part but the configuration space part, this theory solves the problem by putting the wave (and thus matter) in three-dimensional space.

However, these theories are very peculiar, as discussed by Peter Lewis and Michael Esfeld (this volume) among others. Lewis reminds us that the flash ontology is

[9]For more, see Allori [15].

extremely counterintuitive, given that according to this theory matter is mostly empty: for a small macroscopic object made of roughly 10^{19} 'particles,' if each particle undergoes a collapse every 10^{16} s, there is a flash roughly once every 10^{-3} s, and nothing in between. Nevertheless, the gravitational and electromagnetic forces the object is subject to will be continuous in time. Moreover the matter density, being defined in terms of the wavefunction, inherits its tails. Because of this, macroscopic solid objects will have low-matter density tails which other macroscopic solid objects could cross without being subject to any actual interaction, contrary to expectations. In addition, as Esfeld points out, there is a tension between the matter density ontology and the quantum formalism, which is in terms of a finite number of particles. Also, the account of nonlocality may seem more mysterious in this theory than in others, as it implies that the matter density is instantaneously displaced across arbitrary distances.[10]

Regardless of whether or not it is possible to find satisfactory solutions to these challenges,[11] it is interesting to see whether other ontologies could do better. In this regard, the obvious choice would be to try with an ontology of particles. First, as already noticed, there is a sense in which particles are the simplest ontology. Moreover, a particle ontology would increase the explanatory power of the theory: as underlined by Allori [12], Richard Healey [33], and Lewis [34], explanation in the quantum domain is parasitic on classical explanation. Since classical mechanics is about particles, and arguably we can explain the macroscopic feature of matter in terms of such an ontology, there is no in principle reason why this cannot be done also in the quantum domain, if the theory is a particle theory. Finally, as argued in Allori [16] and as discussed in Sect. 6, a particle ontology would also help in

[10]See Egg and Esfeld [30], Esfeld and Deckert [31].

[11]Here's a sketch of some possible responses. To the objection that flashes are counterintuitive one could reply that a satisfactory explanation can lead us far from common sense, so sometimes getting away from commonsensical explanation may be the right thing to do. For instance, while common sense suggests that matter is continuous, atomic theory has shown us that it is not the case: atomic theory, with its in-breath and in-depth explanatory power, is a better explanation of the behavior of matter than our common sense. So, we are justified in accepting atomic theory even if it pictures a world which is distant from what we initially thought. Moreover, the reply goes, in the case of GRWf abandoning common sense for an unfamiliar ontology makes the theory more compatible with relativity, as suggested by Bell (however, see Sect. 6). Also, one could question the fact that the ontology and the explanation is really counterintuitive in a negative sense. It is true that the action of fields is continuous even if there are no flashes. However, this is a problem *only* if the field are intended as material, which in the POA is not necessarily the case: they could be taken to be not generated by the particles, but rather alike to nomological entities, similarly to what happens for the wavefunction in quantum theory (see Allori [14] for a discussion of this). Also, see Esfeld [32] for a defense of GRWf. To the objection that the matter density has tails which could be crossed by other objects without any visible interaction, one could arguably maintain that this is counterintuitive merely when we look at things from a classical perspective: only because classically matter which encounters other matter interacts with it, it does not mean that it has to be the case in the quantum domain. Moreover, the fact that the quantum formalism is in terms of particles does not seem to force us to interpret it as a theory of particles, as one could presumably endorse the formalism of a continuous localization theories (CSL) which does not require particles [32]. Finally nonlocality is a puzzle for all theories, not merely GRWm, so that it is unclear how serious the last objection actually is.

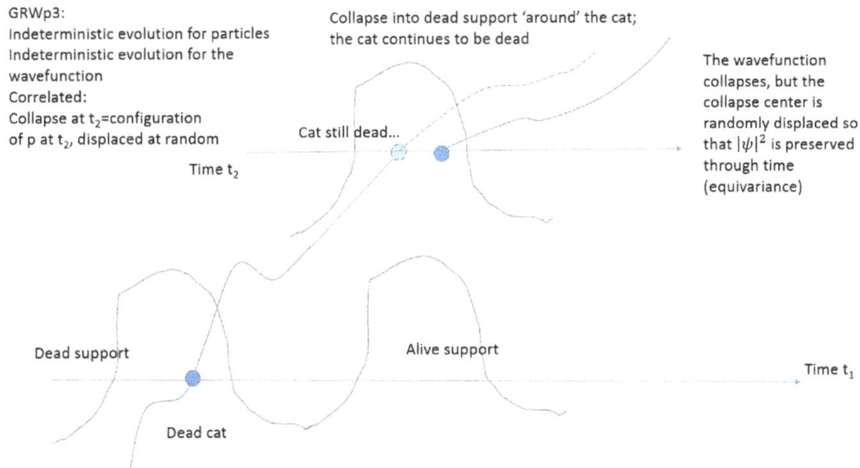

Fig. 1 GRWp3

solving the pessimistic meta-induction argument against scientific realism, since it would provide with an ontological continuity between the classical and the quantum domain.

5 A Particle Ontology for Spontaneous Localization Theories

What would a spontaneous localization theory of particles look like? The possibility of a particle ontology guided by a GRW-evolving wavefunction has been explored in the literature, however never in an exhausting and comprehensive way.[12] Among all the possibilities, I argue that only one survives scrutiny. Let us discuss them in turns.

Daniel Bedingham [35] was the first to propose a particle ontology for a GRW-evolving wave function. This theory has been later dubbed GRWp3 in Allori et al. [10] and Allori [15], given that it is the third GRW particle theory they analyze. In this theory both the particles and the wavefunction evolve stochastically. In particular, the wavefunction evolves in a GRW-fashion while the particles are guided by the same guidance equation of the pilot-wave theory. However, the wavefunction localizes into the *actual* position of the particle at that time (the localization time) but 'displaced' at random (Fig. 1).

Two things need to be noticed. First, the localization of the collapse needs to be anchored to the evolution of the particles appropriately, and that is why it is in the particles configuration at the time of collapse. If one did not require this, the evolution

[12]See Allori et al. [8, 10], Bedingham [35], Allori [15].

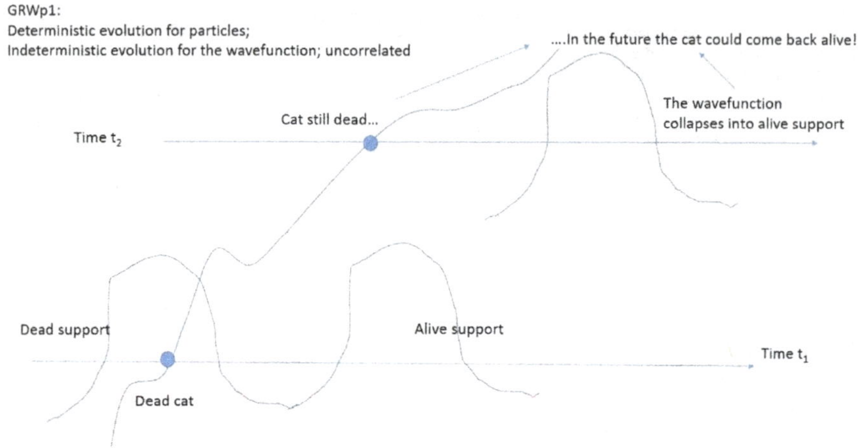

Fig. 2 GRWp1

of the wavefunction and the one of the particles would be uncorrelated. This would lead to a non-empirically adequate theory. In fact, in the POA the status and the behavior of a physical object is determined by the status and the behavior of the PO. The wavefunction is instead part of the dynamical law governing such behavior, so it should determine it appropriately. If the evolution of the PO and the wavefunction are not correlated, as in a theory dubbed GRWp1 in Allori et al. [10], this does not happen (Fig. 2). In fact this could imply, somewhat dramatically, that a cat which has died, and was supposed to stay dead, could come back to life. To see this, assume the configuration of the cat's particles is under the 'dead' support of the wavefunction before the collapse. If so, the cat is, unfortunately, dead, and there's nothing that can be done about it. However, if we do not correlate the two evolutions, then it is possible that the wavefunction collapses into its 'alive' part. For that moment on, therefore it would be this part of the wavefunction which would guide the particles' motion, and this means that the cat could actually come back to life. While this may give some comfort to those who loved the cat, it is not what empirically happens. Instead, if one allows the center the localization of the wavefunction to be the actual position of the corresponding particle at the time of the collapse, then there is the right correlation between the two evolutions. In this way, if the cat is dead before the collapse (that is, the positions of its particles are under the dead support of the wavefunction), then the wavefunction collapses 'around' it, and a dead cat remains dead.

Secondly, it is interesting to notice that the stochastic evolution of the wavefunction does not combine with a deterministic evolution for the particles: in order to have an empirically adequate spontaneous localization particle theory, the particles have to evolve indeterministically as well in order to ensure the equivariance of the theory. In fact a random delocalization of the localization position is required in order

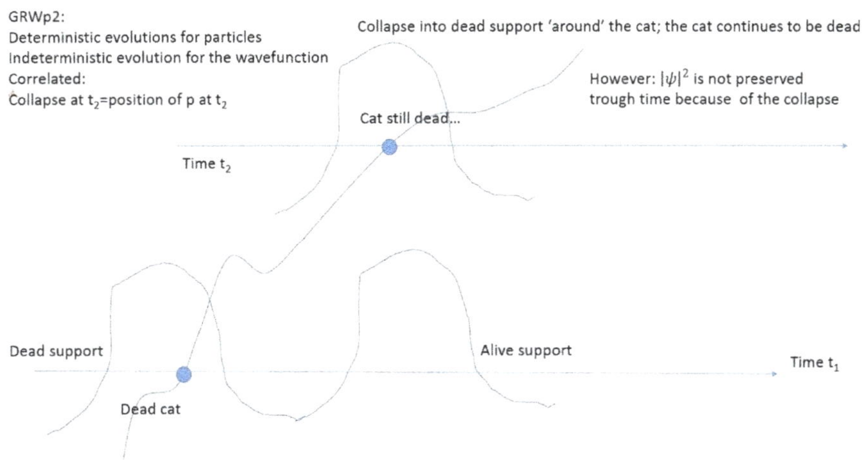

Fig. 3 GRWp2

to 'compensate' the wavefunction collapse so that the empirical distribution remains the one predicted by quantum theory, namely $|\psi|^2$.

This theory, with a deterministic evolution of the particles, was originally[13] proposed by Allori et al. [8], and they called it, rather obviously, GRWp. However, it is empirically inadequate and it was later dubbed GRWp2 in Allori et al. [10] (Fig. 3). The empirical inadequacy of this theory can be seen by noticing that without the displacement, the situation would be almost the same as in the case of the pilot-wave theory, where the wavefunction does not collapse but the particle position is effectively guided only by one of the terms of the superposition, namely the one under which the particles are (Fig. 4).

Thus, there is a sense in which the hallmark of being a GRW-type particle theory is to have *both particles and the wavefunction jump.*[14]

One may think that another possible implementation of this double indeterminism could be accomplished doing the opposite: instead of the wavefunction 'following' the particles and localize where the particles are, one can have the particles 'follow' the wavefunction and jump where the wavefunction localizes. That is, one takes the particles to move as in the pilot-wave theory between localizations, and at the time of collapse all particles jump in the point the wavefunction has collapsed (Fig. 5).

In this theory, dubbed GRWp6 in Allori et al. [10] and Allori [15], the wavefunction takes precedence on the particles. This may already suggest it is going to be a problem, as in the POA the PO is, indeed, primitive. However, before explaining the

---

[13]However, see Bohm and Hiley [36], p. 346.

[14]Other particle theories, aside from the pilot-wave theory, are stochastic mechanics [37], and Bell-type quantum field theories [38–40]. In both theories the wavefunction evolves deterministically, in contrast with GRW-type particle theories. In the former the particles evolve according to a stochastic Markov process, while in the latter the evolution is also stochastic but the particles can also be created and destroyed.

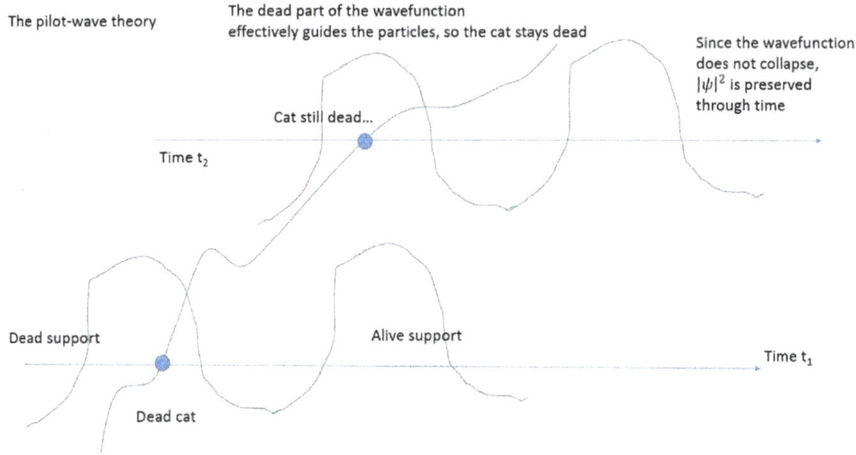

Fig. 4 The pilot-wave theory

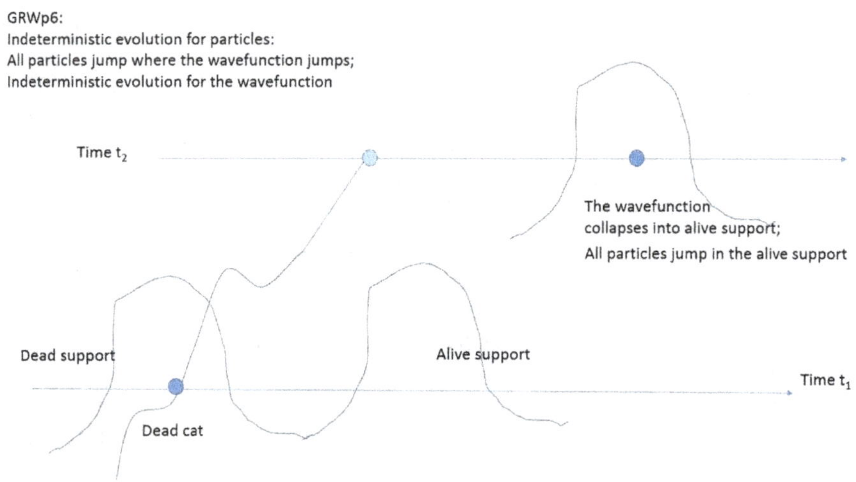

Fig. 5 GRWp6

major drawback of this move, let me first notice that, in order for this theory to get off the ground, one would need all particle to jump together. This is to guarantee that, again, a dead cat would stay dead. In fact if only one particle were to jump, then something similar to what described above is likely to happen again. Consider a situation where the cat is initially dead, and assume that the wavefunction collapses into the 'alive' portion of the wavefunction just after few collapses connected to a few particles. As a consequence, even if the wavefunction indicates 'alive,' most particles would still be 'under' the dead sector of the wavefunction. The 'alive' portion of the wavefunction will soon be dominant over the other part in guiding the evolution of

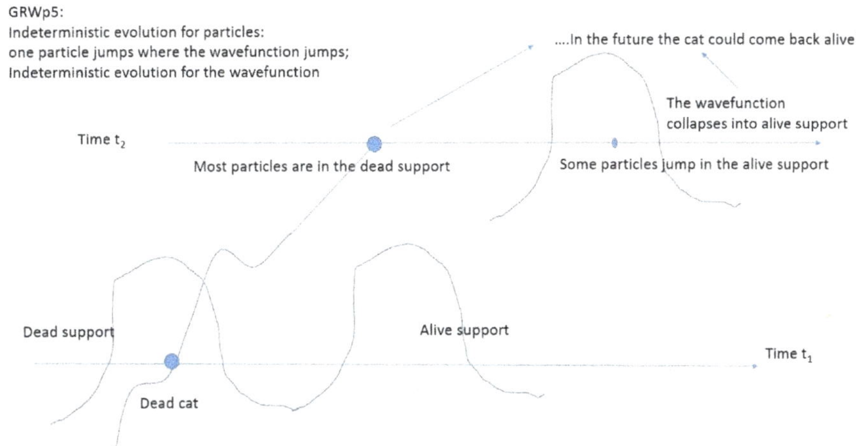

Fig. 6 GRWp5

the particles, therefore opening up for the possibility that the cat will end up alive again. This is what happens in a theory dubbed GRWp5 in Allori et al. [10] (Fig. 6).

In any case, there is another problem: GRWp6 closely resemble a (problematic) many-world theory. In fact this theory can be taken to represent a situation in which there is a world for each term of the superposition of the wavefunction. The problem is that, since all particles jump at the same time when the wavefunction gets localized from one term of the superposition to another, one effectively and instantaneously moves from one world to another. So, GRWp6 is a theory in which many worlds exist, even if not at the same time but one after the other. This makes the theory *empirically incoherent*, namely its truth undermines our empirical justification for believing it to be true.[15] In fact we could instantaneously move from one world in which there are dinosaurs to one in which there aren't any. This implies that our records of the past, including evidence to support the theory, are most likely false: we remember dinosaurs at time t, when we were in world 1, but at time $t + dt$ when we are in world 2 they have disappeared. Similarly, assume we gather some empirical evidence that justifies us in believing in GRWp6 at one time, when we are in world 1; but when we jump into world 2 a second later our memories of that very evidence is most likely false because the two terms of the superpositions describing world 1 and 2 are separated in configuration space and thus describe microscopically different state of affairs. Because of this, presumably, GRWp6 is not a viable theory. This does not happen in GRWp3 because in this theory the particles are 'in charge' and the wavefunction localizes where they actually are located.

Another GRW-like particle theory has been explored by Shan Gao [43]. In his theory particles evolve in random discontinuous motion (RDM) guided by a GRW-evolving wavefunction (Fig. 7). Gao's idea is that the particles spend only an instant at each location, and jump between the different terms of the superposition of the

[15]For more on empirical coherence, see Barrett [41], Huggett and Wüthrich [42].

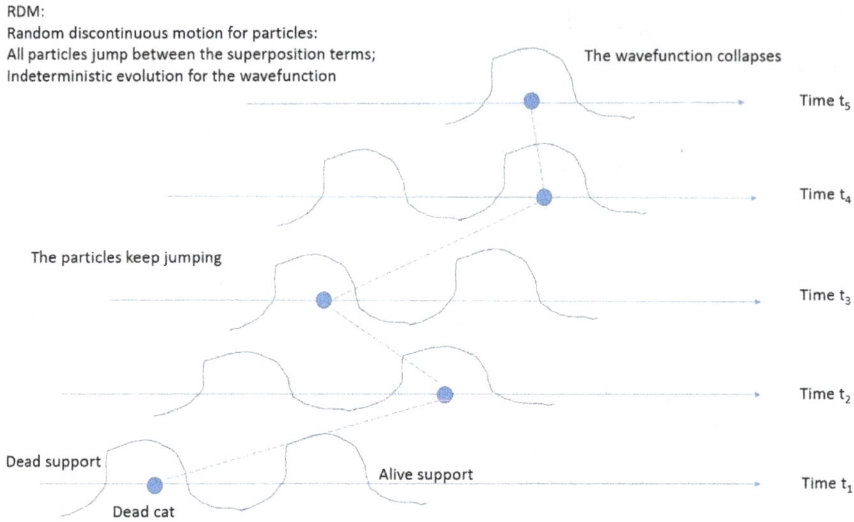

Fig. 7 RDM

wavefunction. This theory is different from GRWp3: in GRWp3 the evolution of the particles is deterministic before and after the localization of the wavefunction around the particles' configuration, while here the particles keep jumping between the different terms of the superpositions before the wavefunction localization. It is perhaps more similar to GRWp6, in that in both theories the wavefunction takes precedence over the PO. However, in GRWp6 the particles jump where the wavefunction localizes but they evolve deterministically before and after, while in RDM the particles keep jumping. Anyway, the collapse of the wavefunction is needed to guarantee that a macroscopic object, say a cat, would not be in a macroscopic superposition of, say, being alive and being dead. After the localization of the wavefunction the particles are confined to move only within the one term of the superposition remaining. Notice that, in order to avoid problems similar to the ones discussed above for GRWp3, one would need all particles to jump together. In any case, the theory seems to suffer from a severe form of empirical incoherence: in GRWp6 the particles were jumping between different words in between collapses; here instead the particles keep jumping between different words at every single instant, before 'setting' for one after the wavefunction collapse.

Interestingly, one may instead read RDM as a theory of flashes, namely of spatiotemporal events, given that particles are usually taken to have a continuous trajectories, unlike what happens in RDM. In any case, it seems interesting to compare RDM with GRWf (Fig. 8). As we said, in RDM the particles keep jumping at every instant, not only at localization times. In contrast, in GRWf each flash corresponds to one of the collapses of the wavefunction, and its space-time location is just the space-time location of that collapse. So, as anticipated, in GRWf matter is mostly empty, because matter exists only at the points of collapse and at the instants of collapse.

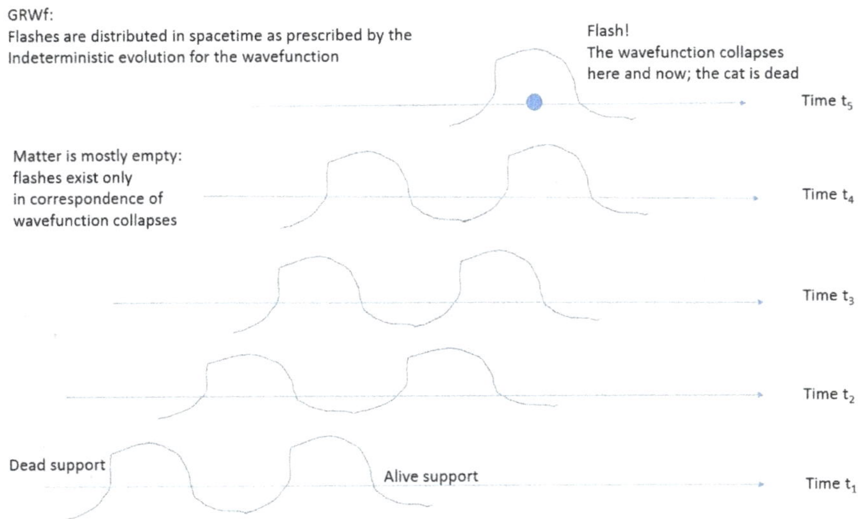

Fig. 8 GRWf

However, it is this feature which allows GRWf to avoid empirical incoherence: in between collapses we are not oscillating between two different words because in between collapses nothing exists.

6 Comparison Among GRW-like Theories: Relativistic GRWp?

In the last section I have discussed how one can construct an empirically adequate theory of particles with a GRW-evolving wavefunction: its name in the literature is GRWp3. However, since this is the only particle theory which survives among the alternatives, I think it is more appropriate to just call it GRWp. As we have seen, this combination requires a stochastic evolution of the particles as well as one for the wavefunction. In this section, I will compare the various GRW-theories with different ontologies: GRWm, GRWp, GRWf. After having done that, I will sketch a proposal for a relativistic particle GRW theory.

When dealing with incompatible alternatives which are empirically equivalent, namely that cannot be set apart by empirical evidence, one usually invokes super-empirical virtues to select one over the other. For instance one could consider parsimony, and argue that the theory which is most parsimonious is the most likely to be true. Assuming that parsimony is indeed a guide to truth, there is however the question of how to define parsimony univocally and objectively. Be that as it may, the matter density seems to score poorly on this criterion when compared with the alternative ontologies. In fact one could argue, as I hinted previously in the text, that the most parsimonious ontology is the one of particles. After all, they need merely a point to be specified, in contrast with the matter density, which is a continuous function. However, one could instead maintain that the flash ontology is the most parsimonious: the flashes are like particles, without trajectories [32]. Nevertheless, as also pointed out in the last section, the flash ontology is more counterintuitive than the particle ontology. As a consequence, the explanatory power of the theory could suffer. Even if explanatory power, like parsimony, is not easy to define univocally, one could maintain that the needed to account for our mistaken intuitions (matter being discontinuous not only in space, like in the case of particles, but also in time in the case of flashes) is a burden for the flashy theory. So, parsimony and explanatory power seem to pull in different directions, and arguably cannot help breaking the tie between GRWf and GRWp, assuming we agree that GRWm scores poorly on these criteria.

Presumably, there are two other considerations that could help in theory selection. One has to do with scientific realism, and the other with relativity. Let us start with the former. As it was clear since the beginning, we are assuming scientific realism, namely we are assuming that these theories can tell us about the nature of reality. However, there is one serious objection to scientific realism: the pessimistic meta-induction argument. This argument goes against the no-miracle argument, which is the main argument for realism: the empirical success of a theory can, and should, be taken as evidence of its truth. The idea behind the pessimistic meta-induction argument is that the conclusion of the no-miracle argument does not follow: since past successful theories turned out to be false, it is unwarranted to believe that our current theories are true simply because they are successful [44]. Since past theories were empirically successful but turned out to be false, it follows that our current theories, even if successful, are more likely to be false than true. One way to respond to the pessimistic meta-induction challenge is to argue that one should be realist about a restricted set of entities, not about the whole theory. Then, if one can show that the entities that are retained in moving from one theory to the next are the ones that are responsible for the empirical success of the theory, the pessimistic meta-induction argument is blocked. In this context, thus, since classically the ontology was the one of particles, a quantum theory with a particle ontology could solve the pessimistic meta-induction argument. Because of this, therefore, one should prefer GRWp over the alternatives GRWm and GRWf: it is the theory that makes scientific realism more plausible in the GRW framework. I have argued [16] that flashes and matter density, in a suitable way, could defeat the pessimistic meta-induction as well, in contrast with GRW0, namely the GRW theory read as a theory about the wavefunction. However, one could maintain that the explanation for how GRWm and GRWf solve this problem diminish their simplicity and explanatory power when compared to GRWp: in GRWp it is obvious that the ontology is preserved from the classical to the quantum domain; this is not so in the case of GRWf and GRWm. So whatever story one has to tell to explain their solution of the pessimistic meta-induction will make the theory not as simple and as explanatory as GRWp.

Finally, another important criterion that could help break the tie between empirically equivalent theories is connected with relativistic invariance. All the GRW theories discussed so far are nonrelativistic, so a natural thought is that if one could extend one of them to the relativistic domain but not the others then that theory should be preferred. For a brief period of time this was indeed the case, as GRWf was the only one of them which had a relativistic extension. This theory, rGRWf, was proposed by Tumulka [45] and provides the probability distribution of the flashes just like GRWf but now it does that with a Dirac evolving wavefunction. Fundamental is the fact that in the construction of the theory no mention of a preferred slicing of spacetime is mentioned, making the theory manifestly Lorentz covariant. This is contrast with relativistic invariant extensions of the pilot-wave theory, which requires a preferred foliation.[16] Arguably, since in the traditional reading of relativity such a preferred foliation does not exist, people have been looking at rGRWf with great interest, as it avoids it entirely. However more recently a relativistic version of GRWm has been proposed by Bedingham et al. [48]. In this theory a Dirac evolving wavefunction defines the matter density field in a Lorentz covariant way, as it only depends on the metric structure, namely the past light-cone in every point. If this is the right way to think about relativistic invariance,[17] then relativistic invariance cannot be used to break the tie between GRWm and GRWf, as both rGRWm and rGRWm exist. However, one may wonder about GRWp: is it possible to construct a relativistic GRWp without the need of a foliation? One may think that the existence of rGRWm, in which there are trajectories for the matter density defined in terms of the wavefunction on the past light-cone, leaves open to the possibility of constructing a relativistic GRWp. One may instead think that this is not possible because just like in the pilot-wave theory in order to define the trajectories, even if discontinuous, one would need a preferred temporal frame. In this regard, it is interesting to consider the attempt by Goldstein and Tumulka [51] to build a pilot-wave theory without a foliation.[18] This theory, let's dub it GT from the names of the authors, specifies covariant particles trajectories with an equation which uses as surfaces of simultaneity the future light cones. In this way, no foliation is needed. The theory is strange, given that it has a microscopic arrow of time pointing towards the past, and also it is not empirically adequate, as it does not have any equivariant measure. However, one could observe the following. In Sect. 5, we dismissed GRWp2, the theory in which the wavefunction jumps where the particles are, because it was not equivariant, while GRWp (in the text previously dubbed GRWp3), in which the particle position was randomly displaced, instead 'regained' equivariance. Perhaps, one could explore the possibility that the lack of equivariance of GT could be 'cured' as we did for GRWp2, and thus define a relativistic particle GRW theory as follows:

---

[16]See Dürr et al. [46, 47].

[17]For a criticism, see Barret [49] and Esfeld and Gisin [50], who argue that these theories, even if they are Lorentz invariant in terms of the overall histories of their PO, are unable to describe single events in a relativistic invariant way.

[18]Another interesting attempt of a relativistic pilot-wave theory without a foliation has been proposed by Sutherland [52, 53], which however involves retrocausation.

1. There are particles evolving according to a suitable pilot-wave-like guidance equation defined with future (or perhaps past) light-cones as simultaneity slices (as in GT),
2. There is a Dirac evolving wavefunction (as in rGRWf and GT),
3. The wavefunction collapses around where the particles are at the time of collapse displaced at random (as in GRWp).

Steps 1 and 2 would guarantee Lorentz covariance without a foliation, Step 3 would instead make the theory equivariant. If this proposal could be made rigorous, then we could have a relativistic GRW theory of particles.

If so, then one could break the tie and select the preferred ontology for GRW-like theories as the one of particles: they are sufficiently simple, they imply less counterintuitive consequences than the alternatives, they help defeating the pessimistic meta-induction argument, and one (presumably) can construct a relativistic GRW theory without a foliation.[19]

7 Conclusion

In the first part of the paper (Sects. 2, 3 and 4) I have argued that the lesson of the POA is that the real problem with quantum theory is not the problem of superpositions, but rather the problem of considering the wavefunction in configuration space as representing physical objects. If so, one should always 'add' something to the wavefunction, regardless of whether one is considering the pilot-wave theory of the spontaneous localization theory. Then in the last part of the paper (Sects. 5 and 6) I have discussed GRW theories of particles, and I have compared them with the other spontaneous localization theories with different ontologies. I finally have argued that, if one were to successfully construct a relativistic spontaneous localization of particles along the lines I have sketched, then this theory would score as the best among the alternatives.

However, if the argument I have advanced in the first part of the paper is sound, then one could wonder: what is the point of considering particle GRW (or many -worlds!) theories at all, in this framework? In fact, if one grants that solving the measurement problem is not sufficient to dissolve the tension between realism and quantum theory, and that one also needs to solve the configuration space problem by postulating some PO in three-dimensional space, then one could also argue that solutions of the measurement problem which take route 2 and 3 (namely many-worlds and spontaneous localization as originally intended) have few chances of being taken seriously. One can turn them into PO theories, but they are doomed to fail because the pilot-wave-theory is *already* the simplest alternative. In other words, let us assume

[19]Of course, if the rGRWp proposed above requires a microscopic inverse arrow of time, then this may be taken to diminish the explanatory power of the theory. However, since the theory has yet to be constructed this kind of considerations seem premature.

that one faces the configuration space problem *before* the measurement problem, rather than after, which one solves by postulating a given PO, may that be of particles, or fields, or strings, or spatiotemporal events. Then the measurement problem does not even arise, as particles never superimpose. If so, *one would hardly think of theories like the spontaneous localization theory as serious option*: the wavefunction solely appears in the law which governs the PO evolution, so why have a stochastic evolution for it if one can obtain perfectly experimentally adequate results with a deterministic one?

My proposal is that looking at theories like the spontaneous localization theory is valuable for a different reason. As we have seen in Sect. 6, theories like rGRWf and rGRWm show genuine relativistic invariance, namely relativistic invariance without a foliation. This is not something that the pilot-wave theory possesses. So, as a matter of an historical accident, namely that people were led to (mistakenly) think that the measurement problem was the one to solve to make quantum theory compatible with scientific realism, they started to develop theories that they otherwise would not have even considered. That is, if the real problem were recognized by everyone to be the configuration space problem, then everyone would *also* agree that its simplest solution is given by the pilot-wave theory, with an ontology of particles and a deterministic evolution for the wavefunction. However, without the other solutions of the measurement problem, presumably people would have just focused on trying to make the pilot-wave theory relativistic invariant, and they perhaps would not have developed theories such as rGRWf or rGRWm which, in contrast with the pilot-wave-theory, are genuinely relativistic. So, this historical accident led us astray in one sense but set us straight in another. That is, on the one hand the measurement problem led us astray because it made us consider theories which are, from the perspective of the solution of the configuration space problem, more complicated than needed. On the other hand it set us straight, because these more complicated theories also show some unique features that make them more amenable to a relativistic extension. In other words, it is valuable to look at these theories even if they are not the simplest solution of the configuration space problem because their stochastic laws may be helpful in solving the tension between quantum theory and relativity.

References

1. Schrödinger, Erwin. 1935. Die gegenwärtige Situation in der Quantenmechanik *Die Naturwissenschaften* 23: 807–812, 823–828, 844–849.
2. Bell, John Stuart. 1987. *Speakable and Unspeakable in Quantum Mechanics.* Cambridge: Cambridge University Press.
3. de Broglie, Louis. 1927. La Nouvelle Dynamique des Quanta. In: *Solvay Conference, Electrons et Photons.* Translated in G. Bacciagaluppi and A. Valentini (2009): *Quantum Theory at the Crossroads: Reconsidering the 1927 Solvay Conference*: 341–371. Cambridge University Press.
4. Bohm, David. 1952. A Suggested Interpretation of the Quantum Theory in Terms of 'Hidden' Variables, I and II. *Physical Review,* 85, 166–193.

5. Everett, Hugh. 1957. Relative State Formulation of Quantum Mechanics. *Review of Modern Physics,* 29, 454–462.
6. Ghirardi, Giancarlo, Rimini, Alberto, and Weber, Tullio. 1986. Unified Dynamics for Microscopic and Macroscopic Systems. *Physical Review D*, 34, 470–491.
7. Dürr, Detlef, Goldstein, Sheldon, and Zanghì, Nino. 1997. Bohmian Mechanics and the Meaning of the Wave Function. In: Cohen, R. S., Horne, M. & Stachel, J. (Eds.), Experimental Metaphysics—Quantum Mechanical Studies for Abner Shimony, Vol. 1; *Boston Studies in the Philosophy of Science* 193: 25–38. Boston: Kluwer Academic Publishers.
8. Allori, Valia, Goldstein, Sheldon, Tumulka, Roderich, and Zanghì, Nino. 2008. On the Common Structure of Bohmian Mechanics and the Ghirardi-Rimini-Weber Theory. *The British Journal for the Philosophy of Science,* 59, 353–389.
9. Allori, Valia, Goldstein, Sheldon, Tumulka, Roderich, and Zanghì, Nino. 2011. Many-Worlds and Schrödinger's First Quantum Theory. *The British Journal for the Philosophy of Science,* 62, 1–27.
10. Allori, Valia, Goldstein, Sheldon, Tumulka, Roderich, and Zanghì, Nino. 2014. Predictions and Primitive Ontology in Quantum Foundations: A Study of Examples. *The British Journal for the Philosophy of Science,* 65, 323–352.
11. Allori, Valia. 2013a. Primitive Ontology and the Structure of Fundamental Physical Theories. In: Albert, David Z., and Ney, Alyssa (eds.). *The Wave Function.* New York: Oxford University Press. 58–75.
12. Allori, Valia. 2013b. On the Metaphysics of Quantum Mechanics. In: Lebihan, Soazig (ed.), *Precis de la Philosophie de la Physique*: 116–151. Vuibert.
13. Allori, Valia. 2015a. Quantum Mechanics and Paradigm Shifts. *Topoi,* 32, 313–323.
14. Allori, Valia. 2015b. Primitive Ontology in a Nutshell. *International Journal of Quantum Foundations* 1 (3): 107–122.
15. Allori, Valia. 2019. Scientific Realism without the Wave-Function: An Example of Naturalized Quantum Metaphysics. In: Saatsi, Juha, French, Steven (eds.) *Scientific Realism and the Quantum,* Oxford University Press.
16. Allori, Valia. 2018. Primitive Ontology and Scientific Realism. Or: The Pessimistic Meta-Induction and the Nature of the Wave Function. *Lato Sensu: Revue de la Société de Philosophie des Sciences* 5 (1): 69–76 (2018).
17. Albert, David Z. 2015. *After Physics.* Cambridge: Harvard University Press.
18. Ney, Alyssa. 2017. Finding the World in the Wave Function: Some Strategies for Solving the Macro-object Problem. *Synthese,* 1–23.
19. Ney, Alyssa. Forthcoming. *Finding the World in the Wave Function.* Oxford University Press.
20. Goldstein, Sheldon, and Teufel, Stefan. 2001. Quantum Spacetime without Observers: Ontological Clarity and the Conceptual Foundations of Quantum Gravity. In Callender, Carig, and Huggett, Nick (eds.), *Physics meets Philosophy at the Planck Scale,* 275–289. Cambridge: Cambridge University Press.
21. Goldstein, Sheldon, and Zanghì, Nino. 2013. Reality and the Role of the Wave Function in Quantum Theory. In Albert, David Z., and Ney, Alyssa (eds.). *The Wave Function.* New York: Oxford University Press: 96–109.
22. Bhogal, Harjit, and Perry, Zee. 2017. What the Humean Should Say About Entanglement. *Noûs,* 1, 74–94.
23. Callender, Craig. 2015. One World, One Beable. *Synthese,* 192, 3153–3177.
24. Esfeld, Michael. 2014. Quantum Humeanism, or: Physicalism without Properties. *The Philosophical Quarterly,* 64, 453–470.
25. Miller, Elisabeth. 2014. Quantum Entanglement, Bohmian Mechanics, and Humean Supervenience. *Australasian Journal of Philosophy,* 92, 567–583.
26. Monton, Bradely. 2013. Against 3 N-Dimensional Space. In: Albert, David Z., and Ney, Alyssa, *The Wave-function: Essays in the Metaphysics of Quantum Mechanics*: 154–167. New York, Oxford University Press.
27. Esfeld, Michael, Lazarovici, Dustin, Huber, Mario, and Dürr, Detlef. 2014. The Ontology of Bohmian Mechanics. *The British Journal for the Philosophy of Science,* 65, 773–796.

28. Suarez, Mauricio. 2015. Bohmian Dispositions *Synthese*, 192 (10): 3203–3228.
29. Bennatti, Fabio, Ghirardi, GianCarlo, Grassi, Renata. 1995. Describing the Macroscopic World: Closing the Circle within the Dynamical Reduction Program. *Foundations of Physics,* 25, 5–38.
30. Egg, Matthias and Esfeld, Michael. 2014. "Non-local common cause explanations for EPR". *European Journal for Philosophy of Science* 4, pp. 181–196.
31. Esfeld, Michael and Deckert, Dirk-André. 2017. *A minimalist ontology of the natural world.* New York: Routledge.
32. Esfeld, Michael. 2019. From the Measurement Problem to the Primitive Ontology Programme.
33. Healey, Richard. 2015 How Quantum Theory Helps us Explain. *The British Journal for the Philosophy of Science* 66: 1–43.
34. Lewis, Peter J. 2019. On Closing the Circle.
35. Bedingham, Daniel. 2011. Relativistic state reduction dynamics. Foundations of. Physics **41**: 686–704..
36. Bohm, David, and Hiley, Basil J. 1993. *The Undivided Universe*, London: Routledge.
37. Nelson, Edward.1985. *Quantum Fluctuations*. Princeton: Princeton University Press.
38. Bell, John S. 1986. Beables for Quantum Field Theory. *Physics Reports*, 137:49–54.
39. Dürr, Detlef, Goldstein, Sheldon, Tumulka, Roderick and Zanghì, Nino. 2004. Bohmian Mechanics and Quantum Field Theory. *Physical Review Letters*, 93:090402.
40. Dürr, Detlef, Goldstein, Sheldon, Tumulka, Roderick and Zanghì, Nino. 2005. Bell-type Quantum Field Theories. *Journal of Physics A: Mathematical and General,* 38(4):R1–R43.
41. Barrett, Jeffrey, A. 1999. *The Quantum Mechanics of Minds and Worlds.* New York: Oxford University Press.
42. Huggett, Nick, and Wüthrich, Christian. 2013. Emergent Spacetime and Empirical (in) coherence. *Studies in Histories and Philosophy of Science* B44: 276–285.
43. Gao, Shan. 2017. *The Meaning of the Wave Function: In Search of the Ontology of Quantum Mechanics*, Cambridge University Press.
44. Laudan, Larry. 1981. A Confutation of Convergent Realism. *Philosophy of Science,* 48, 19–49.
45. Tumulka, Roderich. 2006. A Relativistic Version of the Ghirardi-Rimini-Weber Model. *Journal of Statistical Physics,* 125, 821–840.
46. Dürr, Detlef, Münch-Berndl, Katrin, Goldstein, Sheldon, and Zanghì, Nino. 1999. Hypersurface Bohm-Dirac Models. *Journal of Statistical Physics,* 67, 843–907.
47. Dürr, Detlef, Goldstein, Sheldon, Norsen, Travis, Struyve, Ward, and Zanghì, Nino. 2014. Can Bohmian Mechanics be Made Relativistic? *Proceedings of the Royal Society A* 470(2162): 20130699.
48. Bedingham, Daniel, Dürr, Detlef, Ghirardi, GianCarlo, Goldstein, Sheldon, Tumulka, Roderich and Zanghì, Nino. 2014. "Matter density and relativistic models of wave function collapse". *Journal of Statistical Physics* 154: 623–631.
49. Barrett, Jeffrey A. 2014. Entanglement and disentanglement in relativistic quantum mechanics. *Studies in History and Philosophy of Modern Physics* 47, pp. 168–174.
50. Esfeld, Michael, and Gisin, Nicolas. 2014. The GRW Flash Theory: A Relativistic Quantum Ontology of Matter in Space-Time? *Philosophy of Science* 81 (2):248–264.
51. Goldstein, Sheldon, and Tumulka, Roderich. 2003. Opposite Arrows of Time Can Reconcile Relativity and Nonlocality. Class. Quantum Gravity 20: 557–564.
52. Sutherland, Roderick. 2008. Causally Symmetric Bohm Model. *Studies in History and Philosophy of Modern Physics* 39: 782–805.
53. Sutherland, Roderick. 2017. Lagrangian Description for Particle Interpretations of Quantum Mechanics—Entangled Many-Particle Case. *Foundations of Physics* 47: 174–207.

From the Measurement Problem to the Primitive Ontology Programme

Michael Esfeld

Abstract The paper retraces the development from the measurement problem to the primitive ontology programme. It assesses the contribution of the GRW theory to this programme and discusses the pros and cons of the GRWm matter density ontology and the GRWf flash ontology in comparison to the Bohmian particle ontology. It thereby pursues the evaluation of the proposals for a primitive ontology of quantum physics.

1 The Measurement Problem and the Ontology of Quantum Physics

This paper retraces the development from the measurement problem to the primitive ontology programme in quantum physics and assesses the contribution of the GRW theory to this programme. This section recalls this development. Section 2 discusses the GRWm matter density field ontology, Sect. 3 the GRWf flash ontology, taking the latter—with Bell [1, ch. 22] and *pace* Ghirardi et al. [2]—to be the most important contribution of the GRW theory to the ontology of quantum physics. Section 4 considers the status of the wave function in this context, advocating a wave function realism that does not amount to a dualism of primitive ontology and wave function.

The measurement problem is the central motivation for collapse theories. As set out by Maudlin [3] in what has since become the standard formulation, the measurement problem is the fact that the conjunction of the following three propositions is a contradiction:

1A The wave-function of a system is *complete*, i.e. the wave-function specifies (directly or indirectly) all of the physical properties of a system.
1B The wave-function always evolves in accord with a linear dynamical equation (e.g. the Schrödinger equation).

M. Esfeld (✉)
Department of Philosophy, University of Lausanne, 1015 Lausanne, Switzerland
e-mail: Michael-Andreas.Esfeld@unil.ch

V. Allori et al. (eds.), *Do Wave Functions Jump?*, Fundamental Theories of Physics 198,
https://doi.org/10.1007/978-3-030-46777-7_8

1C Measurements of, e.g., the spin of an electron always (or at least usually) have determinate outcomes, i.e., at the end of the measurement the measuring device is either in a state which indicates spin up (and not down) or spin down (and not up). [3, p. 7]

Accordingly, there are three possiblities to solve the measurement problem:

(1) One can reject (1.A). There is more to the physical systems than what is represented by the wave function. This "more" is traditionally known as "hidden variables" because we do not have experimental access to more than what is extracted from the wave function by means of Born's rule in terms of predictions of measurement outcome statistics. The most prominent theory in this vein is the one going back to de Broglie [4] and Bohm [5]. It was supported by Bell from the 1960s to the 1980s[1] and is today known as Bohmian mechanics.[2] According to this theory, quantum systems always have a determinate value of position, which is not tracked by the wave function. It is, however, misleading to call position a "hidden variable", since all that is ever revealed in measurement outcomes are positions and not wave functions, as pointed out by Bell [1] among others.
(2) One can reject (1.B). In this case, one replaces the Schrödinger dynamics with a dynamics that includes the collapse of the wave function. In the textbook presentations of quantum mechanics, going back to von Neumann [6], this is done in an ad hoc manner, with the wave function being supposed to collapse upon measurement. However, neither are measurements a particular type of interaction—over and above gravitation, electromagnetism and the weak and the strong interaction—that requires a specific law, nor are measurement devices natural kinds on a par with electrons, chemical elements, biological species, etc. The theory of Ghirardi et al. [7] (GRW) improves on this situation by turning the Schrödinger equation into a law for wave function collapse independently of observers, measurements and the like.[3]
(3) One can reject (1.C). In this case, one denies that measurements have outcomes. More precisely, all possible outcomes of any measurement are in fact realized, albeit in different branches of the universe, which do not interfere with one another. This solution goes back to Everett [9]. Consequently, every possible future of a person becomes real in the sense that for every possible future of a person, there is a future self that experiences that future. Hence, there is an obvious problem how to account for probabilities in such a theory, and be it subjective probabilities.

However, at the latest since the seminal paper by Allori et al. [10], it has become clear that describing the situation that we face when it comes to understanding quantum mechanics in terms of these three possibilities to solve the measurement

[1] See in particular Bell [1, chs. 4, 7, 17 and 19].

[2] See Dürr et al. [14].

[3] See also Gisin [8] for a forerunner in that vein.

problem is not the whole story. The reason is that these three possible solutions, thus formulated, are concerned only with the dynamics of the wave function—whether the Schrödinger dynamics is complete so that measurements have no outcomes (not 1.C), whether it has to be amended by wave function collapse (not 1.B) or whether there are additional variables that require an additional dynamics (not 1.A). Insofar as they are only concerned with the dynamics, these solutions do not answer the question of ontology, that is, the question of what the wave function refers to—in other words, the question of what the objects in nature are to which the wave function dynamics relates. This is particularly evident in the case of the collapse dynamics: What are the objects that are subject to this dynamics and what does the collapse of the wave function mean for their behaviour?

When considering the possible answers to the question of the ontology of quantum physics, we are confronted with a division into two principled answers. The one possible answer is what is known as wave function monism. In brief, this is the view that the wave function, conceived as physical object, is the physical reality. The most outspoken advocate of this view is Albert [11, 12, chs. 6–7]. The wave function is defined on configuration space by contrast to three-dimensional space or four-dimensional space-time. For N particles, configuration space has 3 N dimensions such that each point of configuration space represents a possible configuration of the N particles in three-dimensional space. However, if the wave function on configuration space *is* the physical reality, there is no configuration of anything in another, physical space. Consequently, it is misleading to call this space "configuration space".

Be that as it may, the idea of Albert's wave function monism is, in brief, that the wave function undergoes in the space on which it is defined an evolution such that objects that are functionally equivalent to objects in three-dimensional space or four-dimensional space-time come into existence during this evolution. In order to achieve this aim, Albert is sympathetic to the idea that the wave function undergoes collapse in this space. This shows that the GRW dynamics can go together with an ontology that admits only the wave function of the universe on a very high-dimensional space (known as GRW0). Moreover, in general, the solution that rejects 1.C is associated with wave function monism: the idea then is that the wave function of the universe undergoes an evolution of a division into many branches of the universe through decoherence; objects that realize a functional definition of ordinary physical objects come into existence during this evolution, even if configuration space monism is rejected.[4]

The other possible answer is the primitive ontology programme. According to this answer, the wave function does not provide the ontology of quantum mechanics. It plays only a dynamical role. The ontology consists in a configuration of matter in three-dimensional space or four-dimensional space-time. In order to represent this configuration, one has to add a variable to the wave function, namely a variable for the (primitive) ontology. This stance is associated with the solutions that reject 1.B or 1.A (although the rejection of 1.B can, as mentioned, also go together with a wave function only ontology).

[4]See Wallace [13] for a prominent contemporary defense of Everettian quantum mechanics.

Since Bohm's theory has been cast in its contemporary formulation as Bohmian mechanics (BM) by Dürr et al. ([14], ch. 2, originally published 1992), it has been set out as a primitive ontology theory. Indeed, Dürr et al. introduce this term in their 1992 paper (p. 29 in the reprint [14]). BM then is based on four postulates: (i) the primitive ontology of point particles in physical space; (ii) a law that describes the evolution of the configuration of point particles of the universe in which the universal wave function figures, known as guiding equation; (iii) the Schrödinger equation as the law that describes the evolution of the wave function; (iv) a typicality or probability measure in terms of the $|\Psi|^2$ density on the level of the universal wave function; from this measure then follows Born's rule for the prediction of measurement outcome statistics for sub-systems described by their own conditional or effective wave-function.

This structure of a physical theory consisting in a primitive ontology, a law for its evolution, a law for the wave function and a procedure to derive probabilities for measurement outcome statistics applies to any quantum theory that admits a configuration of matter in three-dimensional space or four-dimensional space-time and that assigns to the wave function a dynamical role for the evolution of that configuration. Hence, this structure applies independently of what entities the distribution of matter is taken to consist in (particles or something else) and independently of whether the dynamics for the evolution of the wave function is linear or includes collapse. That is why when taking the ontology of quantum physics into account, the available solutions to the measurement problem come down to two ones: either wave function monism or a primitive ontology of a configuration of matter in ordinary space.

2 The Ontology of GRW I: Matter Density Field

Ghirardi et al. [2] answer the question of ontology by postulating a continuous matter density field that stretches all over space. This theory is known as GRWm, with "m" standing for the matter density variable that is added to the wave function in order to describe the distribution of matter in physical space. Allori et al. characterize the ontology of GRWm in the following manner:

> We have a variable $m(x,t)$ for every point $x \in \mathbb{R}^3$ in space and every time t, defined by
>
> $$m(x,t) = \sum_{i=1}^{N} m_i \int_{\mathbf{R}^{3N}} dq_1 \cdots dq_N \delta(q_i - x)\left|\psi(q_1, \ldots, q_N, t)\right|^2.$$
>
> In words, one starts with the $|\psi|^2$-distribution in configuration space $\mathbb{R}^{3N}$, then obtains the marginal distribution of the ith degree of freedom $q_i \in \mathbb{R}^3$ by integrating out all other variables $q_j, j \neq i$, multiplies by the mass associated with q_i, and sums over i.... The field $m(x, t)$ is supposed to be understood as the density of matter in space at time t. [10, p. 359]

Matter is a continuous, primitive stuff on this view, stretching all over space, in contrast to discrete and thus countable particles. The variable "m" designates matter qua primitive stuff, as again Allori and co-authors make clear:

> Moreover, the matter that we postulate in GRWm and whose density is given by the *m* function does not *ipso facto* have any such properties as mass or charge; it can only assume various levels of density. [15, pp. 331–332]

As in BM, mass, charge, etc. are dynamical variables situated on the level of the wave function that come in through their dynamical role for the evolution of matter; they do not designate intrinsic, essential properties of matter.[5] In both BM and GRWm, matter is characterized by position only. In GRWm, matter is primitive stuff that, moreover, admits of different degrees of density at different points or regions of space, with these degrees of density changing in time. Countenancing a variety of degrees of density as a primitive matter of fact is indispensable in GRWm to account for variation: there is matter all over space with its different degrees of density constituting the variation of matter that we perceive. Thus, for instance, a macroscopic object such as rock or a tree is a concentration of the density of matter in a certain region of space at a time.

Hence, the matter density $m(x,t)$ is an additional variable with respect to the wave function. Consequently, an additional law over and above the collapse law for the evolution of the wave function is needed to establish the link between the wave function and its evolution in configuration space on the one hand and the matter density distribution and its evolution in physical space on the other hand. However, in contrast to the particle positions in BM, the matter density variable is fixed by the wave function. Hence, the dynamics of the wave function keeps track of its evolution. Nonetheless, the matter density variable is "hidden" as well in the sense that it is not fully accessible to an observer. Cowan and Tumulka [19] show that there are facts about the matter density distribution that an observer cannot know. In particular, there are facts about wave function collapse and hence concentration of the matter density in certain points or regions of space that observers cannot measure.

Indeed, Cowan and Tumulka [19] establish that in any primitive ontology theory, the primitive ontology cannot be fully accessible to an observer, whatever it may be (particles, a matter density field, or something else) and independently of whether or not the primitive ontology and its evolution is specified and kept track of by the wave function and its evolution. The reason is the no-signalling theorem: if the primitive ontology were fully accessible, superluminal signalling would be possible in Bell-type experiments. This fact confirms the conclusion of the previous section, namely that the possible stances in the ontology of quantum physics fall into two camps only, that is, either the primitive ontology camp or the wave function monism camp.

The GRWm ontology of a matter density field in physical space is a primitive ontology that is modelled on the wave function. It goes as far as possible in reading the primitive ontology of matter in physical space off from the wave function in configuration space. Of course, there are no superpositions in the sense of any indeterminacy in physical space. The variable $m(x,t)$ always has one definite value. But it

[5] See Brown et al. [16], Pylkkänen et al. [17] and Esfeld et al. [18] for BM.

is modelled on the wave function in the sense that it is a continuous field in physical space as the wave function is a field in configuration space. This close connection raises a number of issues that make it doubtful whether the matter density field is the most convincing proposal for an ontology for the GRW dynamics.

On the one hand, the primitive ontology is one of a field in physical space as the wave function is a field in configuration space. On the other hand, the GRW formalism, as any formalism for quantum mechanics, is a formalism in terms of a finite, determinate and thus countable number of discrete particles, providing a dynamics for these particles, in this case in terms of a collapse or spontaneous localization of the wave function of the *i*th particle. This problem can be remediated by switching to a formalism of a continuous spontaneous localization of the wave function as set out in Ghirardi et al. [20] with the particle labels disappearing.[6] In any case, the fact that a quantum formalism works with a determinate number of particles is no conclusive argument in favour of an ontology of particles.

The more serious problems for the GRWm theory stem from the dynamics. In the first place, the collapse or spontaneous localization of the wave function is achieved mathematically by multiplying the wave function with a Gaussian. Consequently, the collapsed wave function is sharply peaked in a small region of configuration space, but it does not vanish outside that region; it has tails that spread to infinity. This is therefore known as the problem of the tails of the wave function. On its basis, one can object that the GRWm theory does not solve the measurement problem: for instance, in the Schrödinger cat experiment, when the wave function collapses to the outcome dead cat, there then is a high-density dead cat and a low-density live cat. It seems that the low-density cat is just as cat-like (in terms of shape, behaviour, etc.) as the high-density cat, so that there are in fact two cat-shapes in the matter density field, one with a high and another one with a low density. However, one can give arguments against drawing this conclusion so that it is in dispute whether the tails problem implies that the GRWm theory is in trouble solving the measurement problem.[7]

More importantly, quantum non-locality has unpalatable consequences for the GRWm theory. This is already evident from a simple example that involves only a position superposition, but no entanglement. Consider the thought experiment of one particle in a box that Einstein raised at the Solvay conference in 1927:[8] the box is split in two halves that are sent in opposite directions, say from Brussels to Paris and Tokyo. When the half-box arriving in Tokyo is opened and found to be empty, the particle is in the half-box in Paris.

The GRWm account of this experiment is this one: the particle is a matter density field that is split in two halves of equal density when the box is divided; these matter densities travel in opposite directions. Upon interaction with a measurement device, one of these matter densities (say the one in Tokyo) vanishes, while the matter density

[6]See Egg and Esfeld ([21], Sect. 3).

[7]Maudlin [22] argues that it fails to do so. See by contrast Wallace [23], Albert [12] and Egg and Esfeld [21], Sect. 3.

[8]See the account of de Broglie [24, pp. 28–29] and Norsen [25].

in the other half-box (the one in Paris) increases; the whole matter then is concentrated in one of the half-boxes (in Paris in this case). This is to say that the matter density in one of the half-boxes is delocated instantaneously across an arbitrary distance in physical space upon collapse of the wave function in configuration space. It does not travel with any velocity.[9] Even if the collapse of the wave function is conceived as a continuous process, the time it takes for the matter density to disappear in one place and to reappear in another place does not depend on the distance between the two places.

This delocation of matter can with good reason be considered as mysterious, as argued in Esfeld and Deckert [27, pp. 80–81]. Such an account is by no means imposed upon us by quantum non-locality (which can in any case be taken to be counter-intuitive). On BM, for instance, the particles always move on continuous trajectories with a finite velocity. The particle trajectories may be correlated with each other independently of their distance in physical space, thus accounting for quantum non-locality; but nothing is ever delocated spontaneously over arbitrary distances in space.

3 The Ontology of GRW II: Flashes

The problems for the GRWm theory stem from modelling the primitive ontology of matter in physical space on the wave function. However, there is no compelling reason to seek to infer that ontology from the wave function. All that is ever accessible to us in experiments is what is described in terms of the collapse of the wave function in the GRW formalism. This fact suggests the option to consider only the collapses of the wave function as referring to the empirical reality. The resulting primitive ontology then is one of events occurring at space-time points, which are known as flashes, and the theory is known as GRWf.

Such an ontology was proposed by Bell [1, ch. 22] immediately after the publication of the GRW dynamics [7]; the term "flashes" was later coined by Tumulka [28]. The point-events (flashes) are ephemeral. There are no continuous sequences of them, since they occur only when the wave function collapses. Accordingly, this proposal goes together with the original GRW dynamics of an instantaneous collapse of the wave function. Hence, there is no underdetermination of the primitive ontology of the GRW dynamics: the flash ontology is tied to the formalism of a spontaneous localization of the wave function that occurs instantaneously, whereas the matter density ontology goes with the formalism of a continuous spontaneous localization of the wave function. Again, the distribution of flashes (the collapses) is not entirely accessible to an observer and in that sense "hidden", although the flashes are determined by the dynamics of the wave function.[10]

---

[9]See Egg and Esfeld [26].

[10]See Cowan and Tumulka [19].

The flash ontology completes the proposals for a primitive ontology of quantum physics. All the central metaphysical conceptions of objects are realized in these proposals: substances as in Bohmian particles, stuff as in the GRWm matter density field and single events has in the GRWf flashes. Nonetheless, there are further proposals for a primitive ontology of quantum mechanics conceivable, notably also proposals for a primitive particle ontology linked with the GRW dynamics.[11] That notwithstanding, the Bohmian particle ontology, the GRW matter density field ontology and the GRW flash ontology are the only worked out proposals for a primitive ontology of quantum physics that are actually defended in the literature. There is a good reason for this situation: if one starts from the fact that any formulation of non-relativistic quantum mechanics works with a definite number of permanent point particles in its formalism, then BM arguably is the clearest and most simple proposal for a particle ontology and dynamics. If one endorses a collapse dynamics and takes the wave function in that dynamics as guide to the primitive ontology, then one gets to the GRWm ontology. If one lays stress on the fact that only the collapses of the wave function represent observable events, one gets to the GRWf ontology.

The GRWf ontology is not hit by the mentioned objections to the GRWm ontology: there is no problem of particle numbers or discrete objects in the formalism versus a wave function field, since the particle number indicates the number of flashes that can possibly occur at a time, and nothing more than the events described as wave function collapse is empirically accessible anyway. There is no problem of the tails of the wave function, since only the wave function as sharply peaked around a point in configuration space refers to matter in physical space, namely point-events (flashes). And there is nothing mysterious in the account of quantum non-locality, since there are only flashes whose occurrences are correlated with one another in the case of entanglement; but nothing is ever delocated across space.

This latter fact can be taken to suggest that GRWf lends itself to a relativistic extension. Indeed, already Bell [1, pp. 206–212] explored this suggestion and Tumulka [28] worked out a proposal for a relativistic GRWf theory. However, today, it is clear that a simliar result can be achieved in GRWm, as shown by Bedingham et al. [30]. In any case, there are well-grounded doubts voiced notably by Barrett [31] and Esfeld and Gisin [32] whether these models are fully relativistic: in brief, what they describe in a Lorentz-invariant manner are entire possible histories of, in GRWf, flash distributions with probabilities assigned to each such possible entire distribution in space-time. But they are not able to describe the occurrence of individual flashes or contractions of the matter density field in a Lorentz-invariant manner. In other words, they are not able to describe in a relativistic manner how determinate measurement outcomes occur in space-time.

The flash ontology certainly is counter-intuitive, because it refuses to recognize any permanent or persisting material objects. There are only ephemeral flashes with a space-time gap between any two of them.[12] However, these gaps may be so tiny that they cannot be perceived. In other words, the flash ontology is a serious candidate for

[11] See Allori et al. [15] and Allori [29].

[12] Cf. the objection of Maudlin [33].

the correct ontology of our empirical world, as argued, for instance, by Arntzenius [34, ch. 3.15]. The initial conditions and the subsequent development of the real world can be such that the flashes occur in such a dense manner that our experience of a macroscopic world of permanent objects is accounted for. Bell [1, p. 205] regards macroscopic objects as galaxies of flashes. By the same token, an ontology of permanent particles as in BM has to make sure that there are enough particles to account for macroscopic objects that appear as continuous, although there is empty space in the sense of a non-vanishing distance between any two particles.

The flash ontology seems to be distinguished as the most parsimonious one. The reason is that all that to which we have experimental access are the events that are represented by (effective) wave function collapse. The flashes can be conceived as the Bohmian particles deprived of their trajectories or the matter density field with the expansions of the field between the collapse-concentrations deleted. In its light, the Bohmian particle trajectories and the continuous GRWm matter density field look like a surplus structure: they are there, but inaccessible in observation or measurement. However, on closer inspection, it turns out to be questionable how parsimonious the flash ontology really is. The reason is that the flash ontology is committed to a substantival space or space-time in which the flashes occur. Indeed, one can tie the flash ontology to super-substantivalism, that is, the view that an absolute space-time is all there is: space-time flashes occasionally. Again, the flashes are primitive matter, in this case bare particulars: apart from their spatio-temporal location, they do not have any properties. As in the case of the Bohmian particles and the matter density field, mass, charge, etc. are variables that are situated on the level of the wave function and that consist in playing a dynamical role for the evolution of matter instead of being intrinsic, essential properties of material objects.

The commitment to a substantival space-time in the GRWf ontology is a consequence of the fact that there is a gap between any two flashes, while there still is exactly one universe described by exactly one universal wave function and its evolution (instead of each flash being a universe of its own, like a Leibnizean monad). What holds this universe together then is the space-time in which the flashes occur. There may even be times with no flashes at all. Space-time thus takes the position of the substance in which change occurs, in this case change in the number of flashes and the distances between them. This dualism of absolute space-time and matter qua flashes (or qua the flashing versus the empty space-time) has the consequence that this ontology is not so parsimonious, simple or minimal after all, as argued by Esfeld and Deckert [27, pp. 83–84]. The same holds if one were to replace space-time with the wave function, then conceiving the wave function as a sort of substantival stuff in which the flashes occur.

By contrast, neither GRWm nor BM are committed to a substantival space-time, although both are conveniently formulated in terms of a configuration of matter that is inserted into an absolute space and evolving in an absolute time. The GRWm matter density field can be conceived as a substance that stands on its own, without requiring an underlying space or space-time. For instance, Rovelli [35] sets out a proposal for an ontology of fields only without a substantival space-time in the context of the general theory of relativity. By the same token, the permanent Bohmian particles can be conceived as being characterized by their relative distances and their change only instead of positions and trajectories in an absolute space. Indeed, a relationalist formulation of BM has been worked out recently by Dürr et al. [36].[13]

Moreover, the Bohmian particles can be construed as being individuated by the distance relations in which they stand. This is an important advantage of the Bohmian particle ontology when it comes to the metaphysics of matter: no commitment to a primitive stuff with different degrees of density as a primitive matter of fact or bare particulars (as in the case of the GRWf flashes) is called for, since there are relations available that individuate the particles. Consequently, BM turns out to be in fact close to the stance that is known as ontic structural realism in the metaphysics of science according to which there are no objects with intrinsic essences, a primitive stuff substratum or bare particulars (see [27], ch. 2.1).

In any case, we obtain again the result that the dynamical differences—wave function collapse or not—are not central. The central issue is the evaluation of the proposals for a primitive ontology according to criteria such as parsimony or simplicity, coherence and explanatory value.[14]

4 The Status of the Wave Function: Dynamics, not Ontology

"Primitive ontology" does not signify that there also is a non-primitive or secondary ontology (which would in this case apply to the wave function). "Primitive" signifies that the configuration of matter simply exists. It cannot be derived from anything else, notably not from the wave function. Quite to the contrary, if one admits a primitive ontology of a configuration of matter, the wave function then enters the theory only through its dynamical role, namely its role for the evolution of the configuration of matter. In that sense, it is nomological.

An ontological dualism of a primitive ontology of matter on the one hand and a wave function on the other hand would fall victim to the objection of Brown and Wallace [41] according to which the primitive ontology would be superfluous in this case, for everything that is empirically accessible is obtained by applying Born's rule to the wave function. Hence, if the wave function exists as an object of its own, it then contains everything to account for the empirical reality—assuming that wave function monism provides indeed a solution to the measurement problem. However, if one denies this, what accounts for measurement outcomes then is the configuration of the primitive ontology. Consequently, the wave function plays only a dynamical role. There is no reason to admit it to the ontology as an object in addition to the configuration of matter.

If the wave function is nomological in the sense that it plays only a dynamical role, one can adopt with respect to it any one of the three main stances that are available in

[13]See also Vassallo [37], Vassallo and Ip [38] and Koslowski [39].

[14]See Esfeld [40].

the metaphysics of laws of nature. By the same token as with respect to laws, one can maintain that the wave function is a primitive entity of its own (see [42]) or that it is derived from dispositions, powers or structures that are primitive modal entities over and above the primitive ontology (see e.g. [43]). However, the explanatory value of both these stances is doubtful: they reify the wave function without thereby making progress in the explanation of the evolution of the configuration of matter, because the wave function is defined in terms of its dynamical role for the primitive ontology. Furthermore, since they reify the wave function, these stances are not immune to the objection from Brown and Wallace [41].

Given that the wave function enters the theory through its dynamical role for the evolution of the configuration of matter, one can adopt the stance that is known as quantum Humeanism: the wave function is reduced to the evolution of the configuration of matter in the sense that the universal wave function is determined by the overall evolution of the configuration of matter of the universe. Given that entire evolution, the universal wave function is fixed. To my knowledge, the first expression of this stance comes from Bell in "The theory of local beables" (1975):

> One of the apparent non-localities of quantum mechanics is the instantaneous, over all space, 'collapse of the wave function' on 'measurement'. But this does not bother us if we do not grant beable status to the wave function. We can regard it simply as a convenient but inessential mathematical device for formulating correlations between experimental procedures and experimental results, i.e., between one set of beables and another. (Quoted from the reprint in Bell [1, p. 53]

"Beable" is Bell's neo-logism for what exists. His words here have an unnecessarily instrumentalistic tone. The decisive point is that the wave function can be seen as being fixed by correlations between sets of beables—that is, the evolution of the configuration of matter (the "local beables" in Bell's sense). Dowker and Herbauts [44] provide a concrete model of how this can be so in the framework of GRW flashes on a lattice. In recent years, this view has been worked out as a philosophical stance that is inspired by Humean reductionism about laws of nature and therefore known as quantum Humeanism.[15] On the one hand, this stance avoids at its roots any objection against an ontological dualism of a primitive ontology and a wave function. On the other hand, this stance still is a scientific realism with respect to the wave function, since the wave function is anchored in, more precisely determined by what exists, namely the primitive ontology of a configuration of matter and its evolution.[16]

In this context, the seminal contribution of the GRW theory is to make clear that the primitive ontology programme is not limited to Bohmian mechanics. In other words, the alternative to wave function monism going back to Everett [9] is not only Bohm's [5] theory. There is a spectrum of primitive ontology theories that covers all the traditional metaphysical positions about objects and in which there is a lively debate about the best proposal for a primitive ontology and corresponding dynamics. In particular, the GRW flash theory is a serious contender that enters into competition

---

[15] See Miller [45], Esfeld [46], Callender [47] and Bhogal and Perry [48].

[16] See Esfeld and Deckert [27, ch. 2.3] for a detailed argument. See also Allori [29] for scientific realism without ontological commitment to the wave function.

with the Bohmian particle ontology—although, to my mind, at the end of the day, the arguments for a primitive Bohmian particle ontology remain compelling.[17]

Acknowledgments I'm grateful to Valia Allori and Dustin Lazarovici for helpful comments on the draft of this paper.

References

1. Bell, John S. (1987): *Speakable and unspeakable in quantum mechanics.* Cambridge: Cambridge University Press.
2. Ghirardi, GianCarlo, Grassi, Renata and Benatti, Fabio (1995): "Describing the macroscopic world: closing the circle within the dynamical reduction program". *Foundations of Physics* 25, pp. 5–38.
3. Maudlin, Tim (1995): "Three measurement problems". *Topoi* 14, pp. 7–15.
4. de Broglie, Louis (1928): "La nouvelle dynamique des quanta". In: *Electrons et photons. Rapports et discussions du cinquième Conseil de physique tenu à Bruxelles du 24 au 29 octobre 1927 sous les auspices de l'Institut international de physique Solvay.* Paris: Gauthier-Villars. Pp. 105–132. English translation in G. Bacciagaluppi and A. Valentini (2009): *Quantum theory at the crossroads. Reconsidering the 1927 Solvay conference.* Cambridge: Cambridge University Press. Pp. 341-371.
5. Bohm, David (1952): "A suggested interpretation of the quantum theory in terms of 'hidden' variables. I and II". *Physical Review* 85, pp. 166–179, 180–193.
6. von Neumann, Johann (1932): *Mathematische Grundlagen der Quantenmechanik.* Berlin: Springer. English translation *Mathematical foundations of quantum mechanics. Translated by R. T. Beyer.* Princeton: Princeton University Press 1955.
7. Ghirardi, GianCarlo, Rimini, Alberto and Weber, Tullio (1986): "Unified dynamics for microscopic and macroscopic systems". *Physical Review D* 34, pp. 470–491.
8. Gisin, Nicolas (1984): "Quantum measurements and stochastic processes". *Physical Review Letters* 52, pp. 1657–1660.
9. Everett, Hugh (1957): "'Relative state' formulation of quantum mechanics". *Reviews of Modern Physics* 29, pp. 454–462.
10. Allori, Valia, Goldstein, Sheldon, Tumulka, Roderich and Zanghì, Nino (2008): "On the common structure of Bohmian mechanics and the Ghirardi-Rimini-Weber theory". *British Journal for the Philosophy of Science* 59, pp. 353–389.
11. Albert, David Z. (1996): "Elementary quantum metaphysics". In: J. T. Cushing, A. Fine and S. Goldstein (eds.): *Bohmian mechanics and quantum theory: An appraisal.* Dordrecht: Kluwer. Pp. 277–284.
12. Albert, David Z. (2015): *After physics.* Cambridge (Massachusetts): Harvard University Press.
13. Wallace, David (2012): *The emergent multiverse. Quantum theory according to the Everett interpretation.* Oxford: Oxford University Press.
14. Dürr, Detlef, Goldstein, Sheldon and Zanghì, Nino (2013): *Quantum physics without quantum philosophy.* Berlin: Springer.
15. Allori, Valia, Goldstein, Sheldon, Tumulka, Roderich and Zanghì, Nino (2014): "Predictions and primitive ontology in quantum foundations: a study of examples". *British Journal for the Philosophy of Science* 65, pp. 323–352.
16. Brown, Harvey R., Elby, Andrew and Weingard, Robert (1996): "Cause and effect in the pilot-wave interpretation of quantum mechanics". In: J. T. Cushing, A. Fine and S. Goldstein (eds.): *Bohmian mechanics and quantum theory: an appraisal.* Dordrecht: Springer. Pp. 309–319.

[17]See Esfeld [40] and Esfeld and Deckert [27, chs. 2–4].

17. Pylkkänen, Paavo, Hiley, Basil J. and Pättiniemi, Ilkka (2015): "Bohm's approach and individuality". In: A. Guay and T. Pradeu (eds): *Individuals across the sciences*. Oxford: Oxford University Press. Pp. 226–246.
18. Esfeld, Michael, Lazarovici, Dustin, Lam, Vincent and Hubert, Mario (2017): "The physics and metaphysics of primitive stuff". *British Journal for the Philosophy of Science* 68, pp. 133–161.
19. Cowan, Charles Wesley and Tumulka, Roderich (2016): "Epistemology of wave function collapse in quantum physics". *British Journal for the Philosophy of Science* 67, pp. 405–434.
20. Ghirardi, GianCarlo, Pearle, Philip and Rimini, Alberto (1990): "Markov processes in Hilbert space and continuous spontaneous localization of systems of identical particles". *Physical Review A* 42, pp. 78–89.
21. Egg, Matthias and Esfeld, Michael (2015): "Primitive ontology and quantum state in the GRW matter density theory". *Synthese* 192, pp. 3229–3245.
22. Maudlin, Tim (2010): "Can the world be only wavefunction?". In: S. Saunders, J. Barrett, A. Kent and D. Wallace (eds.): *Many worlds? Everett, quantum theory, and reality*. Oxford: Oxford University Press. Pp. 121–143.
23. Wallace, David (2014): "Life and death in the tails of the GRW wave function". http://arxiv.org/abs/1407.4746 [quant-ph].
24. de Broglie, Louis (1964): *The current interpretation of wave mechanics. A critical study*. Amsterdam: Elsevier.
25. Norsen, Travis (2005): "Einstein's boxes". *American Journal of Physics* 73, pp. 164–176.
26. Egg, Matthias and Esfeld, Michael (2014): "Non-local common cause explanations for EPR". *European Journal for Philosophy of Science* 4, pp. 181–196.
27. Esfeld, Michael and Deckert, Dirk-André (2017): *A minimalist ontology of the natural world*. New York: Routledge.
28. Tumulka, Roderich (2006): "A relativistic version of the Ghirardi-Rimini-Weber model". *Journal of Statistical Physics* 125, pp. 825–844.
29. Allori, Valia (2019): "Scientific realism without the wave-function: an example of naturalized quantum metaphysics". In: J. Saatsi and S. French (eds.): *Scientific realism and the quantum*. Oxford: Oxford University Press.
30. Bedingham, Daniel, Dürr, Detlef, Ghirardi, GianCarlo, Goldstein, Sheldon, Tumulka, Roderich and Zanghì, Nino (2014): "Matter density and relativistic models of wave function collapse". *Journal of Statistical Physics* 154, pp. 623–631.
31. Barrett, Jeffrey A. (2014): "Entanglement and disentanglement in relativistic quantum mechanics". *Studies in History and Philosophy of Modern Physics* 47, pp. 168–174.
32. Esfeld, Michael and Gisin, Nicolas (2014): "The GRW flash theory: a relativistic quantum ontology of matter in space-time?". *Philosophy of Science* 81, pp. 248–264.
33. Maudlin, Tim (2011): *Quantum non-locality and relativity. Third edition*. Chichester: Wiley-Blackwell.
34. Arntzenius, Frank (2012): *Space, time and stuff*. Oxford: Oxford University Press.
35. Rovelli, Carlo (1997): "Halfway through the woods: contemporary research on space and time". In: J. Earman and J. Norton (eds.): *The cosmos of science*. Pittsburgh: University of Pittsburgh Press. Pp. 180–223.
36. Dürr, Detlef, Goldstein, Sheldon and Zanghì, Nino (2018): "Quantum motion on shape space and the gauge dependent emergence of dynamics and probability in absolute space and time". http://arxiv.org/abs/1808.06844 [quant-ph].
37. Vassallo, Antonio (2015): "Can Bohmian mechanics be made background independent?". *Studies in History and Philosophy of Science* 52, pp. 242–250.
38. Vassallo, Antonio and Ip, Pui Him (2016): "On the conceptual issues surrounding the notion of relational Bohmian dynamics". *Foundations of Physics* 46, pp. 943–972.
39. Koslowski, Tim (2017): "Quantum inflation of classical shapes". *Foundations of Physics* 47, pp. 625–639.
40. Esfeld, Michael (2014a): "The primitive ontology of quantum physics: guidelines for an assessment of the proposals". *Studies in History and Philosophy of Modern Physics* 47, pp. 99–106.

41. Brown, Harvey R. and Wallace, David (2005): "Solving the measurement problem: de Broglie–Bohm loses out to Everett". *Foundations of Physics* 35, pp. 517–540.
42. Maudlin, Tim (2007): *The metaphysics within physics.* Oxford: Oxford University Press.
43. Suárez, Mauricio (2015): "Bohmian dispositions": *Synthese* 192, pp. 3203–3228.
44. Dowker, Fay and Herbauts, Isabelle (2005): "The status of the wave function in dynamical collapse models". *Foundations of Physics Letters* 18, pp. 499–518.
45. Miller, Elizabeth (2014): "Quantum entanglement, Bohmian mechanics, and Humean supervenience". *Australasian Journal of Philosophy* 92, pp. 567–583.
46. Esfeld, Michael (2014b): "Quantum Humeanism, or: physicalism without properties". *Philosophical Quarterly* 64, pp. 453–470.
47. Callender, Craig (2015): "One world, one beable". *Synthese* 192, pp. 3153–3177.
48. Bhogal, Harjit and Perry, Zee R. (2017): "What the Humean should say about entanglement". *Noûs* 51, pp. 74–94.

Might Laws of Nature 'Ground' Phenomena?

Federico Laudisa

Abstract The present paper focuses on the connection between grounding and laws of nature or, better, on the meaning that the grounding relation might assume if considered by a nomological point of view. In particular, the working hypothesis is that the grounding relation might used to implement the relation between a law of nature and its possible instances. The grounding relation can be distinguished from other modally-loaded relations and thus it might avoid the criticisms of those who are suspicious about the intuition according to which a law somehow 'necessitates' the phenomena falling within its scope. Moreover, if we adopt the above mentioned working hypothesis according to which a law ultimately 'grounds' its instances, then we are more likely to provide the idea of a *governing* law with a more definite meaning: that is, a law of nature would be a 'governing' process with respect a class of given phenomena whenever we are entitled to say that the latter are grounded in the former.

1 Introduction

The theory proposed in 1986 by Ghirardi, Rimini and Weber (GRW henceforth) introduced a quantitatively detailed model of how a state reduction process can be incorporated into quantum mechanics, such that typically quantum phenomena on the microscale coexist with what GRW used to call the *macro-objectification* of physical properties pertaining to apparatuses in measurement interactions. As is well known, the heart of the formulation lies in a nonlinear stochastic modification of the evolution law for wavefunctions, a modification that is supposed to induce spontaneous collapse processes for the wavefunctions themselves [1]. The impressive work developed by Ghirardi and his co-workers along these lines in many years has been accompanied by an increasing awareness concerning the key role that a philosophical reflection should play in the debates on the foundations of quantum mechanics. In particular

F. Laudisa (✉)
Department of Philosophy and Humanities, University of Trento, Via Tommaso Gar 14, 38122 Trento, Italy
e-mail: federico.laudisa@unitn.it

V. Allori et al. (eds.), *Do Wave Functions Jump?*, Fundamental Theories of Physics 198,
https://doi.org/10.1007/978-3-030-46777-7_9

this led the GRW research program in more recent times to take seriously what has been called the *primitive ontology approach*: in this vein, the original GRW model is supposed to evolve into a class of theories endowed with a clear and well-defined ontology (a structure of fundamental physical 'entities'—whatever they are—the theory is supposed to be about), into which the state reduction is to be framed. Two different proposals have been introduced as to the sort of ontology GRW models are held to deal with, the *matter density* ontology and the *flash* ontology. In the former the theory assumes a continuous ontology, consisting essentially in a field on three-dimensional space that, for a given *t*, is to be understood as the density of matter in space at time *t* [2, 3], in the latter the theory assumes a discrete ontology, in which matter is made up by discrete points ('flashes') in spacetime such that to each of these flashes there correspond one of the spontaneous collapses of the wavefunction, and the spacetime location of the flashes is the spacetime location of the collapses. In the flash ontology—whose original proposal is due to J.S. Bell—flashes are the so-called *local beables* of the theory [4].

These more recent years of activity testify then a deep concern for the prospects of a coherent worldview that might be somehow 'extracted' from a suitably interpreted version of quantum mechanics. This worldview aspires to be a metaphysically robust representation of the natural world at the microscale, under the assumption that the above mentioned version of quantum mechanics is held true (in a decent meaning of 'true') and in which we hope to achieve a reasonably clear sense of the basic furniture of the physical world quantum mechanics is supposed to be about. It is to this concern that the present paper is inspired: it is focused on a specific aspect of the notion of *natural law*—a notion that in a metaphysically-loaded worldview is obviously supposed to play a crucial role. I will explore in particular the potential relevance that the metaphysical notion of *grounding* might have for the status of laws, on the background of the debate concerning the 'governing' or 'non-governing' intuition about laws themselves.

2 Grounding: A Nomological Reading?

Within the revival that metaphysics experienced in the last years, grounding has received a considerable attention. Intuitively such attention is very plausible since, when we say in general terms that *x* 'grounds' *y*, we point implicitly to a conceptual region that seems to intersect related, relevant notions both of an ontological and epistemological character—such as causation, explanation, determination, constitution, dependence and so on—without reducing to any of those notions. This kind of interconnection plays also the role of justifying somehow the appeal itself of the notion of grounding. The consideration of how frequently we may find ourselves arguing that *y* occurs—or is what it is—'in virtue of' *x* (one of many possible formulations of a grounding relation) may suggest that such frequency, far from accidental, might have a deeper meaning in a metaphysical context: in fact, a common strategy for defending grounding as an object of investigation takes on the form of a review of cases in which, with different conceptual tools and frameworks, we rely on a grounding intuition (see for instance [5, pp. 812–813]).

Given the several perspectives from which to look at the notion of grounding, in the present paper I will focus on the connection between grounding and laws of nature or, better, on the meaning that the grounding relation might assume if considered by a nomological point of view. By the viewpoint of the relation of metaphysics with science, this focus on laws appears to be well justified, given the role that laws play in specific scientific theories. In particular, the working hypothesis is that the grounding relation might used to implement the relation between a law of nature and its possible instances. The grounding relation can be distinguished from other modally-loaded relations and thus it might avoid the criticisms of those who are suspicious about the intuition according to which a law somehow 'necessitates' the phenomena falling within its scope. Moreover, if we adopt the above mentioned working hypothesis according to which a law ultimately 'grounds' its instances, then we are more likely to provide the idea of a *governing* law with a more definite meaning: that is, a law of nature would be a 'governing' process with respect a class of given phenomena whenever we are entitled to say that the latter are grounded in the former.

The plausibility of the whole project relies on two general assumptions. According to the first the notion of grounding is a fruitful one, so that a possible translation of the intuition of law as a governing process in terms of grounding is not an attempt to explain *obscurum per obscurius*. According to the second—and taking the first assumption for granted—the conceptual option that best suits the governing-grounding translation is primitivism on laws, namely the theory in which laws are basic constituents of the fundamental inventory of the world. Admittedly, both assumptions are strong and engaging. I will return to this point later, but few words of justification are in order. The first assumption undoubtedly presupposes a confidence in the virtues of grounding in metaphysical terms, but it is also true that the whole debate on the status of laws of nature (including the best-system analysis, namely the option which is the most hostile to any kind of governing view of laws) is firmly located within the boundaries of metaphysics: thus if we decide to argue in favour of grounding we do not trivially violate any rule of the game. The second assumption, on the other hand, implies a defence of the primitivist model of law as a plausible model in itself, otherwise it could not play the role that is functional to the first assumption. Now, it is certainly true that the idea of laws as primitives is not the most popular view on laws that I might think of; but, as Tim Maudlin persuasively argued

> Taking laws as primitives may appear to be simple surrender in the face of a philosophical puzzle. But every account must have primitives. The account must be judged on the clarity of the inferences that the primitives warrant and on the degree of systematization they reveal among our pre-analytic inferences [6, p. 15].

My discussion is organized then in the following order. I will first argue in Sect. 3 that the governing intuition about laws is worth defending. I will then argue that primitivism on laws is a viable option in terms of consistency with the governing intuition Sect. 4. Finally in Sect. 5, in response of the alleged vagueness of what it should mean for a type-theory kind of entity such as a primitive law to 'govern' token-type of events, I will propose that the governing intuition might be cashed out in terms of a grounding relation.

3 Laws of Nature: Governing Versus Non-Governing Answers

The debate on the nature of lawhood is a complex, intertwined network of issues. Starting from a very general dichotomy, according to which either laws are *somehow* a part of nature or they pertain *only* to scientific theories (hence—lastly—to *us* as knowing subjects), the controversy proceeds toward further, more specific issues: regularity vs. necessity, the role of Humean supervenience, the relation with the use of laws in scientific practice, the governing vs. the non-governing status of laws, and so on. Let us suppose as a working hypothesis the existence of 'law-like' items in the inventory of the world; a major issue is then of course in what terms should we conceive the relation between such items and the particular phenomena that in some way or another 'fall under' them. It is at this stage that the controversy 'governing versus non-governing' arises.

> [The] Governing answer ... insists that there are genuine laws of nature and furthermore that these laws govern or even produce the events of the world' whereas 'the Non-Governing answer ... has it that there are genuine laws of nature, but that they do not govern or produce the events of the world. The mosaic of events displays certain patterns, and it is in the features of some of these patterns that we find laws.' [7, p. 2].

In a Humean, reductionist perspective, we should be careful in admitting suspect versions of necessity in nature so that, as a consequence, we should not require from laws any 'governing' role as a constitutive feature. In the global metaphysical view of the world that is in the background of (any version of) views like this, any modal, governing feature in mentioning laws must be traced back to *us* as subjects, not to the world, that in itself is nothing but a collection of sparse entities, usually conceived as discrete. In a regularist brand of a reductionist perspective, what is admitted is just the existence of a(n astronomical) number of matters of fact and the existence of a certain number of regularities connecting them [8, p. 32]: although it is far from trivial to specify what a regularity exactly is in non-modal terms, a (naïve) regularity view denies the existence of *necessary* connections or laws whose role in some way or other would be to *ground* the regularities. The more sophisticated versions of a regularity view, summarized in the so-called Mill–Ramsey–Lewis *best-system view*, turn out to be an elaboration of how *we as subjects* should organize regularities, if we want them to play the role that non-Humean views usually associate to the notion of law. The now classic formulation of David Lewis [9, p. 73] makes explicit the disregard for what is often interpreted as the instrinsically modal character of natural laws.

Although according to several philosophers, the failure to capture the modal character of laws of nature is exactly *the* defect of all regularity, best-system approaches to laws (see for instance [10]), objections by Menzies look like a *petitio principii*, since whether laws of nature should be conceived modally or not is just the point under discussion. Moreover, a regularist may well resist the claim that the notion of law is to be *defined* in governing terms: she will try to claim that such move is far from unavoidable and that, as a consequence, a non-governing conception of laws is perfectly consistent. This is what Helen Beebee, for example, tries to accomplish by claiming that one can define law as governing only by assuming an analogy between laws in nature and laws in other domains such as theology, politics, ethics, and so on: but if this assumption is rejected (and it *can* be rejected), the governing feature of laws fails to be a conceptual truth [11].

Now even if we assume the plausibility of these arguments and concede that the analogy between natural laws and other sorts of law does not work, the question remains: what exactly does the regularist view imply about the very notion of law? Beebee characterizes the divide between Humean and anti-Humean stances in these terms:

> For the anti-Humean, laws (unlike accidentally true generalizations) *do* something—they *govern* what goes on in the universe—and they therefore require some sort of ontological basis ... that gives them this ability. Humeans, on the other hand, do not require laws to 'do' anything: like accidentally true generalizations, laws are at bottom merely true descriptions of what goes on. Thus for the Humean there is no need for any ontological distinction between laws and accidents [11, p. 580].

Under this linguistic stipulation concerning *laws*, then, let us focus on what in a Humean framework a purely descriptive view of laws might entail. Humeans deny the need for a search for something that might 'ground' a regularity:

> For the Humean, since the laws are descriptive, what the laws are depends on what the facts are.... Humeans ... do not require laws to 'do' anything: like accidentally true generalizations, laws are at bottom merely true descriptions of what goes on [11, pp. 579–580].

To be true, the expression 'laws are at bottom merely true descriptions of what goes on' is rather vague and, at first sight, it might lend support for the idea that a law might be literally nothing but a *mere collection* of facts, with a possibly paradoxical, set-theoretical kind of consequence. Let us suppose that we have the following collections of states S at their respective times t:

$$F_n = \{\ldots,\ S(t_{n-1}),\ S(t_n)\}$$

$$F_{n+m} = \{\ldots,\ S(t_n),\ \ldots,\ S(t_{n+m-1}),\ S(t_{n+m})\}$$

If we assume that each of these states is a collection of values of a set of relevant physical quantities, each state may well represent a 'fact' in the regularist vein, since each state works as a sort of snapshot at its time $t_{(*)}$ of the physical situation at stake.

Now, intuition tells us that we have here the same law accounting for the evolution of our system at two different times (we can safely assume that the states both in F_n and F_{n+m} are obtained as the computational output of the algorithm implicit in one and the same deterministic dynamical law). But, should a law be nothing but a collection of facts, in the present case we seem to deal with *two* sort-of-laws, since we have *two* different collections! Even worse: at *any* successive instant of time a new law-as-a-collection is generated. But, then, being parsimonious on any alleged modal features of the world seems paradoxically to imply a wild and uncontrollable generation of laws-as-collections over time, a phenomenon in strong tension with an aspiration to a metaphysical economy. On the other hand, if we suppose that there is a unifying principle according to which F_n and F_{n+m} might be shown to be just two instances of one and the same law, this principle could not simply supervene on the states and because of this non-supervenience this account would immediately become a *non-Humean* one.

The counterintuitive character of this objection might simply show that we are not entitled to attack a 'descriptive' or 'Humean' stance by ascribing to it such a thin notion of law in terms of mere collection. For instance, a best-system implementation of the descriptive view can manage the set-theoretic objections, exactly because the 'best' choice in singling out the axioms that are suitable candidates for performing as laws can be done over different sets, that appear as merely different descriptions among which the optimal combination of simplicity and strength is recovered. But in exactly what terms is the notion of law more than a collection of non-nomic facts on one hand, but without a governing character on the other? According to Beebee, the non-governing relation that by a Humean standpoint is supposed to hold between the law and the facts that intuitively are accounted for by the law is expressed concisely and usefully by determinism:

> We can characterize determinism in the following rough and ready way: the state of the universe at any given time together with the laws of nature determines what the state of the universe will be at any future time. But what does 'determines' mean here? For the Humean, the laws and current facts determine the future facts in a purely logical way: you can *deduce* facts from current facts plus the laws. And this is just because laws *are*, in part, facts about the future [11, p. 578].

Now one might object to this Humean understanding of determinism that it overlooks what the specific *mathematical* formulation of a dynamical law implies concerning the 'determining' capability of the law itself. Take for instance Newton's second law. The general form of this dynamical law states a proportionality between force and acceleration: when a specific formulation for a kind of force is inserted into the general schema of the second law, we obtain a mathematical equation that, under non-trivial conditions, turns out to be integrable. This is what justifies us to assume that the knowledge of a given (initial) state and of what is the force (if any) acting on the system makes it possible to determine future states of the system (and also past states, if the evolution satisfies time reversal invariance). So the crucial point is that the determination is possible due to the functional relations among physical quantities, relations that are *encoded* into the mathematical formulation of Newton's law. This encoding does not seem to me easily accounted for in a purely descriptive view of laws, whatever 'purely descriptive' might mean: how can a law play its truly nomic role by being—as a Humean might say—a 'simple, strong summary of the totality of non-nomic facts'?

4 Laws as Primitives

In the above section we have referred just to few of the several reservations that can be made concerning the Humean stance on laws of nature. In more general terms, from the Humean viewpoint it turns out to be far from easy to cope with that seemingly irreducibly modal aspect that informs our explicit and implicit way of employing or referring to laws: for instance, the demand according to which laws must cover *possibilities*, and not just actualities, and the demand according to which, when we ask a law to *explain* facts, we search after what is in virtue of which facts obtain. Moreover, a look at the actual scientific practice when analysing how laws work does nothing but increase the dissatisfaction: one of the most perplexing points of the above discussed features of a Humean stance on laws of nature is that of implying a truly *structureless world*, an implication that seems hard to reconcile with a scientific image of the natural world, even broadly construed.

These criticisms of theHumean stance, albeit not ultimate, look sufficiently plausible to suggest an alternative, the so-called *primitivist* approach,[1] under the assumption that such approach is hospitable to the view of laws as governing items.[2] Basically, primitivism about laws can come in two varieties: in its metaphysical dimension, laws belong to the fundamental inventory of the world, whereas in its conceptual dimension, laws are not to be reduced to more primitive notions: in the words of a notable primitivist, "My analysis of laws is no analysis at all. Rather, I suggest we accept laws as fundamental entities in our ontology. Or, speaking at the conceptual level, the notion of a law cannot be reduced to other more primitive notions." [6, p. 18]. The choice of laws as primitive may have at least two significant strong points. First, the primitive status of laws allows one to have a more effective and stimulating confrontation with the role of laws *within specific scientific theories*: the above reference to actual scientific practice should be read in this sense. Second, the primitive status of laws promises to plausibly accommodate interrelated notions—causation, explanation, counterfactuals—that, together with lawhood, appear to form a true conceptual network. In fact, laws as primitives can be reasonably seen as able to translate causal relations into nomic ones, as grounding counterfactuals if similarity of possible worlds is formulated in terms of compatibility with given laws and,

[1]Under the assumption according to which a descriptive view of laws is a reductionist view in a serious sense, the primitivist approach is an anti-reductionist view which includes similar, but mutually non-equivalent positions such as Carroll [12, 13], Lange [14, 15], and Maudlin [6]: in the present paper we will focus on the Maudlin version.

[2]This clearly need not imply that primitivism is the *only* approach that complies with the assumption.

finally, they can preserve their role in explanation. Let us briefly focus on the two last points.

The connection between laws and counterfactuals is obviously deep, due to the issue of whether modality is intrinsic to the very notion of law or not. As we recalled earlier, laws are intuitively supposed to account not only for the actual phenomena but also for possible ones, and are supposed to account for the uniformity of the natural world in terms of non-contingent, law-governed processes. In the Maudlin version of primitivism, laws as primitives can handle this modal core of the lawhood intuition through their capacity of generating (classes of) models. A given law gives rise to possible worlds to the extent that it sets boundaries to the occurrence of phenomena. All that according to precise prescriptions fixed by the law remains within the boundaries is 'possible': 'the possible worlds consistent with a set of laws are described by the models of a theory that formulates those laws' [6, p. 18]. If setting the boundaries for the validity of a law is what allows for possibility in the primitivist framework, necessity is obtained at a very low cost: since it is the very compatibility with laws that generates a set of models, the laws themselves must hold in all models of the set and therefore display a nomic necessity (in the usual, possible-world language, [6, p. 21]). As to the connection with explanation, the primitive status of laws allows us to select any account of explanation we like in which laws play a sufficiently crucial role, without worrying whether the plausibility of the model of explanation we selected is threatened or not by some more fundamental notion in terms of which the notion of law is reduced: we have explanation of an event f whenever we have a nomic subsumption of f under the relevant set of laws L, namely whenever we may show that, given L, the event f is what we should expect [6, 34ff].

Since no wide-ranging philosophical view concerning such a deep issue as the issue of laws on nature can go unchallenged, let us take into consideration some possible objections to the primitivist account. A first point concerns the status of laws in terms of their alleged 'fundamentality'. At a given stage of development of a scientific theory, we may have reasons to think that a given law is fundamental, a circumstance that seems to go along well with the claim that laws are primitive endowments of the natural world's ontology. History of science, however, has taught us that laws that were supposed to be fundamental turned out to be only special cases of more general laws and still history of science, jointly with philosophy of science, suggests that there are no reasons to think that there is a foreseeable end to this process of ever-increasing generality. How are we to cope with this problem? It hardly looks reasonable to assume that *the whole network of laws* is primitive from the start: would not it be awkward to suppose that, for instance, Kepler's laws are as primitive as Newton's laws, and these in turn as primitive as the Einstein's field equations of general relativity?[3] Although clearly *primitive* and *fundamental* are not equivalent concepts, one might think that the conventionality inherent in

[3]Here Kepler's and Newton's laws are just examples of what might happen to laws that at a certain stage are taken to be fundamental and that, at a later stage, change this status. The charge that some might raise—Kepler's and Newton's laws are not plausible candidates for lawhood anyway because they are in a certain sense 'false'—is irrelevant here: the focus here is on how we should assess the level of fundamentality of certain laws with respect to others, not on their actual truth or falsity.

selecting which laws are supposed to be fundamental and which derivative might be at least disturbing for primitivism. On the other hand, in different areas of science results have been obtained that prove in principle our inability to grasp some kind of knowledge—from the undecidability theorems to the black hole information loss theorems—so that the idea that we might be unable to access part or the totality of the *really fundamental* laws need not contradict the possibly primitive ontological status of some of these laws.

5 Laws as Grounding Items?

On the basis of the arguments defended in previous sections, the proposal I would like to discuss is that a plausible way to implement the intuition that a law is something that 'governs' the phenomena it allegedly covers is in terms of a grounding relation holding between the laws and those phenomena. In this respect, I would like to address two main points, that are both relevant on the background of the Humean versus non-Humean confrontation about laws of nature: the first concerns modality, whereas the second concerns the problem that the difference in ontological category between the law and the law-governed phenomena might represent. As to the first, since it is possible to defend a categorically neutral formulation of grounding, and hence it is also possible to keep separate grounding and modal entailment [16]), a reading of the nomic, governing import of the law in grounding terms need not imply in itself any modal commitment. In any case, even if we opt for the thesis that grounding implies some sort of necessaritarianism [17], the rather innocent sense of necessity implicit in the Maudlin strand of primitivism on laws makes that option more digestible. As to the second, the debate about 'flat' versus 'ordered' conceptions of structures as "target of metaphysical inquiry' [16] might suggest a useful application of an 'ordered' approach to the law-phenomena relation in terms of grounding. Let me elaborate a bit on both.

As to the first point, it is well known that a major issue concerning laws of nature is the status of modality. According to Humeans, as we said, a law is a coincise and balanced summary of local facts whereas for non-Humeans a law is more than that. For the latter camp, the *more-than* part is usually cashed out in modal terms, namely the nomological relation between a law and the phenomena that are supposed to fall under the law is supposed to be a kind of intrinsically modal relation, over and above the 'summary' and some non-Humeans are even prepared to accept nomological modality as itself primitive. Here, however, I would like to recall a serious objections that has been raised to modality (necessity in particular) by Kit Fine. He argues that necessity is unfit to support the metaphysical idea of essence on the basis of claims like the following: While all essential truths about x are necessary truths about x, the converse need not hold, namely not all necessary truths about x need be essential about x. In this sense, Fine argues, modality is too 'coarse-grained' to do a good work in characterizing a metaphysically serious form of dependence [18, 19].

Now, can grounding do better than modality in characterize dependencies that we might find metaphysically relevant, like the nomological ones? If the answer is yes, this can be good news for a non-Humean view of laws. As stated above, the non-Humean view expresses its dissatisfaction toward an anti-necessitarian view of laws by defending the idea that a law is more than a coincise and balanced summary of local facts: in doing this, however, it must solve the problem of somehow regimenting that extra-relation between laws and facts, a problem that non-Humeans usually address by assuming nomicity as a primitive modal kind of relation. But if we suppose that grounding improves over modality in characterizing a metaphysical dependence, a non-Humean might exploit such improvement in favour of her stance, since the recourse to grounding might help to bypass the ordinary anti-modal objections toward a non-Humean view of laws: just like in Fine's view a property may be essential to x (i.e. it may 'ground' x in the sense of *being constitutive* of x) without having modal connotations, similarly a law might 'ground' some phenomena without involving any modal relation between law and phenomena themselves.

Let us see how this suggestion might work with respect to an instance of modal primitivism on laws that, in other respects, seems useful [6, pp. 18–21]. Laws are intuitively supposed to account not only for the actual phenomena but also for possible ones, and are supposed to account for the uniformity of the natural world in terms of non-contingent, law-governed processes: laws as primitives can handle this modal core of the lawhood intuition through their capacity of generating (classes of) models. An additional advantage of the Maudlin version of primitivism seems to be that, if we assume laws as primitives, the principle according to which we generate models only in terms of their compatibility with laws requires – so to speak – a 'minimum' of modality: possibility *is exactly* law-compatibility, whereas necessity is obtained simply from ranging over the worlds so generated. In a Finean perspective, however, this 'minimum' is exactly what is perplexing: a given law L might be 'necessary' in the above sense even if it might have no relevant connection with large classes of phenomena in those worlds. In other words, the bare compatibility with L that in the model-theoretical sense generates the possible-worlds structure is 'insensitive' to the relation between the law L and the events in the worlds that belong to the structure. It is here that the Finean criticism toward modality comes in, and suggests that the assumption of a *grounding* relation between L and the phenomena supposedly covered by it might justify the relevance of the law for the phenomena: L can be plausibly said 'to govern' those phenomena, since the latter are *grounded* in the former.

A last remark concerns the second point raised above, namely the difference in ontological category between the law and the law-governed phenomena. In the debate on grounding, this relation is usually assumed to hold within a selected category—facts or propositions in most cases (see e.g. [20]). Since in my proposal the grounding relation connects a law with phenomena, one may ask whether grounding can be 'cross-categorical', in fact, there seem to be no really compelling reason to prevent in principle the possibility that there might be such cross-categorical grounding relation (see e.g. [16]): on the contrary, in view of the above mentioned possibility of applying these suggestions to issues in foundations of physics, the cross-categoricity appears to be a highly desirable property.

References

1. Ghirardi G.C., Rimini A., Weber T., 1986, "Unified Dynamics for Microscopic and Macroscopic Systems", *Physical Review* D34, pp. 470-496.
2. Bassi A., Ghirardi G.C., 2003, "Dynamical Reduction Models", *Physics Reports* 379, pp. 257-427.
3. Benatti F., Ghirardi G.C., Grassi R., 1995, ""Describing the Macroscopic World: Closing the Circle within the Dynamical Reduction Program", *Foundations of Physics* 25, pp. 5-38.
4. Bell J.S., 1987, *Speakable and Unspeakable in Quantum Mechanics*, Cambridge University Press, Cambridge.
5. Clark M.J., Liggins D. 2012, "Recent Work on Grounding", *Analysis* 72, pp. 812-823.
6. Maudlin, T. 2007. *The Metaphysics within Physics*. New York: Oxford University Press.
7. Cohen, J., Callender C., 2009, "A Better Best System Account of Lawhood" *Philosophical Studies* 145, pp. 1–34.
8. Mumford, S., 2004. *Laws in Nature*. London: Routledge.
9. Lewis, D. K., 1973, *Counterfactuals*. Cambridge, MA: Harvard University Press.
10. Menzies, P. 1993. "Laws, Modality, and Humean Supervenience." In *Ontology, Causality and Mind: Essays in Honour of D. M. Armstrong*, edited by J. Bacon, K. Campbell, and L. Reinhardt, pp. 195–224, Cambridge: Cambridge University Press.
11. Beebee H., 2000, "The Non-governing Conception of Laws of Nature" *Philosophy and Phenomenological Research* 61, pp. 571–594.
12. Carroll, J. W. 1994. *Laws of Nature*. Cambridge: Cambridge University Press.
13. Carroll, J. W. 2008. "Nailed to Hume's Cross?" In *Contemporary Debates in Metaphysics*, edited by J. Hawthorne, T. Sider, and D. W. Zimmerman, pp. 67–81. Oxford: Basil Blackwell.
14. Lange, M. 2000. *Natural Laws in Scientific Practice*. Oxford: Oxford University Press.
15. Lange, M. 2009. *Laws and Lawmakers*. New York: Oxford University Press.
16. Schaffer J. 2009, "What Ground What", in *Metametaphysics: New Essays on the Foundations of Ontology*, edited by Chalmers D., Manley D. and Wasserman R., Oxford Clarendon Press, pp. 347-383.
17. deRosset L., 2010, "Getting Priority Straight", *Philosophical Studies* 149, pp. 73-97.
18. Fine K. 1994, «Essence and modality» *Philosophical Perspectives* 8, pp. 1–16.
19. Wilson J., «No work for a theory of grounding», *Inquiry* 57, 2014, pp.
20. Rosen G. 2010, «Metaphysical dependence: grounding and reduction», in *Modality: Metaphysics, Logic and Epistemology*, ed. by B. Hale, A. Hoffman, Oxford University Press, pp. 109-136.

On Closing the Circle

Peter J. Lewis

Abstract Ghirardi sought to "close the circle"—to find a place for human experience of measurement outcomes within quantum mechanics. I argue that Ghirardi's spontaneous collapse approach succeeds at this task, and in fact does so even without the postulation of a particular account of "primitive ontology", such as a mass density distribution or a discrete "flashes". Nevertheless, I suggest that there is a remaining ontological problem facing spontaneous collapse theories concerning the use of classical concepts like "particle" in quantum mechanical explanation at the micro-level. Neither the mass density nor the flash ontology is any help with this problem.

1 Introduction

I remember the first time I came across Gian Carlo Ghirardi's work. I was a graduate student at U. C. Irvine, and my advisor, Jeff Barrett, sent me to read the original GRW paper [11]. When I was done, I thought to myself "Well, that's it. The physicists have solved the measurement problem, and there's nothing left for us philosophers to do". Fortunately (for me), my initial thought was premature; there was plenty of work left, for both physicists and philosophers, both refining the various spontaneous collapse models, and clarifying the surrounding concepts. Indeed, Ghirardi himself was deeply involved in both sides of this work.

One of the conceptual projects involves what Shimony [15] calls "closing the circle". Physics begins with human experience: we postulate physical theories to explain what we observe. Physical theories, insofar as they are successful, tell us what the world is like. But that world, of course, includes human beings and their experiences. So, for consistency, we need to be able to locate human beings and human experiences within the world-view provided by our physical theories.

Ghirardi recognized the importance of closing the circle in physics, and proposed a particular strategy for doing so in the context of his spontaneous collapse approach to quantum mechanics [9]. What I want to do in this paper is to locate Ghirardi's

P. J. Lewis (✉)
Dartmouth College, Hanover, NH 03755, USA
e-mail: Peter.J.Lewis@Dartmouth.edu

V. Allori et al. (eds.), *Do Wave Functions Jump?*, Fundamental Theories of Physics 198,
https://doi.org/10.1007/978-3-030-46777-7_10

proposal within a broader discussion of what it takes to close the circle. In particular, I will argue that, while some proposals in the foundations of quantum mechanics fail to adequately close the circle, Ghirardi's proposal succeeds; however, it does so by making more physical commitments than are strictly necessary. I advocate a strategy for closing the circle very like Ghirardi's but pared of excess physical structure. Finally, I argue that the more pressing problem for spontaneous collapse theories concerns explanation at the microscopic level, and that here Ghirardi's proposal is little help.

2 How not to Close the Circle

Closing the circle might seem like a trivial exercise. After all, if our physical theories describe the behavior of matter at the smallest scales, and if human beings are just complicated chunks of matter, then our physical theories automatically describe the behavior of human beings, including their eyes and their brains. How, then could a physical theory *fail* to find a place for human experience?

Closing the circle does indeed look trivial from a classical perspective, but quantum mechanics challenges much that we thought we could take for granted. The basic difficulty is just the measurement problem. Take a spin-1/2 particle, and prepare it in a superposition of two z-spin eigenstates: $2^{-1/2}(|\uparrow\rangle_z + |\downarrow\rangle_z)$. Now measure the spin of the particle along the z-axis. In the spirit of closing the circle, take the measuring device to be described by quantum mechanics, where $|\text{up}\rangle_m)$ is an eigenstate in which the measuring device reads "spin-up", and $|\text{down}\rangle_m)$ is an eigenstate in which it reads "spin-down". Quantum mechanics entails that after applying the measuring device to the particle and allowing their states to become correlated, the final state of the particle plus the measuring device is $2^{-1/2}(|\uparrow\rangle_z|\text{up}\rangle_m + |\downarrow\rangle_z|\text{down}\rangle_m)$. It looks like there is nothing in this final state that represents the (unique) outcome of this measurement. And since the measuring device m could include a human observer, there is nothing in the final state that represents the (unique) experience of the observer. So it looks like there is nowhere to locate human experience within the quantum formalism, and closing the circle becomes a *problem*.

Closing the circle is related to von Neumann's *psychophysical parallelism* [16, p. 419], and indeed discussions of closing the circle are often couched in terms of postulating a psychophysical parallelism [14, p. 45], [9, p. 33]. But it important to distinguish psychophysical parallelism from closing the circle, in particular because von Neumann's account of psychophysical parallelism *fails* to close the circle.

Von Neumann notes that there is at best a vague distinction between measured systems and measuring devices, and hence that it is arbitrary where physical analysis stops. When measuring temperature, for example, one can count the thermometer as external to the system, or as *part* of the system and hence subject to physical modelling. The same goes for human observers: one can treat the eye or the brain as external to the physical system, or as part of the physical system. Given this arbitrariness, it shouldn't matter to the predicted outcome of an experiment where we place this "cut". Von Neumann calls the principle that "the boundary between the observed system and the observer can be displaced arbitrarily" the "principle of the psycho-physical parallelism" [16, p. 421].

This is a perfectly good methodological principle, and von Neumann's proof that his "collapse on measurement" formulation of quantum mechanics satisfies it is an important demonstration that it exhibits a particular kind of self-consistency. But it isn't the same as closing the circle. Note in particular his insistence that "we must always divide the world into two parts, the one being the observed system, the other being the observer" [16, p. 420]. The observed part is subject to physical modelling by quantum mechanics; for the observer, physical modelling "is meaningless" [16, p. 420]. This division crucial to his formulation: when the observed part does not interact with the observing part, we should model the observed part using the linear Schrödinger dynamics, but when the two portions interact, we should model the observed part using the non-linear collapse dynamics. In other words, the explanation for the collapse, and thereby for our experience of measurement outcomes, necessarily lies *outside* the system modelled by quantum mechanics. Von Neumann's approach blocks the possibility of closing the circle by design.

One might think that this barrier to closing the circle is inevitable. In introducing the concept, Shimony notes that "the greatest obstacle to "closing the circle" is the ancient one which haunted Descartes and Locke—the mind–body problem," and conjectures that quantum mechanics may be "hospitable to a dualism of mind and body" [15, p. 37]. If the experiencing mind is non-physical, then of course it always lies beyond physical analysis, however far that analysis penetrates the workings of the brain. But positing dualism essentially just amounts to an admission that the circle can't be closed. And it is a peculiarly unmotivated admission: our increasing knowledge of the brain strongly suggests that the varieties of human experience have a *physical* origin. The possibility of a "hard problem" of consciousness is irrelevant here: we know empirically that various experiential states are grounded in particular brain states, even if there is a residual explanatory gap concerning the phenomenal nature of those experiential states. The problem of "closing the circle", then, is the problem of finding a way to explain those *brain states* quantum mechanically. It is here that von Neumann fails, since the quantum mechanical explanation of determinate post-measurement brain states appeals to the *deus ex machina* of interaction with an "observer".

3 Closing the Circle with Quantum Jumps

So Shimony's concern about the mind-body problem is a red herring. Setting this concern aside, Shimony gives an indication of his preferred method for closing the circle, citing the GRW approach as the "most promising to date" [15, p. 35]. By adding a stochastic "collapse" term to the Schrödinger dynamics, the GRW approach can apparently explain measurement outcomes, including human brain states, without appeal to an extra-physical "observer". The collapses ensure that for a system consisting of a large number of well-correlated particles, the wave function is, with high probability, well-localized around one point in configuration space. Since measuring devices and brains consist of large numbers of well-correlated particles, it looks like this is enough to close the circle.

However, there is a remaining gap in the circle, described here by Bell [5, p. 44]:

> There is nothing in this theory but the wavefunction. It is in the wavefunction that we must find an image of the physical world, and in particular of the arrangement of things in ordinary three-dimensional space. But the wavefunction as a whole lives in a much bigger space, of $3N$-dimensions. It makes no sense to ask for the amplitude or phase or whatever of the wavefunction at a point in ordinary space. It has neither amplitude nor phase nor anything else until a multitude of points in ordinary three-space are specified.

The GRW dynamics governs the evolution of the wave function in $3N$-dimensional space; on the face of it, it says *nothing* about a three-dimensional space inhabited by measuring instruments and brains. Certainly it can't be the *wave function* that is well-localized in three-space—that would be a mathematical category mistake. But if the GRW theory says nothing about the contents of three-space, obviously it can't close the circle.

Bell [5, p. 45] proposes a way to plug the gap:

> However, the GRW jumps (which are part of the wavefunction, not something else) are well-localized in ordinary space. Indeed, each is centred on a particular spacetime point $(\mathbf{x}, t)$ … A piece of matter then is a galaxy of such events. As a schematic psychophysical parallelism we can suppose that our personal experience is more or less directly of events in particular pieces of matter, our brains, which events are in turn correlated with events in our bodies as a whole, and they in turn with events in the outer world.

The idea is that the GRW theory does, after all, describe the contents of three-space: it describes discrete, point-like events in three-space, with one such event corresponding to the center-point of each GRW collapse. The GRW collapse dynamics ensures that these point-like "flashes" will pick out a unique reading on the measurement apparatus or a unique brain state for an observer. Note that Bell's invocation of "psycho-physical parallelism" isn't a reference to von Neumann's principle about arbitrarily moving the observer-observed boundary. Bell simply means to point out that experience covaries with a person's brain state, so if the GRW dynamics can ensure determinate brain states, it can ensure determinate experience, and thus close the circle.

Ghirardi sees the same problem, but proposes a different solution. Rather than point-like events, Ghirardi proposes that the GRW theory describes a mass density distribution in three-dimensional space, defined in terms of the configuration-space wave function $\Psi(t)$ by

$$\mathcal{M}(\mathbf{r}, t) = \langle \Psi(t) | M(\mathbf{r}) | \Psi(t) \rangle,$$

where $M(\mathbf{r})$ is the mass density operator for location $\mathbf{r}$ in three-space [9, p. 16]. This proposal has the advantage over Bell's that it is adaptable to continuous variants of the GRW theory (e.g. [10]), in which there are no discrete collapse events. As far as closing the circle goes, though, it proceeds very much like Bell's version: the GRW dynamics (discrete or continuous) makes sure that the mass density distribution picks out a unique reading of the measurement apparatus or a unique brain state of the observer [9, p. 36].

4 Primitive Ontology

Hence we have two distinct proposals for closing the circle within the GRW model, one "flashy" and one "massy". These have been described as distinct *primitive ontologies* for the GRW theory [3, p. 359]. A primitive ontology plays a dual role: it is that which the theory is *about*, and it is that which explains the properties of everyday macroscopic objects [1, p. 60]. That is, the existence of a primitive ontology allows for the possibility of "closing the circle", as it is in terms of the primitive ontology that the connection between the scientific image and the manifest image is spelled out. Indeed, Allori [1, pp. 66–69] takes the problematic nature of quantum mechanics to stem in part from the fact that it was developed *without* a primitive ontology, either because of the conviction that a realist understanding of the theory is impossible, or because of the conviction that it describes the *wave function*, which because of its high-dimensional nature is ill-suited to play the role of primitive ontology. Hence Allori [1, pp. 69–70] concludes that we should supply quantum mechanics with a primitive ontology after the fact—and in the case of the GRW theory, that means either flashes or a mass density distribution.

I do not dispute the need for a theory to "close the circle", and if primitive ontology is ontology such that the circle can be closed, I do not dispute the need for that either. But I have some qualms about the particular proposals on offer. My initial worry is methodological: How do you find out what the primitive ontology of a theory is? Allori [1, p. 63] is surely right that the primitive ontology of a theory can't simply be read off its mathematical formulation. Since the dynamical law of quantum mechanics governs the evolution of the wave function over time, much as Newton's laws govern particle positions, one might think that the ontology of quantum mechanics directly corresponds to the wave function—that quantum mechanics is *about* a wave-like entity inhabiting a high-dimensional configuration space. Allori points out that this is not how we identify primitive ontology. In the case of Newtonian mechanics, an ontology of point masses is *presupposed* as the starting point for physical theory construction, not read off the theory after the fact. Indeed, given that many disparate physical systems can be modelled using the same mathematics, reading the ontology off the mathematics seems doomed to failure. Think of the variety of applications of the mathematics of the simple harmonic oscillator!

So if we can't read the ontology off the theory, how should we proceed? Allori [1, p. 69] suggests that primitive ontologies are "proposals" about how to understand quantum mechanics. That is, we use our ingenuity to come up with an ontology that quantum mechanics *could* be about. This seems very much in the spirit of Ghirardi's proposal of a mass density ontology, and perhaps also in the spirit of Bell's proposal of a flash ontology—although note Bell's [5, p. 45] insistence that the flashes are "part of the wavefunction, not something else". The basic idea is that the primitive ontology of quantum mechanics is a separate *hypothesis*, a hypothesis that is supported to the extent that it can explain the properties of macroscopic objects, including measuring devices and human brains.

Certainly one *can* postulate a primitive ontology for quantum mechanics. But I see danger in this approach. The most obvious danger is underdetermination: there are too many competing proposals, and no way to decide between them. In addition to a mass density ontology and a flash ontology, one might propose that the quantum state describes properties of spacetime regions [17], or a collective property of a set of particles [14]. There are doubtless many other possibilities. How can we determine which is correct? Since each ontology is constructed to be fully consistent with the predictions of quantum mechanics, there is no possibility of an empirical answer. Perhaps extra-empirical virtues like explanatory power can come to the rescue here, but the historical track-record of this approach is debatable.

Furthermore, the explanatory power of the mass density and flash ontologies can be called into question. Consider, for example, a solid object whose quantum wave function is well localized in a particular region of configuration space, with "tails" extending elsewhere. The corresponding mass density distribution is large in the relevant region of 3-space, and small elsewhere. Now consider a second object passing through the region where the mass density of the first is small. How should we expect it to behave? The mass density picture suggests that the second object moves through a "sea" of rarefied matter, resulting in a small but constant force. But spontaneous collapse quantum mechanics tells us that there is no such force; instead, the small "tails" on the wave function tell us the probability of a collapse in which the first object moves discontinuously, say into the path of the second object. Rather than a small constant force, there is zero force, with a small probability of a large force.

The problem, then, is that the mass density ontology suggests a continuous effect where the reality is discontinuous. The flash ontology suffers from the opposite problem: it suggests discontinuity even in continuous cases. Consider, for example, a small object consisting of around 10^{19} particles. If each particle suffers a GRW collapse every 10^{16} s, there is a flash along the trajectory of the object roughly once every millisecond. Between these times, there is no ontology corresponding to the object whatsoever. Nevertheless, despite the discontinuity of the primitive ontology, the gravitational and electromagnetic forces exerted by the object on surrounding objects will be *continuous* in time.

One might object to both these examples that they ignore the full explanatory apparatus of quantum mechanics: quantum mechanics can explain both the discontinuous behavior in the first case and the continuous behavior in the second. That is correct, but the point is that the explanation in each case is given by the wave function and the Born rule, not the primitive ontology. The wave function may be explanatorily suspect because it inhabits a high-dimensional space [1, p. 59], but if the idea is that the primitive ontology can "provide an explanatory scheme derived along the lines of the classical one" [1, p. 70], neither the mass density ontology nor the flash ontology clearly meets the classical explanatory standard.

5 Wave Function as Structure

I have argued that simply *positing* a primitive ontology for quantum mechanics is a risky business, both because of the potential of radical ontological underdetermination, and because the proposed ontologies may fail to do the requisite explanatory work. How *should* we identify the appropriate primitive ontology, then? Allori [1, p. 63] suggests a historical approach:

> The mathematical formalism of a theory has a *history* that constrains the interpretation of its formalism: the theory started with a metaphysical position and its appropriate mathematical representation, and it continued with the implementation of the suitable mathematical apparatus necessary to determine how the primitive ontology evolves.

So, for example, Newton *begins* with an ontological posit—that there are massive objects moving in a three-dimensional space—and *then* constructs the relevant mathematical tools to represent the ontology and its temporal evolution. Rather than trying to divine the appropriate ontology by gazing at the mathematics of the final theory, we should look to the interpretation intended by the developers of that theory.

Unfortunately, though, as Allori [1, p. 67] is keenly aware, the development of quantum mechanics doesn't seem to fit this model. Although Schrödinger began with a particular interpretation of the wave function in mind—a three-dimensional field—he abandoned this interpretation when he realized that the wave function for multi-particle systems is defined on configuration space, not 3-space. With the blessing of Bohr and Heisenberg, quantum mechanics forged ahead *without* any conception of the ontology described by the mathematics.

Does this mean that we are forced to posit a primitive ontology for quantum mechanics after the fact, to remedy the oversight of its developers? Perhaps not: I think we can make some progress by considering other historical precedents. While quantum mechanics may be unique in being a theory developed in the *absence* of a primitive ontology, there are a number of historical examples of theories developed on the basis of a *mistaken* primitive ontology. Consider, for example, Fresnel's wave theory of light. Fresnel begins with an ontological posit—an all-pervasive elastic solid—and constructs a mathematical theory of transverse waves in this medium. We think Fresnel's mathematical theory was essentially correct, even though we now think there is no such elastic solid.

What should we make of cases like this? Worrall [18] takes them as evidence for *structural realism*: scientific theories tell us about the structure of the world, but not in general about what instantiates that structure. This view has a good deal of plausibility. What do we know about the ontological nature of *mass* or *charge* or *spin*, over and above the mathematical structures of the physical theories containing those terms? "Nothing" seems like an appropriate answer.

Suppose, then, that we take a structural approach to the wave function. What would that mean? It would mean endorsing the claim that the wave function correctly describes the structure of physical systems in certain contexts, but without endorsing any particular account of the kind of thing that instantiates this structure. This is not the same as wave function realism—the position that the wave function describes a fundamental entity in a high-dimensional space. Rather, the wave function describes the structure instantiated by whatever fundamental entities there may be in ordinary three-dimensional space: particles, fields, flashes, mass density, or something else entirely. A structure is not in itself an *object*, but rather a way that objects relate to each other.

Of course, "structure" is a rather vague term, and it is reasonable to ask for more details about the sense of the term "structure" as it is used here. I wish I had more to say, but for now I only have a negative characterization to give. One kind of structure the world exhibits is *nomological* structure: events exhibit regularities, and those regularities are (or are produced by) *laws*. Dürr et al. [8] suggest that wave function structure is nomological structure. But the main motivation for taking wave function structure as nomological is that there is a sense in which it can be taken to govern the dynamical evolution of the primitive ontology. The wave function fixes the motion of the particles in Bohm's theory, the evolution of the mass density distribution in mass-density GRW, and the probability distribution of flashes in flashy GRW. If one withholds from endorsing any particular primitive ontology, then there is no particular reason to think that the relationship between the wave function and the ontology is best characterized as nomological.

Furthermore, even if we endorse one of the existing proposals concerning primitive ontology, there are well-known reasons to resist thinking of the wave function as nomological: the form of the wave function depends on the nature of the system under consideration, and it changes over time [6, p. 533]. We don't usually conceive of laws this way. Of course, we can always extend our conception of law to include contingent, time-evolving laws [7, p. 3157], but such a move threatens to elide an important distinction. Suppose we apply quantum mechanics to the motion of a set of charged particles. These particles exert forces on each other according to an inverse square law, and this law is reflected in the Hamiltonian term appearing in the Schrödinger equation. The inverse square law is neither contingent nor time-evolving. One could always propose that there are two basic kinds of law, but it does less damage to standard physical thought to conceive of the wave function as a summary of the relations between the three-dimensional entities involved (whatever they may be), rather than a law governing those entities.

6 Closing the Circle—and Opening Another

My proposal, then, is that we should think of the wave function as a structure, but withhold commitment to any particular account of the ontology that instantiates this structure. And my claim is that this is enough to close the circle, at least for a spontaneous collapse theory. To appreciate what it takes to close the circle, consider again the case of Fresnel's wave theory of light. Poisson famously derived from this theory the existence of a bright spot in the center of a circular shadow. Mathematically speaking, what he actually derived was a region of high-amplitude wave structure surrounded by a region of low-amplitude wave structure. This was enough to underwrite the existence of the bright spot, even without any hypothesis concerning the nature of the ontology that instantiates the wave structure. That is, Fresnel's wave theory closes the circle: it enables us to locate observable experimental outcomes within the framework of the theory.

We can do the same with a spontaneous collapse theory. Consider a measurement of the spin of a spin-1/2 particle in which it is deflected by a magnetic field and then to one of a pair of suitably-positioned detectors, each of which responds to detection by raising a flag. If the particle is initially in a symmetric superposition of spins along the measurement direction, then the wave function of the particle plus detection apparatus evolves to a symmetric superposition in configuration space, but one term in this superposition is rapidly made many orders of magnitude larger than the other by the spontaneous collapse mechanism. That is, the vast majority of the wave function amplitude at the end of the measurement is concentrated around a particular small region of configuration space. Even in the absence of a preferred account of the underlying ontology, we know how to interpret this structure: it is the structure of things in three-space, whatever their underlying ontology may turn out to be. Ordinary objects like flags are made out of this three-dimensional ontology. That is, the high-amplitude region of configuration space tells us the locations of the two flags in three-space: it is either a structure in which the "up" flag is raised or a structure in which the "down" flag is raised.

Hence there is no need to endorse any particular account of primitive ontology so that a spontaneous collapse theory can close the circle. For macroscopic systems, spontaneous collapse picks out a particular small region of configuration space, and that region specifies the locations of ordinary objects in 3-space. Thus we can find the unique observed outcome of a measurement within the theory. More directly, since neuroscience suggests that our experience supervenes on the electrochemical configuration of our brains, the specification of a small region of configuration space also specifies human experiences.

Can we remain agnostic about primitive ontology then? I am not sure. I suspect that spontaneous collapse theories do face an explanatory problem, but that it concerns the micro-world rather than human experience. We have no serious difficulty locating measurement results and our experience of them within the structural framework of spontaneous collapse theories. The things that are more difficult to locate are the explanatory entities of classical physics—particles and fields. As Healey [12] has forcefully argued, quantum explanation is parasitic on classical concepts. We measure the spin of a spin-1/2 *particle* by passing it through a magnetic *field*. The Hamiltonian term in the Schrödinger equation contains a term corresponding to the interaction of this point-particle with the field. But how do we find these particles and fields in the wave function structure?

One might think that it is here that the proposals for primitive ontology might do some work. The flash ontology, though, is clearly of no use: microscopic systems produce no flashes over reasonable time-scales, and hence correspond to no ontology whatsoever according to the flash proposal. The mass density ontology is more promising, the obvious approach being to treat a particle as a localized region of high mass density. Ghirardi cautions against such an interpretation, though. Essentially, the problem is the microscopic analog of the explanatory worry for the mass density ontology explained in Sect. 4. When the mass density for a particle is spread out over a large region, a second particle passing through the region won't experience a continuous, small force, as the spread-out mass density would lead you to expect, but rather zero force, with a small probability of a large force ("collision"). Hence Ghirardi et al. [9, p. 18] construct a criterion for deciding when the mass density distribution is "objective", one that typically applies to macroscopic objects but not to single particles.

One could reject this latter move, and hold that microscopic systems have an associated mass density that is just as objective as that of macroscopic systems [13]. On this account, a single "particle" is really a localized region of high mass density, albeit one that can split in two or spread out. One might quite reasonably think that this is just how quantum "particles" behave: sometimes they spread out, and act more like waves. Concerning Ghirardi's worry about the behavior of regions of low mass density, Monton responds that the anomalous behavior is explained by the wave function alone, not the mass density. Monton is happy to concede that "mass density is epiphenomenal" [13, p. 419]. But while this may be acceptable for Monton's purposes, clearly mass density is not functioning as primitive ontology here.

So neither the mass density ontology nor the flash ontology adequately explains the role of particles in quantum mechanics. One possible move at this point would be to insist that the primitive ontology *is* just particles in three-dimensional space. The primitive ontology approach is flexible: in addition to the obvious particle-based quantum theory (Bohm's theory), versions of the GRW theory can be constructed that have a primitive ontology of particles [2, 4]. The challenges facing this approach are well known—most notably that it is hard to square the law governing the evolution of the particles with relativity. But this looks to me like the direction to take. That is, I submit that the real ontological puzzle of the quantum world doesn't concern human experience, but rather concerns how our physical theories—quantum, relativistic, and classical—hang together.

7 Conclusion

Ghirardi's physical and philosophical insight ran deep. He realized the need for an account of the quantum world in which it is possible to locate our experience of measurement outcomes, and unlike von Neumann, he succeeded at producing one. In fact, matters are simpler than he realized: there is no need for a mass density distribution, or any other particular account of primitive ontology, in order to close the circle, since the wave function, understood as a structure of three-dimensional things, can do the job by itself.

However, this doesn't mean that there is no further work to be done. Quantum mechanics does not only have to be hospitable to human experience; it also needs to be hospitable to classical explanation, since quantum explanation is parasitic on classical. Ghirardi's mass density ontology is of little help here. Unless, with Healey [12, p. 11], we give up thinking of quantum mechanics as *descriptive* at all, we need to look elsewhere for an understanding of the ontology of the quantum world.

References

1. Allori, Valia (2013), "Primitive ontology and the structure of fundamental physical theories," in A. Ney and D. Z. Albert (eds.), *The Wave Function: Essays on the Metaphysics of Quantum Mechanics*. New York: Oxford University Press, 58–75.
2. Allori, Valia (2019), "Scientific Realism without the wave-function: an example of naturalized quantum metaphysics," In: J. Saatsi and S. French (eds.), *Scientific Realism and the Quantum*. Oxford: Oxford University Press.
3. Allori, Valia, Sheldon Goldstein, Roderich Tumulka, and Nino Zanghí (2008), "On the common structure of Bohmian mechanics and the Ghirardi-Rimini-Weber theory," *British Journal for the Philosophy of Science* 59: 353–389.
4. Allori, Valia, Sheldon Goldstein, Roderich Tumulka, and Nino Zanghí (2014) "Predictions and primitive ontology in quantum foundations: a study of examples," *British Journal for the Philosophy of Science* 65 (2): 323–352.
5. Bell, J. S. (1987), "Are there quantum jumps?" in C. W. Kilmister (ed.), *Schrödinger: Centenary of a Polymath*, Cambridge University Press, 41–52. Reprinted in Bell, J. S. (2004), *Speakable and Unspeakable in Quantum Mechanics*, Second Edition. Cambridge: Cambridge University Press, 201–212.
6. Brown, Harvey R., and David Wallace (2005), "Solving the Measurement Problem: De Broglie-Bohm Loses Out to Everett," *Foundations of Physics* 35: 517–540.
7. Callender, Craig (2015), "One world, one beable," *Synthese* 192: 3153–3177.
8. Dürr, Detlef, Sheldon Goldstein, and Nino Zanghì (1992), "Quantum equilibrium and the origin of absolute uncertainty," *Journal of Statistical Physics* 67: 843–907.
9. Ghirardi, G. C., R. Grassi, and F. Benatti (1995), "Describing the macroscopic world: closing the circle within the dynamical reduction program," *Foundations of Physics* 25: 5–38.
10. Ghirardi, G. C., P. Pearle, and A. Rimini (1990), "Markov processes in Hilbert space and continuous spontaneous localization of systems of identical particles," *Physical Review A* 42: 78–89.
11. Ghirardi, G. C., A. Rimini and T. Weber (1986), "Unified dynamics for microscopic and macroscopic systems," *Physical Review D* 34: 470–491.
12. Healey, Richard A. (2015) "How quantum theory helps us explain," *British Journal for the Philosophy of Science* 66: 1–43.
13. Monton, Bradley (2004), "The problem of ontology for spontaneous collapse theories," *Studies in History and Philosophy of Modern Physics* 35: 407–421.
14. Monton, Bradley (2013), "Against 3N-dimensional space," in A. Ney and D. Z. Albert (eds.), *The Wave Function: Essays on the Metaphysics of Quantum Mechanics*. New York: Oxford University Press, 154–167.
15. Shimony, A. (1989), "Search for a World View that will Accommodate our Knowledge of Microphysics," in J. Cushing and E. McMullin (eds.), *Philosophical Consequences of Quantum Theory: Reflections on Bell's Theorem*, pp. 25–37. Notre Dame: Notre Dame University Press.

16. Von Neumann, J. (1955), *Mathematical Foundations of Quantum Mechanics*, Princeton University Press. Originally published (1932) as *Mathematische Grundlagen der Quantenmechanik*, Springer.
17. Wallace, David, and Christopher G. Timpson (2010), "Quantum mechanics on spacetime I: Spacetime state realism," *British journal for the philosophy of science* 61: 697–727.
18. Worrall, John (1989), "Structural realism: The best of both worlds?" *Dialectica* 43: 99–124.

Mathematical Physics

On the Complete Positivity of the Ghirardi-Rimini-Weber Model

F. Benatti and F. Gebbia

Abstract We study the complete positivity of the standard, Markovian Ghirardi-Rimini-Weber model, propose an explicitly time-dependent generalization and show that in some cases one can have complete positivity even in presence of negative-rates.

1 Introduction

From a mathematical physics point of view, the master equation associated to the so-called Ghirardi-Rimini-Weber (GRW) [1] model falls within the family of master equations that generate semigroups of completely positive and trace-preserving maps on the state space of density matrices of open quantum systems. These master equations are characterised by generators of Gorini-Kossakowski-Sudarshan-Lindblad (GKSL) form [2–5] . The main feature of the GRW model is that the noise contribution to its generator implements Gaussian spatial localizations. Far from being a simplifying mathematical nicety, the property of complete positivity, ensured by GKSL generators, is necessary for the full physical consistency of any open quantum dynamics. Indeed, it is imposed by the existence of entangled states describing physically plausible contexts where the open quantum system comes together with an arbitrary ancillary system [6, 7]. There, the request that the open system dynamics, let it be denoted by Λ_t, preserves the positivity of all time-evolving open system states (density matrices) ρ is not enough. Indeed, dynamical maps of the form $\Lambda_t \otimes \mathrm{Id}_n$ must also preserve the positivity of all entangled states of the open system statistically coupled to any ancillary n-level system, the latter being unaffected by the trivial

F. Benatti · F. Gebbia (✉)
Department of Physics, University of Trieste, Trieste, Italy

INFN, Trieste, Italy
e-mail: francesca.gebbia@ts.infn.it
e-mail: benatti@ts.infn.it

V. Allori et al. (eds.), *Do Wave Functions Jump?*, Fundamental Theories of Physics 198,
https://doi.org/10.1007/978-3-030-46777-7_11

dynamics Id_n [4, 6]. This request amounts to asking that, for all $n \geq 1$, $\Lambda_t \otimes \mathrm{Id}_n$ map any positive operator-valued matrix of the form

$$\left[X_i^\dagger X_j\right]_{i,j=1}^n = \sum_{i,j=1}^n X_i^\dagger X_j \otimes |i\rangle\langle j| \tag{1}$$

into a positive operator-valued matrix

$$\left[\Lambda_t\left[X_i^\dagger X_j\right]\right]_{i,j=1}^n = \sum_{i,j=1}^n \Lambda_t\left[X_i^\dagger X_j\right] \otimes |i\rangle\langle j| , \tag{2}$$

where X_i are any choice of n operators acting on the Hilbert space $\mathcal{H}$ of the open quantum system and $\{|i\rangle\}_{i=1}^n$ is a suitably chosen orthonormal basis in the n-level ancilla Hilbert space $\mathbb{C}^n$ [8]. By Kraus-Stinespring theorem, this property is fulfilled if and only the map Λ_t can be represented as

$$\Lambda_t[X] = \sum_{\ell\in L} L_\ell(t)\, X\, L_\ell^\dagger(t) , \tag{3}$$

with operators $L_\ell(t)$ on $\mathcal{H}$ such that $\sum_\ell L_\ell^\dagger(t)\, L_\ell(t)$ converges in a suitably chosen operator topology. When the master equation satisfied by the open system states is

$$\partial_t \rho_t = \mathbb{L}[\rho_t] , \tag{4}$$

with time-independent generator $\mathbb{L}$, then the generated dynamical maps, formally $\Lambda_t = \exp(t\mathbb{L})$, compose as a forward-in-time Markovian semigroup. If the generator is norm-bounded when acting on the algebra of bounded operators on $\mathcal{H}$, then the GKSL theorem states that the complete positivity of Λ_t is ensured if and only if the generator has the GKSL form [2, 3, 8]

$$\mathbb{L}[\rho_t] = -\frac{i}{\hbar}\left[\hat{H}\,,\,\rho_t\right] + \sum_{\theta\in\Theta}\left(K_\theta\,\rho_t\,K_\theta^\dagger - \frac{1}{2}\left\{K_\theta^\dagger\,K_\theta\,,\,\rho_t\right\}\right) , \tag{5}$$

for suitably chosen Kraus K_θ on $\mathcal{H}$.

Interestingly, for time-dependent generators $\mathbb{L}_t$, the theory of non-Markovian open dynamics [10–13] has taught that one may get completely positive solutions of

$$\partial_t \rho_t = \mathbb{L}_t[\rho_t] \tag{6}$$

even when the generator, besides the "positive" dissipative, time-dependent contribution

$$+\sum_{\theta\in\Theta}\left(K_\theta(t)\,\rho_t\,K_\theta^\dagger(t) - \frac{1}{2}\left\{K_\theta^\dagger(t)\,K_\theta(t)\,,\,\rho_t\right\}\right) \tag{7}$$

also contains a "negative" contribution of the form

$$-\sum_{\xi \in \Xi} \left(\widetilde{K}_{\xi}(t)\, \rho_t\, \widetilde{K}_{\xi}^{\dagger}(t) - \frac{1}{2} \left\{ \widetilde{K}_{\xi}^{\dagger}(t)\, \widetilde{K}_{\xi}(t)\, ,\, \rho_t \right\} \right) . \tag{8}$$

Unfortunately no general constraints on the time-dependent Kraus operators $K_{\theta}(t)$ in the "positive" dissipative contribution and $\widetilde{K}_{\xi}(t)$ in the "negative" one are known that ensure the complete positivity of Λ_t. At present, only sufficient conditions are available that stem from concrete examples.

In the following, we first generalize the GRW model inserting a particular time-dependence in its dissipative term that introduces "negative" contributions and explicitly solve it in the case of a particle in one-dimension characterized by a time-independent quadratic Hamiltonian. Then, we discuss the complete positivity of the solutions by addressing the standard, strictly Markovian, GRW model case and a time-dependent version with particularly chosen "negative" contributions.

2 GRW-Model

The so-called GRW model for one quantum particle in one spatial dimension described by position and momentum operators $\hat{q}$, $\hat{p}$ with $[\hat{q}\ ,\ \hat{p}] = i\hbar$, corresponds to the master equation

$$\partial_t \rho_t = -\frac{i}{\hbar} \left[\hat{H}, \rho_t \right] + \lambda \sqrt{\frac{\alpha}{\pi}} \int_{-\infty}^{+\infty} dx\, \mathrm{e}^{-\frac{\alpha}{2}(\hat{q}-x)^2}\, \rho_t\, \mathrm{e}^{-\frac{\alpha}{2}(\hat{q}-x)^2} - \lambda \rho_t\ . \tag{9}$$

The generator on the right hand side consists of the usual commutator of the open quantum system density matrix ρ_t with the Hamiltonian $\hat{H}$ plus a noise term

$$N(\rho_t) \equiv \lambda \sqrt{\frac{\alpha}{\pi}} \int_{-\infty}^{+\infty} dx\, e^{-\frac{\alpha}{2}(\hat{q}-x)^2}\, \rho_t\, e^{-\frac{\alpha}{2}(\hat{q}-x)^2}\ , \tag{10}$$

and a damping term $-\lambda\rho_t$ which ensures the preservation of the overall probability: $\partial_t\, \mathrm{tr}(\rho_t) = 0$. In the master equation there appear two phenomenological parameters: $\alpha^{-1/2} \geq 0$, a typical localisation length and $\lambda \geq 0$, the frequency of occurrence of the localisation processes $N[\rho_t]$; indeed, the spatial matrix elements

$$\langle q_1 | N(\rho_t) | q_2 \rangle = \lambda\, \mathrm{e}^{-\frac{\alpha}{4}(q_1-q_2)^2}\, \langle q_1 | \rho_t | q_2 \rangle\ , \tag{11}$$

where $\hat{q}|q_{1,2}\rangle = q_{1,2}|q_{1,2}\rangle$, are strongly suppressed when $|q_1 - q_2| \geq 1/\sqrt{\alpha}$.

Notice that the above expression is a continuous form of the GKSL equations (5) where the index set Θ and the running index $\theta \in \Theta$ have become $\mathbb{R}$ and $x \in \mathbb{R}$ while the sum has turned into an integral with $K_{\theta} = \mathrm{e}^{-\alpha(\hat{q}-x)^2/2}$.

Before seeking explicitly the dynamical maps Λ_t solutions to (9), let us observe that, by Fourier transforming,

$$\mathrm{e}^{-\frac{\alpha}{4}(q_1-q_2)^2} = \frac{1}{\sqrt{\pi}}\int_{-\infty}^{+\infty} du\, \mathrm{e}^{i\,u\,\sqrt{\alpha}(q_1-q_2)}\, \mathrm{e}^{-u^2},$$

one rewrites the noise term as

$$N(\rho_t) = \frac{\lambda}{\sqrt{\pi}}\int_{-\infty}^{+\infty} du\, \mathrm{e}^{-u^2}\, \mathrm{e}^{i\,\sqrt{\alpha}\,u\,\hat{q}}\, \rho_t e^{-i\,\sqrt{\alpha}\,u\,\hat{q}}\,. \tag{12}$$

The spatial localization can then be interpreted as the effect of momentum translations by a normally distributed random momentum $\hbar\sqrt{\alpha}\,u$. Such an expression of $N[\rho_t]$ suggests a possible generalization of the GRW model that may account for non-Markovian, memory effects. Indeed, by substituting $\lambda\,\frac{1}{\sqrt{\pi}}\,\mathrm{e}^{-u^2}$ with a real function $\lambda_t(u)$ explicitly depending on both time t and u such that $\lambda_t := \int_{-\infty}^{+\infty} du\,\lambda_t(u)$ exists, the GRW master equation becomes

$$\partial_t\rho_t = -\frac{i}{\hbar}\left[\hat{H},\rho_t\right] + \int_{-\infty}^{+\infty} du\,\lambda_t(u)\,\mathrm{e}^{i\,\sqrt{\alpha}\,u\,\hat{q}}\,\rho_t\,\mathrm{e}^{-i\,\sqrt{\alpha}\,u\,\hat{q}} - \lambda_t\rho_t\,. \tag{13}$$

In order to preserve hermiticity, one has to impose the condition

$$\lambda_t(u) = \lambda_t(-u) \qquad \forall\, t \geq 0\,. \tag{14}$$

In order to keep the analytical difficulties to a minimum, we shall also ask $\lambda_t(u)$ to possess a well-defined Fourier transform

$$\widetilde{\lambda}_t(v) := \int_{-\infty}^{+\infty} du\,\mathrm{e}^{i\,u\,v}\,\lambda_t(u)\,, \tag{15}$$

which is then real for all $t \geq 0$. Of course, one recovers the standard Markovian GRW model (9) by eliminating the explicit time-dependence and setting

$$\lambda_t(u) = \frac{\lambda}{\sqrt{\pi}}\,\mathrm{e}^{-u^2}\,. \tag{16}$$

In the generalized GRW model, $\lambda_t(u)$ need not in principle be a positive definite function of t and u; however, one has to check that the maps Λ_t solutions to (13) are positive and also, as briefly sketched in the Introduction, completely positive. Already the first request limits the choice of possible non-positive frequency-like functions $\lambda_t(u)$ as indicated by the following finite-dimensional example.

Example 1 (Non-Markovian Interlude) Consider the one-qubit, purely dissipative master equation

$$\partial_t \rho_t = \lambda(t)\,(\sigma_z\,\rho_t\,\sigma_z \;-\; \rho_t) \qquad \rho_t = \frac{1}{2}\left(\hat{\mathbb{I}} + \boldsymbol{r}_t \cdot \hat{\boldsymbol{\sigma}}\right) \tag{17}$$

where $\lambda(t)$ is a smooth function of time and $\boldsymbol{r}_t$ is the Block vector representing the qubit density matrix ρ_t by a point within the unit sphere in $\mathbb{R}^3$, while $\boldsymbol{\sigma} = (\sigma_1, \sigma_2, \sigma_3)$ is the vector of Pauli matrices. The time dependence of $\boldsymbol{r}_t$ follows from the equations

$$\dot{r}_{1,2}(t) \;=\; -\,2\,\lambda(t)\,r_{1,2}(t)\;, \quad \dot{r}_3(t) = 0\;, \tag{18}$$

by insertion of the Block form of ρ_t into the master equation. They are readily solved by

$$r_{1,2}(t) \;=\; r_{1,2}\,\mathrm{e}^{-2L(t)}\;, \quad r_3(t) = r_3\;, \quad L(t) := \int_0^t \mathrm{d}s\,\lambda(s)\;, \tag{19}$$

so that

$$\rho \mapsto \rho_t = \Lambda_t[\rho] = \frac{1}{2}\left(\hat{\mathbb{I}} + \mathrm{e}^{-2L(t)}(r_1\,\hat{\sigma}_1 + r_2\,\hat{\sigma}_2) + r_3\,\hat{\sigma}_3\right)\;. \tag{20}$$

Since the Block vector must remain of norm ≤ 1, one must impose $L(t) \geq 0$ for all $t \geq 0$, whereas the function $\lambda(t)$ may also be negative. By inspecting the eigenvalues of the Choi-Jamiolkowski matrix [14]

$$M_t := \Lambda_t \otimes \mathrm{Id}[P_{sym}] = \frac{1}{2}\begin{pmatrix} 1 & 0 & 0 & \mathrm{e}^{-2L(t)} \\ 0 & 0 & 0 & 0 \\ 0 & 0 & 0 & 0 \\ \mathrm{e}^{-2L(t)} & 0 & 0 & 1 \end{pmatrix}\;, \tag{21}$$

where

$$P_{sym} = \frac{1}{2}\left(\hat{\mathbb{I}} \otimes \hat{\mathbb{I}} + \hat{\sigma}_1 \otimes \hat{\sigma}_1 - \hat{\sigma}_2 \otimes \hat{\sigma}_2 + \hat{\sigma}_3 \otimes \hat{\sigma}_3\right)\;,$$

one sees that M_t is positive semi-definite and thus Λ_t completely positive. In the standard semigroup setting of the GKSL theorem, $\lambda(t) = \lambda \geq 0$ is necessary and sufficient for Λ_t being completely positive, in the non-Markovian setting, $\lambda(t)$ can be negative and yet Λ_t may be completely positive.

In the example above, the positive rates $\lambda(t)$ are interpreted as the (time-dependent) frequencies at which the system loses coherence because of the presence of the environment. Such loss of coherence corresponds to information about the open system which is lost into the environment; therefore, negative $\lambda(t)$ are associated to a gain in coherence which is due to information back-flowing from the environment into the open system [15–17]. A similar interpretation is applicable to the "positive" and "negative" contributions to the generators in (7), respectively (8).

3 Explicit Solution to the Generalized GRW Model

In order to solve explicitly the generalized GRW model (13), we restrict to quadratic Hamiltonians,

$$\hat{H} = A\hat{p}^2 + B\,\hat{p}\hat{q} + B^*\,\hat{q}\hat{p} + C\hat{q}^2\,; \qquad A, C \in \mathbb{R}\,,\ B \in \mathbb{C}\,. \tag{22}$$

They generate unitary Heisenberg time-evolutions $\hat{U}_t = \mathrm{e}^{-i\,\hat{H}\,t/\hbar}$ such that

$$\begin{cases} \hat{q}_t = \hat{U}_t^\dagger\,\hat{q}\,\hat{U}_t = a_t\hat{q} + b_t\hat{p} \\ \hat{p}_t = \hat{U}_t^\dagger\,\hat{p}\,\hat{U}_t = c_t\hat{q} + d_t\hat{p} \end{cases} \quad \text{with initial conditions} \quad \begin{cases} a_0 = 1,\ b_0 = 0 \\ c_0 = 0,\ d_0 = 1 \end{cases} \tag{23}$$

and $a_t d_t - b_t c_t = 1$. We remove the Hamiltonian contribution to the GRW generator by defining

$$\widetilde{\rho}_t = \hat{U}_t^\dagger\,\rho_t\,\hat{U}_t\,, \tag{24}$$

whence the GRW equation (13) becomes:

$$\partial_t \widetilde{\rho}_t = -\lambda_t\,\widetilde{\rho}_t + \int_{-\infty}^{+\infty} du\,\lambda_t(u)\,\hat{W}(\hbar\sqrt{\alpha}\,u\,b_t, -\hbar\sqrt{\alpha}\,u\,a_t)\,\widetilde{\rho}_t\,\hat{W}^\dagger(\hbar\sqrt{\alpha}\,u\,b_t, -\hbar\sqrt{\alpha}\,u\,a_t)\,, \tag{25}$$

where we have introduced the Weyl operators

$$\hat{W}(x, y) = \mathrm{e}^{\frac{i}{\hbar}(x\hat{p} - y\hat{q})}\,, \qquad x\,,\ y \in \mathbb{R}\,, \tag{26}$$

and used that the quadratic Hamiltonian is such that

$$\hat{U}_t^\dagger\,\hat{W}(x, y)\,\hat{U}_t = \hat{W}(xd_t - yb_t, ya_t - xc_t)\,, \tag{27}$$

namely, such that the unitary dynamics maps Weyl operators into Weyl operators.

3.1 Solution in the Heisenberg Picture

Under the generalized GRW model (13), a generic initial Weyl operator $\hat{W}(x, y)$ evolves in time into $\widetilde{W}_t(x, y)$, in general not a Weyl operator, by means of the dual version of (25) which is obtained from the duality relation

$$\mathrm{tr}\left(\widetilde{\rho}_t\,\hat{W}(x, y)\right) = \mathrm{tr}\left(\rho\,\widetilde{W}_t(x, y)\right)\,, \tag{28}$$

Notice that (24) gives $\widetilde{\rho}_{t=0} = \rho$ and that the dynamics $\hat{W}(x, y) \mapsto \hat{W}_t(x, y)$ of the Weyl operators under the dual of the master equation (13) follows from

$$\hat{W}_t(x, y) = \hat{U}_t^\dagger \, \widetilde{W}_t(x, y) \, \hat{U}_t \, , \tag{29}$$

where $\widetilde{W}_t(x, y)$ is solution to the master equation

$$\partial_t \widetilde{W}_t(x, y) = -\lambda_t \widetilde{W}_t(x, y) + \int_{-\infty}^{+\infty} du \, \lambda_t(u) \, \hat{W}(-\hbar\sqrt{\alpha}\, u \, b_t, \hbar\sqrt{\alpha}\, u \, a_t) \, \widetilde{W}_t(x, y) \, \hat{W}^\dagger(-\hbar\sqrt{\alpha}\, u \, b_t, \hbar\sqrt{\alpha}\, u \, a_t) \, . \tag{30}$$

In the Appendix we show that the solution $\widetilde{W}_t(x, y)$ is given by

$$\widetilde{W}_t(x, y) = \exp\Big(-\kappa_t + \int_0^t ds \, \widetilde{\lambda}_s(b_s y + a_s x)\Big) \, \hat{W}(x, y) \quad \text{where} \tag{31}$$

$$\kappa_t = \int_0^t ds \, \lambda_s = \int_0^t ds \int_{-\infty}^{+\infty} du \, \lambda_s(u) \, , \quad \widetilde{\lambda}_s(b_s y + a_s x) := \int_{-\infty}^{+\infty} du \, \lambda_s(u) \, \mathrm{e}^{i\sqrt{\alpha}\, u \, (b_s y + a_s x)} \, , \tag{32}$$

or, equivalently,

$$\widetilde{W}_t(x, y) = \int_{\mathbb{R}^2} \frac{d\overline{x} \, d\overline{y}}{(2\pi\hbar)^2} \, \widetilde{G}_t(\overline{x}, \overline{y}) \, \hat{W}^\dagger(\overline{x}, \overline{y}) \, \hat{W}(x, y) \, \hat{W}(\overline{x}, \overline{y}) \quad \text{where} \tag{33}$$

$$\widetilde{G}_t(\overline{x}, \overline{y}) := \int_{\mathbb{R}^2} \frac{du \, dv}{(2\pi\hbar)^2} \, \mathrm{e}^{\frac{i}{\hbar}(\overline{y}u - \overline{x}v)} \, G_t(u, v) \, . \tag{34}$$

It thus follows from (29) that the full GRW dynamics of Weyl operators reads

$$\hat{W}(x, y) \longmapsto \hat{W}_t(x, y) = \exp\Big(-\kappa_t + \int_0^t ds \, \widetilde{\lambda}_S(b_s y + a_s x)\Big) \, \hat{U}_t^\dagger \, \hat{W}(x, y) \, \hat{U}_t \, . \tag{35}$$

Together with (27), the expression (35) shows that Weyl operators are transformed into Weyl operators multiplied by the exponential function

$$G_t(x, y) := \exp\Big(-\kappa_t + \int_0^t ds \, \widetilde{\lambda}_s(b_s y + a_s x)\Big) \, . \tag{36}$$

Notice that, given the Heisenberg dynamics of Weyl operators, the Schrödinger time-evolution of density matrices solution to (9) is given, via (28), by

$$\rho_t = \int_{\mathbb{R}^2} \frac{d\overline{x} \, d\overline{y}}{(2\pi\hbar)^2} \, \widetilde{G}_t(\overline{x}, \overline{y}) \, \hat{W}(\overline{x}, \overline{y}) \, \hat{U}_t \, \rho \, \hat{U}_t^\dagger \, \hat{W}^\dagger(\overline{x}, \overline{y}) \, . \tag{37}$$

Since the unitary maps $\hat{W}(x, y) \mapsto \hat{U}_t^\dagger \, \hat{W}(x, y) \, \hat{U}_t$ are completely positive and any composition of completely positive maps is completely positive, from now on we focus the attention on the maps

$$\Lambda_t : \hat{W}(x, y) \mapsto \Lambda_t[\hat{W}(x, y)] := \widetilde{W}_t(x, y) = G_t(x, y) \, \hat{W}(x, y) \, . \tag{38}$$

4 Complete Positivity of the Generalized GRW Model

As much as (33), expression (37) is a continuous version of the Kraus-Stinespring decomposition of completely positive maps as in (3) only if the function $\widetilde{G}_t(\overline{x}, \overline{y})$ is non-negative almost everywhere so that one can absorb $\sqrt{\widetilde{G}_t(\overline{x}, \overline{y})}$ into the Weyl operators and thus get a "positive" continuous diagonal expression similar to (7). Furthermore, $G_t(x, y)$ goes into $\mathrm{e}^{-\kappa_t}$ when $x, y \to \infty$; therefore, its Fourier transform $\widetilde{G}_t(\overline{x}, \overline{y})$ does not exist as a function, but has a meaning as a distribution. It is then the integral of $\widetilde{G}_t(\overline{x}, \overline{y})$ over suitable positive test functions that must return non negative values.

Before explicitly checking whether Λ_t is completely positive or not, one first ascertains that Λ_t is unital as it indeed should be: namely, $\Lambda_t[\hat{\mathbb{I}}] = \hat{\mathbb{I}}$. Also, like in the qubit case of Example 1, the function κ_t in (32) must satisfy $\kappa_t \geq 0$ for all $t \geq 0$. Indeed, in order to be completely positive, the maps Λ_t must be Schwartz-positive [8]; namely, for all $\hat{X}$ in the Weyl algebra, it must hold that

$$\Lambda_t\left[\hat{X}^\dagger \hat{X}\right] \geq \Lambda_t\left[\hat{X}^\dagger\right] \Lambda\left[\hat{X}\right]. \tag{39}$$

With $\hat{X} = \hat{W}(x, y)$, from (38) one gets

$$\Lambda_t\left[\hat{W}^\dagger(x, y)\, \hat{W}(x, y)\right] = \Lambda_t\left[\hat{\mathbb{I}}\right] = \hat{\mathbb{I}} \geq \Lambda_t\left[\hat{W}^\dagger(x, y)\right] \Lambda_t\left[\hat{W}(x, y)\right] = G_t^2(x, y)\,\hat{\mathbb{I}}\,. \tag{40}$$

Hence $G_t(x, y)$ must satisfy $G_t(x, y) \leq 1$ whence

$$\kappa_t \geq \int_0^t ds\, \widetilde{\lambda}_s\left(a_s x + b_s y\right) \qquad \forall x, y \in \mathbb{R}\,. \tag{41}$$

If κ_t is allowed to become negative, then, because of the Riemann-Lebesgue lemma [18] ($\lambda_t(u)$ has indeed been assumed to be square-integrable), $\lim_{x\to\pm\infty} \widetilde{\lambda}_t(x) = 0$, whence the inequality above can be violated for suitably large x or y or both.

According to Eq. (2) in the Introduction and using the fact that any operator on $\mathcal{H}$ can be written as a linear combination of Weyl operators, complete positivity can be reduced to asking that for all $n \geq 1$ and all choices of $(x_i, y_i) \in \mathbb{R}^2$,

$$\Lambda_t \otimes \mathrm{Id}_n\,[\mathcal{W}_n] \quad \text{with} \quad \mathcal{W}_n = \left[\hat{W}^\dagger(x_i, y_i)\hat{W}(x_j, y_j)\right]_{i,j=1}^n \tag{42}$$

be a positive matrix. Notice that the entries of $\mathcal{W}_n$ are elements of the Weyl algebra and act on vectors $|\Psi\rangle \in \mathcal{H} \otimes \mathbb{C}^n$ with n components $|\psi_i\rangle \in \mathcal{H}$, the Hilbert space of square integrable functions on $\mathbb{R}$. Explicitly, the entries of $\Lambda_t \otimes \mathrm{Id}_n\,[\mathcal{W}_n]$ have the form

$$\Lambda_t\left[\hat{W}^\dagger(x_i,y_i)\hat{W}(x_j,y_j)\right] = \mathrm{e}^{\frac{i}{2\hbar}(x_iy_j-x_jy_i)}\,\Lambda_t\left[\hat{W}(x_j-x_i,y_j-y_i)\right]$$
$$= G_t(x_j-x_i,y_j-y_i)\,\hat{W}^\dagger(x_i,y_i)\hat{W}(x_j,y_j)\,. \quad (43)$$

Taking the expectation of $\Lambda_t\otimes\mathrm{Id}_n\,[\mathcal{W}_n]$ with respect to any $|\Psi\rangle\in\mathcal{H}\otimes\mathbb{C}^n$, complete positivity entails

$$\sum_{i,j} G_t(x_j-x_i,y_j-y_i)\,\langle\psi_i|\hat{W}^\dagger(x_i,y_i)\hat{W}(x_j,y_j)|\psi_j\rangle \geq 0 \quad (44)$$

for any number and choice of $|\psi_j\rangle\in\mathcal{H}$ and $(x_j,y_j)\in\mathbb{R}^2$. Inserting the explicit expression of $G_t(x,y)$ into (44), we have thus to check whether

$$\Delta_\psi(t) := \sum_{i,j}\exp\left(\int_0^t ds\,\widetilde{\lambda}_s\Big(a_s(x_j-x_i)+b_s(y_j-y_i)\Big)\right)\langle\psi_i|\hat{W}^\dagger(x_i,y_i)\hat{W}(x_j,y_j)|\psi_j\rangle \geq 0\,. \quad (45)$$

By means of the exponential series and by making explicit the Fourier transforms $\widetilde{\lambda}_s\Big(a_s(x_j-x_i)+b_s(y_j-y_i)\Big)$, one rewrites

$$\Delta_\psi(t) = \sum_{k=0}^\infty\frac{1}{k!}\prod_{\ell=1}^k\int_0^t ds_\ell\int_{\mathbb{R}} du_\ell\,\lambda_{s_\ell}(u_\ell)\,H(\mathbf{x}_n,\mathbf{y}_n,\mathbf{s}_k,\mathbf{u}_k,\Psi_n) \quad (46)$$

$$H(\mathbf{x}_n,\mathbf{y}_n,\mathbf{s}_k,\mathbf{u}_k,\Psi_n) := \left\|\sum_j\exp\left(i\sqrt{\alpha}\left(y_j\sum_{p=1}^k u_p b_{s_p}+x_j\sum_{p=1}^k u_p a_{s_p}\right)\right)\hat{W}(x_j,y_j)|\psi_j\right\|^2 \geq 0\,, \quad (47)$$

where $\mathbf{x}_n := (x_1,x_2,\ldots,x_n)$ and analogously for $\mathbf{y}_n$, $\mathbf{s}_k := (s_1,s_2,\ldots,s_k)$ and analogously for $\mathbf{u}_k$, while Ψ_n stands for the vector $|\Psi\rangle\in\mathcal{H}\otimes\mathbb{C}^n$.

Because of the integration of the multiple products of the "non-positive" rate functions $\lambda_{s_\ell}(u_\ell)$, controlling the sign of the series in (46) is an extremely difficult task: in the following we consider a few cases that, besides allowing for definite answers, permit to shed light on some salient features of the problem.

4.1 Standard GRW Model for a Free Particle

We restrict to the free Hamiltonian $\hat{H}=\dfrac{\hat{p}^2}{2m}$; then, in (23), $a_t=d_t=1$, $b_t=\dfrac{t}{m}$, $c_t=0$ and

$$\widetilde{W}_t(x,y) = G_t(x,y)\,\hat{W}(x,y)\,,\quad G_t(x,y) = \exp\left(-\kappa_t+\int_0^t ds\,\widetilde{\lambda}_s\left(x+\frac{y}{m}s\right)\right). \quad (48)$$

First, we focus upon the standard GRW model by choosing $\lambda_t(u)$ as in (16), whence

$$G_t(x, y) = \exp\left(-\lambda t + \lambda \int_0^t ds\, \mathrm{e}^{-\frac{\alpha}{4}\left(x+\frac{y}{m}s\right)^2}\right) . \tag{49}$$

It follows that Schwartz positivity (40) is fulfilled; indeed, $\kappa_t = \lambda\, t \geq 0$ for all $t \geq 0$ and

$$t - \int_0^t ds\, e^{-\frac{\alpha}{4}\left(x+\frac{y}{m}s\right)^2} \geq 0 \qquad \forall x, y \in \mathbb{R} . \tag{50}$$

On the other hand,

$$\Delta_\psi(t) = \sum_{k=0}^{\infty} \frac{\lambda^k}{k!} \left(\prod_{\ell=1}^{k} \int_{\mathbb{R}} du_\ell\, \frac{\mathrm{e}^{-u_\ell^2}}{\sqrt{\pi}}\right) \int_0^t ds_1 \cdots \int_0^t ds_k\, H(\mathbf{x}_n, \mathbf{y}_n, \mathbf{s}_k, \mathbf{u}_k, \Psi_n) , \tag{51}$$

is positive for all $|\psi_i\rangle$, $|\psi_j\rangle \in \mathcal{H}$ and (x_i, y_i), $(x_j, y_j) \in \mathbb{R}^2$ since $H(\mathbf{x}_n, \mathbf{y}_n, \mathbf{s}_k, \mathbf{u}_k, \Psi_n) \geq 0$, whence the GRW dynamics of a free particle is explicitly completely positive.

4.2 Generalized GRW Model for a Free Particle: Singular Rate Function

The existing literature on non-Markovian dynamics makes it clear how difficult it is to arrive at general prescriptions for negative rates in master equations as (13) that nevertheless ensure the complete positivity of the generated maps Λ_t. In this section we content ourselves with applying the approach developed in the previous section for particular functions $\lambda_t(u)$ that however illustrate some of the features of the problem.

In order to appreciate the main changes with respect to the argument developed for the Markovian case above when negative rates are present, we stay with the free Hamiltonian and choose

$$\lambda_t(u) = \lambda_t\, \frac{\mathrm{e}^{-\frac{u^2}{\alpha\hbar^2}}}{\sqrt{\pi\alpha\hbar^2}} , \tag{52}$$

with rate function consisting of one positive spike of strength Δ_1 followed by a negative one of strength Δ_2:

$$\lambda_t = \lambda\, \Delta_1\, \delta(t - T_1) - \lambda\, \Delta_2\, \delta(t - T_2) , \quad T_2 > T_1 , \quad \Delta_{1,2} > 0 , \quad \lambda > 0 . \tag{53}$$

Then, $\kappa_t = \lambda(\Delta_1 - \Delta_2)$ and the function $G_t(x, y)$ in (49) reads

$$G_t(x, y) = \exp\left(-\lambda(\Delta_1 - \Delta_2) + \lambda\, \Delta_1 \exp\left(\mathrm{e}^{\frac{-\alpha}{4(x+yT_1/m)^2}}\right) - \lambda\, \Delta_2 \exp\left(\mathrm{e}^{\frac{-\alpha}{4(x+yT_2/m)^2}}\right)\right) . \tag{54}$$

Certainly, in order to fulfill $\kappa_t \geq 0$ for large x and y, one has to choose $\Delta_1 \geq \Delta_2$. However, despite of this constraint, choosing $x = -yT_1/m$ yields

$$G_t(-yT_1/m, y) = \exp\left(\lambda\,\Delta_2\left(1 - \mathrm{e}^{-\alpha\, y^2 (T_2 - T_1)^2/(4m^2)}\right)\right) \geq 1\,, \tag{55}$$

thus contradicting condition (41) which necessarily follows from complete positivity. In fact, the generated dynamical maps Λ_t are not even positive. Indeed, positivity of Λ_t requires that

$$\begin{aligned} A &:= \langle\psi|\,\Lambda_t\left[\left(1 + \mathrm{e}^{i\phi}\,\hat{W}(x, y)\right)^{\dagger}\left(1 + \mathrm{e}^{i\phi}\,\hat{W}(x, y)\right)\right]|\psi\rangle \\ &= 2\left(1 + G_t(x, y)\,\mathrm{Re}\left\{\mathrm{e}^{i\phi}\,\langle\psi|\,\hat{W}(x, y)\,|\psi\rangle\,\right\}\right) \geq 0 \end{aligned} \tag{56}$$

for all Hilbert space vectors $|\psi\rangle$ and for all choices of $\phi \in [0, 2\pi]$ and $(x, y) \in \mathbb{R}^2$. With $x = -yT_1/m$,

$$A = 2\left(1 + \mathrm{e}^{\lambda\Delta_2\left(1 - \mathrm{e}^{-\frac{\alpha}{4m^2}y^2(T_2 - T_1)^2}\right)}\,\mathrm{Re}\left\{\mathrm{e}^{i\phi}\,\langle\psi|\,\hat{W}(-yT/m, y)\,|\psi\rangle\,\right\}\right)\,. \tag{57}$$

Finally, choosing $|\psi\rangle$ corresponding to the standard Gaussian $\psi(x) = \mathrm{e}^{-x^2/2}/\sqrt[4]{\pi}$ and suitably setting ϕ, one derives the necessary condition

$$A = 2\left(1 - \mathrm{e}^{\lambda\,\Delta_2\left(1 - \mathrm{e}^{-\frac{\alpha}{4m^2}\,y^2(T_2 - T_1)^2}\right)}\,\mathrm{e}^{-y^2\frac{T_1^2\hbar^2 + m^2}{4m^2\hbar^2}}\right) \geq 0\,. \tag{58}$$

Choosing y very small and expanding to first order in y^2, we obtain the approximation

$$A \simeq \frac{(y_1 - y_2)^2}{2m^2\hbar^2}\left(T_1^2\hbar^2 + m^2 - \lambda\Delta_2\alpha\hbar^2\,(T_2 - T_1)^2\right)\,, \tag{59}$$

whose right-hand side can be made negative by choosing either T_2 or Δ_2 large enough (and accordingly setting y in order to perform the expansion). This example shows how delicate is the trade-off between the time-integral of the time-dependent spatial Gaussian localisations and the width of the time-intervals where the rate function becomes negative together with the amount of such negativity.

4.3 Generalized GRW Model for a Free Particle: Small α Limit

The second example uses again (52) but with the following rate function:

$$\lambda_t = \begin{cases} \lambda \geq 0 \ : & 0 \leq t < T - \Delta \\ -\lambda \leq 0 : & T - \Delta \leq t \leq T + \Delta \\ \lambda \geq 0 \ : & t > T + \Delta \end{cases} . \tag{60}$$

Namely, according to the comment after Example 1, the localisation processes always dissipate information about the system into the environment responsible for them, apart from an interval of time $[T - \Delta, T + \Delta]$ when information comes back from the environment into the system. In order to make the complete positivity issue addressable, we consider the approximation to the the generalised GRW-model that follows by considering a very large localization length, namely a very small parameter α which yields

$$G_t(x, y) = \exp\left(-\frac{\alpha}{4}\int_0^t ds\,\lambda_s\left(x + \frac{y}{m}s\right)^2\right) . \tag{61}$$

Notice that the small α expansion of the generalised GRW model (12) yields

$$\partial_t \rho_t = -\frac{i}{\hbar}\left[\hat{H}\,,\,\rho_t\right] - \frac{\lambda_t\,\alpha}{4}\,[\hat{q}\,,\,[\hat{q}\,,\,\rho_t\,]] . \tag{62}$$

Then, inserting the expression for $G_t(x, y)$ in (61) into (37), $\Delta_\psi(t)$ in (45) relative to the solution of the above master equation reduces to

$$\Delta_\psi(t) := \int_{\mathbb{R}^4} \frac{dx\,dy\,du\,dv}{(2\pi\hbar)^2}\,e^{i/\hbar(yu-xv)}\,e^{-\alpha/4\,g_{0,t}(u,v)}\left\|\sum_j e^{-i/\hbar(yx_j-xy_j)}\,\hat{W}(x_j, y_j)|\psi_j\rangle\right\|^2 , \tag{63}$$

whence no series expansion is necessary and one can directly focus on the quadratic form

$$g_{0,t}(u, v) := A_{0,t}\,u^2 + C_{0,t}\,v^2 + 2\,B_{0,t}\,u\,v , \tag{64}$$

where, with λ_t as in (60),

$$A_{0,t} := \int_0^t \mathrm{d}s\,\lambda_s = \lambda\,t\,, \quad B_{0,t} := \frac{1}{m}\int_0^t \mathrm{d}s\,s\,\lambda_s = \frac{\lambda\,t^2}{2m}\,, \quad C_{0,t} := \frac{1}{m^2}\int_0^t \mathrm{d}s\,s^2\,\lambda_s = \frac{\lambda\,t^3}{3m^2} . \tag{65}$$

Observe that integrating over intervals $0 \leq t_1 \leq t_2$ where λ does not change sign yields

$$A_{t_1,t_2} = \lambda\,(t_2 - t_1)\,, \quad B_{t_1,t_2} = \frac{\lambda\,(t_2^2 - t_1^2)}{2m}\,, \quad C_{t_1,t_2} = \frac{\lambda(t_2^3 - t_1^3)}{3m^2} . \tag{66}$$

Then, the determinant of the quadratic form

$$g_{t_1,t_2}(u,v) := A_{t_1,t_2}\,u^2 + C_{t_1,t_2}\,v^2 + 2\,B_{t_1,t_2}\,u\,v \tag{67}$$

is always positive, while the trace has the sign of λ. It thus follows that $g_{0,t}(u,v)$ is positive semi-definite for $0 \leq t \leq T - \Delta$, while it decreases in the interval $T - \Delta \leq t \leq T + \Delta$ where $\lambda < 0$ and increases again for $t > T + \Delta$. Therefore, $g_{0,t}(u,v) \geq 0$ at all $t \geq 0$ if $g_{0,T+\Delta}(u,v) \geq 0$. One computes

$$A_{0,T+\Delta} = \lambda\,(T - 3\Delta)\,,\quad B_{0,T+\Delta} = \frac{\lambda}{m}\left(T^2 + \Delta^2 - 6T\Delta\right),\quad C_{0,T+\Delta} = \frac{\lambda}{3m^2}\left(T^3 - 3\Delta^3 + 3T\Delta^2 - 9T^2\Delta\right). \tag{68}$$

Clearly, for sufficiently far away negativity interval ($T \gg 1$) or sufficiently low negativity ($\Delta \ll 1$), trace and determinant become positive. Then, $g_{0,T+\Delta}(u,v)$ and thus all $g_{0,t}(u,v)$, $t \geq 0$, are positive semi-definite with positive Gaussian Fourier transforms yielding $\Delta_\psi(t) \geq 0$ for all $t \geq 0$ and thus entailing the complete positivity of all generated maps Λ_t.

5 Conclusions

In the spirit of non-Markovian open quantum systems that evolve according to physically consistent completely positive maps and, at the same time, show negative rates and back-flow of information from environment to system, we have proposed a time-dependent generalisation of the Ghirardi-Rimini-Weber model and set the technical framework for discussing the complete positivity of the dynamical maps it generates. After explicitly proving the latter property in the case of the standard, Markovian GRW model, we have presented an example of a singular negative rate consisting of one positive spike and a subsequent negative one that does not allow either for completely positive or positive solutions. Instead, we considered the small α expansion of the generalised GRW model that yields completely positive solutions, thus providing a scenario where to study a *bona fide* back-flow of information in the context of spatial localisations.

Acknowledgments The author F.B. acknowledges that his research has been conducted within the framework of the Trieste Institute for Theoretical Quantum Technologies.

Appendix

Products of Weyl operators are proportional to Weyl operators,

$$\hat{W}(x_1, y_1)\hat{W}(x_2, y_2) = e^{-\frac{i}{2\hbar}(x_1 y_2 - x_2 y_1)}\,\hat{W}(x_1 + x_2, y_1 + y_2)\,. \tag{69}$$

Thus, they span linearly the so-called Weyl algebra whose closure in the strong-operator topology yields the algebra of bounded operators on the Hilbert space $\mathcal{H}$ of square-integrable functions [19]. We shall then seek the solution $\widetilde{W}_t(x, y)$ to the master equation (30) as a linear combination of Weyl operators,

$$\widetilde{W}_t(x, y) = \int_{-\infty}^{+\infty} d\overline{x}\, d\overline{y}\, F_t^{xy}(\overline{x}, \overline{y})\, \hat{W}(\overline{x}, \overline{y}) , \tag{70}$$

where the scalar function $F_t^{xy}(\overline{x}, \overline{y})$ becomes the unknown to be found. Insertion of such an expression into the left and right hand sides of (30) yields

$$\int_{-\infty}^{+\infty} d\overline{x}\, d\overline{y} \left(\partial_t F_t^{xy}(\overline{x}, \overline{y}) + \lambda_t F_t^{xy}(\overline{x}, \overline{y}) - \widetilde{\lambda}_t(b_t\overline{y} + a_t\overline{x})\, F_t^{xy}(\overline{x}, \overline{y})\right) \hat{W}(\overline{x}, \overline{y}) = 0 ,$$

where we have used the fact that (69) implies

$$\hat{W}(x_1, y_1)\hat{W}(x, y)\hat{W}^\dagger(x_1, y_1) = \mathrm{e}^{\frac{i}{\hbar}(xy_1 - yx_1)}\, \hat{W}(x, y) , \tag{71}$$

and we have set (see (15))

$$\widetilde{\lambda}_t(b_t\overline{y} + a_t\overline{x}) := \int_{-\infty}^{+\infty} du\, \lambda_t(u)\, \mathrm{e}^{i\sqrt{\alpha}\, u\, (b_t\overline{y} + a_t\overline{x})} .$$

Finally, using that $\mathrm{tr}\left(\hat{W}(x, y)\hat{W}(-\overline{x}, -\overline{y})\right) = 2\pi\hbar\, \delta(x - \overline{x})\delta(y - \overline{y})$, one gets the following differential equation for the unknown function $F_t(x, y)$,

$$\partial_t F_t^{xy}(\overline{x}, \overline{y}) = -\lambda_t F_t^{xy}(\overline{x}, \overline{y}) + \widetilde{\lambda}_t(b_t\overline{y} + a_t\overline{x})\, F_t^{xy}(\overline{x}, \overline{y}) ,$$

with initial condition $F_{t=0}^{xy}(\overline{x}, \overline{y}) = \delta(x - \overline{x})\delta(y - \overline{y})$. Then,

$$F_t^{xy}(\overline{x}, \overline{y}) = \exp\left(-\int_0^t ds \left(\lambda_s - \widetilde{\lambda}_s(b_s\overline{y} + a_s\overline{x})\right)\right) F_0^{xy}(\overline{x}, \overline{y}) .$$

Once the previous expression is substituted into (70), one finally finds

$$\widetilde{W}_t(x, y) = \exp\left(-\kappa_t + \int_0^t ds\, \widetilde{\lambda}_s(b_s y + a_s x)\right) \hat{W}(x, y) ,$$

with $\kappa_t = \int_0^t ds\, \lambda_s = \int_0^t ds \int_{-\infty}^{+\infty} du\, \lambda_s(u)$. Further, using (71) again and by means of two Dirac deltas, one rewrites

$$
\begin{aligned}
\widetilde{W}_t(x,y) &= \int_{\mathbb{R}^2} du\,dv\,\delta(u-x)\delta(v-y)\,G_t(u,v)\,\hat{W}(x,y) \\
&= \int_{\mathbb{R}^4} \frac{du\,dv\,d\overline{x}\,d\overline{y}}{(2\pi\hbar)^2}\,\mathrm{e}^{\frac{i}{\hbar}(\overline{y}(u-x)-\overline{x}(v-y))}\,G_t(u,v)\,\hat{W}(x,y) \\
&= \int_{\mathbb{R}^2} \frac{d\overline{x}\,d\overline{y}}{(2\pi\hbar)^2}\,\widetilde{G}_t(\overline{x},\overline{y}))\,\hat{W}^\dagger(\overline{x},\overline{y})\,\hat{W}(x,y)\,\hat{W}(\overline{x},\overline{y}) \qquad \text{where} \\
\widetilde{G}_t(\overline{x},\overline{y}) &:= \int_{\mathbb{R}^2} \frac{du\,dv}{(2\pi\hbar)^2}\,\mathrm{e}^{\frac{i}{\hbar}(\overline{y}u-\overline{x}v)}\,G_t(u,v)\,.
\end{aligned}
$$

References

1. G. C. Ghirardi, A. Rimini, T. Weber, Phys. Rev. **D 34**, 470 (1986)
2. A. Gorini, A. Kossakowski, and E.C.G. Sudarshan, J. Math. Phys. **17**, 821 (1976)
3. G. Lindblad, Comm. Math. Phys. **48**, 119 (1976)
4. R. Alicki, K. Lendi, *Quantum Dynamical Semigroups and Applications* (Springer, Berlin, 1987)
5. H.-P. Breuer, F. Petruccione, *The Theory of Open Quantum Systems* (Oxford University Press, Oxford, 2007)
6. F. Benatti, R. Floreanini, Int. J. Mod. Phys. **B 19**, 3063 (2005)
7. F. Benatti, R. Floreanini, R. Romano, J. Phys. A: Math. Gen. **35**, L351 (2002)
8. R. Alicki, M. Fannes, *Quantum Dynamical Systems* (Oxford University Press, Oxford, 2001).
9. H.-P. Breuer, E.-M. Laine, J. Piilo, B. Vacchini, Rev. Mod.Phys. **88**, 021002 (2016).
10. D. Chruściński and A. Kossakowski, J. Phys. B: At. Mol. Opt. Phys. **45** , 154002 (2012)
11. D. Chruściński and S. Maniscalco, Phys. Rev. Lett. **112**, 120404 (2014)
12. D. Chruściński and F. A. Wudarski, Phys. Lett. A **377**, 1425 (2013)
13. D. Chruściński and F. A. Wudarski, Phys. Rev. A **91**, 012104 (2015)
14. R. Horodecki, P. Horodecki, M. Horodecki, and K. Horodecki, Rev. Mod. Phys. **81**, 865 (2009)
15. E. Laine, H. Breuer, J. Piilo,Phys. Rev. A, **81**, 062115 (2010)
16. Á. Rivas, S.F. Huelga, M.B. Plenio, Rep. Progr. Phys., **77**, 9 (2014)
17. H. Breuer, E. Laine, J. Piilo, Phys. Rev. Lett., **103**, 210401 (2009)
18. W. Rudin, *Functional Analysis* (McGraw-Hill, New York, 1991)
19. O. Bratteli, D.W. Robinson, *Operator Algebras and Quantum Statistical Mechanics II* (Springer-Verlag, Heidelberg, 1981)

Energy-Lifetime Relations

Robert Grummt and Nicola Vona

Abstract We studied an explicit model of alpha decay, for which we could calculate all quantities involved in the linewidthlifetime relation and in the energy-time uncertainty relation. The former is often regarded as a consequence of the latter, but we show that it cannot be the case, as it is possible to adjust the potential and the initial state in such a way that the linewidth-lifetime product gets arbitrarily close to 1, while at the same time the energy-time uncertainty product gets arbitrarily large. Additionally, this implies that Var E is a physically irrelevant quantity: Wave functions that differ only minimally in the tails are physically indistinguishable and can be produced by the same experiment, yet they can have radically different values of Var E, that therefore cannot be considered as a characteristic quantity of the physical system.

1 Introduction

We write these lines in honour of Gian-Carlo Ghirardi, who worked on the description of unstable systems from early on in his career [1]. Prof. Ghirardi made it a priority to create possibilities for people to meet, discuss and collaborate: To him the people behind the work were most important. The Summer School he co-organized regularly in Sesto, Italy, has contributed greatly to our education, and we have wonderful memories of it. He was also a very active member of the European COST project on "Fundamental Problems in Quantum Physics," whose sole purpose was to bring

All authors contributed equally to this work.

R. Grummt · N. Vona (✉)
Mathematisches Institut der LMU München, Theresienstr. 39, 80333 München, Germany
e-mail: nicola.vona@me.com

R. Grummt
e-mail: robert@templex.net

V. Allori et al. (eds.), *Do Wave Functions Jump?*, Fundamental Theories of Physics 198,
https://doi.org/10.1007/978-3-030-46777-7_12

together physicists, mathematicians and philosophers of all generations. Thank you for all the moments you have shared with us.

For an unstable nucleus undergoing exponential decay, the *linewidth-lifetime relation*

$$\Gamma\tau = 1 \tag{1}$$

connects the lifetime τ of the nucleus to the spread of the energy of the decay products, expressed as the full width at half maximum Γ of the probability density function of the energy.

Since τ expresses an uncertainty on time and Γ one on energy, this relation is often presented as an instance of the energy-time uncertainty relation (see for example [6]). Nevertheless, Fock and Krylov [4] argued that these two relations are indeed independent. They observed that many initial states differing only for the tails of the energy density can exhibit similar values of Γ, τ, and $\operatorname{Var} T$, still having completely different values of $\operatorname{Var} E$. The latter is very sensitive to the tails of the energy density, while Γ is related to the shape of the energy density around its maximum; The lifetime τ depends mostly on the maximum of the energy density, and in an exponential distribution the mean is directly related to the variance, so also $\operatorname{Var} T$ depends mostly on the maximum of the energy density. Therefore, for a fixed shape of the energy distribution around its maximum, Γ, τ, and $\operatorname{Var} T$ are fixed, while $\operatorname{Var} E$ can be changed at will by modifying the tails of the distribution, so that the product in (1) stays constant and the product $\operatorname{Var} E \operatorname{Var} T$ can have arbitrary values.

Here we present a rigorous proof of the arguments by Fock and Krylov. We considered an explicit model of alpha decay, for which we could calculate all implicated quantities. We show that it is possible to adjust the potential and the initial state in such a way that the product $\Gamma\tau$ gets arbitrarily close to 1, while at the same time the product $\operatorname{Var} E \operatorname{Var} T$ gets arbitrarily large. Therefore, *the linewidth-lifetime relation cannot be a consequence of the time-energy uncertainty relation.*

2 Gamow's Model of Alpha Decay and Skibsted's Variation of It

The theoretical study of exponential decay, and in particular of alpha decay, goes back to Gamow [2]. We will summarize his key insight for the three dimensional Schrödinger equation with rotationally symmetric potential V, having compact support in $[0, R_V]$. This model can be analyzed mathematically with reasonably simple tools as shown by Skibsted in [7]. We will only be concerned with the case of zero angular momentum to avoid the angular momentum barrier potential, which would not have compact support. In this case the Schrödinger equation is equivalent to the one dimensional problem

$$i\partial_t\psi = \left(-\partial_r^2 + V(r)\right)\psi =: H\psi \tag{2}$$

Gamow's key insight was to consider eigenfunctions $f(k_0, r)$ of the stationary Schrödinger equation

$$\left(-\partial_r^2 + V(r)\right) f(k, r) = k^2 f(k, r) \tag{3}$$

that satisfy the boundary conditions $f(k_0, r) = e^{ik_0 r}$ for $r \geq R_V$ and $f(k_0, 0) = 0$, and have complex eigenvalue k_0 such that

$$k_0 = \alpha - i\beta, \tag{4}$$

for some $\alpha, \beta > 0$. The function $f(k_0, r)$ yields a solution

$$f_t(k_0, r) := e^{-ik_0^2 t} f(k_0, r) \tag{5}$$

to the time-dependent Schrödinger equation

$$i\partial_t f_t(k_0, r) = \left(-\partial_r^2 + V(r)\right) f_t(k_0, r) = k_0^2 f_t(k_0, r). \tag{6}$$

Letting

$$E - i\frac{\gamma}{2} := k_0^2 = \alpha^2 - \beta^2 - i2\alpha\beta, \tag{7}$$

with $E, \gamma > 0$, we see that

$$|f_t(k_0, r)|^2 = e^{-\gamma t}|f(k_0, r)|^2, \tag{8}$$

i.e. $f_t(k_0, r)$ decays exponentially in time with lifetime $1/\gamma$.

In the sequel we will refer to $f(k_0, r)$ as Gamow function. Note that both boundary conditions on the Gamow function are natural: the condition $f(k_0, r) = e^{ik_0 r}$ for $r \geq R_V$ means that $f(k_0, r)$ is purely outgoing, which is reasonable for states describing decay, while the condition $f(k_0, 0) = 0$ means that no probability should enter the region $r < 0$, which is the standard condition on physical states expressed in spherical coordinates.

Clearly, Gamow's description does not immediately connect with quantum mechanics because it contains complex eigenvalues and exponentially increasing eigenfunctions, that are not square integrable. Skibsted analyzed in [7] the sense in which Gamow's model of alpha decay carries over to quantum mechanics. There, the meta-stable state is modelled via the truncated Gamow function

$$f_R := \mathbf{1}_R f(k_0, \cdot) \tag{9}$$

For rotationally symmetric potentials V that are compactly supported in $[0, R_V]$, with $\|rV(r)\|_1 < \infty$ and $R_2(t) := 2\alpha t + R$ he has shown that

$$e^{-iHt} f_R \approx e^{-ik_0^2 t} f_{R_2(t)} \tag{10}$$

if $\beta \ll 1$ is small enough. He also assumed that the potential has neither bound nor virtual states, that irrelevant for studying exponential decay for long lived particles. Skibsted essentially showed in [7] that the velocity with which the alpha-particle escapes the nucleus is 2α, while the lifetime of the meta-stable state is $(4\alpha\beta)^{-1}$. Comparison with empirical data shows that $\alpha \approx 1$, while the lifetime is very large and therefore $\beta \ll 1$.

To show the above result, he expressed the solution of the time dependent Schrödinger equation in terms of generalized eigenfunctions $\psi^+(k, r)$, like

$$e^{-iHt}\psi(r) = \int_0^\infty \hat{\psi}(k)\psi^+(k, r)e^{-ik^2 t}\, dk, \quad \text{with} \quad \hat{\psi}(k) = \int_0^\infty \psi(r)\bar{\psi}^+(k, r)\, dr. \tag{11}$$

For a good introduction into generalized eigenfunctions ψ^+, especially for the potentials under consideration, see Chap. 12 of [5]. We will also use the generalized eigenfunctions in the next section.

3 The Energy Time Uncertainty Relation and the Linewidth-Lifetime Relation Are Different

In [3, 8] we used Skibsted's model to show that the product $\Gamma\tau$ and the product $\text{Var}\, E\, \text{Var}\, T$ behave very differently for meta-stable states with large enough lifetime (see Theorem 4.2 in [3]). This section summarises the line of argumentation.

Skibsted calculated the energy distribution of the truncated Gamow f_R in Lemma 3.2 of [7], and found that it is approximately a Breit-Wigner distribution with linewidth $4\alpha\beta$. Unfortunately, the Breit-Wigner distribution has fat tails that result in infinite energy variance, so we can not look at $\text{Var}\, E$ right away. The modification applied by Skibsted to the Gamow wave function is enough to recover square integrability in space and produce a physical state, but it is not enough to regularise the energy tails. The energy tails are strongly influenced by the hard cut-off in space in eq. (9), therefore in [3, 8] we considered a Gaussian cut-off with width σ in space, that regularises the energy tails and results in a finite energy variance. To reduce the number of free parameters, we set $\sigma = \beta$, knowing that we are interested in the limit $\beta \to 0$. With this choice, we find that

$$\text{Var}\, E = O(\beta^{-2}), \quad \text{for } \beta \to 0 \tag{12}$$

(see Lemmas 4.4 and 4.9 in [3]). The energy variance grows when the cut-off gets steeper, getting infinite for the sharp cut-off.

For $\sigma = \beta$ small enough the time evolved wave function with Gaussian cut-off stays close to the one of Skibsted, that satisfies eq. (10), therefore $\tau \approx 1/4\alpha\beta$.

For Γ, consider that the generalised Fourier transform $\hat{\psi}$ of the wave function is the sum of the one of the Skibsted wave function and the one of the Gaussian tails; the tails amount for a small proportion of the L_2 norm of the wave function, therefore they cannot change the overall shape of $\hat{\psi}$. Along these lines, Lemma 4.11 of [3] shows that for β small enough $\Gamma \approx 4\alpha\beta$, from which follows that $\Gamma\tau \approx 1$ (Theorem 4.2 of [3]).

We are now left with calculating Var T, but to do so we first need to specify how we model the time measurement. To this end, we used the flux of the probability current through a far away detecting surface as probability density function for the arrival time of the alpha particle at the detector. The flux of the probability current in general does not have the properties needed to be a probability density function; Nevertheless, its use in this case is justified by the fact that the distance between the detector and the decaying nucleus is much bigger than the nucleus itself, therefore the measurement is practically performed under scattering conditions.[1] We consider the detector to be a sphere of radius R around the origin. Note that the position of the cut-off in eq. (9) is equal to the detector radius R, that is appropriate to model all experiments that start with a bulk of material, and the only information available is that the decay products did not hit the detector yet. The probability current is non-zero only in the radial direction, and letting

$$\psi_t(r) := e^{-iHt}\psi(r), \tag{13}$$

one readily finds that its flux Φ through the detector is

$$\Phi(t) = \frac{2}{\|\psi\|_2^2} \operatorname{Im}\left[\bar{\psi}_t(R)\,(\partial_r\psi_t)\,(R)\right]. \tag{14}$$

To define the arrival time probability density Π_T, we normalise the flux (14) through the detector surface to one on the time interval $(0, \infty)$, getting

$$\Pi_T(t) = \frac{\Phi(R,t)}{\int_0^\infty \Phi(R,t')dt'}. \tag{15}$$

Now, we can calculate the mean arrival time as

$$\langle t\rangle := \int_0^\infty t\,\Pi_T(t)\,dt \tag{16}$$

[1] For a general discussion of the role of the probability current in the description of time measurements see [9, 10].

and similarly the time variance. To calculate such integrals, we have to get a handle on the time evolved wave function for all times. For this purpose, in [3, 8] we divided the time evolution in two parts: the exponential decay regime and the scattering regime. Skibsted's results, summarised in eq. (10), yield the necessary control over the exponential decay regime. To treat the scattering regime, we used the stationary phase argument, that in essence is based on the following partial integration trick, that starts from eq. (11)

$$e^{-iHt}\psi(r) = \int_0^\infty \hat{\psi}(k)\psi^+(k,r)e^{-ik^2t}\,dk \tag{17}$$

$$= \left[\frac{i\hat{\psi}(k)\psi^+(k,r)}{2kt}e^{-ik^2t}\right]_0^\infty - \int_0^\infty \partial_k\left[\frac{i\hat{\psi}(k)\psi^+(k,r)}{2kt}\right]e^{-ik^2t}\,dk. \tag{18}$$

The generalized eigenfunctions ψ^+ are well understood for the potentials under consideration, and Chap. 12 of [5] provides the relation

$$\psi^+(k,r) = \frac{1}{2i}(S(k)f(k,r) - f(-k,r)), \tag{19}$$

where $S(k)$ denotes the S-matrix element for zero angular momentum. We see that we can get bounds on the scattering regime by means of bounds on the derivatives of the S-Matrix. Following this route, in [3, 8] we found explicit bounds on the overall time evolution (see Corollary 4.1 in [3]), from which we proved that

$$\text{Var}\,T = (4\alpha\beta)^{-2}\cdot(1+\varepsilon)\,, \tag{20}$$

where $\varepsilon \to 0$ for $\beta \to 0$. (see Lemma 4.9 in [3], and Lemma 4.3 for an explicit bound).

Summarising, we found that when $\beta \to 0$ then

$$\text{Var}\,E = O(\beta^{-2}) \qquad \Gamma \approx 4\alpha\beta \tag{21}$$

$$\text{Var}\,T \approx (1/4\alpha\beta)^2 \qquad \tau \approx 1/4\alpha\beta \tag{22}$$

$$\text{Var}\,E\,\text{Var}\,T = O(\beta^{-4}) \qquad \Gamma\tau \approx 1 \tag{23}$$

from which follows

$$\text{Var}\,E\,\text{Var}\,T \to \infty \tag{24}$$

while

$$\Gamma\tau \to 1, \tag{25}$$

that shows that the linewidth-lifetime relation cannot be an instance of the energy-time uncertainty relation.

4 Conclusion

We considered a variation of Skibsted's model of a long lived meta-stable state, for which we could calculate explicitly all quantities involved in the linewidth-lifetime relation and in the energy-time uncertainty relation, proving that the former cannot be a consequence of the latter.

Notice also that the behaviour of Var E is not only incompatible with the linewidth-lifetime relation, it also means that Var E is a physically irrelevant quantity: Wave functions with very similar cut-off σ are physically indistinguishable and can be produced by the same experiment, yet they can have radically different values of Var E, that therefore cannot be considered as a characteristic quantity of the physical system.

Acknowledgments Nicola Vona Gratefully Acknowledges the Financial Support of the Elite Network of Bavaria.

References

1. L. Fonda, G. C. Ghirardi, and Rimini A. Decay theory of unstable quantum systems. *Rep. Prog. Phys.*, 41:587–631, 1978.
2. G. Gamow. Zur Quantentheorie des Atomkernes. *Zeitschrift für Physik*, 51(3-4):204–212, 1928.
3. R. Grummt. *On quantum mechanical decay processes*. PhD thesis, LMU München, Feb 2014. http://edoc.ub.uni-muenchen.de/16621.
4. N. S. Krylov and V. A. Fock. On the uncertainty relation between time and energy. *Journal of Physics USSR*, 11:112–120, 1947.
5. R.G. Newton. *Scattering Theory of Waves and Particles*. International series in pure and applied mathematics. McGraw-Hill, 1966.
6. J.W. Rohlf. *Modern Physics from $\alpha\alpha$ to Z*. Wiley, 1994.
7. E. Skibsted. Truncated Gamow functions, α-decay and the exponential law. *Comm. Math. Phys.*, 104(4):591–604, 1986.
8. N. Vona. *On Time in Quantum Mechanics*. PhD thesis, LMU München, Feb 2014. http://edoc.ub.uni-muenchen.de/16620.
9. Nicola Vona and Detlef Dürr. The role of the probability current for time measurements. In Philippe Blanchard and Jürg Fröhlich, editors, *The Message of Quantum Science – Attempts Towards a Synthesis*. Springer, 2014. preprint available at http://arxiv.org/abs/1309.4957.
10. Nicola Vona, Günter Hinrichs, and Detlef Dürr. What does one measure when one measures the arrival time of a quantum particle? Phys. Rev. Lett., 111:220404, Nov 2013.

On the Continuum Limit of the GRW Model

Günter Hinrichs

Abstract We consider the relation between time-discrete and continuous models for wave function collapse. In the special case of the original GRW model and the Diósi model, it can be made mathematically precise how the latter arises as a scaling limit of the former.

1 Discrete and Continuous Collapse Models

The Ghirardi-Rimini-Weber theory (shortly GRW, see [7], [3] and, for a mathematically rigorous treatment, [11]) is the earliest prominent attempt to solve the so-called measurement problem in quantum mechanics by including a well-defined random collapse mechanism in the Schrödinger evolution of the wavefunction. More precisely, the wavefunction undergoes the usual Schrödinger evolution, but in addition, at random times given by a Poisson process, it is multiplied by a suitably localized Gaussian function ("collapsed"). Such an evolution is clearly well-defined whenever the Schrödinger evolution is. Apart from being nonrelativistic, a physical deficiency that limits the possible range of validity of GRW considerably is the fact that symmetry of the wavefunction is not preserved. Nevertheless, it is still of interest e.g. in the discussion of conceptual questions due to its transparency. In the physically more refined (but still nonrelativistic) CSL model [8], a Gaussian field in three-dimensional space couples to the wavefunction, effecting a collapse continuously in time. In this way, symmetry can naturally be preserved, but the mathematics becomes much more involved and not even the existence of solutions seems to have been worked out.

Although they share some formal structures, CSL is not a continuous version of GRW, but a new model. In addition, GRW has a continuum limit, namely the QMUPL model [4, 5] (going back to Diósi). It does not preserve symmetry, but is mathematically much better understood than CSL because it is described by a

G. Hinrichs (✉)
Mathematisches Institut, Ludwig-Maximilians Universität, Theresienstr. 39, 80333 München, Germany
e-mail: guenter.hinrichs@math.uni-augsburg.de

V. Allori et al. (eds.), *Do Wave Functions Jump?*, Fundamental Theories of Physics 198,
https://doi.org/10.1007/978-3-030-46777-7_13

stochastic differential equation driven by a finite number of Wiener processes. The limit from GRW to QMUPL is performed by increasing the frequency of the collapses and diminishing their strengh (making the Gaussians broader), keeping the order of magnitude of the total effect constant, and has been derived rigorously in [6]. Apart from clarifying the relation between the two models GRW and QMUPL, the structure of the proof might also be interesting in dealing with CSL because attempts to solve the equation would probably most straightforwardly start with similar discrete approximations. In the passage from GRW to QMUPL, use was made of the linear structure behind the models and a Trotter-like product formula for linear stochstic differential equations. In the following, the approach from [6] will be reformulated in a way that makes this clearer and some central steps will be recalled.

2 The Linear Structure Behind the GRW Model

For simplicity, we consider a one-particle wavefunction $\varphi_0 \in L^2(\mathbb{R}^3, \mathbb{C})$ (with L^2-norm $\|\cdot\|$) without physical constants, of which the pure Schrödinger evolution would be $e^{-itH}\varphi_0$ with some self-adjoint Hamiltonian. We collect the randomness in a sample space $\Omega := \mathbb{R}^{\mathbb{N}} \times (\mathbb{R}^3)^{\mathbb{N}}$ with coordinate projections $X = (X_1, X_2, \dots)$ on the first factor (which is to desribe the times between two collapses) and $Y = (Y_1, Y_2, \dots)$ on the second factor (describing the collapse centers). We define $\mathbb{P}(X \in \cdot)$ to be a countable product of standard exponential distributions. The evolution of φ_0 collapsing at times $T_{n,\mu} := \sum_{k=1}^{n} \frac{X_k}{\mu}$ $(n \in \mathbb{N})$, where the model constant $\mu > 0$ is a frequency parameter, is defined recursively by

$$\psi^{\alpha,\mu}_{T_{n,\mu}}(Y_1, \dots, Y_n, X, x) := \left(\frac{\alpha}{\pi}\right)^{\frac{3}{4}} e^{-\frac{\alpha}{2}|x-Y_n|^2} e^{-\frac{i}{\mu}X_n H} \varphi^{\alpha,\mu}_{T_{n-1,\mu}}(Y_1, \dots, Y_{n-1}, X, x),$$

$$\varphi^{\alpha,\mu}_{T_{n,\mu}} := \frac{\psi^{\alpha,\mu}_{T_{n,\mu}}}{\|\psi^{\alpha,\mu}_{T_{n,\mu}}\|}$$

and

$$\begin{aligned}\mathbb{P}_{\alpha,\mu}(Y_n \in A \mid Y_1, \dots, Y_{n-1}, X) &:= \int_A \left(\frac{\alpha}{\pi}\right)^{\frac{3}{2}} \int e^{-\alpha|x-y|^2} |(e^{-\frac{i}{\mu}X_n H} \varphi^{\alpha,\mu}_{T_{n-1,\mu}})(Y, X, x)|^2 \mathrm{d}x\mathrm{d}y \\ &= \int_A \|\psi^{\alpha,\mu}_{T_{n,\mu}}(Y_1, \dots, Y_{n-1}, y, X, \cdot)\|^2 \mathrm{d}y,\end{aligned} \tag{2.1}$$

i. e. at each collapse time, the wavefunction is multiplied by a Gaussian with some prescribed inverse variance α, the center of which is a random variable distributed according to $|\varphi|^2$ convoluted with the normalized squared Gaussian. The sequence of wavefunctions is extended continuously according to

$$\varphi^{\mu}_{T_{n,\mu}+s} := e^{-isH} \varphi^{\alpha,\mu}_{T_{n,\mu}} \text{ for } 0 < s < X_{n+1}$$

between the collapse times. By this and the preceding equations, the stochastic process $(\varphi_t^{\alpha,\mu})_{t\geq 0}$ is specified.

Explicitely, one calculates

$$\mathbb{P}_{\alpha,\mu}(Y_1 \in A_1, Y_2 \in A_2 \mid X) = \int_{\{Y_1 \in A_1\}} \mathbb{P}_{\alpha,\mu}(Y_2 \in A_2 \mid Y_1, X) \mathrm{d}\mathbb{P}_{\alpha,\mu}(\cdot \mid X)$$

$$= \int_{\{Y_1 \in A_1\}} \int_{A_2} \left(\frac{\alpha}{\pi}\right)^{\frac{3}{2}} \int e^{-\alpha|x-y_2|^2} |(e^{-\frac{i}{\mu}X_2 H} \frac{\psi_{T_{1,\mu}}^{\alpha,\mu}}{\|\psi_{T_{1,\mu}}^{\alpha,\mu}\|})(Y_1, X, x)|^2 \mathrm{d}x \mathrm{d}y_2 \mathrm{d}\mathbb{P}_{\alpha,\mu}(\cdot \mid X)$$

$$= \int_{A_1} \int_{A_2} \left(\frac{\alpha}{\pi}\right)^{\frac{3}{2}} \int e^{-\alpha|x-y_2|^2} |(e^{-\frac{i}{\mu}X_2 H} \psi_{T_{1,\mu}}^{\alpha,\mu})(y_1, X, x)|^2 \mathrm{d}x \mathrm{d}y_2 \frac{1}{\|\psi_{T_{1,\mu}}^{\alpha,\mu}(y_1, X, \cdot)\|^2}$$

$$\times \|\psi_{T_{1,\mu}}^{\alpha,\mu}(y_1, X, \cdot)\|^2 \mathrm{d}y_1$$

$$= \int_{A_1} \int_{A_2} \left(\frac{\alpha}{\pi}\right)^3 \int e^{-\alpha|x-y_2|^2} |(e^{-\frac{i}{\mu}X_2 H} e^{-\frac{\alpha}{2}|\cdot-y_1|^2} e^{-\frac{i}{\mu}X_1 H} \varphi_0)(x)|^2 \mathrm{d}x \mathrm{d}y_2 \mathrm{d}y_1 .$$

One observes that the norm term cancels. (The analogous effect arises in deriving the equation for the statistical operator $\rho_t^{\alpha,\mu}(x, y) := \mathbb{E}_{\alpha,\mu} \varphi_t^{\alpha,\mu}(x) \overline{\varphi_t^{\alpha,\mu}(y)}$, resulting in a simple closed equation for ρ_t in so-called Lindblad form. One chooses in (2.1) the convolution instead of the exact $|\varphi|^2$ distribution in order to achieve precisely this.) Inductively,

$$\mathbb{P}_{\alpha,\mu}(Y_1 \in A_1, \ldots, Y_n \in A_n \mid X)$$

$$= \int_{A_1 \times \cdots \times A_n} \left(\frac{\alpha}{\pi}\right)^{\frac{3n}{2}} \int |(e^{-\frac{\alpha}{2}|\cdot-y_n|^2} e^{-\frac{i}{\mu}X_n H} \cdots e^{-\frac{\alpha}{2}|\cdot-y_1|^2} e^{-\frac{i}{\mu}X_1 H} \varphi_0)(x)|^2 \mathrm{d}x \mathrm{d}(y_1, \ldots, y_n)$$

$$= \int_{A_1 \times \cdots \times A_n} \left(\frac{\alpha}{\pi}\right)^{\frac{3n}{2}} \int |e^{\alpha x \cdot y_n - \frac{\alpha}{2}|x|^2} e^{-\frac{i}{\mu}X_n H} \cdots e^{\alpha x \cdot y_1 - \frac{\alpha}{2}|x|^2} e^{-\frac{i}{\mu}X_1 H} \varphi_0(x)|^2 \mathrm{d}x$$

$$\times e^{-\alpha(|y_1|^2 + \cdots + |y_n|^2)} \mathrm{d}(y_1, \ldots, y_n)$$

The last step is convenient because now the y-integral can be read as a mean value w. r. t. independent Gaussian random variables. We thus arrive at the following equivalent formulation of GRW, in which an auxiliary wavefunction $\psi_t^{\alpha,\mu}$, starting from $\psi_0^{\alpha,\mu} = \varphi_0$, undergoes a random linear evolution w. r. t. an auxiliary probability measure $\mathbb{Q}$ for all $t \in \mathbb{R}$ and is then subject to nonlinear modifications in order to get the physically relevant process $(\varphi_t^{\alpha,\mu})$ and the physically relevant measure $\mathbb{P}_{\alpha,\mu}$. (It corresponds to what Giancarlo Ghirardi used to call "linear plus cooking formulation" in the context of CSL.)

Given parameters $\alpha, \mu > 0$, let $(X_n)_{n\in\mathbb{N}}$ be random variables and $(Y_n)_{n\in\mathbb{N}}$ three-dimensional random vectors on some probability space $(\Omega, \mathfrak{A}, \mathbb{Q})$ which are independent of each other, all X_n having standard exponential distributions, all Y_n $N(0, \frac{1}{2\alpha}\mathbb{1})$ normal distributions. Take the Poisson process

$$N_\mu(t) := \max\left\{k \in \mathbb{N} : \sum_{j=1}^{k} \frac{X_j}{\mu} \leq t\right\}$$

and its jump times

$$T_{n,\mu} := \sum_{k=1}^{n} \frac{X_k}{\mu} \quad (n \in \mathbb{N})$$

and set

$$\psi_t^{\alpha,\mu}(x) := e^{i(t-T_{N_\mu(t),\mu})H} \prod_{n=1}^{\kappa_\mu(t)} e^{\alpha x \cdot Y_n - \frac{\alpha}{2}|x|^2} e^{iX_n H} \varphi_0(x) \tag{2.2}$$

where the factors have to be arranged from right to left. Then

$$\mathbb{P}_{\alpha,\mu}(Y_1 \in A_1, \ldots, Y_n \in A_n \mid X) := \mathbb{E}_{\mathbb{Q}}(\mathbb{1}_{\{Y_1 \in A_1, \ldots, Y_n \in A_n\}} \|\psi_{T_{n,\mu}}^{\alpha,\mu}\|^2 \mid X) \tag{2.3}$$

together with the prescription $\mathbb{P}_{\alpha,\mu}(X \in \cdot) := \mathbb{Q}(X \in \cdot)$ defines a new measure on $\mathfrak{F} := \sigma((X_n), (Y_n))$.

$$\varphi_t^{\alpha,\mu} := \frac{\psi_t^{\alpha,\mu}}{\|\psi_t^{\alpha,\mu}\|}$$

as a stochastic process w. r. t . $\mathbb{P}_{\alpha,\mu}$ is the GRW process (coinciding in distribution with the one from the first definition).

3 The Continuum Limit of GRW and the Stochastic Trotter Formula

From time 0 to sufficiently large t, there are on average $t\mu$ collapses. If $H = 0$, they result in the multiplication of φ_0 with a single Gaussian function with inverse variance $\frac{1}{2}t\mu\alpha$. Therefore, one might guess that also in the case $H \neq 0$ the order of magnitude of the total effect of the collapses up to a fixed time t should only depend on $\lambda := \frac{\alpha\mu}{2}$. For a continuum limit, it is therefore natural to prescribe a $\lambda > 0$, set $\alpha = \frac{2\lambda}{\mu}$ – this will be done from now on – and let the mean frequency μ of collapses go to ∞.

On the technical side, one should first observe that, if $(\Omega, \mathfrak{A}, \mathbb{Q})$ is chosen such that it admits a three-dimensional standard Wiener process $(\xi_t)_{t\geq 0}$, then, for any choice of λ and μ, independent Y_n with $N(0, \frac{1}{2\alpha}\mathbb{1}) = N(0, \frac{\mu}{4\lambda})$ distribution can be realized via its increments e. g. as

$$Y_n := \frac{\mu}{2\sqrt{\lambda}}\left(\xi_{\frac{n}{\mu}} - \xi_{\frac{n-1}{\mu}}\right). \tag{3.1}$$

Consequently, the collapse part of the evolution of ψ_t can be viewed as a discretization of the continuous evolution $\varphi_0 \mapsto \psi_t = A_{0,t}\varphi_0$ with

$$A_{s,t}\psi(x) := e^{\sqrt{\lambda}x\cdot(\xi_t-\xi_s)-\lambda|x|^2(t-s)}\psi(x)$$

for $0 \leq s \leq t < \infty$. Namely, (2.2) becomes

$$\psi_t^{\alpha,\mu} = e^{i(t-T_{N_\mu(t),\mu})H} \prod_{n=1}^{N_\mu(t)} A_{\frac{n-1}{\mu},\frac{n}{\mu}} e^{iX_n H}\varphi_0 \,. \tag{3.2}$$

The constants in (3.1) could have been put in different ways in front of the brackets and in the indices. Our choice is convenient because the indices of $A_{s,t}$ appear fron a technical point of view on the same level as a "time parameter" X_n in (3.2), even if they do not have such a physical meaning. Consequently, the difference of two successive indices is chosen as $\frac{1}{\mu}$, which is the mean time between two successive collapses.

(3.2) recalls the Trotter product formula: Two different evolutions alternate with increasing frequency. Ito calculus shows that $\psi_t := A_{0,t}\varphi_0$ satisfies the stochastic differential equation

$$\mathrm{d}\psi_t = \sqrt{\lambda}\psi_t x \cdot \mathrm{d}\xi_t - \frac{\lambda}{2}|x|^2\psi_t \mathrm{d}t \,,$$

so the Trotter product formula suggests that $\psi^{\frac{2\lambda}{\mu},\mu}$ converges for $\mu \to \infty$ in some sense to a solution of a combination of the latter and the Schrödinger equation, namely of

$$d\psi_t(x) = -iH\psi_t \mathrm{d}t + \sqrt{\lambda}\psi_t(x)x \cdot \mathrm{d}\xi_t - \frac{\lambda}{2}|x|^2\psi_t(x)\mathrm{d}t \text{ with } \psi_0 = \varphi_0 \,. \tag{3.3}$$

In [6], this has been worked out. If one uses this equation for a similar change of measure as (2.3), namely

$$\mathbb{P}_\lambda(A) := \mathbb{E}_\mathbb{Q}(\mathbb{1}_A\|\psi_t\|^2) \text{ for all } A \in \sigma(\{\xi_s \mid s \leq t\}) \tag{3.4}$$

(which can be shown to be a consistent definition of a measure), then, considering the stochastic process given by $\varphi_t := \frac{\psi_t}{\|\psi_t\|}$ under $\mathbb{P}_\lambda$, one arrives at the QMUPL model announced in the introduction. The measure changes (2.3) and (3.4) fit well together with each other and with the product formula and one finally arrives at the following result from [6]:

Theorem 1 *Let $H = -\frac{1}{2}\Delta + V$ with a bounded potential V having bounded first and second derivatives. Then, for all $t_1, \ldots, t_n \geq 0$, in the sense of weak convergence of measures,*

$$\lim_{\mu\to\infty} \mathbb{P}_{\frac{2\lambda}{\mu},\mu} \circ (\varphi_{t_1}^{\frac{2\lambda}{\mu},\mu}, \ldots, \varphi_{t_n}^{\frac{2\lambda}{\mu},\mu})^{-1} = \mathbb{P}_\lambda \circ (\varphi_{t_1}, \ldots, \varphi_{t_n})^{-1}.$$

In the proof of the product formula, one first shows that $(\psi_t^{\frac{2\lambda}{\mu},\mu})$ forms a Cauchy sequence in μ in a suitable sense. This part of the proof, in which one only needs to work with the "simple" discrete model, could potentially be transferred e. g. to suitable approximations of CSL and, if this turned out to be successful, could even be used as the mathematical definition of such a model. The next part of the proof, showing that $\psi_t^{\frac{2\lambda}{\mu},\mu} \to \psi_t$ in a suitable sense, makes use of specific mathematical results for the QMUPL model (to be found e. g. in [1, 2, 9, 10]), whereas the last step, the "nonlinear part", is quite straightforward once the previous work is done.

4 Giancarlo Ghirardi

Do wavefunctions jump? I have never been quite convinced of that, at present find it not fruitful to pursue this question further and tend to be tired of being confronted again and again with this subject just because I happened to deal with collapse models in my diploma thesis... The more I would like to emphasize that this feeling does not in the least refer to Prof. Ghirardi or his work in collapse theory. In conferences, I found it very pleasant to see his modest and winning way of presenting things. He used to emphasize that what he said might be wrong, should not be overestimated and anyway, (according to John Bell,) he gained his living most importantly with his teaching and not his research. A conference in which everyone takes such an attitude towards his positions is still to be held... In this way, he brought people together and, even being sceptic, one was more induced to take into account that he might be right than to stick to his admissions of maybe being wrong. And, after all, he might be right...

Requiescat in pace!

References

1. A. Bassi. Collapse models: analysis of the free particle dynamics. *J. Phys. A*, 38(14):3173–3192, 2005.
2. A. Bassi, D. Dürr, and Martin Kolb. On the long time behavior of free stochastic Schrödinger evolutions. *Rev. Math. Phys.*, 22(1):55–89, 2010.
3. A. Bassi and G.C. Ghirardi. Dynamical reduction models. *Phys. Rep.*, 379(5-6):257–426, 2003.
4. L. Diósi. Continuous quantum measurement and Itô formalism. *Phys. Lett. A*, 129(8-9):419–423, 1988.

5. L. Diósi. Models for universal reduction of macroscopic quantum fluctuations. *Phys. Rev. A*, 40(3):1165–1174, Aug 1989.
6. D. Dürr, G. Hinrichs and M. Kolb. On a Stochastic Trotter Formula with Application to Spontaneous Localization Models. *J. Stat. Phys.*, 143(6):1096–1119, 2011.
7. G. C. Ghirardi, A. Rimini, and T. Weber. Unified dynamics for microscopic and macroscopic systems. *Phys. Rev. D (3)*, 34(2):470–491, 1986.
8. Gian Carlo Ghirardi, Philip Pearle, and Alberto Rimini. Markov processes in Hilbert space and continuous spontaneous localization of systems of identical particles. *Phys. Rev. A (3)*, 42(1):78–89, 1990.
9. C. M. Mora and Rolando Rebolledo. Regularity of solutions to linear stochastic Schrödinger equations. *Infin. Dimens. Anal. Quantum Probab. Relat. Top.*, 10(2):237–259, 2007.
10. C. M. Mora and Rolando Rebolledo. Basic properties of nonlinear stochastic Schrödinger equations driven by Brownian motions. *Ann. Appl. Probab.*, 18(2):591–619, 2008.
11. R. Tumulka. The point processes of the GRW theory of wave function collapse. *Rev. Math. Phys.*, 21(2):155–227, 2009.

Continuous Collapse Models on Finite Dimensional Hilbert Spaces

Antoine Tilloy

Collapse models come in many flavors, with varying levels of complexity. Yet even the simplest physically realistic models have a phenomenology that is non-trivial to study rigorously, if only because continuous space imposes an infinite dimensional Hilbert space. Here, we would like to focus on toy models, that apply to finite dimensional Hilbert spaces, that can be efficiently simulated, and are amenable to a precise and to some extent rigorous study. We shall mostly be interested in collapse models for qubits, *i.e.* with $\mathscr{H} = \mathbb{C}^2$, that is for the simplest quantum mechanical system one can think of.

The prototypical equation we will discuss gives the dynamics of a probability (or population) $p_t = |\langle\psi_t|+\rangle_z|^2 \in [0, 1]$ for a qubit to be in one state (say its ground state, or spin-up state $|+\rangle_z$) which is a fixed point of the collapse process. It is an Itô stochastic differential equation that reads:

$$\mathrm{d}p_t = \underbrace{\lambda\,(p_{\mathrm{eq}} - p_t)\,\mathrm{d}t}_{\text{"regular" dynamics}} + \underbrace{\sqrt{\gamma}\; p_t(1 - p_t)\,\mathrm{d}W_t}_{\text{collapse dynamics}}, \tag{1}$$

where $p_{\mathrm{eq}} \in]0, 1[$ is a constant equilibrium probability in the absence of collapse, W_t is a Wiener process (Brownian motion), λ is the rate or frequency associated to the dynamics in absence of collapse and γ is the rate[1] associated to collapse dynamics. We shall explain later how this equation is obtained, but let us briefly give an intuition for its phenomenology.

[1]Note that the inverse time scale γ appears with a square root, intuitively because $\mathrm{d}W$ scales like $\sqrt{\mathrm{d}t}$.

A. Tilloy (✉)
Max-Planck-Institut für Quantenoptik, Hans-Kopfermann-Straße, 85748 Garching, Germany
e-mail: antoine.tilloy@mpq.mpg.de

V. Allori et al. (eds.), *Do Wave Functions Jump?*, Fundamental Theories of Physics 198,
https://doi.org/10.1007/978-3-030-46777-7_14

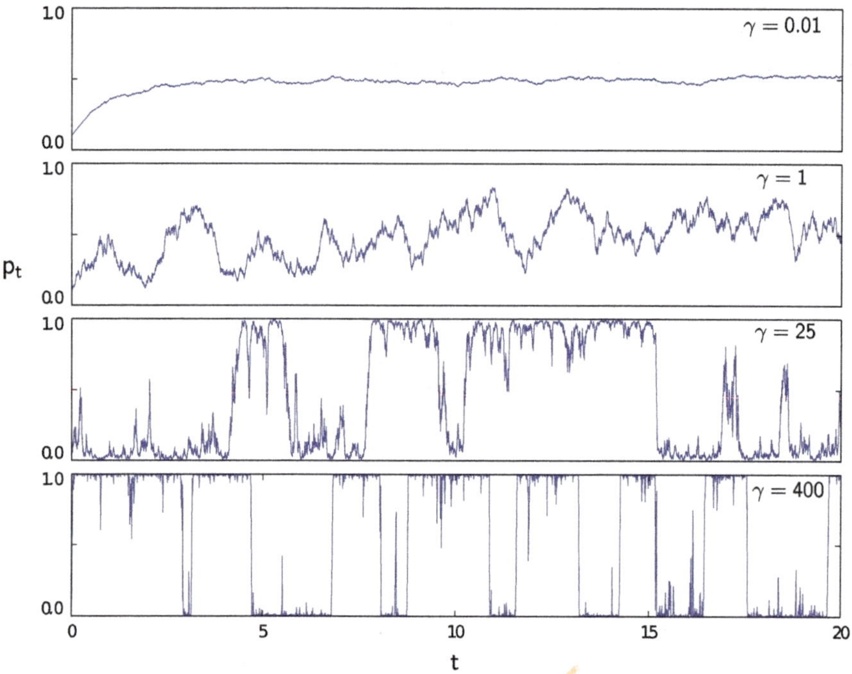

Fig. 1 Trajectories p_t from $\mathrm{d}p_t = \lambda\,(p_{\mathrm{eq}} - p_t)\,\mathrm{d}t + \sqrt{\gamma}\,p_t(1-p_t)\,\mathrm{d}W_t$ for increasing values of γ—Typical trajectories are shown for $\lambda = 1$, $p_{\mathrm{eq}} = 0.5$, and $\gamma = \{0.01, 1, 25, 400\}$. They are obtained through a naive Euler discretization of (1) with $\mathrm{d}t = 5.10^{-4}$. Far smarter discretization schemes can be used for this particular type of stochastic differential equations (see *e.g.* [1]), and they confirm this qualitative behavior. In particular, the sharp, almost punctual excursions decorating the jump process in the last plot are *not* numerical artifacts

The dynamics in absence of collapse, if taken alone, yields an exponential convergence (controlled by the rate λ) to the equilibrium probability p_{eq}. It is typically the dynamics one obtains by coupling a qubit to a thermal bath. On the other hand, the collapse term induces an inhomogeneous diffusion with a coefficient that vanishes in $p = 0$ and $p = 1$. Hence, under this dynamics, the probability wanders in an unbiased way until it reaches one of these two fixed points where the dynamics freezes. In brief, thermalization dynamics deterministically drives the probability to $p = p_{\mathrm{eq}} \in]0, 1[$, while collapse stochastically drives it to $p = 0$ or $p = 1$. It is from this competition that rich dynamics can emerge.

Before saying more, it is instructive to look at a typical trajectory of the stochastic process as the collapse rate γ is progressively increased. The results are shown in Fig. 1. One sees that upon increasing the value of γ, there is a crossover from a continuous diffusion to a jump dynamics (up to some subtleties). Interestingly, the solutions of (1) seem to converge, in some non-trivial sense, when $\gamma \to +\infty$. It is this limit we shall be mostly interested in understanding precisely here.

1 Setup

1.1 The Stochastic Schrödinger Equation and Its Origin

We consider a spontaneous collapse model for a quantum state $|\psi\rangle \in \mathscr{H} = \mathbb{C}^D$ with $D < +\infty$. The dynamics is postulated to be given by the *stochastic Schrödinger equation* (SSE):

$$\mathrm{d}|\psi_t\rangle = \left\{-iH\mathrm{d}t + \sqrt{\gamma}\,(\mathcal{O} - \langle\mathcal{O}\rangle_t)\,\mathrm{d}W_t - \frac{\gamma}{2}\left[\mathcal{O}^\dagger\mathcal{O} - 2\,\langle\mathcal{O}^\dagger\rangle_t\mathcal{O} + \langle\mathcal{O}^\dagger\rangle_t\langle\mathcal{O}\rangle_t\right]\mathrm{d}t\right\}|\psi_t\rangle, \quad (2)$$

where $\mathcal{O}$ is a generic operator,[2] $\langle\mathcal{O}\rangle_t = \langle\psi_t|\mathcal{O}|\psi_t\rangle$, W_t is a Wiener process (Brownian motion), γ is the collapse strength (or rate), and H is the system Hamiltonian independent of the collapse process. This stochastic differential equation with multiplicative noise is to be understood in the Itô convention [2]. As an illustration, taking D large and $\mathcal{O}$ to be a discretized version of the position operator X, (2) would yield an approximation of the "*Quantum Mechanics with Universal Position Localization*" (QMUPL) model [3, 4] in one space dimension. More complicated setups can easily be considered, where many operators (possibly non-commuting) are being continuously collapsed simultaneously, but we will stick to this simple dynamics (2) in what follows.

Where is such a stochastic differential equation coming from? There are at least 3 ways to motivate it:

1. *From collapse models in continuous space*—Starting from a collapse model in continuous space like the continuous spontaneous localization model (CSL), one may derive an effective collapse equation on a smaller Hilbert space. This happens if one considers degrees of freedom that are intrinsically discrete (for example spin in a Bell or EPR experiment), or if only a few states can be reached by the dynamics (for example if the potential $V(\hat{X})$ appearing in the Schrödinger equation has a few deep minima).
2. *From consistency requirements*—One may ask what is the most general collapse equation that (i) yields a linear evolution for the density matrix averaged over the noise $\bar{\rho} = \mathbb{E}[|\psi\rangle\langle\psi|]$ (ii) is Markovian (iii) preserves state purity. It turns out that all equations with these properties essentially have the same form as (2) up to some additional phase factors (see [5, 6] and references therein).
3. *From continuous measurement theory*—Since their inception in the eighties, collapse models have been developed alongside the theory of continuous measurement [7, 8]. The latter aims to describes the continuous monitoring of quantum systems within orthodox quantum theory. In this context, one simply pushes the use of the collapse postulate sufficiently far away from the system studied to avoid problems or ambiguities for all practical purposes. A continuous measurement or monitoring is then obtained in a proper limit where infinitely weak

[2] A non-Hermitian $\mathcal{O}$ can be used to obtain a so called "dissipative" collapse model, but we will focus here mostly on the Hermitian case which already yields rich dynamics.

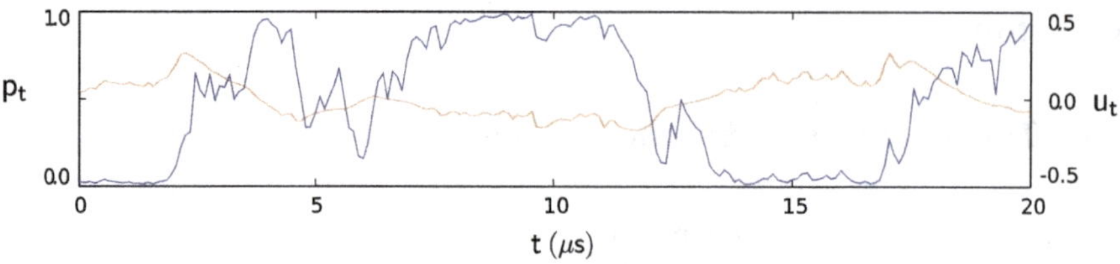

Fig. 2 **Quantum trajectory of a continuously monitored transmon qubit** (from *real* experimental data). In blue, p_t is the population in the z basis, and in light orange $u_t = u_t^*$ is the non-diagonal coefficient of ρ in the z basis (see (8)). The stochastic master equation (see 1.2) describing the evolution is slightly more complicated than the idealized ones we consider in subsequent examples and reads [9]: $\mathrm{d}\rho_t = -i[\Omega\sigma_y, \rho_t]\mathrm{d}t + \sum_j \mathcal{D}[L_j](\rho_t)\mathrm{d}t + \sqrt{\eta_d}\,\frac{\Gamma_d}{2}\mathcal{H}[\sigma_z](\rho_t)\mathrm{d}W_t$, with $j = \{u, v, w, \varphi\}$, $L_u = \sqrt{\Gamma_1/2}\sigma_-$, $L_v = i\sqrt{\Gamma_1/2}\,\sigma_-$, $L_w = \sqrt{\Gamma_d/2}\,\sigma_z$, $L_\varphi = \sqrt{\Gamma_\varphi/2}\,\sigma_z$, $\Gamma_d = (0.9\mu s)^{-1}$, $\Gamma_\varphi = (17.9\mu s)^{-1}$, $\Omega = 2\pi/(5.2\mu s)$, $\Gamma_1 = (765.3\mu s)^{-1}$ and $\eta_d = 34\%$. Experimental data courtesy of Benjamin Huard and Quentin Ficheux of École Normale Supérieure. For more detail, see Ficheux's thesis [10]

measurements are carried infinitely frequently. It turns out that the equations one obtains in this context are exactly the same as those of continuous collapse models.[3] The interpretation is of course different, but the formalism is identical (Fig. 2).

This latter motivation from continuous measurement theory is crucial for us, as it provides most of the intuition for the results. Furthermore, considering a finite dimensional Hilbert space is more common and natural on the continuous measurement side, where the systems monitored are typically effective qubits or few level systems, not fundamental constituents of nature. Finally, whilst the stochastic trajectories of (2) are not observable in the collapse context, they can be reconstructed in the continuous measurement context where the noise is knowable a posteriori, as it is a function of the (random) measurement results.

1.2 The Stochastic Master Equation

In practice it is more convenient to work with the equation for $\rho_t = |\psi_t\rangle\langle\psi_t|$, which makes the general structure more manifest. Using the Itô formula,[4] we get the *stochastic master equation* (SME):

[3]This equivalence holds only for Markovian collapse models. For colored noise (or non-Markovian collapse models), there is no longer a simple continuous measurement interpretation.

[4]In this context, "using the Itô formula" simply means writing

$$\mathrm{d}\rho_t = \mathrm{d}|\psi_t\rangle\ \langle\psi_t| + |\psi_t\rangle\ \mathrm{d}\langle\psi_t| + \mathrm{d}|\psi_t\rangle\ \mathrm{d}\langle\psi_t|,$$

using the formal rule $\mathrm{d}W_t \mathrm{d}W_t = \mathrm{d}t$ and keeping terms of order one in $\mathrm{d}t$ and $\mathrm{d}W_t$.

$$\mathrm{d}\rho_t = \mathscr{L}(\rho_t)\,\mathrm{d}t + \gamma\mathcal{D}[\mathcal{O}](\rho_t)\,\mathrm{d}t + \sqrt{\gamma}\,\mathcal{H}[\mathcal{O}](\rho_t)\,\mathrm{d}W_t \tag{3}$$

where $\mathscr{L}(\rho) = -i[H, \rho]$ and we have used the continuous measurement theory notations:

$$\mathcal{D}[\mathcal{O}](\rho) = \mathcal{O}\rho\mathcal{O}^\dagger - \frac{1}{2}\{\mathcal{O}^\dagger\mathcal{O}, \rho\} \tag{4}$$

$$\mathcal{H}[\mathcal{O}](\rho) = \mathcal{O}\rho + \rho\mathcal{O} - \mathrm{tr}\left[(\mathcal{O} + \mathcal{O}^\dagger)\rho\right]\rho. \tag{5}$$

This SME (3) is equivalent with the SSE (2) if the initial state ρ_t is pure (= rank 1), but it is more flexible, allowing $\mathscr{L}$ that are not Hamiltonian flows and do not preserve purity. Its first term $\mathcal{D}[\mathcal{O}]$ is linear and is simply a Lindblad operator: it encodes the decoherence associated with the collapse process and remains upon averaging over the noise. The second term $\mathcal{H}$ is a non-linear map on ρ. It is responsible for the collapse[5] and it would disappear upon noise averaging. Indeed, $\bar{\rho}_t = \mathbb{E}[\rho_t]$ is straightforward to compute because Itô integrals against the Wiener process have zero average. The average density matrix $\bar{\rho}$ ve rifies the master equation (ME):

$$\frac{\mathrm{d}}{\mathrm{d}t}\bar{\rho}_t = \mathscr{L}(\bar{\rho}_t) + \gamma\mathcal{D}[\mathcal{O}](\bar{\rho}_t). \tag{6}$$

The fact that it is linear is fundamental and insures the consistency of the collapse model and prevents problems with signalling or the probabilistic interpretation of the quantum state [11, 12]. This linearity is very natural in the measurement context, where averaging over the randomness of the measurement results is equivalent to tracing over an environment and thus preserves linearity.

2 Pure Collapse

Let us first consider (3) in the limit when there is no additional dynamics (*i.e.* $\mathscr{L} = 0$) and for a qubit (*i.e.* $\mathscr{H} = \mathbb{C}^2$):

$$\mathrm{d}\rho_t = \gamma\mathcal{D}[\mathcal{O}](\rho_t)\,\mathrm{d}t + \sqrt{\gamma}\,\mathcal{H}[\mathcal{O}](\rho_t)\,\mathrm{d}W_t. \tag{7}$$

To make things specific, we take $\mathcal{O} = \sigma_z/2$ where $\sigma_x, \sigma_y, \sigma_z$ are the 3 Pauli matrices. We introduce a parameterization of the qubit density matrix:

$$\rho_t = \begin{pmatrix} p_t & u_t \\ u_t^* & 1 - p_t \end{pmatrix}, \tag{8}$$

[5]In the context of continuous measurement theory, it encodes the progressive acquisition of information and is sometimes called the "stochastic innovation" term.

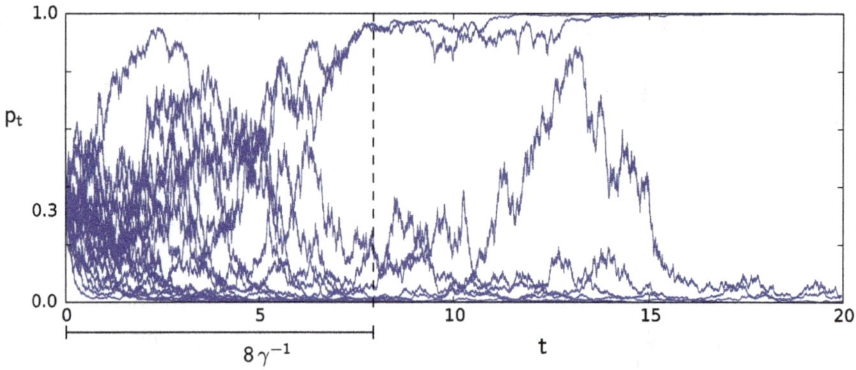

Fig. 3 A few realizations of the stochastic process $[\mathrm{d}p_t = \sqrt{\gamma}\, p_t(1-p_t)\,\mathrm{d}W_t]$ – for $\gamma = 1$ and $p_0 = 0.3$. Trajectories converge to $p = 1$ or $p = 0$ exponentially fast on average, with characteristic timescale $\tau \propto \gamma^{-1}$

where $p_t \in [0, 1]$ is the probability to be in (or population in) the state $|+\rangle_z$, *i.e.* $p_t = \langle +|\rho_t|+\rangle_z$, and u_t is a complex phase. We can expand (7) to obtain an equation for p and u. The one for the phase is

$$\mathrm{d}u_t = -\frac{\gamma}{8} u_t\,\mathrm{d}t + \frac{\sqrt{\gamma}}{2}(2p_t - 1)u_t\,\mathrm{d}W_t. \tag{9}$$

The stochastic trajectory of the phase depends on the population, but its average obeys an autonomous equation: writing $\bar{u}_t = \mathbb{E}[u_t]$ we have $\frac{\mathrm{d}}{\mathrm{d}t}\bar{u}_t = -(\gamma/8)\bar{u}_t$, and $\bar{u}_t = \bar{u}_0 \mathrm{e}^{-\gamma t/8}$. Hence, on average, collapse dynamics induces exponential decoherence in the eigenbasis of the collapse operator. This is expected. The equation for p_t is more interesting:

$$\mathrm{d}p_t = \sqrt{\gamma}\, p_t(1-p_t)\,\mathrm{d}W_t. \tag{10}$$

It contains no deterministic part and is a pure "martingale" (*i.e.* unbiased on average). A few realizations of this stochastic process are plotted in Fig. 3. The diffusion is inhomogeneous and makes p_t converge exponentially fast[6] (on average) to 0 or 1. This is easily seen by considering $\Delta_t = \sqrt{p_t(1-p_t)}$ which measures the distance from the final state. Using the Itô formula, one obtains:

[6]The probability p_t never touches 0 or 1 exactly. This behavior is to be contrasted from that of the Wright-Fisher equation encountered in population dynamics. The latter is similar but for a crucial square root:

$$\mathrm{d}x_t = \sqrt{x_t(1-x_t)}\,\mathrm{d}W_t, \tag{11}$$

and with this modification the boundaries $x = 0$ or $x = 1$ are reached almost surely in finite time.

$$\mathrm{d}\Delta_t = -\frac{\gamma}{8}\Delta_t\,\mathrm{d}t + \frac{\sqrt{\gamma\Delta_t}}{2}(1-2p_t)\,\mathrm{d}W_t. \tag{12}$$

Hence, $\frac{\mathrm{d}}{\mathrm{d}t}\bar{\Delta}_t = -(\gamma/8)\bar{\Delta}_t$, and $\bar{\Delta}_t = \bar{\Delta}_0\,\mathrm{e}^{-\gamma t/8}$.

Now we may wonder if the collapse towards $p = 1$ or $p = 0$ is done according to the Born rule. This is indeed the case:

$$\mathbb{P}\left[|\psi_t\rangle \underset{t\to+\infty}{\longrightarrow} |+\rangle_z\right] = |\langle\psi_0|+\rangle_z|^2 \text{ or, equivalently } \mathbb{P}\left[p_t \underset{t\to+\infty}{\longrightarrow} 1\right] = p_0. \tag{13}$$

This is seen easily by exploiting the fact that p_t is a martingale, a property which explains most of the features of the collapse process (see *e.g.* [13, 14]). Using (10), we have simply that $\frac{\mathrm{d}\bar{p}_t}{\mathrm{d}t} = 0$, hence $\bar{p}_t \equiv p_0$, and $\lim_{t\to+\infty}\bar{p}_t = p_0$. The later limit is:

$$\lim_{t\to+\infty}\bar{p}_t = \mathbb{P}\left[p_t \underset{t\to+\infty}{\longrightarrow} 1\right]\times 1 + \mathbb{P}\left[p_t \underset{t\to+\infty}{\longrightarrow} 0\right]\times 0 = \mathbb{P}\left[p_t \underset{t\to+\infty}{\longrightarrow} 1\right], \tag{14}$$

where we have used the fact that p_t converges almost surely to 0 or 1.

Hence the effect of our stochastic process is (i) to shrink the non-diagonal coefficients in the basis of $\mathcal{O}$, i.e. *decohere* (ii) make the diagonal coefficients all converge towards one, i.e. on has a *collapse* as expected. Both happen exponentially fast, with a rate controlled by γ, and with the expected probability. This generalizes trivially from $\mathscr{H} = \mathbb{C}^2$ to $\mathbb{C}^D$ for $D < \infty$.

3 Jumps

Now that we understand the dynamics induced by a continuous collapse process, we can add intrinsic dynamics of the system. In the limit where this dynamics is slow compared to the continuous collapse process, we will see the emergence of quantum jumps. This may be understood as a form of semiclassical limit: how does a quantum system behave when collapse is so fast that the state is almost always well localized?

3.1 Qubit with Dissipative Dynamics

The simplest setup we can consider is that of a qubit coupled to a thermal bath which induces a relaxation in the energy basis. Namely, we consider the evolution:

$$\mathrm{d}\rho_t = \mathscr{L}_{\text{thermal}}(\rho_t)\,\mathrm{d}t + \gamma\mathcal{D}[\mathcal{O}](\rho_t)\,\mathrm{d}t + \sqrt{\gamma}\,\mathcal{H}[\mathcal{O}](\rho_t)\,\mathrm{d}W_t. \tag{15}$$

with the Lindblad operator:

$$\mathscr{L}_{\text{thermal}}(\rho) = \lambda_\uparrow \mathcal{D}[\sigma_+](\rho) + \lambda_\downarrow \mathcal{D}[\sigma_-](\rho), \tag{16}$$

where $\sigma_+ = |+\rangle\langle-|_z = \sigma_-^\dagger$ and $\lambda_{\uparrow/\downarrow}$ represent the excitation and de-excitation induced by the bath. The complete evolution preserves diagonal density matrices: as in the pure collapse case, the non-diagonal coefficients shrink exponentially without any feedback on the diagonal ones. As a result, we consider only the evolution of p_t. Expanding (15) yields:

$$\mathrm{d}p_t = \lambda\,(p_{\text{eq}} - p_t)\,\mathrm{d}t + \sqrt{\gamma}\,p_t(1-p_t)\,\mathrm{d}W_t, \tag{17}$$

with $\lambda = \lambda_\downarrow + \lambda_\uparrow$ and $p_{\text{eq}} = \lambda_\downarrow/\lambda$. This is the equation we advertised in the introduction with its non-trivial competition between collapse driving p_t to 0 or 1 and relaxation driving it to p_{eq}.

For large γ, the stochastic process p_t seems to converge (in a weak sense) to a Markovian jump process (see 1). Intuitively, the dominant noise term $\sqrt{\gamma}\,p_t(1-p_t)\mathrm{d}W_t$ forces p_t to be almost always 0 or 1 and the subleading deterministic term induces Markovian transitions between these boundary values.

For this simple example, one can easily characterize the emerging jump process quantitatively provided one accepts that the large γ limit is indeed a Markov process. Such a Markov process would be characterized by two jump rates $M_{(+)\leftarrow(-)}$ and $M_{(-)\leftarrow(+)}$. In the large γ limit, p_t becomes a Markov chain between 0 and 1 and we its average value $\bar{p}_t$ will thus obey:

$$\frac{\mathrm{d}}{\mathrm{d}t}\bar{p}_t = -M_{(-)\leftarrow(+)}\bar{p}_t + M_{(+)\leftarrow(-)}(1-\bar{p}_t) \tag{18}$$

But using (17) we have that for all γ (and not just γ infinite):

$$\frac{\mathrm{d}}{\mathrm{d}t}\bar{p}_t = -\lambda_\uparrow p_t + \lambda_\downarrow(1-\bar{p}_t). \tag{19}$$

Hence we simply read that $M_{(-)\leftarrow(+)} = \lambda_\uparrow$ and $M_{(+)\leftarrow(-)} = \lambda_\downarrow$. In this very simple example, the jump rates can be read straightforwardly from the averaged master equation and it is only the very emergence of the jump process that is less trivial and requires the stochastic description.

3.2 Qubit with Coherent Dynamics

We now consider a second example where continuous collapse competes with a non-commuting unitary evolution. Namely, we choose a Hamitlonian $H = (\omega/2)\sigma_y$ while still collapsing with the operator $\mathcal{O} = \sigma_z/2$. The SME reads

$$\mathrm{d}\rho_t = -i\,\frac{\omega}{2}\,[\sigma_y, \rho_t]\,\mathrm{d}t + \gamma\mathcal{D}[\mathcal{O}](\rho_t)\,\mathrm{d}t + \sqrt{\gamma}\,\mathcal{H}[\mathcal{O}](\rho_t)\,\mathrm{d}W_t. \tag{20}$$

As before, it can be expanded into a pair of stochastic differential equations for p_t and u_t which parameterize the density matrix. However, this time the two equations are coupled, and the non-diagonal coefficient u_t does not shrink to zero. In fact, real and pure density matrices are preserved by the evolution (20) so that there is still only one dynamical parameter (which is an angle in the Bloch sphere). But the discussion remains easier with p_t and u_t.

Intuitively, what do we expect will happen? In the z basis, the unitary evolution with σ_x creates Rabi oscillations, hence $p_t \sim \cos(\omega t)$ in the absence of collapse. On the other hand, when continuous collapse dominates, we expect p_t to spend most of its time near 0 or 1 as before. There is indeed some non-trivial competition between the two. Let us just see how the trajectories look in Fig. 4. As before, we observe an emergent jump behavior in the large γ limit, starting from a completely different evolution for γ small (Rabi oscillations versus thermal relaxation in the previous case). However, there an important difference: as γ increases, the jumps get sharper and more discontinuous, but at the same time their frequency decreases (in $1/\gamma$). This is a signature of the Zeno effect: a coherent transition is slowed down by collapse (or measurement).

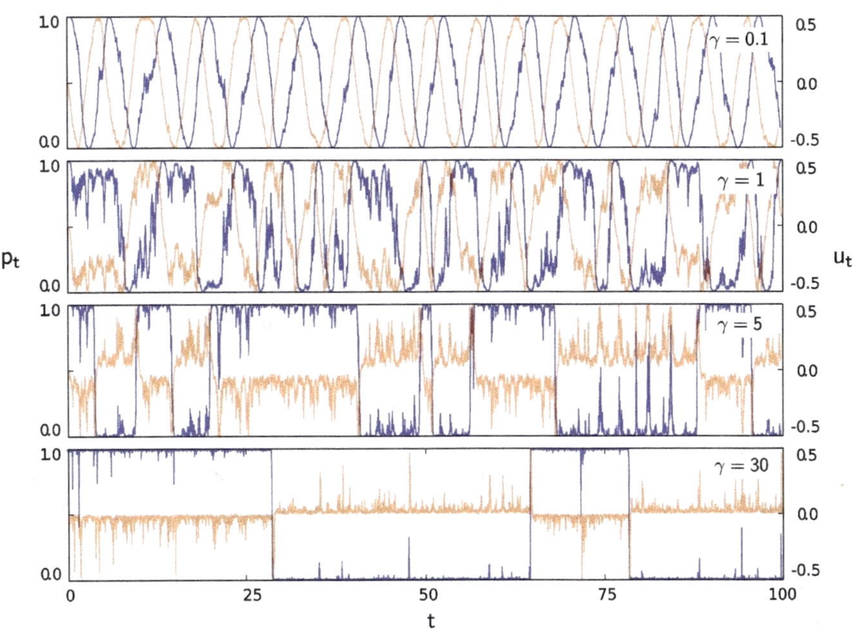

Fig. 4 **Trajectories** p_t (dark blue) **and** u_t (light orange) from – Typical trajectories are shown for $\omega = 1$, $p_{\text{eq}} = 1.0$, and $\gamma = \{0.1, 1, 5, 30\}$. They are obtained through a naive Euler discretization of (1) with $\mathrm{d}t = 10^{-4}$. u_t is real. For $\gamma \gg 100$ we would no longer see any jump in the figure and p_t would appear stuck in either 0 or 1

In this example, one can also prove rigorously the emergence of jumps and compute their rate explicitly. However, it is easier to go directly to the general case, which makes the perturbative reasoning more transparent.

3.3 General Case

We saw in the two examples before that the emergence of jumps seemed ubiquitous in the strong measurement limit, but that their rate depends on their origins: jumps mediated by a bath have a fixed rate independent of the measurement strength whereas "unitary" jumps are Zeno suppressed. For the latter, the jump rates vanish for large γ, and would thus be zero if the limit were taken too brutally. It means that to obtain a non-trivial limit, we need to adequately rescale the system dynamics (given by the Liouvillian $\mathscr{L}$) while the measurement strength γ is sent to $+\infty$. Up to this subtlety, we will show, or rather suggest, that the jump limit is ubiquitous, and there is generically a transition from continuous diffusive dynamics to discrete jump dynamics in the fast collapse limit (see Fig. 5).

We recall the setup, following the derivation in [15]. We continuously collapse a certain self-adjoint operator $\mathcal{O} = \sum_k \nu_k |k\rangle\langle k|$ at a rate γ, where the ν_k are real and, we assume, all different. We have a system evolution in the absence of collapse given by $\mathscr{L}_\gamma$ which depends on γ because we allow ourselves to rescale the part of the dynamics yielding jumps that would otherwise be Zeno suppressed. The evolution of the density matrix reads

$$\mathrm{d}\rho_t = \mathscr{L}_\gamma(\rho_t)\,\mathrm{d}t + \gamma\mathcal{D}[\mathcal{O}](\rho_t)\,\mathrm{d}t + \sqrt{\gamma}\,\mathcal{H}[\mathcal{O}](\rho_t)\,\mathrm{d}W_t. \tag{21}$$

We now need to parameterize the Liouvillian more explicitly. To this end, we write $[\mathscr{L}(\rho)]^{ij} = L_{kl}^{ij}\rho^{kl}$ with summation on repeated indices and postulate the scaling:

$$\begin{aligned}
L_{ll}^{ii} &= A_l^i + o(1)\\
L_{kl}^{ii} &= \sqrt{\gamma}\,B_{kl}^i + o(\sqrt{\gamma}) \quad \text{for } k \neq l\\
L_{ll}^{ij} &= \sqrt{\gamma}\,C_l^{ij} + o(\sqrt{\gamma}) \quad \text{for } i \neq j\\
L_{kl}^{ij} &= \gamma\,D_{kl}^{ij} + o(\gamma) \quad \text{for } i \neq j \text{ and } k \neq l \text{ and } D_{kl}^{ij} = -d_{kl}\delta_k^i\delta_l^j.
\end{aligned} \tag{22}$$

The justification for this scaling comes naturally when calculating the jump rates. Since we shall not carry the proof here, let us just say that the A term corresponds to incoherent contributions like those of the first example, and thus has to be taken fixed. The B and C terms essentially correspond to a Hamiltonian contribution like that of the second example, and need to be enhanced as γ is increased to get a non-zero jump rate. Finally, the diagonal part of the D term has an effect similar to that of the collapse on the average density matrix, and thus needs to be scaling like γ to remain relevant in the limit.

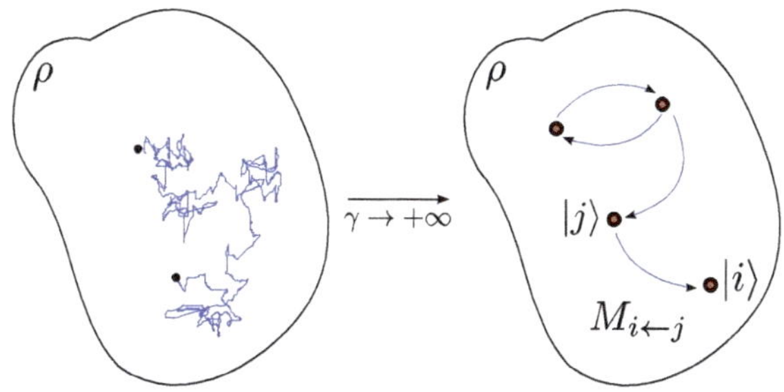

Fig. 5 Illustration of the jump theorem – For finite γ, the state ρ diffuses in Hilbert space. When the collapse rate is sent to infinity, the state spends most of its time near the eigenvectors of the collapse operator. It becomes a Markov chain, randomly jumping from pointer to pointer with a rate $M_{i\leftarrow j}$ that can be computed explicitly

The jump theorem then gives [15] (see also [16]) that in the large γ limit, ρ becomes a Markov chain between the projectors $|k\rangle\langle k|$ with jump rates (or Markov matrix):

$$M_{i\leftarrow j} = A^i_j + 2\,\Re\mathrm{e}\sum_{k<l}\frac{B^i_{kl}C^{kl}_j}{\Delta_{kl}} \tag{23}$$

with $\Delta_{kl} = \frac{1}{2}|\nu_k - \nu_l|^2 + d_{kl}$. To give some intuition about the result we consider a slightly less general situation where $\mathscr{L}(\rho) = A(\rho) - i[H,\rho]$ with A acting diagonally as in (22). Then the jump rates simplify to:

$$M_{i\leftarrow j} = \overbrace{L^{ii}_{jj}}^{\text{“incoherent” contribution}} + \underbrace{\frac{4}{\gamma}\left|\frac{H_{ij}}{\nu_i - \nu_j}\right|^2}_{\text{“coherent” contribution}} . \tag{24}$$

So again, if H is not rescaled $\propto \sqrt{\gamma}$, the coherent contribution is suppressed in the limit. Note the interesting form of this coherent term: it depends not only on the collapse basis but also on the eigenvalues of the collapse operator. This is a very distinct behavior from the one obtained from a projective measurement or instantaneous collapse to a pointer, where nothing physical can depend on the eigenvalues.

There are two strategies to derive (23), a quick and dirty method using the master equation, and a more rigorous one using the evolution for the probability distribution of ρ:

1. One can accept, from the pure collapse discussion, that the collapse will make the state stay near the eigenvectors of the collapse operator most of the time, and further assume the transitions from pointer to pointer will be Markovian. Then, to compute their rate, one only needs to study the master equation for $\bar{\rho}$ (averaged over the collapse noise):

$$\partial_t \bar{\rho}_t = \left(A + \sqrt{\gamma}\,(B + C) + \gamma\,(D + \mathcal{D}[\mathcal{O}])\right)(\bar{\rho}_t). \tag{25}$$

On then carries perturbation theory to second order in γ to find a closed master equation for the diagonal part of ρ: $\partial_t \text{diag}(\bar{\rho}_t) = M\, \text{diag}(\bar{\rho}_t)$, where $\text{diag}(\bar{\rho}_t)$ is written as a column vector. The matrix M is then identified as the Markov matrix of (23). This is the strategy followed in [17]: it is simple as one only needs the master equation (and not the stochastic master equation) but requires one to assume that the limit is indeed a Markov process between pointers.[7]

2. A more rigorous method consists in going one step more abstract. The idea is to study not only the average of the state $\bar{\rho}$, but rather its full probability distribution $\mathbb{P}_t[\rho\,|\,\rho_0]$. From the stochastic master equation (21), one can find the second order Fokker-Planck operator $\mathfrak{D}$ such that

$$\partial_t \mathbb{P}_t[\rho\,|\,\rho_0] = \mathfrak{D}\mathbb{P}_t[\rho\,|\,\rho_0]. \tag{26}$$

With the scaling we have chosen, this differential operators admits the expansion $\mathfrak{D} = \mathfrak{D}_0 + \gamma\mathfrak{D}_1$, hence $\mathbb{P}_t = \exp\left(t\mathfrak{D}_0 + t\gamma\mathfrak{D}_1\right)\mathbb{P}_0$. One then notes that for large γ, probability distributions which survive are in the kernel of $\mathfrak{D}_1$ because it is a negative operator. One then shows that these probability distributions are Dirac measures on the eigenvectors of the collapse operator. A perturbative expansion around these stable points in the space of probability distributions gives the Markovian transitions between them [15]. Hence this method allows to prove that the large γ limit is indeed that of a Markovian jump process between pointers at the same time as it allows to compute the jump rates.

In what sense do we have convergence towards the jump process in the large γ limit? Actually, we have no more than a convergence in law, that is, expressions of the form $\mathbb{E}[f(\rho_{t_1}, \ldots, \rho_{t_N})]$ converge to the ones computed with the limiting jump process in the large γ limit. Could we hope to prove more? No, because a fine grained structure, that is not captured by the jump process, survives in the limit: the *quantum spikes*.

4 Spikes

4.1 A First Observation

To understand the phenomenon of spikes, we will restrict our analysis to the simplest instance in which they appear, in the context of the scalar stochastic differential equation (17)

$$\mathrm{d}p_t = \lambda\,(p_{\text{eq}} - p_t)\,\mathrm{d}t + \sqrt{\gamma}\,p_t(1 - p_t)\,\mathrm{d}W_t. \tag{27}$$

[7] To see that this is not obvious, note that there exist other unravelings of the master equation (25), *i.e.* different stochastic master equations giving the same average master equation, that do not give jumps between pointers in the limit. Hence the jump limit really is a feature of the stochastic description.

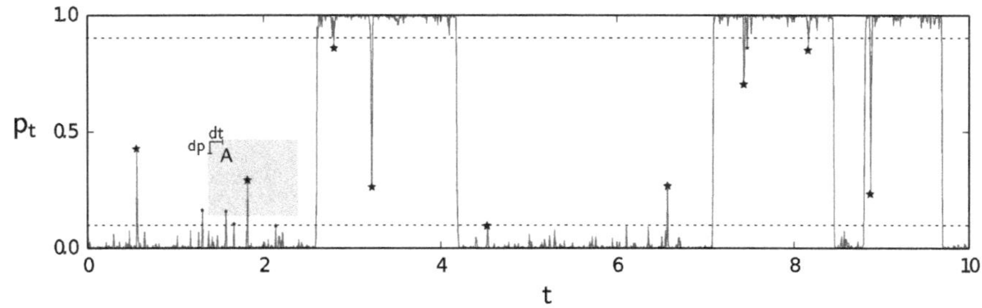

Fig. 6 Trajectories p_t for $\gamma \gg \lambda$ – A typical trajectory for $\lambda = 1$, and $\gamma = 400$, is visually very close to what one would get for $\gamma = +\infty$. Spikes above a certain threshold (here 10%) are marked with stars. To quantify spikes, one considers a domain A of the plane (t, p) in a region without jumps. The number of spikes ending in A is a Poisson random variable with an intensity given in (29)

Already from Fig. 1, the careful reader will have noticed that p_t does not quite converge to a jump process. There seem to be sharp excursions decorating the jump process which one would almost dismiss as numerical artifacts. We call these seemingly instantaneous excursions *quantum spikes* and highlight them them in Fig. 6.

Quantum spikes are fast in the sense that they take a time $\propto \gamma^{-1}$ and thus appear discontinuous when $\gamma \to +\infty$. However, their height remains of order 1 in the limit. Thus, while they disappear in the sense of Lebesgue measure in the fast collapse limit (and thus in quantities like $\mathbb{E}[f(p_{t_1}, \ldots, p_{t_N})]$), they remain if one considers instead first passage times or statistics of local extrema.

It is rather obvious to see what a spike is from a plot like that of Fig. 6, but it is important (and less trivial) to define spikes more precisely. For simplicity, we consider upward spikes starting from 0 (downward spikes, starting from $p = 1$ are treated in the same way). Let us give ourselves two fixed thresholds $\delta \ll \varepsilon \ll 1$. We call a spike an excursion, or piece of trajectory, starting from ε and eventually reaching δ. Because excursions away from 0 become instantaneous in the large γ limit, the only thing we see from them is a vertical line from ε up to the maximum value reached during the excursion and down to δ, hence the name spike (see Fig. 7). Once we have sent γ to $+\infty$ and spikes are effectively instantaneous, we can lower the thresholds ε and δ arbitrarily close to 0 so that the statistics of spikes do not depend on them.

With this definition, spikes can be given a precise characterization. Namely, the number of spikes ending in a finite domain A of the plane (p, t) (see Fig. 6) is a *Poisson process* of intensity μ:

$$\mathbb{P}\left[n \text{ spikes ending in} A\right] = \frac{e^{-\mu}\mu^n}{n!} \quad \text{with } \mu = \int_A d\nu(p, t) \tag{28}$$

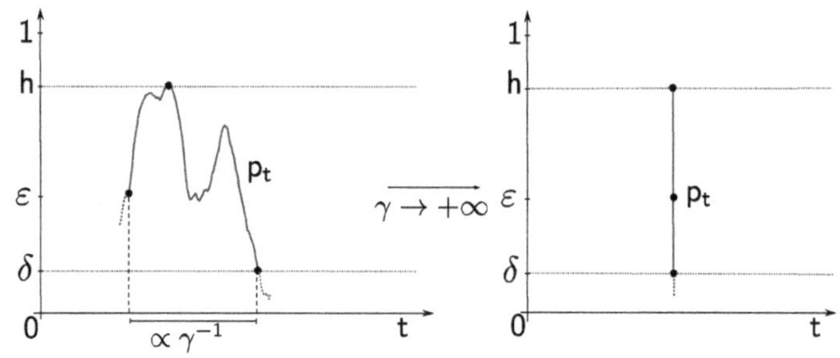

Fig. 7 For γ finite, one needs two thresholds $0 < \delta < \varepsilon \ll 1$ to define a spike (starting from 0). We consider that an excursion starts when the process reaches ε from below, and then stops when it hits δ. In the middle, the excursion reached a maximum here written $h > \varepsilon$. When $\gamma \to +\infty$, the excursion becomes instantaneous, all that remains is a vertical line up to h, which we call a spike. Once γ has been sent to $+\infty$, the two thresholds ε and δ can be sent to zero, and spikes of arbitrarily small size may be considered

The density ν is then given by the following (truncated) power laws:

$$\mathrm{d}\nu_0(p,t) = \mathrm{d}t\,\mathrm{d}p\,\frac{\lambda\, p_{eq}}{p^2} \quad \text{for spikes starting from 0} \tag{29}$$

$$\mathrm{d}\nu_1(p,t) = \mathrm{d}t\,\mathrm{d}p\,\frac{\lambda\,(1-p_{eq})}{(1-p)^2} \quad \text{for spikes starting from 1.} \tag{30}$$

Importantly, γ appears nowhere, the limiting distribution is well defined for γ infinite. Further the integrated density diverges for small spikes $\int_{]0,\varepsilon]\times\Delta t} \mathrm{d}\nu_0(p,t) = +\infty$, and there are thus infinitely many of them.

4.2 Martingale Intuition

The essence of the reason for the existence of spikes is the following:

> As the process p_t is a martingale away from the boundaries when $\gamma \to +\infty$, if there are jumps, there must be aborted jumps (or spikes) as well.

Let us make this argument more precise. Away from the boundaries $p = 0$ or $p = 1$, and when γ is large, the collapse term dominates and $\mathrm{d}p_t \simeq \sqrt{\gamma}\, p_t(1-p_t)\,\mathrm{d}W_t$. Importantly, this means that p_t is approximately a martingale, in particular:

$$\forall T \geq t,\ \mathbb{E}\big[\, p_T \mid p_t = \varepsilon \,\big] \simeq \varepsilon. \tag{31}$$

Let us imagine the process started near 0 at t_0 and has increased to a small value $p_{t_0} = \varepsilon$. What is the probability that the jump completes before p goes below δ?

Let us write $\tau \geq t$ the time when p hits δ (and the jump is considered aborted) or hits $1-\delta$ (and the jump is considered completed). This random variable is a so called stopping time, which implies that equation (31) holds if T is replaced by τ [2]. Furthermore, by definition we have:

$$\mathbb{P}[\text{jump completes}|p_{t_0}=\varepsilon]\times(1-\delta)+\mathbb{P}[\text{jump aborts}|p_{t_0}=\varepsilon]\times\delta=\mathbb{E}[p_\tau|p_{t_0}]. \tag{32}$$

The latter term is just $p_{t_0}=\varepsilon$ because the process is a martingale. Hence

$$\mathbb{P}[\text{jump completes}|p_{t_0}=\varepsilon]=\varepsilon+\text{negligible corrections }\mathcal{O}(\delta), \tag{33}$$

i.e. the probability that a jump completes is equal to how far it already went! In turn, this means there are jumps that do not complete, excursions that reach a certain value p (for example $1/2$), and then go back to their initial value.

4.3 Sketch of a Proof

The previous "martingale" argument for the existence and even necessity of spikes is almost sufficient to compute their distribution. As before, we look only at spikes starting near 0, and we neglect subleading terms $\mathcal{O}(\delta)$. However, instead of looking at the jump completion, we can look at the stopping time τ_h for a given fixed height h such that $\varepsilon<h<1$. This random variable just gives the time when p reached h or the δ-neighbourhood of 0 after starting in ε. We have:

$$\mathbb{P}[p_{\tau_h}=h|p_{t_0}=\varepsilon]\times h+\mathbb{P}[p_{\tau_h}=0|p_{t_0}=\varepsilon]\times 0=\mathbb{E}[p_{\tau_h}|p_{t_0}=\varepsilon]=\varepsilon. \tag{34}$$

Hence,

$$\mathbb{P}[p_{\tau_h}=h|p_{t_0}=\varepsilon]=\frac{\varepsilon}{h}. \tag{35}$$

This probability is also the probability that the maximum p reaches before going back to the δ-neighbourhood of 0 is superior to h:

$$\mathbb{P}\left[\max_{u<\tau}(p_u)\geq h \mid p_{t_0}=\varepsilon\right]=\frac{\varepsilon}{h}. \tag{36}$$

Therefore, we have in differential form:

$$\mathrm{d}\mathbb{P}\left[\max_{t_0<u<\tau}(p_u)=h<1-\delta \mid p_{t_0}=\varepsilon\right]=\varepsilon\,\frac{\mathrm{d}h}{h^2}. \tag{37}$$

This explains the $1/p^2$ in the density of spike maxima (29). Note in passing that there is an additional term for $h = 1$ (again, up to a δ neighbourhood), because $\mathbb{P}\left[\max_{u\leq T}(p_u) = 1 \mid p_{t_0} = \varepsilon\right] = \varepsilon$. Hence

$$\mathrm{d}\mathbb{P}\left[\max_{t_0<u\leq T}(p_u) = h \leq 1 \mid p_{t_0} = \varepsilon\right] = \varepsilon\left[\frac{\mathrm{d}h}{h^2} + \delta(1-h)\,\mathrm{d}h\right]. \tag{38}$$

How often do we get to try to jump, *i.e.* how often does the process reach at least ε in any small time interval Δt? To answer this question, we can use a simple consistency argument, namely that the probability to reach ε is related to the jump rate which we know from the previous Sect. 3.

Let us consider a time interval Δt such that $\varepsilon\lambda^{-1} \ll \Delta t \ll \gamma^{-1}$. The probability that a jump from 0 to 1 occurs during Δt is simply $\lambda p_{\mathrm{eq}}\Delta t$. This jump probability can be decomposed into the probability to reach at least ε and then complete a jump,[8] which reads, from the previous discussion $\mathbb{P}[\{p_u\}_{t\leq u\leq t+\Delta t}$ reaches at least $\varepsilon] \times \varepsilon$. Hence we have

$$\mathbb{P}[\{p_u\}_{t\leq u\leq t+\Delta t} \text{ reaches at least } \varepsilon] = \frac{\lambda p_{\mathrm{eq}}}{\varepsilon}\Delta t. \tag{39}$$

As a result, in every small interval Δt, the probability density that there is an excursion reaching a maximum h is

$$\mathrm{d}\mathbb{P}\left[\max_{t<u\leq t+\Delta t}(p_u) = h < 1 \mid p_{t_0} = \varepsilon\right] = \frac{\Delta t\lambda p_{\mathrm{eq}}}{h^2}\mathrm{d}h. \tag{40}$$

Because $\Delta t \gg \gamma^{-1}$, the statistics of spikes from two different intervals are independent and we thus get that the maxima of the excursions are given by the Poisson process we advertised in (29).

This sketch of proof, relying almost only on the martingale property, follows closely the way spikes were first characterized [18]. This line of argument has been made rigorous by Kolb & Lisenfeld [19]. Other proof strategies exist, notably one exploiting a time reparameterization of the dynamics transforming p_t into a reflected Brownian motion (see [20] for a physicist explanation of the argument, and [21, 22] for mathematically rigorous derivations). The case of the qubit with a coherent evolution discussed in 3.2 can also be treated, and the spikes have the same power law statistics (up to a different prefactor). However, although spikes show up in numerical simulations of dynamics in larger Hilbert spaces and seem ubiquitous, no theoretical characterization is known beyond the qubit case.

---

[8] Note that we do not consider the probability that ε is reached more than once during Δt and thus that there could be several jump attempts. This is because the probability to reach ε during Δt is already much smaller than 1 for our choice of Δt and thus probabilities of having more than one attempt per Δt are subleading.

4.4 Are Spikes Real?

Now that we have precisely characterized them and are sure of their mathematical existence, we should ask ourselves whether spikes are relevant. Are spikes real, out there in the world, or just a modeling artifact? In fact, this question is far subtler than it seems, and the answer depends on what ontological commitments one makes.

Classical spikes and hidden Markov models – A first question we could ask is to know if spikes could appear classically, merely as the result of imperfect knowledge of an underlying (well defined) jump process. This is indeed the case for the equation we have focused on

$$\mathrm{d}p_t = \lambda\,(p_{\mathrm{eq}} - p_t)\,\mathrm{d}t + \sqrt{\gamma}\,p_t(1-p_t)\,\mathrm{d}W_t \tag{41}$$

which can be obtained as the real time probability of a hidden Markov model [18]. More precisely, consider a (classical) Markov process R_t (R for real) that can jump randomly from the value 0 to the value 1 in such a way that, on average:

$$\frac{\mathrm{d}}{\mathrm{d}t}\bar{R}_t = \lambda(p_{eq} - \bar{R}_t) \;\text{ with }\; \bar{R}_t = \mathbb{E}[R_t]. \tag{42}$$

One can construct a model of (classical) continuous imperfect observation of this process (say with repeated blurry pictures) and consider the *filtered* probability $p_t^{\mathrm{f}} = \mathbb{P}[R_t = 1|\text{blurry pictures up to } t]$. This filtered probability p_t^{f} encodes the knowledge we have of the Markov process position R_t at time t using all past blurry pictures. There exists a particular classical imperfect observation scheme such that p_t^{f} verifies exactly equation (41) [18]. Hence, in this context, spikes, which are still somehow unexpected, can be explained as an artifact of our residual ignorance of the underlying jump process. Even with the optimal filter, we cannot know the system arbitrarily well because it jumps instantly and we *know* that it does! At the Bayesian optimum, provided by the filter, one gets (perhaps surprisingly) a lot of false alerts (the spikes).

More generally, if the density matrix remains diagonal in the collapse basis (which is the case for incoherent transitions), there exists a classical hidden Markov model whose filtered probability verifies exactly the stochastic master equation we put forward. In fact, in the continuous measurement context, SMEs can be understood as the generalization to non-diagonal matrices of the Kushner-Stratonovich filtering equations used in the context of classical estimation. Hence, so long as one sticks with diagonal density matrices, one can always interpret the collapse as a Bayesian updating of a state of knowledge about a well defined classical variable that evolves independently of the collapse/measurement (that is, without mechanical back-action). Naturally, this equivalence breaks down if the density matrix is not diagonal during the evolution (for example if it evolves unitarily) as is the case in the second example we considered in 3.2.

Quantum spikes and hidden variable theories – In the general case, the convenient classical interpretation can no longer explain the spikes away. One could still construct hidden variables[9] doing discrete jumps (without spikes) but then their jump probabilities would depend on the quantum state [17], and thus the spikes would appear to be physical. Hence, while there are spikes that can be explained classically, it seems there exists genuinely quantum spikes as well, which cannot be dismissed as easily as an artifact of Bayesian updating.

A matter of ontology—In the end, to know if spikes would be real or not in the context of a given continuous collapse model, one needs to say precisely what is real in the first place, *i.e.* what the ontology of the theory is. A popular choice is to take some expectation value over the state,[10] *i.e.* $\langle\psi_t|\mathcal{O}|\psi_t\rangle$. For such a choice, spikes are unequivocally real. But there are other possibilities, for example flashes (or their continuous equivalent sometimes called "signal" in the continuous measurement context). These latter ontologies are convenient in some cases, as they allow to consistently couple the collapse model describing quantum matter with a classical sector (for example gravity [23, 24]). Further, for what interests us here, these ontologies do not have spikes.

Trimming spikes by knowing the future—There is a third option, also inspired from continuous measurement theory but which, to our knowledge, has never been considered in the context of collapse models: forward-backward estimates (or rather their quantum version). In the classical case, we saw that a filtered probability

$$p_t^f := \mathbb{P}\left[R_t = 1|\{\text{observations before } t\}\right], \tag{43}$$

encoding the knowledge one has in real time about a classical Markov process had spikes. Another quantity, quite natural in the classical context, is the forward-backward or a smoothed probability:

$$p_t^{f.b.} := \mathbb{P}\left[R_t = 1|\{\text{all past and future observations}\}\right]. \tag{44}$$

This quantity can only be computed after all the observations have been carried, and not in real time. Intuitively, it is easy to know a posteriori that a spike was just a spike and that no real jump was about to happen, and thus the forward-backward probability $p_t^{f.b.}$ should not have spikes. This intuition is confirmed by numerical simulations: $p_t^{f.b.}$ is smoother (differentiable) and without spikes [18]. In the quantum context, there is no unambiguous definition of a density matrix conditioned on the past and the future. Different notions have been put forward, like the *past quantum state* [25] and the *smoothed quantum state* [25, 26]. For the former, spikes disappear, but the resulting state generically loses its density operator properties (which does not matter if one just aims to define an ontology from an expectation value) while in the former the output is still a *bona fide* quantum state $\rho_t^{f.b.}$ but spikes are generically not tamed.

[9]Such hidden variable theories are easy to construct, and are essentially the discrete version of Bohmian mechanics introduced by Bell in the context of quantum field theory. Including a continuous collapse/measurement on top of such dynamics is done *e.g.* in [17].

[10]In the context of physically realistic collapse models, the operator is typically position dependent and proportional to the regularized mass density.

Summary—The reality of spikes depends on the choices we make. One of the equations we obtained in the quantum context and that shows spikes can be obtained from a (classical) hidden Markov model. In the latter, spikes live in our minds only: nothing is spiky in Nature, but our best *real time* knowledge is. Spikes vanish once we look back and are only asked to tell a posteriori where the process was. This makes it a bit unsatisfying to have the spikes be real in the quantum context, especially if it implies giving them a reality as well in the cases that could be just as well described by a classical hidden Markov model. This very minor aesthetic criterion could help compare different collapse model ontologies.

5 Generalization and Open Problems

Let us summarize what is known on the mathematical front. For a finite dimensional Hilbert space and a generic continuous collapse process, one can easily prove a convergence towards pointer states as predicted by the Born rule. When a small additional dynamics is added, we see the emergence of jumps in the fast collapse regime. The statistics of these jumps can be computed in full generality. Another feature, spikes, seem ubiquitous. However, they are quantitatively understood only in the qubit case.

A first possible generalization is to go from a continuous collapse model to a discrete one, and replace the diffusive equations we had with jump ones. Note that there the jumps we would see are not the same as the emerging jumps between pointers, but could be far smaller jumps in Hilbert space formally equivalent to weak measurements as in the Ghirardi-Rimini-Weber model [27]. So long as the collapses are not exactly projective, they introduces a new timescale, just like the γ we had in the continuous case. When the frequency of discrete collapses is sent to infinity, one obtains jumps between pointers that can be quantified exactly just as before [16, 17]. However, while spikes are numerically present as well in this context and seem to have the exact same power law statistics, no proof is known even for the qubit case.

A second generalization would be to characterize spikes precisely for Hilbert spaces of arbitrary finite dimension. This is a surprisingly non trivial task. One reason is that in higher dimensions, it is unclear to know on which submanifold of the Hilbert space the spikes happen, as there is no longer a single path connecting pointers. In general, knowing on which submanifold the trajectories stay for a given collapse operator is a hard question, which has been solved only in simple cases [28].

Finally, it would be interesting to rigorously extend the results we presented here to more realistic situations with Hilbert spaces of infinite dimension. A lot of progress was made on the mathematical physics front in the recent years, by Ballesteros et al. [29] in the pure collapse setup and Bauer et al. [30] in the case where collapse competes with other simple dynamics.[11] However, the general case and an extension to the even larger Hilbert spaces of quantum field theories remain an open-problem.

6 Summary and Conclusion

Continuous collapse models on finite dimensional Hilbert spaces give rise to rich dynamics, already in the simplest case yielding the stochastic differential equation

$$\mathrm{d}p_t = \lambda\,(p_{\mathrm{eq}} - p_t)\,\mathrm{d}t + \sqrt{\gamma}\,p_t(1 - p_t)\,\mathrm{d}W_t. \tag{45}$$

When pure collapse is considered [$\lambda = 0$ in (45)], one obtains a progressive reduction of the quantum state to one (random) eigenvector (or pointer) of the collapse operator [$p = 0$ or 1 in (45)]. This happens because collapse acts as a pure noise term which vanishes only on these eigenvectors, which are thus fixed points of the dynamics.

The fact that this reduction is progressive and not instantaneous allows to compare its rate γ to other system dynamics (unitary or dissipative). Generically, when the collapse process is much faster than other dynamics [$\gamma \gg \lambda$ in (45)], we see the (expected) emergence of quantum jumps between the pointers. The latter are not strictly instantaneous, and take roughly γ^{-1} to complete. These jumps can be characterized precisely and we note two important facts:

1. The jump rates decrease as a function of the collapse rate γ when their origin is a Hamiltonian coupling (Zeno effect), whereas they do not depend on γ for dissipative transitions (no Zeno effect), as is the case for (45),
2. The jump rates generically depend on the eigenvalues of the collapse operator and not just on the eigenvectors as one would expect for projective measurements.

Finally, the jumps are not the whole story. Perhaps surprisingly, they come decorated with spikes, thin excursions that never complete into jumps. Spikes are power law distributed and prevent any strong form of convergence towards the jump process. They seem ubiquitous and the proof of their existence and rigorous characterization has been done in simple cases. However, their ontological status (are spikes real or just in our mind?) is subtle, especially when collapse models are compared with hidden Markov models.

In the end, it is not entirely clear if the fine characterization of collapse models in the overly simple context we have discussed has any physical relevance. It is possible that spikes, for example, will remain a mathematical curiosity and nothing more. Nonetheless, even then, this study will have put into light a stochastic differential equation (45) with a surprisingly rich behavior, that sparked interest in mathematics [19, 22], and even finance [33].

---

[11] An example of such competition in the continuous case is given by the QMUPL model for a free particle where $H \propto \hat{P}^2$ and $\mathcal{O} \propto \hat{X}$. For this particular model, a lot is known rigorously (see *e.g.* [31, 32] and references therein) and the dynamics is very rich, with behavior already qualitatively distinct from the simpler discrete setting we considered.

References

1. P. Rouchon and J. F. Ralph, Phys. Rev. A **91**, 012118 (2015) https://doi.org/10.1103/PhysRevA.91.012118
2. B. Øksendal, *Stochastic differential equations* (Springer, Berlin Heidelberg, 2003)
3. L. Diósi, Phys. Rev. A **40**, 1165 (1989) https://doi.org/10.1103/PhysRevA.40.1165
4. A. Bassi, K. Lochan, S. Satin, T. P. Singh, and H. Ulbricht, Rev. Mod. Phys. **85**, 471 (2013 a) https://doi.org/10.1103/RevModPhys.85.471
5. A. Bassi, D. Dürr, and G. Hinrichs, Phys. Rev. Lett. **111**, 210401 (2013 b) https://doi.org/10.1103/PhysRevLett.111.210401
6. L. Diósi, Phys. Rev. Lett. **112**, 108901 (2014) https://doi.org/10.1103/PhysRevLett.112.108901
7. K. Jacobs and D. A. Steck, Contemp. Phys. **47**, 279 (2006) https://doi.org/10.1080/00107510601101934
8. H. M. Wiseman and G. J. Milburn, *Quantum measurement and control* (Cambridge university press, Cambridge UK, 2009)
9. Q. Ficheux, S. Jezouin, Z. Leghtas, and B. Huard, Nat. Comm. **9**, 1926 (2018) https://doi.org/10.1038/s41467-018-04372-9
10. Q. Ficheux, *Quantum Trajectories with Incompatible Decoherence Channels*, Ph.D. thesis, school École normale supérieure Paris (2018) DOI: https://tel.archives-ouvertes.fr/tel-02098804
11. N. Gisin, Helv. Phys. Acta **62**, 363 (1989)
12. A. Bassi and K. Hejazi, Eur. J. Phys. **36**, 055027 (2015) https://doi.org/10.1088/0143-0807/36/5/055027
13. S. L. Adler, D. C. Brody, T. A. Brun, and L. P. Hughston, J. Phys. A: Math. Gen. **34**, 8795 (2001) https://doi.org/10.1088/0305-4470/34/42/306
14. M. Bauer and D. Bernard, Phys. Rev. A **84**, 044103 (2011) https://doi.org/10.1103/PhysRevA.84.044103
15. M. Bauer, D. Bernard, and A. Tilloy, J. Phys. A: Math. Theor. **48**, 25FT02 (2015) https://doi.org/10.1088/1751-8113/48/25/25ft02
16. M. Ballesteros, N. Crawford, M. Fraas, J. Fröhlich, and B. Schubnel, Annales Henri Poincaré **20**, 299 (2019) https://doi.org/10.1007/s00023-018-0741-z
17. A. Tilloy, *Mesure continue en mécanique quantique: quelques résultats et applications*, Ph.D. thesis, school Paris Sciences et Lettres (2016) DOI: https://doi.org/10.pdf
18. A. Tilloy, M. Bauer, and D. Bernard, Phys. Rev. A **92**, 052111 (2015) https://doi.org/10.1103/PhysRevA.92.052111
19. M. Kolb and M. Liesenfeld, Annales Henri Poincaré (2019), https://doi.org/10.1007/s00023-019-00772-9
20. M. Bauer, D. Bernard, and A. Tilloy, J. Phys. A: Math. Theor. **49**, 10LT01 (2016) https://doi.org/10.1088/1751-8113/49/10/10lt01
21. M. Bauer and D. Bernard, Annales Henri Poincaré **19**, 653 (2018) https://doi.org/10.1007/s00023-018-0645-y
22. C. Bernardin, R. Chetrite, R. Chhaibi, J. Najnudel, and C. Pellegrini, arXiv:1810.05629 (2018)
23. A. Tilloy, Phys. Rev. D **97**, 021502 (2018) https://doi.org/10.1103/PhysRevD.97.021502
24. A. Tilloy, arXiv:1903.01823 (2019
25. S. Gammelmark, B. Julsgaard, and K. Mølmer, Phys. Rev. Lett. **111**, 160401 (2013) https://doi.org/10.1103/PhysRevLett.111.160401
26. I. Guevara and H. Wiseman, Phys. Rev. Lett. **115**, 180407 (2015) https://doi.org/10.1103/PhysRevLett.115.180407
27. G. C. Ghirardi, A. Rimini, and T. Weber, Phys. Rev. D **34**, 470 (1986) https://doi.org/10.1103/PhysRevD.34.470
28. A. Sarlette and P. Rouchon, J. Math. Phys. **58**, 062106 (2017) https://doi.org/10.1063/1.4984587

29. M. Ballesteros, N. Crawford, M. Fraas, J. Fröhlich, and B. Schubnel, in *Mathematical Problems in Quantum Physics*, Vol. 717, edited by F. Bonetto, D. Borthwick, E. Harrell, and M. Loss (American Mathematical Society, 2018) p. 241 https://doi.org/10.1090/conm/717/14452
30. M. Bauer, D. Bernard, and T. Jin, SciPost Phys. **5**, 37 (2018) https://doi.org/10.21468/SciPostPhys.5.4.037
31. A. Bassi, J. Phys. A: Math. Gen. **38**, 3173 (2005) https://doi.org/10.1088/0305-4470/38/14/008
32. A. Bassi, D. Dürr, and M. Kolb, Rev. Math. Phys. **22**, 55 (2010) https://doi.org/10.1142/S0129055X10003886
33. C. Henkel, Physica A **469**, 447 (2017) https://doi.org/10.1016/j.physa.2016.11.125

Theoretical Physics

Collapse Models, Relativity, and Discrete Spacetime

Daniel J. Bedingham

Abstract In collapse models, the induced narrowing of the wavefunction typically leads to an increase in energy. For realistic non-relativistic models, the parameters of the model can be set such that the energy increase is small enough to be within experimental bounds. However, for relativistic versions of collapse models the energy increase is divergent. Here we show how to regulate this divergent behaviour by formulating a collapse model on a discrete Lorentzian spacetime. The result is a relativistic collapse model with finite energy production. This energy increase can be made sufficiently small with a reasonable choice for the discreteness scale of spacetime.

1 Introduction

In practical applications of quantum theory there is a quantum state with a rule for how it changes with time when the system is left alone (unitary dynamics described by the Schrödinger equation), along with a collapse rule to use when a measurement is made. The collapse rule determines, from the quantum state, the various probabilities for the possible measurement outcomes, and updates the state according to which one was realised. The fundamental problem is that both of these rules involve a change to the state, but the question of when to use one or the other is determined by a judgement as to whether the process involved is a measurement or not. This is no good for a supposedly fundamental theory. Admittedly, for most practical applications, whether or not something is a quantum measurement is clear, but as we try to make larger quantum systems (or smaller measuring devices) the distinction will become less clear.

D. J. Bedingham (✉)
Department of Physics, Royal Holloway, University of London, Egham YW20 0EX, UK
e-mail: daniel.bedingham@rhul.ac.uk

V. Allori et al. (eds.), *Do Wave Functions Jump?*, Fundamental Theories of Physics 198,
https://doi.org/10.1007/978-3-030-46777-7_15

There are various responses to the measurement problem, none of which can claim general acceptance. Collapse models are one such response [1–5], the aim being to find a unified dynamical rule which reproduces the usual two rules of standard quantum theory (unitary state dynamics or state collapse during measurement) approximately in situations where they would respectively be expected. This removes the need to form a view about whether a processes is a measurement in order to explain what happens.

Collapse models have certain advantages. In the orthodox picture of quantum theory the state describes the various possibilities and their probabilities of occurring, and measurement outcomes are our direct contact with the quantum world. Collapse models closely resemble this picture (but without the need to distinguish measurements), as will be explained in Sect. 2 where we outline the general structure of collapse models. Also, collapse models can be made to satisfy the principle of relativity, meaning that they can be expressed in terms of covariant equations with no dependence on a particular frame of reference. This will be outlined in Sect. 3.

In Sect. 3 it is also shown that basic relativistic collapse models suffer from divergent behaviour. This is addressed in the remainder of the article, Sects. 3 and 4, where we construct a collapse model on a discrete Lorentzian spacetime. Specifically we use a scalar quantum field theory defined on a causal set spacetime structure. Causal sets as a description of spacetime are primarily motivated by attempts to combine quantum theory with gravity. For us, the discrete spacetime structure regulates the divergent behaviour of the collapse model.

2 Collapses Are Like Measurements

In a generalised quantum measurement (see Ref. [6]) there are a collection $\{\hat{M}_m\}$ of measurement operators. The index m refers to the measurement outcome. If the quantum state is $|\psi\rangle$, then the probability of the measurement outcome m is given by

$$p(m) = \frac{\langle\psi|\hat{M}_m^\dagger \hat{M}_m|\psi\rangle}{\langle\psi|\psi\rangle}, \tag{1}$$

and the (unnormalised) state of the system after the measurement with outcome m changes by

$$|\psi\rangle \to \hat{M}_m|\psi\rangle. \tag{2}$$

In order that the probabilities for the different outcomes sum to 1, the measurement operators must satisfy the completeness relation

$$\sum_m \hat{M}_m^\dagger \hat{M}_m = 1. \tag{3}$$

This definition encompasses familiar projective measurements (where $\hat{M}_m = \hat{P}_m$ with $\hat{P}_i\hat{P}_j = \delta_{ij}\hat{P}_i$), and weak measurements. For example, for a qubit state with basis vectors $|0\rangle$ and $|1\rangle$ we can form a set of measurement operators $\{\hat{M}_0 = (|0\rangle\langle 0| + K|1\rangle\langle 1|)/(1+K^2), \hat{M}_1 = (K|0\rangle\langle 0| + |1\rangle\langle 1|)/(1+K^2)\}$ with $0 < K < 1$. Given an initial state of the form $|\psi\rangle = a|0\rangle + b|1\rangle$, and a measurement outcome $m = 0$, the resultant state using Eq. (2), is

$$|\psi\rangle \to \hat{M}_0|\psi\rangle \propto a|0\rangle + Kb|1\rangle. \tag{4}$$

This measurement does not precisely determine the basis state to which the qubit belongs, correspondingly it has a milder impact on the state than a projective measurement. The effect in this case is to enhance the amplitude of $|0\rangle$ relative to $|1\rangle$.[1]

The structure of a collapse model generally involves a sequence of operations of this kind occurring spontaneously. The sense in which this happens is that at certain times, the state is randomly impacted in the same way as described by Eq. (2) with probability of the particular outcome m given by Eq. (1). There is no measurement as such. The times at which these collapses take place may be random and uniformly distributed in time, or regular and evenly spaced, or various other possibilities.

The well-known Ghirardi-Rimini-Weber (GRW) model [3] takes this form. As originally presented, the model concerns a set of distinguishable particles, labelled by i. For each particle there is an independent set of random collapse times with Poisson (uniform) distribution of rate λ. For particle i the collapse (or measurement) operator takes the form

$$\hat{L}_i(\mathbf{z}_t) = \frac{1}{(2\pi\sigma)^{3/4}} e^{-\frac{(\hat{\mathbf{x}}_i - \mathbf{z}_t)^2}{4\sigma^2}}. \tag{5}$$

Here $\hat{\mathbf{x}}_i$ is the (3-dimensional) position operator for particle i and $\mathbf{z}_t \in \mathbf{R}^3$ is a random variable which plays the role of the measurement outcome. It is easy to check that this operator satisfies the completeness relation

$$\int d^3\mathbf{z}\hat{L}_i^2(\mathbf{z}) = 1, \tag{6}$$

equivalent to Eq. (3), but for a Hermitian operator with a continuously valued random outcome.

The way the model works is that when the collapse operator spontaneously acts at time t say, the effect on the state is equivalent to that of performing a weak measurement of the particle's position. This has the effect via Eq. (2) of localising the state of the particle to within a distance σ of the random outcome $\mathbf{z}_t$. The variable $\mathbf{z}_t$ can, by the measurement analogy, be thought of as a weak estimate of the particle's position as this time.

---

[1]Weak measurements such as this can be brought about by the use of a projective measurement on an auxiliary quantum system which interacts with the system of interest.

By having a very low rate of collapse for individual particles λ it is possible for these collapses to have a negligible effect on small quantum systems for which unitary state dynamics already provide an excellent description (the likelihood of a collapse occurring when we look is negligible). However, for a large scale superposition of the form '*all particles close to* x_1 *or all particles close to* x_2', the states of the individual particles are entangled and it only takes the collapse of one constituent particle to collapse the entire bulk object. If there are a lot of particles, the probability of one of them experiencing a collapse in a short time can be large. This means that, for example, a macroscopic pointer in a quantum measurement will rapidly collapse to one position or another. The GRW model can therefore describe both the unitary behaviour of small systems and the collapse behaviour of quantum measurement in a single dynamical framework.

There remains the question of how to make a correspondence between this model and the physical world of our experience. In order to do this we must be able to determine definite locally defined quantities from the model which represent observable features of the world. Although various ways of teasing out a physical world from this model have been proposed, the most natural one (in my view) takes into account the analogy between collapses and generalised quantum measurements to treat the random outcomes $\mathbf{z}_t$ (which in the generalised quantum measurement case would be viewed as the weakly measured position outcomes) as representative of the location of matter. The collection of all these collapse locations $\{\mathbf{z}_t\}$ for all particles, each with an independent Poisson distributed sequence of collapse times, describes the world: where these collapses are dense there is matter (see Ref. [7]).

To be clear, in this picture the $\{\mathbf{z}_t\}$ describe the physical world, and the quantum state is a means of estimating those $\{\mathbf{z}_t\}$ in the future based on what we know about the past (which means those $\{\mathbf{z}_t\}$ in the past).

The GRW model is not the only collapse model. For other models, these same ideas can be translated. We note in particular that continuous collapse models (which employ a continuous diffusion process) can be viewed as a limit case of a discrete collapse process.

3 Constructing a Relativistic Collapse Model

For a relativistic collapse model [8–10] our primary demand is that the description should be covariant, that is we should be able to write down the equations in a form that doesn't depend on the frame of reference (or more generally the coordinates on a given patch of spacetime). This is more or less guaranteed if we can express the collapse process in tensor form. Suppose we consider a relativistic quantum field with state $|\Psi\rangle$. In the interaction picture the unitary dynamics of the state is described by

$$|\Psi_{\sigma'}\rangle = \hat{U}(\sigma', \sigma)|\Psi_\sigma\rangle, \tag{7}$$

with

$$\hat{U}(\sigma', \sigma) = Te^{-i\int_{\sigma}^{\sigma'} dV \hat{\mathcal{H}}_{\text{int}}(x)}, \tag{8}$$

where $\hat{\mathcal{H}}_{\text{int}}(x)$ is the interaction Hamiltonian density, dV is the spacetime volume measure, and T is the time ordering operator. The unitary operator $\hat{U}(\sigma', \sigma)$ describes the change in state from spacelike hypersurface σ to spacelike hypersurface σ'. The spacelike hypersurface generalises the idea of a timeslice. The state is given with reference to some σ since we are interested in the state at some point in time. It should be the case that nowhere is σ' to the past of σ—the unitary dynamics represent a change of the state as we move forward in time.

Provided that $\hat{\mathcal{H}}_{\text{int}}(x)$ is a Lorentz-scalar operator then the equations are covariant. Furthermore, in order for Eq. (7) to be unambiguously defined, we require that $[\hat{\mathcal{H}}_{\text{int}}(x), \hat{\mathcal{H}}_{\text{int}}(y)] = 0$ for spacelike separated x and y. This ensures that the ordering of spacelike separated interactions has no effect on the overall outcome and allows us to combine unitary operators, for example

$$\hat{U}(\sigma'', \sigma) = \hat{U}(\sigma'', \sigma')\hat{U}(\sigma', \sigma), \tag{9}$$

provided σ' is nowhere to the past of σ and nowhere to the future of σ''. The result is that the dynamics is foliation independent, meaning that when we advance the state from σ to σ'', the result is independent of any intermediate state σ' we may choose to pass through.

In order to construct a satisfactory relativistic model of collapses we first assume that the collapse events are associated to points in spacetime (rather than just points in time as with GRW) and that they are distributed uniformly across spacetime (i.e. Poisson distributed over invariant spacetime volume). This gives an invariant distribution of collapse events. We assume that if there is a collapse event at point x, then when the state advances to a surface σ to which the point x belongs it changes by

$$|\Psi_\sigma\rangle \rightarrow \hat{J}_x(z_x)|\Psi_\sigma\rangle, \tag{10}$$

where $\hat{J}_x$, the collapse operator at point x, is a Lorentz-scalar operator, and z_x is the random outcome. The outcome should be drawn from a probability distribution

$$p(z_x) = \frac{\langle\Psi_\sigma|\hat{J}_x(z_x)^\dagger \hat{J}_x(z_x)|\Psi_\sigma\rangle}{\langle\Psi_\sigma|\Psi_\sigma\rangle}, \tag{11}$$

and therefore the collapse operators must satisfy

$$\int dz \hat{J}_x(z)^\dagger \hat{J}_x(z) = 1 \tag{12}$$

(for continuously-valued z). In addition we must stipulate that

$$[\hat{J}_x(z_x), \hat{J}_y(z_y)] = 0; \quad [\hat{J}_x(z_x), \hat{\mathcal{H}}_{\text{int}}(y)] = 0, \tag{13}$$

for spacelike separated x and y. Again this ensures that the ordering of spacelike separated interactions and collapse events has no effect on the outcome.

Given these rules it can be shown [10] that the joint distribution for the outcomes of a set of collapses occurring between σ and σ' (with σ' nowhere to the past of σ), is independent of the foliation of spacetime used to pass between these surfaces. In addition, given the outcomes for this set of collapses, then the final state associated to the hypersurface σ' is unambiguously defined.

The construction is therefore independent of frame or foliation apart from the need to specify initial and final hypersurfaces σ and σ'. However, this is no problem: the initial surface σ is there to specify the known history that we take to define the initial state (with our interpretation of the $\{z_x\}$ as determining local physical quantities, by this we mean that we know $\{z_x\}$ to the past of σ sufficiently well to determine the state on σ); the final surface σ' simply sets the region over which collapse outcomes are of interest.

The challenge is then to find a collapse operator which satisfies these rules. For a real scalar quantum field with field operator $\hat{\phi}(x)$ this is not hard, for example [11]

$$\hat{J}_x(z_x) = \left(\frac{\beta}{\pi}\right)^{1/4} e^{-\frac{\beta}{2}(\hat{\phi}(x)-z_x)^2}, \tag{14}$$

where β is some fixed parameter which we can call the collapse strength (note that there is also another fixed parameter representing the density of collapse events in spacetime, call it μ). With a standard interaction Hamiltonian of the form $\hat{\phi}^4(x)$, all commutation properties above follow from the fact that the field operator commutes with itself at spacelike separation.

The way that this model works is that the $\{z_x\}$ are equivalent to the results of weak measurements of $\hat{\phi}(x)$. This gives a measure of the amount of matter at a given point in spacetime. The $\{z_x\}$ therefore represent the distribution of matter throughout spacetime.

The problem arises when we examine the energy increase caused by these collapse events [11]. It is a general feature of collapse models that as collapse leads to a narrowing of the wave function, there is an increase in the average energy of the state. In the present case, as collapse results in a narrowing of the state of $\hat{\phi}(x)$, there is an increase in the variance of the conjugate momentum and therefore an average increase in energy. This would all be fine if this energy increase were sufficiently small to be consistent with experimental bounds, but it turns out that there is an infinite rate of increase of energy density. To see this calculate the expected change of energy for a collapse event at point x in the limit that β is small:

$$\Delta E = \int dz \frac{\langle \Psi_\sigma | \hat{J}_x(z)[\hat{H}, \hat{J}_x(z)] | \Psi_\sigma \rangle}{\langle \Psi_\sigma | \Psi_\sigma \rangle} = \frac{\beta \delta^3(\mathbf{x}=0)}{4} \tag{15}$$

where $\hat{H}$ is the free field Hamiltonian, and $\delta^3(\mathbf{x}=0) = 1/(2\pi)^3 \int d^3\mathbf{p} e^{i\mathbf{x}\cdot\mathbf{p}}|_{\mathbf{x}=0}$.

For the remainder of this article we present a possible solution to this problem which is to regulate this energy increase by working in a discrete form of Lorentzian spacetime. The basic argument is that if spacetime is discrete, then there is some fundamental length scale a which means that $\delta^3(\mathbf{x}=0) \sim a^{-3}$.[2] This is potentially a very large number, yet it leaves the possibility of setting the remaining parameters of the model (the collapse strength β; and the density of collapse events in spacetime μ) such that energy increases are small yet collapse effects sufficient to account, through $\{z_x\}$, for the macro world of our experience.

4 Discrete Spacetime, Quantum Fields, and Collapses

A causal set [12, 13] is a set of points with a partial order relation $\preceq$ such that if $x \preceq y$ then the point x is understood to be in the causal past of y (including the possibility that $x = y$). That the ordering is partial means that two points x and y do not have to satisfy either $x \preceq y$ or $y \preceq x$. In these cases it is understood that the points are not causally related. We impose three conditions on the causal set: (i) Reflexivity—($x \preceq x$); (ii) Antisymmetry—($x \preceq y \preceq x \implies x = y$); and (iii) Transitivity—($x \preceq y \preceq z \implies x \preceq z$). These conditions would be satisfied by points on a Lorentzian manifold with no closed causal curves. Finally, the discreteness of spacetime is expressed by a further condition that if $x \preceq y$, then there are a finite number of points z such that $x \preceq z \preceq y$. Note that if $x \preceq y$ and $x \neq y$ we can write $x \prec y$.

Comparison between a causal set and a Lorentzian manifold is made using the idea of a faithful embedding: a causal set can be faithfully embedded in a Lorentzian manifold if the causal set can be mapped onto points in the manifold in such a way that the partial order relations match the causal relations between the points on the manifold and that the points are uniformly distributed on average over spacetime

[2]This can be motivated by considering a one-dimensional space of length L as a regular array of N points. The kronecker delta function can be written

$$\delta_{nm} = \frac{1}{N} \sum_{k=1}^{N} e^{2\pi i \frac{k}{N}(n-m)}, \tag{16}$$

and so if we define $x_n = Ln/N$ and $p_k = 2\pi k/L$ so that $\Delta x = a = L/N$, and $\Delta p = 2\pi/L$, then

$$a^{-1}\delta_{nm} = \frac{1}{2\pi} \sum_{k=1}^{N} \Delta p e^{ip_k(x_n - x_m)}. \tag{17}$$

This is the discretised version of the Dirac delta function $\delta(x_n - x_m)$ in one dimension. Setting $m = n$ results in $\delta(x=0) \sim a^{-1}$. We also note that, by the same argument $\delta(p=0) \sim L/(2\pi)$.

volume. It is a conjecture of the causal set approach to discrete spacetime that a given causal set cannot be faithfully embedded in two Lorentzian manifolds that are dissimilar on scales greater that the discreteness scale (taken from the number of points per unit volume of spacetime). The idea is that the manifold picture of spacetime can then be abandoned in favour of the more fundamental casual set picture. In this new picture, the causal structure of spacetime is determined by the partial ordering of the causal set, and each point in the causal set can be considered to contribute a unit spacetime volume element.

The immediate challenge for our purpose is to define a quantum field on a causal set. In order to do this we follow Ref. [14] (see also Ref. [15]), here providing only a bare outline. To begin we review some features of scalar quantum field theory in Minkowski space. For a scalar quantum field with mass m, the advanced and retarded propagators are defined by

$$\begin{aligned} iG_{\text{adv}}(x-y) &= \langle 0| \left[\hat{\phi}(x), \hat{\phi}(y)\right] |0\rangle \Theta(y_0 - x_0) \\ -iG_{\text{ret}}(x-y) &= \langle 0| \left[\hat{\phi}(x), \hat{\phi}(y)\right] |0\rangle \Theta(x_0 - y_0), \end{aligned} \tag{18}$$

where $|0\rangle$ is the vacuum state and Θ is the Heaviside step function. The Pauli-Jordan function is defined by

$$i\Delta(x-y) = \langle 0| \left[\hat{\phi}(x), \hat{\phi}(y)\right] |0\rangle, \tag{19}$$

and so

$$\Delta(x-y) = G_{\text{adv}}(x-y) - G_{\text{ret}}(x-y). \tag{20}$$

The field operator can be written in terms of creation and annihilation operators as $\hat{\phi}(x) = \int d^3\mathbf{p}/[(2\pi)^3\sqrt{2\omega_\mathbf{p}}]\left[e^{-ip\cdot x}\hat{a}(\mathbf{p}) + e^{ip\cdot x}\hat{a}^\dagger(\mathbf{p})\right]$ with $p\cdot x = \omega_\mathbf{p}x_0 - \mathbf{p}\cdot\mathbf{x}$, $\omega_\mathbf{p} = \mathbf{p}^2 + m^2$, $[\hat{a}(\mathbf{p}), \hat{a}^\dagger(\mathbf{q})] = (2\pi)^3\delta^{(3)}(\mathbf{p}-\mathbf{q})$, $[\hat{a}(\mathbf{p}), \hat{a}(\mathbf{q})] = 0$, $\hat{a}(\mathbf{p})|0\rangle = 0$, and so the Pauli-Jordan function is

$$i\Delta(x-y) = \int \frac{d^3\mathbf{p}}{(2\pi)^3 2\omega_\mathbf{p}}\left[e^{-ip\cdot(x-y)} - e^{ip\cdot(x-y)}\right]. \tag{21}$$

Now

$$\int d^4y\, i\Delta(x-y)e^{\pm iq\cdot y} = \mp\frac{2\pi\delta(p_0=0)}{2\omega_\mathbf{q}}e^{\pm iq\cdot x}, \tag{22}$$

where $\delta(p_0=0) = 1/(2\pi)\int dx_0 e^{ip_0x_0}|_{p_0=0}$ (to make sense of this we can think of spacetime as a regular lattice of finite size, see footnote[2]). We can therefore regard the mode functions $e^{\pm iq\cdot x}$ as the eigenfunctions of the Pauli-Jordan function with eigenvalue $\mp 2\pi\delta(p_{0=0})/2\omega_\mathbf{q}$.

On the other hand, for a causal set with N elements which can be faithfully embedded into 4D Minkowski space (note that such a set is easily found by *sprinkling* points in Minkowski space uniformly over spacetime volume and determining the partial ordering from the causal relations between points) we can define an $N \times N$ matrix [16]

$$K_{\text{ret}} = aL(I - abL)^{-1}, \quad a = \frac{\sqrt{\mu}}{2\pi\sqrt{6}}, \quad b = -\frac{m^2}{\mu}, \tag{23}$$

where μ is the density of points sprinkled in Minkowski space, I is the $N \times N$ identity matrix, and L is the $N \times N$ link matrix defined such that

$$L_{xy} = \begin{cases} 1 & \text{if } x \prec y \text{ and there are no points } z \text{ such that} x \prec z \prec y \\ 0 & \text{otherwise.} \end{cases} \tag{24}$$

The matrix K_{ret} represents a weighted sum over paths through the causal set and with the given choices of parameters a and b, it is found to closely resemble the retarded propagator G_{ret} for a scalar field on a 4D Minkowski space [16]. We therefore regard it as the causal set equivalent of the retarded propagator. We can define an advanced propagator as the transpose of the retarded propagator

$$K_{\text{adv}} = K_{\text{ret}}^T, \tag{25}$$

and the causal set analogue of the Pauli-Jordan function

$$\Delta = K_{\text{ret}} - K_{\text{adv}}. \tag{26}$$

Since the matrix $i\Delta$ is skew-symmetric and Hermitian, then its rank is even and its non-zero eigenvalues appear in pairs related by a change of sign. We can then divide the eigenfunctions into pairs u_i, v_i such that

$$i\Delta u_i = \lambda_i u_i, \quad i\Delta v_i = -\lambda_i v_i, \tag{27}$$

with eigenvalues $\lambda_i > 0$ and with $i = 1, \ldots, s$ where $2s$ is the rank of $i\Delta$. We are free to set $u_i = v_i^*$, $u_i^\dagger u_j = v_i^\dagger v_j = \delta_{ij}$, and $u_i^\dagger v_j = 0$. Referring to the continuum case (22) we therefore treat the u_i, v_i as mode functions. We use them to construct a field operator as follows: define a vacuum state $|0\rangle$; for each mode i define creation and annihilation operators $\hat{a}_i$, $\hat{a}_i^\dagger$ which satisfy $[\hat{a}_i, \hat{a}_j^\dagger] = \lambda_i \delta_{ij}$, $[\hat{a}_i, \hat{a}_j] = 0$; basis states take the form $(\hat{a}_1^\dagger)^{n_1}(\hat{a}_1^\dagger)^{n_1} \cdots (\hat{a}_s^\dagger)^{n_s}|0\rangle$; construct a scalar field operator as

$$\hat{\phi}_x = \sum_{i=1}^{s} \left[(u_i)_x \hat{a}_i + (v_i)_x \hat{a}_i^\dagger \right]. \tag{28}$$

Note that here, $(u_i)_x$ means the xth element of the N vector u_i.

To see that this construction gives the correct result in the continuum limit we label the eigenstates by $\mathbf{p}$ and set

$$(u_{\mathbf{p}})_x = \frac{e^{-ip\cdot x}}{[(2\pi)^4\delta^4(p=0)]^{1/2}}; \quad \lambda_{\mathbf{p}} = \frac{2\pi\delta(p_0=0)}{2\omega_{\mathbf{p}}}; \quad \hat{a}(\mathbf{p}) = \hat{a}_{\mathbf{p}}\left[\frac{(2\pi)^3\delta^3(\mathbf{p}=0)}{\lambda_{\mathbf{p}}}\right]^{1/2}, \tag{29}$$

(the eigenfunction/eigenvalues are taken from (22) with the eigenfunctions now normalised such that $u_{\mathbf{p}}^\dagger u_{\mathbf{q}} = v_{\mathbf{p}}^\dagger v_{\mathbf{q}} = \delta_{\mathbf{pq}}$). This results in annihilation and creation operators satisfying $[\hat{a}(\mathbf{p}), \hat{a}^\dagger(\mathbf{q})] = (2\pi)^3\delta^3(\mathbf{p}-\mathbf{q})$, $[\hat{a}(\mathbf{p}), \hat{a}(\mathbf{q})] = 0$; and the scalar field operator becomes

$$\begin{aligned}\hat{\phi}_x &= \sum_{\mathbf{p}}\left[(u_{\mathbf{p}})_x\hat{a}_{\mathbf{p}} + (v_{\mathbf{p}})_x\hat{a}_{\mathbf{p}}^\dagger\right] \\ &= \sum_{\mathbf{p}}\left[\frac{e^{-ip\cdot x}}{[(2\pi)^3\delta^3(\mathbf{p}=0)\sqrt{2\omega_{\mathbf{p}}}]}\hat{a}(\mathbf{p}) + \frac{e^{ip\cdot x}}{[(2\pi)^3\delta^3(\mathbf{p}=0)\sqrt{2\omega_{\mathbf{p}}}]}\hat{a}^\dagger(\mathbf{p})\right] \\ &= \int\frac{d^3\mathbf{p}}{[(2\pi)^3\sqrt{2\omega_{\mathbf{p}}}]}\left[e^{-ip\cdot x}\hat{a}(\mathbf{p}) + e^{ip\cdot x}\hat{a}^\dagger(\mathbf{p})\right],\end{aligned} \tag{30}$$

where we have used $d^3\mathbf{p} \sim \delta^{-3}(\mathbf{p}=0)$ (see footnote [2]). The scalar quantum field constructed on the causal set therefore has the correct continuum limit.

Given a causal set with scalar quantum field operator defined in the way outlined above we envisage an interaction picture as follows: start with some initial state $|\Psi\rangle$; impose some total ordering on the points of the causal set which respects the partial ordering (i.e. if we denote the total order relation by $<$, then $x \prec y \implies x < y$); go through the points in order, at point x, for a $\hat{\phi}^4$-type interaction, the state changes according to

$$|\Psi\rangle \to \hat{U}_x|\Psi\rangle, \quad \hat{U}_x = e^{-ig\hat{\phi}_x^4}, \tag{31}$$

where g is a coupling constant. The overall outcome will be independent of the total ordering (provided that partial ordering is maintained) if $[\hat{\phi}_x, \hat{\phi}_y] = 0$ whenever $x \nprec y$ and $y \nprec x$. To prove this use

$$\begin{aligned}[\hat{\phi}_x, \hat{\phi}_y] &= \sum_{ij}\left\{(u_i)_x(v_j)_y[\hat{a}_i, \hat{a}_j^\dagger] + (v_i)_x(u_j)_y[\hat{a}_i^\dagger, \hat{a}_j]\right\} \\ &= \sum_i\left\{\lambda_i(u_i)_x(v_i)_y - \lambda_i(v_i)_x(u_i)_y\right\} \\ &= \sum_i\left\{\lambda_i(u_i)_x(u_i^\dagger)_y - \lambda_i(v_i)_x(v_i^\dagger)_y\right\} = i\Delta_{xy} = i(K_{\text{ret}})_{xy} - i(K_{\text{adv}})_{xy}.\end{aligned} \tag{32}$$

Now by construction $(K_{\rm ret})_{xy} = 0$ if $y \not\prec x$ and $(K_{\rm adv})_{xy} = 0$ if $x \not\prec y$. This proves the result.

To make the theory into a collapse theory we use the same idea. As we go through the totally ordered points, at point x the state changes according to

$$|\Psi\rangle \to \hat{J}_x(z_x)|\Psi\rangle, \quad \hat{J}_x(z_x) = \left(\frac{\beta}{\pi}\right)^{1/4} e^{-\frac{\beta}{2}(\hat{\phi}_x - z_x)^2}, \tag{33}$$

where β is the collapse strength and z_x is a random variable occurring with probability

$$p(z_x) = \frac{\langle\Psi|\hat{J}_x(z_x)^\dagger \hat{J}_x(z_x)|\Psi\rangle}{\langle\Psi|\Psi\rangle}. \tag{34}$$

The variable z_x is equivalent to the result of a weak measurement of $\hat{\phi}_x$ and so represents the 'amount of matter' at point x. It is straightforward to show that the joint probability for a set of collapses is independent of the total ordering (provided that the partial ordering is maintained); and also that for a given sequence of points $\{x\}$ with given collapse outcomes $\{z_x\}$, the final state is independent of the total ordering (cf. Ref. [10]). When the causal set is faithfully embedded into a Lorentzian manifold, these statements translate into the fact that the dynamical process is independent of frame or foliation.

The total particle number operator is $\hat{N} = \sum_i \hat{a}_i^\dagger \hat{a}_i / \lambda_i$. During a collapse event the change in total particle number is found to be

$$\Delta N = \int dz \frac{\langle\Psi|\hat{J}_x(z)[\hat{N}, \hat{J}_x(z)]|\Psi\rangle}{\langle\Psi|\Psi\rangle} = \frac{\beta}{2}\sum_{i=1}^{s} \lambda_i (u_i)_x (u_i^\dagger)_x. \tag{35}$$

This can be expected to be finite.

5 Discussion

Rewriting physics within this new fundamental picture presents severe challenges. Ideally one should be able to derive all features starting with only the causal set and any additional structure that has been added (in our case the scalar quantum field). Let us reasonably assume that the picture can be well approximated by a scalar quantum field theory on a Lorentzian manifold. Let us also reasonably assume that in the non-relativistic limit, this collapse model can be well approximated by one of the candidate non-relativistic collapse models: GRW or the Continuous Spontaneous Localisation (CSL) model. This being the case, we should expect that the collapse lengthscale of the non-relativistic model, which represents the lengthscale over which the effects of a collapse event are correlated, should be given by the natural lengthscale of the discrete space (this is μ^{-4} where μ is the density of causal set points embedded within the corresponding manifold).

In reference [17] it is argued that in order to be compatible with experimental evidence (large scale interferometry, spontaneous heating of matter) and to be philosophically reasonable (i.e. to predict collapse of perceptible objects in an imperceptible time scale), the collapse lengthscale in either GRW or CSL models should be greater than approximately 10^{-10} metres. For a discreteness scale this seems to be quite large.[3] It is certainly much larger than the expected discreteness scale, the Planck length, 10^{-35} m. However, even with a Planck length discreteness scale, it would be possible to define a field $\hat{\phi}_x$ on only a uniformly distributed subset of points from the full causal set (with density $< (10^{-10}\text{m})^{-4}$). The discreteness scale of this field would then be $> 10^{-10}$ m making it viable collapse model.

The general construction is not limited to Minkowski spacetimes. For general curved spacetimes we could ambitiously envisage that the curvature is sourced by the matter/energy distribution described by $\{z_x\}$ thereby removing the need for a quantum gravity [18]. In the context of the causal set construction this would mean that when adding points to the future frontier of a causal set (this should be done in such a way that no new point can precede an existing point), the order relations made with existing points should be dependent in some way (perhaps stochastically) on $\{z_x\}$. This is complicated by the fact that in order to determine the state and the $\hat{\phi}_x$ operators we should know the full causal set so that we can write down the complete set of available modes. Therefore, in order to describe the matter content of space, we must first know the complete structure of spacetime. This may be a problem for practical determination of the unknown future but is not necessarily a problem for consistency of the theory: we need only establish a correlation between matter/energy distribution $\{z_x\}$ and spacetime structure. One might hope that this could be done in such a way as to correspond to the Einstein equation.

References

1. A. Bassi and G. Ghirardi, Physics Reports 379, 257 (2003).
2. A. Bassi, K. Lochan, S. Satin, T. P. Singh, and H. Ulbricht, Reviews of Modern Physics 85, 471 (2013).
3. G. C. Ghirardi, A. Rimini, and T. Weber, Physical Review D 34, 470 (1986).
4. P. Pearle, Physical Review A 39, 2277 (1989).
5. G. C. Ghirardi, P. Pearle, and A. Rimini, Physical Review A 42, 78 (1990).
6. M. A. Nielsen and I. L. Chuang, Quantum computation and quantum information (Cambridge University Press, 2010).
7. J. S. Bell, John S. Bell on the foundations of quantum mechanics , 172 (2001).
8. D. J. Bedingham, Foundations of Physics 41, 686 (2011).
9. D. Bedingham, in Journal of Physics: Conference Series, Vol. 306 (IOP Publishing, 2011) p. 012034.
10. D. Bedingham, S. K. Modak, and D. Sudarsky, Phys. Rev. D 94, 045009 (2016).

[3] I am unaware of any argument or experimental result ruling out such a large discreteness scale.

11. G. C. Ghirardi, R. Grassi, and P. Pearle, Foundations of Physics 20, 1271 (1990).
12. L. Bombelli, J. Lee, D. Meyer, and R. D. Sorkin, Physical review letters 59, 521 (1987).
13. S. Surya, "The causal set approach to quantum gravity," arXiv:1903.11544 [gr-qc] (2019).
14. S. Johnston, Phys. Rev. Lett. 103, 180401 (2009).
15. R. D. Sorkin, International Journal of Geometric Methods in Modern Physics 14, 1740007-3127 (2017), arXiv:1703.00610 [gr-qc].
16. S. Johnston, Classical and Quantum Gravity 25, 202001 (2008).
17. W. Feldmann and R. Tumulka, Journal of Physics A: Mathematical and Theoretical 45, 065304 (2012).
18. D. Bedingham, "Collapse models and spacetime symmetries," in Collapse of the Wave Function: Models, Ontology, Origin, and Implications (Cambridge University Press, 2018) pp. 74–94.

Opto-Mechanical Test of Collapse Models

Matteo Carlesso and Mauro Paternostro

The gap between the predictions of collapse models and those of standard quantum mechanics widens with the complexity of the involved systems. Addressing the way such gap scales with the mass or size of the system being investigated paves the way to testing the validity of the collapse theory and identify the values of the parameters that characterize it.

Despite increasing sensitivities are taking experiments closer to working points where the potential differences between collapse-based formulations and standard quantum theory should become apparent, the task of finding the precise value of the parameters of a given collapse models is nevertheless difficult. In fact, environmental decoherence—having at the statistical level the same signature as collapse models—could mask any collapse-induced effect, thus biasing the interpretation of related experimental observations.

The current efforts aimed at the test of collapse models can be notionally split into two broad classes: interferometric and non-interferometric tests. The former, which aim at directly probe the validity of the quantum superposition principle, provide a natural test for any collapse model. They rely on the creation of a spatial superposition and, after a suitable time of free evolution—necessary for the propagation of the collapse effects—on the subsequent measurement of its interference contrast. The comparison of such contrast, which is weakened by the environmental and collapse noises, with the predictions of quantum mechanics provides experi-

M. Carlesso (✉)
Department of Physics, University of Trieste, Strada Costiera 11, 34151 Trieste, Italy
e-mail: matteo.carlesso@ts.infn.it

Istituto Nazionale di Fisica Nucleare, Trieste Section, Via Valerio 2, 34127 Trieste, Italy

M. Paternostro
Centre for Theoretical Atomic, Molecular and Optical Physics, School of Mathematics and Physics, Queen's University, Belfast BT7 1NN, UK
e-mail: m.paternostro@qub.ac.uk

V. Allori et al. (eds.), *Do Wave Functions Jump?*, Fundamental Theories of Physics 198,
https://doi.org/10.1007/978-3-030-46777-7_16

mental upper bounds to the collapse parameters. The most successful experiments in this context have been performed using matter-wave interferometry and are extensively discussed elsewhere [cf. Chap. 26]. Here we focus on the second class of experimental assessments, namely the non-interferometric one, with the declared goal of illustrating their potential for the successful falsification of collapse models in close-to-state-of-the-art platforms.

The remainder of this Chapter is organised as follows: In Sect. 1 we review the recently proposed non-interferometric approach to the testing of collapse models. Sect. 2 specialises our assessment to the opto-mechanical platform. In particular, we focus on the description of two recent thought experiments, which have paved the way to the design of experimental routes to the falsification of collapse mechanisms. In Sect. 3 we assess quantitative bounds provided by a set of experiments that broadly fall into the category of non-interferometric settings. Finally, Sects. 4 and 5 address the open questions linked to plausible extensions of standard and nearly canonical formulations of collapse theories and the use of rotational degrees of freedom of mechanical rotors as ultra-sensitive tools for the inference of the minuscule effects of collapse models.

1 Non-interferometric Experiments: A New Perspective in Collapse Model Testing

Differently from interferometric tests, where a superposition needs to be created, sustained and finally measured, non-interferometric assessments tests do not rely on the availability of high-quality non-classical resource states. A plethora of different experiments fall in this class, from those involving the x-ray radiation spontaneously emitted from Germanium (see Chaps. 18 and 28) to those focussing on the change of the internal energy of matter-like systems [1–3], from the monitoring of the free expansion of cold atoms [4] to experiments based on the dynamics of opto-mechanical systems, which are currently considered to be one of the most promising platforms for the delicate discrimination between collapse-based models an standard quantum mechanics.

Here, we review the proposals put forward in Ref. [5–7], which have planted the seeds for the opto-mechanical exploration of collapse models via non-interferometric approaches. For concreteness, we will focus on the Continuous Spontaneous Localization (CSL) model [8–10], which is characterized by parameters λ and r_C: the first is the collapse rate, while the second is its correlation distance.

To introduce the effects induced by the CSL model, we consider a confined system of mass m whose dimensions are, for the sake of simplicity, point-like. The system is initially in thermal equilibrium at temperature T, which we shall assume to be small so as to make thermal fluctuations irrelevant. The confining mechanism is then switched off and the system is let to freely evolve for a time t, when measurement of the position of the system is performed. During the free evolution, the effect of the CSL mechanism can be read out in the spread of the position, which reads

$$\langle \hat{\mathbf{x}}^2(t) \rangle = \langle \hat{\mathbf{x}}^2(t) \rangle_{\mathrm{QM}} + \frac{\lambda \hbar^2 t^3}{2 m_0^2 r_{\mathrm{C}}^2}, \tag{1}$$

where m_0 is the mass of a nucleon, $\langle \hat{\mathbf{x}}^2(t) \rangle_{\mathrm{QM}}$ gives the contribution due to quantum mechanics, and the last term is due to the CSL effect. There is a qualitative difference between the evolution of the spread due to quantum mechanics (which is $\sim t^2$) and the contribution arising from the collapse mechanism ($\sim t^3$). The diffusion induced by the environment has a behaviour similar to the one due the collapse mechanism [11]. On this basis, a way to extrapolate the parameters of CSL would pass through the observation of the diffusive Brownian process and the consequent establishment of bounds on the collapse parameters. This idea was put forward in Ref. [12], which considered a levitated charged nanosphere in a Paul trap supported by an optical cavity [the latter being needed for passive cooling of the system, cf. Fig. 1a]. Clearly, the standard decoherence sources, such as thermal photon emission, absorption and scattering as well as the collision with the residual gas particles, would also contribute to the diffusive motion of the system. The analysis performed in [12, 13] is, in this context, particularly useful as reporting a comparison between possible diffusive contributions from collapse models and analogous terms resulting from standard decoherence mechanisms. By following ideas akin to those pursued in Ref. [12], quantitative bounds on the CSL parameters were derived from a cold atom experiment

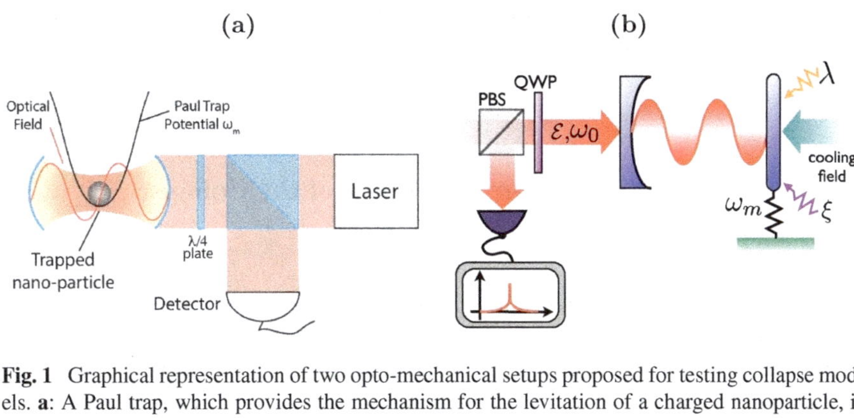

Fig. 1 Graphical representation of two opto-mechanical setups proposed for testing collapse models. **a**: A Paul trap, which provides the mechanism for the levitation of a charged nanoparticle, is supported by an optical cavity, required for the particle cooling. Picture taken from Ref. [12]. **b**: End-cavity opto-mechanical setup as proposed in Refs. [5–7] : the cavity field is sustained by an external laser at frequency ω_0. The end mirror resonates at frequency ω_{m} and is subject to environmental noise—described as Brownian motion at non-zero temperature and associated with the noise operator ξ—and collapse noise (described by the operator λ). Picture taken from Ref. [5]. The detection scheme is the same in both the setups: a quarter-wave-plate (λ/4-plate or QWP) and a polarizing beam splitter (PBS) are used to redirect the light leaving the cavity to a detector for the reconstruction of the optical DNS

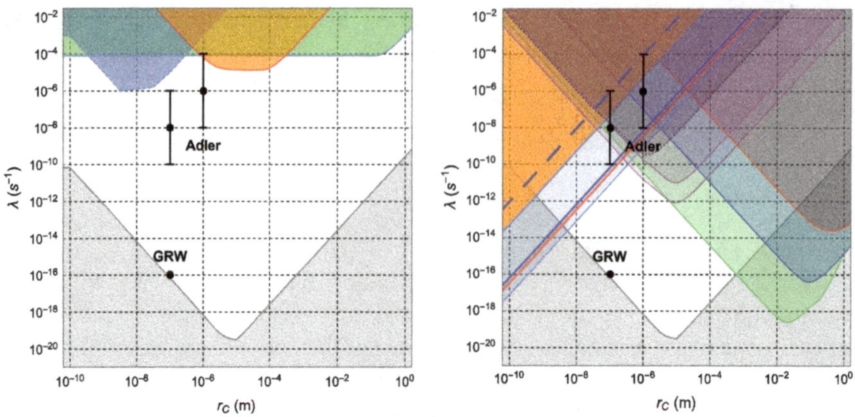

Fig. 2 Exclusion plots for the CSL parameters with respect to the GRW's and Adler's theoretical values [10, 20]. *Left panel*—Excluded regions from interferometric experiments: molecular interferometry [21, 22] (blue area), atom interferometry [23] (green area) and experiment with entangled diamonds [24] (orange area). *Right panel*—Regions of the parameter space of CSL excluded by a set of non-interferometric experiments: AURIGA, LIGO and LISA Pathfinder [25, 26] (red, blue and green areas, respectively), cold atoms [27] (orange area), phonon excitations in crystals [1] (red line), blackbody radiation from the neutron star PSR J 1840-1419 and from Neptune [3] (dashed and continuous blue lines, respectively), X-ray measurements [14, 28–31] (light blue area) and nanomechanical cantilever and its improved version [32, 33] (purple areas with dashed and continuous bound). The grey color highlights the region excluded on the basis of theoretical arguments [22]

[4], where the free expansion of the gas cloud was characterized and compared with the collapse-induced diffusion. The corresponding upper bounds are reported in Fig. 2.

2 Opto-Mechanical System as a Probe of the Collapse Mechanism

Let us now turn to the role played by opto-mechanical in the assessment of collapse models. They focus on an indirect effect provided by the collapse mechanism, which is an extra Brownian-like motion of the center of mass of the mechanical component of an opto-mechanical system. Such motion leads to an extra diffusion mechanism that can be detected through standard experimental techniques and, under suitable conditions, provide information on the undergoing collapse mechanism. In order to fix the ideas, we assume a single-sided Fabry-Perot cavity endowed with an end-cavity mechanical oscillator and driven by an external laser, which also provides the mechanism for the measurement of the mechanical motion [cf. Fig. 1b]. The latter is influenced by a phononic environment (at non-zero temperature) and, allegedly, the CSL-like collapse noise. The action of the latter can be added to the Langevin equations governing the opto-mechanical motion, which read [5]

$$\frac{\mathrm{d}\hat{x}_t}{\mathrm{d}t} = \frac{\hat{p}_t}{m}, \text{ and } \frac{\mathrm{d}\hat{p}_t}{\mathrm{d}t} = -m\omega_{\mathrm{m}}^2\hat{x}_t + \hbar\chi\hat{a}_t^\dagger\hat{a}_t - \gamma_{\mathrm{m}}\hat{p}_t + \hat{\xi}_t + \hat{F}_t^{\mathrm{CSL}}, \tag{2}$$

where ω_{m} and γ_{m} are the harmonic frequency of the mirror and its damping constant, χ denotes the coupling of the mechanical oscillator with the cavity field, whose creation and annihilation operators are $\hat{a}^\dagger$ and $\hat{a}$ respectively. Here, $\hat{\xi}_t$ and $\hat{F}_t^{\mathrm{CSL}}$ denote the stochastic forces due to the environment and the collapse mechanism, respectively. Indeed, the collapse action can be mimicked by adding to the Schrödinger equation a stochastic potential $\hat{V}_{\mathrm{CSL}}$, whose corresponding force is given by $\hat{F}_{\mathrm{CSL}}(t) = \frac{i}{\hbar}[\hat{V}_{\mathrm{CSL}}(t), \hat{p}]$. In the case of CSL we have [14]

$$\hat{V}_{\mathrm{CSL}} = -\hbar\sum_j \frac{m_j}{m_0}\int \mathrm{d}\mathbf{x}\, \hat{\Psi}_j^\dagger(\mathbf{x}, t)\hat{\Psi}_j(\mathbf{x}, t)N(\mathbf{x}, t), \tag{3}$$

where $\hat{\Psi}_j^\dagger(\mathbf{x}, t)$ and $\hat{\Psi}_j(\mathbf{x}, t)$ are respectively the creation and annihilation operators of a j-type particle of mass m_j, and $N(\mathbf{x}, t)$ is the a stochastic noise inducing the collapse, whose mean and correlator are

$$\mathbb{E}[N(\mathbf{x}, t)] = 0, \text{ and } \mathbb{E}[N(\mathbf{x}, t)N(\mathbf{y}, s)] = \lambda\delta(t - s)G(\mathbf{x} - \mathbf{y}), \tag{4}$$

with $\mathbb{E}$ the stochastic average over the noise and $G(\mathbf{x}) = e^{-\mathbf{x}^2/4r_{\mathrm{C}}^2}$. Equation (4) gives a clear interpretation of λ and r_{C} as, respectively, the collapse rate and the noise correlation distance.

The signatures of the collapses of the mechanical motion can be tracked through the density noise spectrum (DNS), whose definition reads

$$\mathcal{S}_{xx}(\omega) = \int \frac{\mathrm{d}\Omega}{4\pi}\,\mathbb{E}\left[\langle\{\tilde{x}(\omega), \tilde{x}(\Omega)\}\rangle\right], \tag{5}$$

where $\tilde{x}(\omega)$ is the Fourier transform of the fluctuations of $\hat{x}_t$. Following the derivation in Ref. [15], one finds

$$\mathcal{S}_{xx}(\omega) = \frac{2\hbar^2|\alpha|^2\kappa\chi^2}{m^2\left[\kappa^2 + (\Delta - \omega)^2\right]|d(\omega)|^2} + \frac{\hbar m\gamma_{\mathrm{m}}\omega\coth\left(\frac{\hbar\omega}{2k_{\mathrm{B}}T}\right) + \mathcal{S}_{\mathrm{CSL}}}{m^2|d(\omega)|^2}, \tag{6}$$

where $|\alpha|^2$ denotes the intensity of the intra-cavity laser, Δ is the laser-cavity detuning, T is the environmental temperature, and κ is the cavity dissipation rate. Moreover we have introduced the susceptibility function $1/|d(\omega)|^2$ with

$$|d(\omega)|^2 = (\omega_{\mathrm{m,eff}}^2(\omega) - \omega^2)^2 + \gamma_{\mathrm{m,eff}}^2(\omega)\omega^2. \tag{7}$$

Here, $\omega_{\text{m,eff}}(\omega)$ and $\gamma_{\text{m,eff}}(\omega)$ denote the effective mechanical frequency and damping rate, respectively. Finally, $\mathcal{S}_{\text{CSL}}$ quantifies the action of CSL noise, which can be obtained from $\mathbb{E}[\langle \hat{F}_{\text{CSL}}(t)\hat{F}_{\text{CSL}}(t')\rangle] = \mathcal{S}_{\text{CSL}}\delta(t-t')$ with [16]

$$\mathcal{S}_{\text{CSL}} = \frac{\hbar^2 \lambda r_C^3}{\pi^{3/2} m_0^2} \int \mathrm{d}\mathbf{k}\, |\tilde{\mu}(\mathbf{k})|^2 e^{-\mathbf{k}^2 r_C^2} k_x^2, \tag{8}$$

where $\tilde{\mu}(\mathbf{k})$ is the Fourier transform of the mass density. Here, due to the presence of the latter, two aspects can be considered. First, $\mathcal{S}_{\text{CSL}}$ is proportional to the square of the mass m of the system. Thus, heavier masses can provide a stronger signature of the collapse mechanism. Second, Eq. (8) strongly depends on the geometry of the system and in particular on the ratio between its size L and r_C. Indeed, in the limit of $r_C \ll L$ the collapse noise will act incoherently on parts of the system which are distant more that r_C, while for $r_C \sim L$ such action will be coherent. Finally, for $r_C \gg L$, the collapse action will be still coherent but unfocused on the system, thus effectively loosing strength. The dependence of $\mathcal{S}_{\text{CSL}}$ on the geometry of the system is clearly visible in the shape of the corresponding upper bounds on the collapse parameters. Indeed, as it is shown in Fig. 1, once the dimensions L of the system are fixed, one has the strongest bound on λ for the value of $r_C \sim L$. This reflects in the characteristic V-shaped form of the bounds of the CSL parameters.

Equation (6) gives insight in the collapse action on the mechanical oscillator. This is the change of the equilibrium temperature of the system from the environmental one T to an enhanced effective one. Indeed, in the limit for high temperatures of the environment this reads [16]

$$\hbar m \gamma_m \omega \coth\left(\frac{\hbar\omega}{2k_B T}\right) + \mathcal{S}_{\text{CSL}} \rightarrow 2m\gamma_m k_B (T + \Delta T_{\text{CSL}}) \tag{9}$$

with

$$\Delta T_{\text{CSL}} = \frac{\mathcal{S}_{\text{CSL}}}{2m\gamma_m k_B}. \tag{10}$$

One should notice that, here, another parameter of the opto-mechanical setup plays an important role, namely the damping rate γ_m that quantifies mechanical dissipation. Clearly, the more the system dissipates, the faster the thermalization process to the environmental temperature, and the smaller the collapse contribution. On the contrary, in the limit of no dissipation (i.e. for $\gamma_m \rightarrow 0$), ΔT_{CSL} diverges: this is exactly what should be expected from the model, whose collapse noise can be associated to an infinite-temperature bath. In passing, we remark that generalizations of collapse models have been proposed [17–19] where the noise inducing collapse is associated with a finite temperature T_{CSL} and an ensuing dissipative process. We refer to Sect. 4 for details on such models.

As underlined in Ref. [6], the thermal noise, proportional to $\coth(\frac{\hbar\omega}{2k_B T})$, is not the only limitation in detecting the collapse-induced diffusion. Indeed, also the measurement process contributes to enhancing the noise in the readout signal, thus screening the signal from the collapse mechanism. Clearly, a precise characterization of the thermal effects and the measurement backaction would provide stronger upper bounds to the collapse parameters.

3 Experimental Bounds

The first application that we consider is the one reported in Ref. [25], where three experiments—LIGO, AURIGA and LISA Pathfinder—have been considered. The first two are gravitational wave detectors, while the last one is only a prototype of a future gravitational wave detector. In all such experiments, a mechanical resonator is monitored through optical techniques. Due to the mass of the systems (~2 kg for LISA Pathfinder, ~40 kg for LIGO and ~2300 kg for AURIGA), the back-action of the optics can be neglected, and one considers only the last term in Eq. (6), which depends explicitly on the experiment considered. The single arm of LIGO and LISA Pathfinder consists of two masses, modelled as harmonic oscillators, whose relative distance is monitored. Conversely, AURIGA is a resonant bar whose elongation is measured. For the latter, one can model the system as two half-mass harmonic oscillators oscillating in counterphase. Thus, the modelling is the same for all three experiments. Equation (8) is consequently modified to read

$$\mathcal{S}_{\text{CSL}} = \frac{\hbar^2 \lambda r_C^3}{2\pi^{3/2} m_0^2} \int \mathrm{d}\mathbf{k}\ |\tilde{\mu}(\mathbf{k})|^2 e^{-\mathbf{k}^2 r_C^2} k_x^2 (1 - e^{iak_x}), \tag{11}$$

where a is the distance between the two masses. Such systems are well outside the quantum realm due to their masses, which also prevent their use in interferometric experiments. However, they set important bounds on the collapse parameters, which are here reported in Fig. 1.

The second application that we aim at covering is that reported in Refs. [32, 33], where a heavy micrometrical sphere is attached to a silicon cantilever, which acts as a mechanical resonator. As the sphere is ferromagnetic, in place of the optics, a low noise SQUID can be employed to monitor the mechanical motion of the cantilever. The system is placed in high vacuum and low temperature to minimize the thermal action of the environment. Moreover, in order to better characterize the thermal component of the noise, different measurements of the DNS of the system were performed at different temperatures of the environment, ranging from 11 mK to ~ 1 K. Thus, by exploiting Eq. (10), one can determine upper bounds on the collapse parameters λ and r_C, which are reported in Fig. 1.

4 Testing of the Dissipative and Colored CSL Models

The CSL model have two weaknesses [10]. The first is the steady increase in the energy of any (free) system in time, e.g. an hydrogen atom is heated by $\simeq 10^{-14}$ K per year taking the values $\lambda = 10^{-16}\,\text{s}^{-1}$ and $r_C = 10^{-7}$ m. Although the increment is small, it is not realistic feature even for a phenomenological model. On the other hand, one expects that, through a dissipative mechanisms, the system will eventually termalize to the finite temperature of the collapse noise. Although there are theoretical arguments suggesting the value of such a temperature to be $T_{\text{CSL}} \simeq 1$ K [17, 19], one needs to validate them. While an interferometric investigation was performed in Ref. [22, 34], and a non-interferometric measurement of the free expansion of a cold-atom cloud was studied in Ref. [4], the theoretical setting for an opto-mechanical test of the dissipative extension of the CSL model was proposed in Ref. [18]. Figure 3 shows how the experimental bounds change when the dissipation is explicitly considered in the collapse mechanism for two values of the T_{CSL}.

The second weakness of the CSL model is that its noise has a white spectrum. This is clearly an approximation as no physical noise can be perfectly white. Conversely, one expects the existence of a cutoff frequency Ω_C above which the collapse mechanism is negligible. Theoretical arguments suggest $\Omega_C \sim 10^{12}$ Hz [35, 36]. The introduction of the cutoff changes the predictions of the model: the correlations of the noise in Eq. (4) are modified in $\mathbb{E}[N(\mathbf{x},t)N(\mathbf{y},s)] = \lambda f(t-s)G(\mathbf{x}-\mathbf{y})$, where $f(t)$ describes the time correlations of the collapse noise. Correspondingly, the DNS in an opto-mechanical system becomes $\mathcal{S}_{\text{cCSL}}(\omega) = \mathcal{S}_{\text{CSL}} \times \tilde{f}(\omega)$, where $\tilde{f}(\omega)$ is the Fourier transform of $f(t)$ [37]. Bounds on the CSL parameters for colored noise were studied in detail in Ref. [4, 22, 37]. In particular, upper bounds from

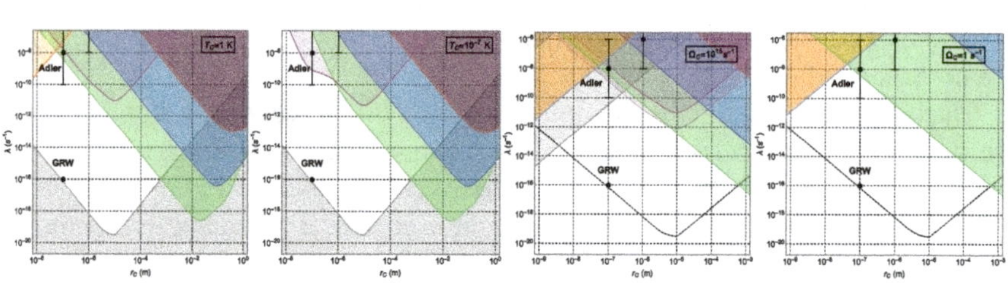

Fig. 3 **First and second panels**: Upper bounds on the dissipative CSL parameters λ and r_C for two values of the CSL noise temperature: $T_{\text{CSL}} = 1$ K (first panel) and $T_{\text{CSL}} = 10^{-7}$ K (second panel). Picture taken from [18]. **Third and fourth panels**: Upper bounds on the colored CSL parameters λ and r_C for two values of the frequency cutoff: $\Omega_c = 10^{15}$ Hz (third panel) and $\Omega_c = 1$ Hz (fourth panel). Picture taken from [37]. Red, blue and green lines (and respective shaded regions): Upper bounds (and exclusion regions) from AURIGA, LIGO and LISA Pathfinder, respectively [25]. Purple region: Upper bound from cantilever experiment [33]. Orange and grey top regions: Upper bound from cold atom experiment [4, 27] and from bulk heating experiments [1]. The bottom area shows the excluded region based on theoretical arguments [22]

high frequencies experiments (or involving small time scales) are weakened when moving to small value of Ω_C. Figure 3 shows the upper bounds to the colored CSL extension for two values of Ω_C.

5 Proposals for Future Testing

Opto-mechanical proposals have been put forward aimed at strengthening the current upper bounds on the collapse parameters. A first one consists in the modification of the cantilever experiment in Ref. [33], where the homogeneous mass is substituted with one made of several layers of two different materials [38]. This will increment the effect of the CSL noise for the values of r_C of the order of the thickness of the layers. The hypothetical upper bounds that can be inferred from such scheme are shown in Fig. 4.

A second possible test focuses on the rotational degrees of freedom in place of the vibrational ones [16, 39]. The former can quantify the CSL action in a form similar to that in Eq. (10), where the collapse-induced contribution to the temperature is that related to the rotational degrees of freedom and reads $\Delta T_{\mathrm{CSL}}^{\mathrm{rot}} = \mathcal{S}_{\mathrm{CSL}}^{\mathrm{rot}}/2k_{\mathrm{B}}D_\phi$, where D_ϕ is the rotational damping rate and

$$\mathcal{S}_{\mathrm{CSL}}^{\mathrm{rot}} = \frac{\hbar^2\lambda r_C^3}{\pi^{3/2}m_0^2}\int \mathrm{d}\mathbf{k}\ \left|k_y\partial_{k_z}\tilde{\mu}(\mathbf{k}) - k_z\partial_{k_y}\tilde{\mu}(\mathbf{k})\right|^2 e^{-r_C^2\mathbf{k}^2}. \tag{12}$$

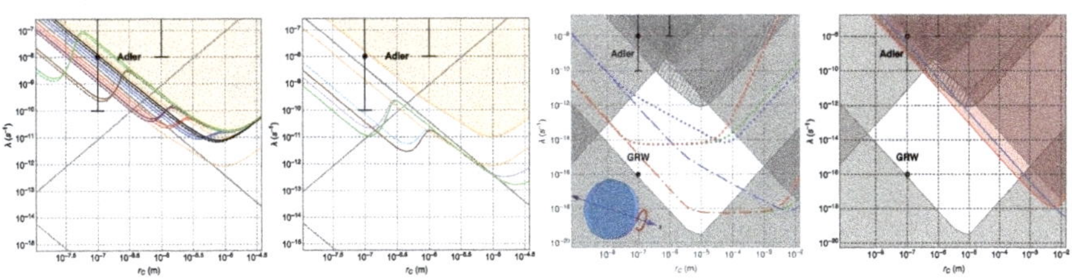

Fig. 4 Exemplification of two possible experimental tests of collapse models. **First panels**: Hypothetical upper bounds obtained from substituting the sphere attached to the cantilever used in [33] with a multilayer cuboid of the same mass for various thickness of the layers [38]. The bounds are compared with that from the improved cantilever experiment [33] shown in orange. Picture taken from [38]. **Second panel**: Same as the first panel, but with a mass ten times larger. Picture taken from [38]. **Third panel**: Results of the analysis proposed in [16, 39] where the rotational degrees of freedom of a cylinder are studied. The red line denotes the upper bound that can be obtained from the constrains given by the rotational motion, compared with those from the translations (blue and green lines). Picture taken from [16]. **Fourth panel**: Red shaded area highlights the hypothetical excluded value of the collapse parameters that could be to derived from the conversion of the translational noise of LISA Pathfinder to rotational one [16]. This is compared to the new (old) upper bounds from the translational motion shown with the blue line [16] (grey area [25]). Picture taken from [16]

Equation (12) quantifies the stochastic torque induced on the system by the collapse noise. When such a scheme is applied to macroscopic systems, it can provide a sensible improvement of the bounds on the collapse parameters, cf. Fig. 4. A direct application was considered in [16], where the bound from LISA Pathfinder [25] can be significantly improved by considering also the rotational degrees of freedom.

The above are only two of several proposals [12, 13, 40–42] suggested over the past few years aimed to push the exploration of the CSL parameter space. More will be discussed in Chaps. 25, 27 and 29.

Acknowledgments The authors acknowledge the invaluable collaboration with A. Bassi, H. Ulbricht, S. Donadi, A. Vinante on the topics of this work. They are grateful acknowledge support from the H2020 Collaborative project TEQ (grant agreement 766900) and COST Action CA15220. MC acknowledges support from INFN. MP acknowledges support from the DfE-SFI Investigator Programme (grant 15/IA/2864), the Royal Society Wolfson Research Fellowship ((RSWF R3 183013) and the Leverhulme Trust Research Project Grant (grant nr. RGP-2018-266).

References

1. S. L. Adler and A. Vinante Phys. Rev. A **97** 052119 (2018) https://doi.org/10.1103/PhysRevA.97.052119
2. M. Bahrami Phys. Rev. A **97** 052118 (2018) https://doi.org/10.1103/PhysRevA.97.052118
3. S. L. Adler, A. Bassi, M. Carlesso, and A. Vinante Phys. Rev. D **99** 103001 (2019) https://doi.org/10.1103/PhysRevD.99.103001
4. M. Bilardello, S. Donadi, A. Vinante, and A. Bassi Physica A **462** 764 (2016) DOI: http://www.sciencedirect.com/science/article/pii/S0378437116304095
5. M. Bahrami *et al.* Phys. Rev. Lett. **112** 210404 (2014) https://doi.org/10.1103/PhysRevLett.112.210404
6. S. Nimmrichter, K. Hornberger, and K. Hammerer Phys. Rev. Lett. **113** 020405 (2014) https://doi.org/10.1103/PhysRevLett.113.020405
7. L. Diósi Phys. Rev. Lett. **114** 050403 (2015) https://doi.org/10.1103/PhysRevLett.114.050403
8. G. C. Ghirardi, P. Pearle, and A. Rimini Phys. Rev. A **42** 78 (1990) https://doi.org/10.1103/PhysRevA.42.78
9. P. Pearle Phys. Rev. A **39** 2277 (1989) https://doi.org/10.1103/PhysRevA.39.2277
10. A. Bassi and G. C. Ghirardi Phys. Rep. **379** 257 (2003) DOI: http://www.sciencedirect.com/science/article/pii/S0370157303001030
11. O. Romero-Isart Phys. Rev. A **84** 052121 (2011) DOI https://doi.org/10.1103/PhysRevA.84.052121
12. D. Goldwater, M. Paternostro, and P. F. Barker https://doi.org/10.1103/PhysRevA.94.010104 Phys. Rev. A **94** 010104(R) (2016)
13. S. McMillen *et al.* Phys. Rev. A **95** 012132 (2017) https://doi.org/10.1103/PhysRevA.95.012132
14. S. L. Adler, A. Bassi, and S. Donadi J. Phys. A **46** 245304 (2013) DOI: http://stacks.iop.org/1751-8121/46/i=24/a=245304
15. M. Paternostro *et al.* New J. Phys. **8** pages 107 (2006) DOI: http://stacks.iop.org/1367-2630/8/i=6/a=107
16. M. Carlesso, M. Paternostro, H. Ulbricht, A. Vinante, and A. Bassi New J. Phys. **20** 083022 (2018a) https://doi.org/10.1088/1367-2630/aad863/meta
17. A. Smirne and A. Bassi Sci. Rep. **5** 12518 (2015) https://doi.org/10.1038/srep12518
18. J. Nobakht, M. Carlesso, S. Donadi, M. Paternostro, and A. Bassi Phys. Rev. A **98** 042109 (2018) https://doi.org/10.1103/PhysRevA.98.042109

19. A. Smirne, B. Vacchini, and A. Bassi https://doi.org/10.1103/PhysRevA.90.062135 Phys. Rev. A **90** 062135 (2014)
20. S. L. Adler J. Phys. A **40** 2935 (2007) DOI: http://stacks.iop.org/1751-8121/40/i=12/a=S03
21. S. Eibenberger *et al.* Phys. Chem. Chem. Phys. **15** 14696 (2013) https://doi.org/10.1039/C3CP51500A
22. M. Toroš, G. Gasbarri, and A. Bassi Phys. Lett. A **381** 3921 (2017) DOI: http://www.sciencedirect.com/science/article/pii/S0375960117309465
23. T. Kovachy *et al.* Nature **528** 530 (2015a) https://doi.org/10.1038/nature16155
24. K. C. Lee *et al.* Science **334** pages 1253 (2011) DOI: http://science.sciencemag.org/content/334/6060/1253
25. M. Carlesso, A. Bassi, P. Falferi, and A. Vinante Phys. Rev. D **94** 124036 (2016) DOI: https://doi.org/10.1103/PhysRevD.94.124036
26. M. Armano *et al.* Phys. Rev. Lett. **120** 061101 (2018) DOI https://doi.org/10.1103/PhysRevLett.120.061101
27. T. Kovachy *et al.* Phys. Rev. Lett. **114** 143004 (2015b) https://doi.org/10.1103/PhysRevLett.114.143004
28. C. E. Aalseth *et al.* (collaboration The IGEX Collaboration) Phys. Rev. C **59** 2108 (1999) https://doi.org/10.1103/PhysRevC.59.2108
29. A. Bassi and S. Donadi Phys. Lett. A **378** 761 (2014) DOI: http://www.sciencedirect.com/science/article/pii/S0375960114000073
30. S. Donadi, D.-A. Deckert, and A. Bassi Ann. Phys. **340** 70 (2014) DOI: http://www.sciencedirect.com/science/article/pii/S0003491613002443
31. K. Piscicchia *et al.* Entropy **19** (2017) DOI: http://www.mdpi.com/1099-4300/19/7/319
32. A. Vinante *et al.* Phys. Rev. Lett. **116** 090402 (2016) https://doi.org/10.1103/PhysRevLett.116.090402
33. A. Vinante *et al.* Phys. Rev. Lett. **119** 110401 (2017) https://doi.org/10.1103/PhysRevLett.119.110401
34. M. Toroš and A. Bassi J. Phys. A **51** 115302 (2018) DOI: http://stacks.iop.org/1751-8121/51/i=11/a=115302
35. S. L. Adler and A. Bassi J. Phys. A **40** 15083 (2007) DOI: http://stacks.iop.org/1751-8121/40/i=50/a=012
36. S. L. Adler and A. Bassi J. Phys. A **41** 395308 (2008) DOI: http://stacks.iop.org/1751-8121/41/i=39/a=395308
37. M. Carlesso, L. Ferialdi, and A. Bassi The European Physical Journal D **72** 159 (2018b) https://doi.org/10.1140/epjd/e2018-90248-x
38. M. Carlesso, A. Vinante, and A. Bassi Phys. Rev. A **98** 022122 (2018c) https://doi.org/10.1103/PhysRevA.98.022122
39. B. Schrinski, B. A. Stickler, and K. Hornberger J. Opt. Soc. Am. B **34** C1 (2017) DOI: http://josab.osa.org/abstract.cfm?URI=josab-34-6-C1
40. B. Collett and P. Pearle Found. Phys. **33** 1495 (2003) https://doi.org/10.1023/A:1026048530567
41. R. Kaltenbaek *et al.* Eur. Phys. J. Quant. Tech. **3** 5 (2016) https://doi.org/10.1140/epjqt/s40507-016-0043-7
42. R. Mishra, A. Vinante, and T. P. Singh Phys. Rev. A **98** 052121 (2018) https://doi.org/10.1103/PhysRevA.98.052121

Two Invariant Surface-Tensors Determine CSL of Massive Body Wave Function

Lajos Diósi

Abstract Decoherence of massive body wave function under Continuous Spontaneous Localization is reconsidered. It is shown for homogeneous probes with wave functions narrow in position and angle that decoherence is a surface effect. Corresponding new surface integrals are derived as the main result. Probe's constant density and two completely geometric surface-dependent invariant tensors encode full dependence of positional and angular decoherence of masses, irrespective of their microscopic structure. The two surface-tensors offer a new insight into CSL and a flexible approach to design laboratory test masses.

1 Introduction

Spontaneous decoherence and collapse models, reviewed e.g. by [1, 2] share the form of modified von Neumann equation of motion for the quantum state $\hat{\rho}$:

$$\frac{d\hat{\rho}}{dt} = -\frac{\mathrm{i}}{\hbar}[\hat{H}, \hat{\rho}] + \mathcal{D}\hat{\rho}, \tag{1}$$

where $\hat{H}$ is the many-body Hamiltonian of masses m_a with positions $\hat{\mathbf{x}}_a$ and momenta $\hat{\mathbf{p}}_a$, resp., for $a = 1, 2, \ldots$. The term of spontaneous decoherence takes this generic form:

$$\mathcal{D}\hat{\rho} = -\int\int D(\mathbf{r}-\mathbf{r}')[\hat{\varrho}(\mathbf{r}), [\hat{\varrho}(\mathbf{r}'), \hat{\rho}]]\mathrm{d}\mathbf{r}\mathrm{d}\mathbf{r}', \tag{2}$$

containing the mass density operator at location $\mathbf{r}$:

L. Diósi (✉)
Wigner Research Center for Physics, H-1525 Budapest 114, P.O.Box 49, Budapest, Hungary
e-mail: diosi.lajos@wigner.hu

V. Allori et al. (eds.), *Do Wave Functions Jump?*, Fundamental Theories of Physics 198,
https://doi.org/10.1007/978-3-030-46777-7_17

Fig. 1 For a shape (e.g. a cuboid) lacking rotational symmetry, both position and angle are localized since both of them alter the surface (left). For a sphere, angle does not alter the surface, hence position is localized but angle is not (middle). If we carve N transversal gaps into the cuboid (right), to multiply the surface then we enhance the localization rate by a factor about $N+1$ in the longitudinal direction (horizontal, in our case).

$$\hat{\varrho}(\mathbf{r}) = \sum_a m_a \delta(\mathbf{r} - \hat{\mathbf{x}}_a). \tag{3}$$

The non-negative decoherence kernel $D(\mathbf{r}-\mathbf{r}')$ is model dependent.

In a conference talk [4], I compared some characteristic features of the two leading proposals, the Continuous Spontaneous Localization (CSL) of Ghirardi, Pearle, and Rimini, and the model of Penrose and myself [5, 6] called DP-model after the two independent proponents. I claimed and gave examples (Fig. 1) for CSL in particular that the *surfaces of homogeneous massive bodies are the only subjects of localization*. My observation has been waiting for mathematical formulation until now.

In recent literature, the central mathematical object is the *geometric factor* of decoherence:

$$\mu_{\mathbf{k}} = \sum_a m_a \mathrm{e}^{-\mathrm{i}\mathbf{k}\mathbf{r}_a}, \tag{4}$$

defined in the c.o.m. frame, introduced by [7], also discussed by [8] in this volume. This object is the Fourier-transform of the classical mass density in the c.o.m. frame:

$$\mu(\mathbf{r}) = \sum_a m_a \delta(\mathbf{r} - \mathbf{r}_a). \tag{5}$$

Usually, the contribution of the geometric factor is evaluated in the Fourier-representation. I am going to show that working in the physical space instead of Fourier's is not only possible but even desirable.

In Sect. 2 we recapitulate the decoherence of c.o.m. motion in terms of the geometric factor. For constant density probes, Sect. 3 derives a new practical expression of the decoherence in terms of a simple surface integral, the method is applied for angular (rotational) decoherence in Sect. 4. Possible generalizations towards probes with unsharp edges and for wider superpositions are outlined in Sect. 5, while Sect. 6 is for conclusion and outlook.

2 Center-of-mass Decoherence

The standard CSL model [1] introduces two universal parameters, collapse rate $\lambda = 10^{-17}s^{-1}$, localization $\sigma = 10^{-5}$ cm, and it contains the nuclear mass m_N. The decoherence kernel $D(\mathbf{r}-\mathbf{r}')$ is a Gaussian whose nonlocal effect can be absorbed by a Gaussian smoothening of the mass density $\hat{\varrho}(\mathbf{r})$. The key quantity is the σ-smoothened mass distribution operator:

$$\hat{\varrho}_\sigma(\mathbf{r}) = \sum_a m_a G_\sigma(\mathbf{r}-\hat{\mathbf{x}}_a), \tag{6}$$

where $G_\sigma(\mathbf{r})$ is the central symmetric Gaussian distribution of width σ. Then the decoherence term (2) becomes a single-integral:

$$\mathcal{D}\hat{\rho} = -\frac{4\pi^{3/2}\lambda\sigma^3}{m_N^2}\int [\hat{\varrho}_\sigma(\mathbf{r}),[\hat{\varrho}_\sigma(\mathbf{r}),\hat{\rho}]]\mathrm{d}\mathbf{r}. \tag{7}$$

Inserting Eq. (6), Fourier-representation yields this equivalent form:

$$\mathcal{D}\hat{\rho} = -\frac{\lambda\sigma^3}{2\pi^{3/2}m_N^2}\int \mathrm{e}^{-\mathbf{k}^2\sigma^2}\sum_{a,b} m_a m_b[\mathrm{e}^{\mathrm{i}\mathbf{k}\hat{\mathbf{x}}_a},[\mathrm{e}^{-\mathrm{i}\mathbf{k}\hat{\mathbf{x}}_b},\hat{\rho}]]\mathrm{d}\mathbf{k}. \tag{8}$$

We are interested in the c.o.m. dynamics of the total mass $M=\sum_a m_a$:

$$\frac{d\hat{\rho}_{\rm cm}}{dt} = -\frac{\mathrm{i}}{\hbar}[\hat{H}_{\rm cm},\hat{\rho}_{\rm cm}] + \mathcal{D}_{\rm cm}\hat{\rho}_{\rm cm}, \tag{9}$$

where $\hat{\mathbf{X}}, \hat{\mathbf{P}}$ will stand for the c.o.m. coordinate and momentum. To derive the c.o.m. decoherence term (and also the rotational decoherence term later on in Sect. 4), substitute

$$\hat{\mathbf{x}}_a = \hat{\mathbf{X}} + \mathbf{r}_a + \hat{\boldsymbol{\varphi}}\times\mathbf{r}_a \tag{10}$$

in (8), where $\mathbf{r}_a$ are the constituent coordinates in the c.o.m. frame in rigid body approximation; $\hat{\boldsymbol{\varphi}}$ is the vector of angular rotation, assuming $\langle\hat{\boldsymbol{\varphi}}\rangle$, $\Delta\varphi \ll \pi$. Then Eq. (8), by taking trace over the rotational degrees of freedom, reduces to the following c.o.m. decoherence term:

$$\mathcal{D}_{\rm cm}\hat{\rho}_{\rm cm} = -\frac{\lambda\sigma^3}{\pi^{3/2}m_N^2}\int \mathrm{e}^{-\mathbf{k}^2\sigma^2}|\mu_\mathbf{k}|^2\left(\mathrm{e}^{\mathrm{i}\mathbf{k}\hat{\mathbf{X}}}\hat{\rho}_{\rm cm}\mathrm{e}^{-\mathrm{i}\mathbf{k}\hat{\mathbf{X}}}-\hat{\rho}_{\rm cm}\right)\mathrm{d}\mathbf{k}, \tag{11}$$

where we recognize the presence of the geometric factor $\mu_\mathbf{k}$. At small quantum uncertainties, when $\Delta\mathbf{X} \ll \sigma$, we use the momentum-diffusion equation as a good approximation:

$$\mathcal{D}_{\rm cm}\hat{\rho}_{\rm cm} = -\frac{\lambda\sigma^3}{2\pi^{3/2}m_N^2}\int \mathrm{e}^{-\mathbf{k}^2\sigma^2}|\mu_\mathbf{k}|^2[\mathbf{k}\hat{\mathbf{X}},[\mathbf{k}\hat{\mathbf{X}},\hat{\rho}_{\rm cm}]]\mathrm{d}\mathbf{k}. \tag{12}$$

This equation describes position-decoherence, together with momentum-diffusion, both of them being non-isotropic in the general case. We are going to concentrate on the evaluation of the tensorial coefficient of decoherence on the r.h.s. of (12).

3 Invariant Surface-Tensor for C.O.M. Decoherence

As we see, the geometric factor $\mu_\mathbf{k}$ itself does not matter but its squared modulus does. We consider the approximation (12) which allows for a spectacular simple geometric interpretation of the relevant structure:

$$\int \mathrm{e}^{-\mathbf{k}^2\sigma^2}|\mu_\mathbf{k}|^2(\mathbf{k}\circ\mathbf{k})\mathrm{d}\mathbf{k} = (2\pi)^3\int \nabla\mu_\sigma(\mathbf{r})\circ\nabla\mu_\sigma(\mathbf{r})\,\mathrm{d}\mathbf{r}. \tag{13}$$

We can recognize $\mu_\sigma(\mathbf{r})$ as the σ-smoothened mass density in the c.o.m. frame.[1] This latter form becomes amazingly useful if the bulk is much larger than σ and possesses constant density ϱ when averaged over the scale of σ. If, furthermore, we assume the density drops sharply from ϱ to zero through the surface then $\nabla\mu_\sigma(\mathbf{r})$ is vanishing everywhere but in about a σ-layer around the surface. Let $\mathbf{n}$ stand for the normal vector of the surface at a given point $\mathbf{r}$ and let h be the height above the surface, then

$$\nabla\mu_\sigma(\mathbf{r}+h\mathbf{n}) = -\varrho\mathbf{n}g_\sigma(h), \tag{14}$$

$g_\sigma(h)$ is the central Gaussian of width σ. The volume integral can be rewritten, with good approximation, as an integral along h and a subsequent surface integral:

$$\begin{aligned}(2\pi)^3\int \nabla\mu_\sigma(\mathbf{r})\circ\nabla\mu_\sigma(\mathbf{r})\mathrm{d}\mathbf{r} &= (2\pi)^3\varrho^2\oint \mathbf{n}\circ\mathbf{n}\left(\int g_\sigma^2(h)\mathrm{d}h\right)\mathrm{d}S\\ &= \frac{(2\pi)^3\varrho^2}{2\pi^{1/2}\sigma}\oint(\mathbf{n}\circ\mathbf{n})\mathrm{d}S. \end{aligned} \tag{15}$$

If the prove has cavities in it, and the characteristic sizes of the probe and cavities keep to be much larger than σ, then the surface integral must be extended for the surfaces of the cavities as well. Using Eqs. (13) and (15), the decoherence term (12) obtains the attractive form

---

[1]Previous works, like e.g. [10] and Supplemental Material (S11) of [11], used the double-integral:

$$\pi^{3/2}\sigma^{-3}\iint \exp\left(-|\mathbf{r}-\mathbf{r}'|^2/(4\sigma^2)\right)\nabla\mu(\mathbf{r})\circ\nabla\mu(\mathbf{r}')\mathrm{d}\mathbf{r}\mathrm{d}\mathbf{r}',$$

without deriving the equivalent single-integral as of the r.h.s. of Eq. (13).

$$\mathcal{D}_{\mathrm{cm}}\hat{\rho}_{\mathrm{cm}} = -\frac{2\pi\lambda\sigma^2\varrho^2}{m_N^2}\oint[\mathbf{n}\hat{\mathbf{X}},[\mathbf{n}\hat{\mathbf{X}},\hat{\rho}_{\mathrm{cm}}]]\mathrm{d}S. \tag{16}$$

This is our main result. It shows that the c.o.m. decoherence is completely determined by the constant density ϱ and the shape of the body, through the *surface-tensor*

$$\mathrm{S}_{\mathrm{cm}} =: \oint(\mathbf{n}\circ\mathbf{n})\mathrm{d}S. \tag{17}$$

In CSL, at small quantum uncertainties $\Delta\mathbf{X} \ll \sigma$, the c.o.m. decoherence of homogeneous sharp-edged bulks is a *surface effect*!

Recall that the main result (16) remains valid if the probe has cavities and we integrate over the surfaces of the cavities as well. This allows us to multiply the CSL decoherence by carving cavities inside the otherwise homogeneous probe, CSL decoherence can be multipled (cf. Fig. 1). This explains the reason of enhanced decoherence in layered structures, proposed by [9].

The heating rate, coming from the decoherence term in (12), is defined by the Heisenberg derivative $\Gamma_{\mathrm{cm}} = \mathcal{D}_{\mathrm{cm}}(\hat{\mathbf{P}}^2/2M)$. Now easy is to write it in a more explicite form than before. Reading $\mathcal{D}_{\mathrm{cm}}^\dagger = \mathcal{D}_{\mathrm{cm}}$ off from (16), one immediately obtains

$$\Gamma_{\mathrm{cm}} = \frac{2\pi\lambda\sigma^2\varrho^2}{m_N^2}\frac{S}{M} = \frac{2\pi\lambda\sigma^2\varrho}{m_N^2}\frac{S}{V}, \tag{18}$$

where S is the total surface (including cavities' internal surfaces) and V is the total volume. Note that Γ_{cm} is the same if we start from the general dynamics (11) not restricted by $\Delta\mathbf{X} \ll \sigma$. [It does not matter if we calculate the Heisenberg derivative of the quadratic $\hat{\mathbf{P}}^2$ by $\mathcal{D}_{\mathrm{cm}}$ in (11) or, alternatively, by the $\hat{\mathbf{X}}$-quadratic approximation of $\mathcal{D}_{\mathrm{cm}}$ in (12).] Interestingly, c.o.m. heating is inverse proportional to the size of the bulk. Recall the total heating rate

$$\Gamma = \mathcal{D}\sum_a \frac{\hat{\mathbf{p}}^2}{2m_a} = \frac{3\hbar^2\lambda}{2m_N^2\sigma^2}M, \tag{19}$$

always much larger than the c.o.m. heating. For a sphere of radius R we get $\Gamma_{\mathrm{cm}}/\Gamma = 3(\sigma/R)^4$.

Examples. Consider the longitudinal motion of a cylinder, Eq. (16) reduces to

$$\mathcal{D}_{\mathrm{cm}}\hat{\rho}_{\mathrm{cm}} = -\frac{2\pi\lambda\sigma^2\varrho^2}{m_N^2}S_\perp[\hat{x},[\hat{x},\hat{\rho}_{\mathrm{cm}}]], \tag{20}$$

where $S_\perp$ is the total surface perpendicular to the motion (i.e.: the area of both faces of the cylinder). At a given constant density ϱ, the decoherence is independent of the length of the cylinder. It can be squeezed to become a plate or elongated to become a rod. This invariance of the decoherence offers a fair guidance when we design laboratory probes. However, the same invariance may raise conceptual questions as well. With increasing length of the rod while decoherence rate remains constant, CSL might leave the longitudinal superposition of our massive rod with counter-intuitive long coherence times. An other remarkable feature of the surface-tensor S is that spontaneous decoherence in one direction can be decreased by tilted edges instead of perpendicular ones. If the faces of the above cylinder are replaced by cones of apex angle θ then the two factors $\mathbf{n}\hat{\mathbf{X}}$ in Eq. (16) get a factor $\sin(\theta/2)$ each while the surface of the cones becomes $\sin^{-1}(\theta/2)$-times larger than $S_\perp$. The spontaneous longitudinal decoherence becomes suppressed by the factor $\sin(\theta/2)$. E.g.: sharp pointed needles become extreme insensitive to longitudinal CSL.

4 Rotational Decoherence

Rotational decoherence of objects under CSL has recently been discussed by [12, 13]. Derivation of our main result (16) on decoherence of lateral superpositions tells us how to express this time the decoherence of angular superpositions in terms of a surface integral. We outline the steps, without the details. After substituting $\hat{\mathbf{x}}_a$ by Eq. (10) into Eq. (8), we trace over the c.o.m. motional d.o.f., yielding

$$\mathcal{D}_{\mathrm{rot}}\hat{\rho}_{\mathrm{rot}} = -\frac{\lambda\sigma^3}{2\pi^{3/2}m_N^2}\int \mathrm{e}^{-\mathbf{k}^2\sigma^2}\sum_{a,b} m_a m_b[\mathrm{e}^{\mathrm{i}\mathbf{k}(\mathbf{r}_a+\hat{\varphi}\times\mathbf{r}_a}, [\mathrm{e}^{-\mathrm{i}\mathbf{k}(\mathbf{r}_b+\hat{\varphi}\times\mathbf{r}_b}, \hat{\rho}_{\mathrm{rot}}]]\mathrm{d}\mathbf{k}. \tag{21}$$

If $\Delta(\varphi\times\mathbf{r}_a)\ll\sigma$ for all a, we approximate the integral as follows:

$$\int \mathrm{e}^{-\mathbf{k}^2\sigma^2}\sum_{a,b} m_a m_b \mathrm{e}^{\mathrm{i}\mathbf{k}(\mathbf{r}_a-\mathbf{r}_b)}[\hat{\boldsymbol{\varphi}}\mathbf{k}\mathbf{r}_a, [\hat{\boldsymbol{\varphi}}\mathbf{k}\mathbf{r}_b, \hat{\rho}_{\mathrm{rot}}]]\mathrm{d}\mathbf{k}, \tag{22}$$

where we define the triple scalar product by $\mathbf{abc}=\mathbf{a}(\mathbf{b}\times\mathbf{c})$. This integral is equivalent to the following volume integral:

$$(2\pi)^3\int[\hat{\boldsymbol{\varphi}}\mathbf{r}\nabla\mu_\sigma(\mathbf{r}), [\hat{\boldsymbol{\varphi}}\mathbf{r}\nabla\mu_\sigma(\mathbf{r}), \hat{\rho}_{\mathrm{rot}}]]\mathrm{d}\mathbf{r}. \tag{23}$$

Applying the arguments and approximations as in Sect. 3, we rewrite this volume integral as a surface integral:

$$\frac{(2\pi)^3\varrho^2}{2\pi^{1/2}\sigma}\oint[\hat{\boldsymbol{\varphi}}\mathbf{r}\mathbf{n}, [\hat{\boldsymbol{\varphi}}\mathbf{r}\mathbf{n}, \hat{\rho}_{\mathrm{rot}}]]\mathrm{d}S. \tag{24}$$

Using this form for the integral in Eq. (21), the rotational decoherence term takes the following ultimate form:

$$\mathcal{D}_{\mathrm{rot}}\hat{\rho}_{\mathrm{rot}} = -\frac{2\pi\lambda\sigma^2\varrho^2}{m_N^2}\oint[\hat{\boldsymbol{\varphi}}\mathbf{rn}, [\hat{\boldsymbol{\varphi}}\mathbf{rn}, \hat{\rho}_{\mathrm{rot}}]]\mathrm{d}S. \tag{25}$$

The rotational decoherence is determined by the constant density ϱ and the *rotational surface-tensor*:

$$\mathrm{S}_{\mathrm{rot}} =: \oint(\mathbf{r}\times\mathbf{n})\circ(\mathbf{r}\times\mathbf{n})\mathrm{d}S, \tag{26}$$

where, as before, $\mathbf{r}$ is the coordinate of a surface point in the c.o.m. frame and $\mathbf{n}$ is the corresponding normal vector to the surface. Remember, the validity of (25) was limited by $\Delta(\varphi\times\mathbf{r}_a)\ll\sigma$ for all a. In terms of the locations $\mathbf{r}$, the condition becomes $\Delta(\varphi\times\mathbf{r})\ll\sigma$ for all surface points $\mathbf{r}$.

Calculation of the spontaneous heating rate of the rotational degrees of freedom is straightforward, yielding

$$\Gamma_{\mathrm{rot}} = \frac{2\pi\lambda\sigma^2\varrho}{m_N^2}\mathrm{Tr}(\mathrm{I}^{-1}\mathrm{S}_{\mathrm{rot}}), \tag{27}$$

where $\mathrm{I}=\int(\mathbf{r}\circ\mathbf{r})\mathrm{d}\mathbf{r}$ is the inertia tensor of the probe.

Examples. Consider the rotation of a long cylindric rod of length L and radius $R\ll L$, around a perpendicular axis $\mathbf{n}_{\mathrm{rot}}$ through its center. All along the rod —except for its short middle part of size $\sim R$— the expression $\mathbf{rnn}_{\mathrm{rot}}=r\sin(\Phi)$ is a good approximation where $r\in(-L/2, L/2)$ is the axial coordinate and Φ is the azimuthal angle of the surface position $\mathbf{r}$. Using this approximation, we can easily evaluate the axial element of the rotational surface-tensor $\mathrm{S}_{\mathrm{rot}}$ that controls the angular decoherence (25):

$$\oint(\mathbf{rnn}_{\mathrm{rot}})^2\mathrm{d}S = \frac{\pi RL^3}{12}. \tag{28}$$

As another example, consider our cylinder rotating around its axis of symmetry: CSL predicts zero decoherence (cf. Fig. 1). But we introduce a small elliptical eccentricity $e\ll 1$ of the cross section. In leading order, we have $\mathbf{rnn}_{\mathrm{rot}}=\frac{1}{2}Re^2\sin(2\Phi)$, yielding the following contribution of the shape to the strength of angular decoherence:

$$\oint(\mathbf{rnn}_{\mathrm{rot}})^2\mathrm{d}S = \frac{e^4}{4}\pi R^2L, \tag{29}$$

that is $e^4/4$ times the volume of the cylinder. Recall that $e^2=2\Delta R/R$ where ΔR is the small difference between the main diameters of the elliptic cross section. The obtained result may raise the same conceptual problem that we mentioned for the longitudinal superposition of the massive rod/needle: azimuthal superpositions of massive cylinders of low eccentricity may become practically insensitive to CSL.

5 Outlines of Generalizations

That in CSL the c.o.m and rotational decoherences are surface effects for homogeneous probes has been explicitly shown in Sects. 3 and 4 for ideal sharp edges and for spatial superpositions much smaller than σ. Both of the latter restrictions can be relaxed and $\mathcal{D}_{\mathrm{cm}}$ still remains a surface integral.

The case of unsharp edges is not much different from the ideal case. Let $H(h)\varrho$ be the profile of how the density drops from the constant ϱ down to zero through a thin layer defining the surface where the layer's thickness is small w.r.t. the sizes of the probe. Then the following generalization of Eq. (14) helps:

$$\nabla\mu_\sigma(\mathbf{r}+h\mathbf{n}) = \varrho\mathbf{n}\int g_\sigma(h-h')\mathrm{d}H(h'). \tag{30}$$

The rest of constructing the surface integral is the same as for Eq. (14) which described the special case where H was the (descending) step function.

The case of not necessarily small quantum positional uncertainties was described by Eq. (11). It takes an equivalent closed form in coordinate representation:

$$\mathcal{D}_{\mathrm{cm}}\hat{\rho}_{\mathrm{cm}}(\mathbf{X},\mathbf{Y}) = -\frac{\lambda\sigma^3}{\pi^{3/2}m_N^2}(2\pi)^3\int\Big[\mu_\sigma(\mathbf{r}+\mathbf{X})\mu_\sigma(\mathbf{r}+\mathbf{Y}) - \mu_\sigma^2(\mathbf{r})\Big]\mathrm{d}\mathbf{r}\;\hat{\rho}_{\mathrm{cm}}(\mathbf{X},\mathbf{Y}). \tag{31}$$

The relevant structure is the integral, which we write as

$$(2\pi)^3\int\left[\mu_\sigma(\mathbf{r}+\mathbf{X}-\mathbf{Y}) - \mu_\sigma(\mathbf{r})\right]\mu_\sigma(\mathbf{r})\mathrm{d}\mathbf{r}. \tag{32}$$

As long as the quantum uncertainty $|\mathbf{X}-\mathbf{Y}|$ is much smaller than the sizes of the probe, but not necessarily smaller then σ, the integral is vanishing everywhere in the bulk except for a thin layer of thickness $\sim|\mathbf{X}-\mathbf{Y}|$ below the surface. Accordingly, we incline to anticipate CSL decoherence remains a surface effect and, investing some harder mathematical work, $\mathcal{D}_{\mathrm{cm}}$ as well as $\mathcal{D}_{\mathrm{rot}}$ would take a form of surface integral, generalizing (16) and (25) beyond their quadratic approximations in $\hat{\mathbf{X}}$ and $\hat{\varphi}$.

6 Concluding Remarks

We have discussed CSL for constant density test masses and proved that spontaneous decoherence of both translational and rotational motion is determined by the density ϱ and by two invariant surface-tensors of the bodies:

$$\mathrm{S}_{\mathrm{cm}} = \oint(\mathbf{n}\circ\mathbf{n})\mathrm{d}S,$$

$$\mathrm{S}_{\mathrm{rot}} = \oint (\mathbf{r} \times \mathbf{n}) \circ (\mathbf{r} \times \mathbf{n}) \mathrm{d}S.$$

These two fully encode the relevant features of the probe's geometry. Previously, these features were encoded by the so-called geometric factor

$$\mu_{\mathbf{k}} = \varrho \int \mathrm{e}^{-\mathrm{i}\mathbf{k}\mathbf{r}} \mathrm{d}\mathbf{r},$$

an integral over the probe's volume and a function of the wave number **k**. In case of general heavily inhomogeneous test masses the necessity of using the geometric factor is certainly doubtless. But for homogeneous probes, the surface-tensors should take over the role.

Important is the new insight into the physics of CSL in motion of a general massive bulk as a whole. First, microscopic structure is totally irrelevant, only the σ-smoothened density matters. Furthermore, displacements of homogeneous regions are not decohered at all. Only the displacements of inhomogeneities are decohered. The sharper the inhomogeneity, the stronger the decoherence it induces. In a constant density probe, the only inhomogeneous part is its surface, hence is CSL decoherence a surface effect for it—that we have here exploited. The same is true for layered probes where mass density jumps—through surfaces (walls) between the layers—contribute to the decoherence tensors. Inhomogeneities other than the said two-dimensional inhomogeneous regions around surfaces may rarely be sharp and fat enough to contribute to c.o.m. or rotational decoherence. Decoherence of probes with smooth material inhomogeneities may remain dominated by the said surfaces, our method of surface-tensors might extend for them!

Acknowledgements The author thanks the National Research Development and Innovation Office of Hungary Projects Nos. 2017-1.2.1-NKP-2017-00001 and K12435, and the EU COST Action CA15220 for support.

References

1. A. Bassi and G.C. Ghirardi, *Phys. Rep.* **379**, 257 (2003)
2. A. Bassi , K. Lochan, S. Satin, T.P. Singh, and H. Ulbricht, *Rev. Mod. Phys.* **85**, 471 (2013)
3. G. Ghirardi, P. Pearle, and A. Rimini, *Phys. Rev.* **A 42**, 78 (1990)
4. L. Diósi, *Spontaneous quantum measurement of mass distribution: DP and CSL models* (Castiglioncello, Sept. 2014). https://wigner.mta.hu/~diosi/slides/dice2014.pdf
5. L. Diósi, *Phys. Lett.* **A 120**, 377 (1987)
6. R. Penrose, *Gen. Rel. Grav.* **28**, 581 (1996)
7. S. Nimmrichter, K. Hornberger, and K. Hammerer, *Phys. Rev. Lett.* **113**, 020405 (2014)
8. S.L. Adler, A. Bassi, and M. Carlesso, *The CSL Layering Effect from a Lattice Perspective*, **198** arXiv:1907.11598
9. M. Carlesso, A. Vinante, and A. Bassi, *Phys. Rev.* **A 98**, 022122 (2018)
10. M. Bahrami, M. Paternostro, A. Bassi, and H. Ulbricht, *Phys. Rev. Lett.* **112**, 210404 (2014)
11. A. Vinante, M. Bahrami, A. Bassi, O. Usenko, G. Wijts, and T.H. Oosterkamp, *Phys. Rev. Lett.* **116**, 090402 (2016)

12. B. Schrinski, B. A. Stickler, and K. Hornberger, *J. Opt. Soc. Am.* **B34**, C1 (2017)
13. M. Carlesso, M. Paternostro, H. Ulbricht, A. Vinante, and A. Bassi, *New J. Phys.*, **20**, 083022 (2018)

Collapse and Charged Particles

Sandro Donadi

Abstract In this chapter, the radiation emission in collapse models is discussed. The basic idea is that, in these models, the noise responsible for the collapse also randomly accelerate protons and electrons in matter, inducing an emission of radiation which is not predicted by standard quantum theory. This offers a way to test collapse models and set bounds on their free parameters. Here we focus our attention on the theoretical calculations required to compute the radiation emission rate from a charged system, which is the quantity measured in the experiments. As we will discuss, the use of perturbative techniques requires some care in order to get the correct results.

1 Introduction

Prof. GianCarlo Ghirardi has been a very inspiring figure for my scientific career. When I was at the 3rd year of my bachelor in physics at Trieste University, he was teaching the course of Quantum Mechanics, and I really appreciated his classes as well as his own notes on the topic, which are for me one of the best references for Quantum Mechanics. Even before being my professor, Prof. Ghirardi inspired me through his book "Un'occhiata alle carte di Dio" (Sneaking a Look at God's Cards) [1], in my opinion the best and most complete popular book on foundations of Quantum Mechanics. It was by reading this book that I became aware and interested on foundations of Quantum Mechanics, and led me to choose to purse my research career on this topic, with special focus on collapse models [2, 3], a line of research started by Prof. Ghirardi, together with Prof. A. Rimini and Prof. T. Weber in their seminal paper [4].

S. Donadi (✉)
Frankfurt Institute for Advanced Studies (FIAS), Ruth-Moufang-Straße 1, 60438 Frankfurt am Main, Germany
e-mail: donadi@fias.uni-frankfurt.de

V. Allori et al. (eds.), *Do Wave Functions Jump?*, Fundamental Theories of Physics 198,
https://doi.org/10.1007/978-3-030-46777-7_18

In this chapter, we discuss the calculation of the radiation emission rate in collapse models as a consequence of the interaction between charged particles with the collapse-inducing noise. This phenomenon have been studied in great details for the Continuous Spontaneous Localizations (CSL) [5], and the Quantum Mechanics with Universal Position Localizations (QMUPL) collapse models [6, 7]. The goal of this chapter is to summarize the main theoretical results.

2 First Calculations of the Collapse-Induced Radiation Emission

The first calculation of the collapse-induced radiation emission in the CSL model was given by Fu [8]. The starting point, which is also taken in all the subsequent articles on the topic, is to replace the non-linear dynamics of the CSL model with the unitary and linear evolution given by the stochastic Schrödinger equation[1]:

$$i\hbar\frac{d|\phi(t)\rangle}{dt} = \left[H - \hbar\sqrt{\gamma}\int d\mathbf{x}\, w(\mathbf{x},t)\psi^{\dagger}(\mathbf{x})\psi(\mathbf{x})\right]|\phi(t)\rangle. \tag{1}$$

where H describes the standard quantum Hamiltonian, which we write explicitly below, while the second term describes the effect of the CSL collapse noise. The coupling constant $\gamma = \lambda 8\pi^{3/2} r_C^3$ depends on the two fundamental free parameters of the CSL model: λ which sets the strength of the collapse, and r_C which gives the spatial correlation of the noise (see Eq. (2) below). The operators $\psi(\mathbf{x})$ ($\psi^{\dagger}(\mathbf{x})$) are the annihilation (creation) operators of a particle in the point "$\mathbf{x}$",[2] and $w(\mathbf{x}, t)$ is a classical noise field with zero average and correlation

$$\mathbb{E}[w(\mathbf{x},t)w(\mathbf{y},s)] = \delta(t-s)F(\mathbf{x}-\mathbf{y}), \qquad F(\mathbf{x}) = \frac{1}{(\sqrt{4\pi}r_C)^3}e^{-\mathbf{x}^2/4r_C^2}. \tag{2}$$

Equation (1) does *not* describe a collapse dynamics, being a standard Schrödinger equation with a random potential. However, it can be easily proved that it leads to the same master equation as the non-linear CSL equation [9]. Therefore, as far as we are concerned in computing average values of observables, which is our case, this dynamics and the true (non-linear) CSL dynamics lead to the same predictions.

We are working with the second quantization formalism, and we want to compute the emission rate of photons. The Hamiltonian is

---

[1] In Eq. (1) we introduce only one matter field $\psi(\mathbf{x})$. The equation can be easily generalized to the case where different kind of particles are considered, which is however not required for the purposes of this chapter.

[2] In principle, in the CSL model there is also a dependence on the particles spins, which however is irrelevant for the phenomenon we are discussing, so it can be neglected.

$$H = \int d\mathbf{x} \; [\mathcal{H}_S(\mathbf{x}) + \mathcal{H}_R(\mathbf{x}) + \mathcal{H}_{INT}(\mathbf{x})] \tag{3}$$

with the first term

$$\mathcal{H}_S(\mathbf{x}) = \frac{\hbar^2}{2m} \vec{\nabla}\psi^\dagger(\mathbf{x}) \cdot \vec{\nabla}\psi(\mathbf{x}) + \; V(\mathbf{x})\psi^\dagger(\mathbf{x})\psi(\mathbf{x}) \tag{4}$$

being the Hamiltonian density of the system, with the potential V which in the rest of the chapter will be taken equal to zero (free particle) or harmonic. The second term is the Hamiltonian density of the free of the electromagnetic (EM) field:

$$\mathcal{H}_R(\mathbf{x}) = \frac{1}{2}\left(\varepsilon_0 \mathbf{E}_\perp^2(\mathbf{x}) + \frac{\mathbf{B}^2(\mathbf{x})}{\mu_0}\right), \tag{5}$$

where $\mathbf{E}_\perp$ is the transverse part of the electric component and $\mathbf{B}$ is the magnetic component. Finally, $\mathcal{H}_{INT}$ contains the standard interaction between the quantized electromagnetic field and the non-relativistic Schrödinger field:

$$\mathcal{H}_{INT}(\mathbf{x}) = \; i\frac{\hbar e}{m}\psi^\dagger(\mathbf{x})\mathbf{A}(\mathbf{x}) \cdot \vec{\nabla}\psi(\mathbf{x}) + \frac{e^2}{2m}\mathbf{A}^2(\mathbf{x})\psi^\dagger(\mathbf{x})\psi(\mathbf{x}). \tag{6}$$

The electromagnetic potential $\mathbf{A}(\mathbf{x})$ takes the form:

$$\mathbf{A}(\mathbf{x}) \; = \sum_{\mathbf{k},\mu} \alpha_k \left[\vec{\epsilon}_{\mathbf{k},\mu}\, a_{\mathbf{k},\mu} e^{i\mathbf{k}\cdot\mathbf{x}} + \vec{\epsilon}^{\,*}_{\mathbf{k},\mu}\, a^\dagger_{\mathbf{k},\mu} e^{-i\mathbf{k}\cdot\mathbf{x}}\right] \tag{7}$$

where $\alpha_k = \sqrt{\hbar/2\varepsilon_0\omega_k L^3}$ with $\omega_k = kc$, $\vec{\epsilon}_{\mathbf{k},\mu}$ are the polarization vectors of the EM field and $a_{\mathbf{k},\mu}$ ($a^\dagger_{\mathbf{k},\mu}$) are the annihilation (creation) operators of a photon with wave-vector $\mathbf{k}$ and polarization μ.

The goal of the calculation is to compute the radiation emission rate defined as

$$\frac{d\Gamma_k}{dk} = k^2 \int d\Omega_{\mathbf{k}} \sum_{\mu} \frac{d}{dt}\mathcal{P}_{\mathbf{k},\mu}(t) \tag{8}$$

with

$$\mathcal{P}_{\mathbf{k},\mu}(t) := \sum_f |\langle f, \mathbf{k}, \mu | U(t) | i, 0_{ph}\rangle|^2 \tag{9}$$

giving the probability that a photon with wave-vector $\mathbf{k}$ and polarization μ is emitted at time t. In Eq. (9), $U(t) = \mathcal{T}\exp\left(\frac{i}{\hbar}\int_0^t dt'\, H_{\text{tot}}(t')\right)$ is the time evolution operator, with "$\mathcal{T}$" denoting the time ordering and $H_{\text{tot}}(t')$ the total Hamiltonian in the squared bracket of Eq. (1), $|i, 0_{ph}\rangle$ is the initial state of the system and the EM field in its vacuum state, while $|f, \mathbf{k}, \mu\rangle$ is the final state of the system and that of an emitted photon with wave vector $\mathbf{k}$ and polarization μ. Since we are interested in the emission

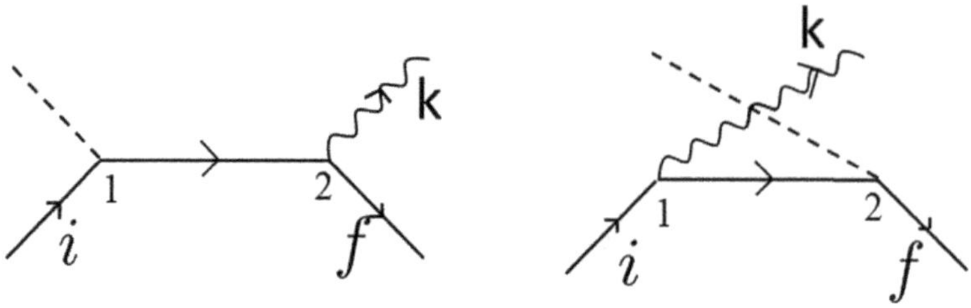

Fig. 1 Lowest order relevant Feynman diagrams describing the emission of a photon (wavy line) from an electron (continuous lines) induced by the interaction with the collapse noise (dashed line)

rate, a time derivative is taken in Eq. (8), and because the final state of the system is not measured, as well as the polarization and directions of the emitted photons, the sum over these degrees of freedoms is taken in Eqs. (8) and (9).

In [8], given the stochastic Schrödinger equation (1) describing the dynamics, the computation of the emission rate and in particular of the matrix elements in Eq. (9) was performed using the standard perturbative approach: first, one moves to interaction picture with respect to the free Hamiltonian $H_0 = \int d\mathbf{x}\ [\mathcal{H}_S(\mathbf{x}) + \mathcal{H}_R(\mathbf{x})]$; then, the Dyson expansion is done and the relevant contributions are computed. The perturbative approach is justified by the fact that both the EM interaction and noise interaction can be treated perturbatively. In terms of Feynman diagrams, the two relevant contribution considered by Fu are given by the diagrams in Fig. 1.

The calculation, long but straightforward, gives as a result the emission rate formula[3]:

$$\frac{d\Gamma_k}{dk} = \frac{\lambda \hbar e^2}{4\pi^2 \varepsilon_0 c^3 m^2 r_C^2 k}. \tag{10}$$

where $\hbar$ and c have the usual meaning, ε_0 is the vacuum permittivity, e and m are respectively the charge and the mass of an electron, and $k = \omega_k / c$ the module of the wave vector of the emitted photons.

If a mass proportional version of the CSL model is considered, which as discussed briefly in the next section and more in details in chapter 28 is suggested precisely by this experiment, the coupling constant has to be replaced as $\sqrt{\gamma} \to \sqrt{\gamma} m/m_0$, with m_0 being a reference mass taken equal to the mass of a nucleon. Then Eq. (10) becomes

$$\frac{d\Gamma_k}{dk} = \frac{\lambda \hbar e^2}{4\pi^2 \varepsilon_0 c^3 m_0^2 r_C^2 k}. \tag{11}$$

The calculation of Fu was later generalized by Adler and Ramazanoglu in [10] where, among other results, the analysis was extended to the case of the non-

[3]Equation (10) appears different to the final result reported in [8] only because here the SI system of units is used while in [8] the CSG system of units was considered. If one takes Fu's result and maps $e^2 \to e^2/(4\pi)$, Eq. (10) is consistently obtained.

Markovian CSL model, and the radiation emitted from an Hydrogen atom was explicitly computed. Besides recovering Fu result of Eq. (11), they found that when a non-white noise extension of CSL is considered, i.e. when the noise correlation (2) is replaced by

$$\mathbb{E}[w(\mathbf{x},t)w(\mathbf{y},s)] = f(t-s)F(\mathbf{x}-\mathbf{y}), \tag{12}$$

the emission rate in Eq. (11) gets multiplied by a factor $\tilde{f}(\omega_k)$ which is the Fourier transform of the function $f(t)$ in Eq. (12) computed at the frequency of the emitted photon. Regarding the emission from the hydrogen atom, it was shown that for high energy photons the emission is the twice that given by Eq. (11) (the electron and the proton emissions add incoherently), while for small energy photons the emission is suppressed (the contribution from the electron tends to cancel that from the proton).

3 Comparison with the Experimental Data

The predicted emitted rate in Eq. (10) was compared to data from an experiment measuring the photons emitted form Germanium in [8]. By considering only the emission from the 4 outer electrons of Germanium (which were approximated as free being weakly bounded to the Germanium nucleus), and considering the values $r_C = 10^{-7}$ m and $\lambda = 10^{-16}$ s^{-1}, the radiation emitted was lager than that predicted by the original CSL model. However, if the mass proportional CSL model is considered, the emitted rate given now by Eq. (11) is weakened by 6 orders of magnitude, leading to a bound $\lambda \lesssim 10^{-10}$ s^{-1}. This proves how efficient the phenomenon of emission of radiation is in setting constrains to the CSL model.

The same analysis has been carried on more in detail recently in [11], using data from more recent experiments and a more detailed statistical analysis. This is discussed in detail in chapter 28, and it leads to the stronger bound $\lambda \leq 6.8 \times 10^{-12}$ s^{-1} for $r_C = 10^{-7}$ m. Letting also the r_C as a free parameter, the bounds are reported in the exclusion plot in Fig. 2 of chapter 28 .

4 A Discrepancy Among Different Calculations of the Emission Rate

A puzzling situation appeared in 2009, when the calculation of the radiation emission was repeated using the QMUPL model by Bassi and Dürr [12]. In the calculation performed using the QMUPL model, the starting point is the Schrödinger equation

$$i\hbar\frac{d|\phi(t)\rangle}{dt} = \left[H - \hbar\sqrt{\lambda_q}\,\mathbf{q}\cdot\mathbf{w}(t)\right]|\phi(t)\rangle \tag{13}$$

where H here is the same as that defined in Section 2 just written in the first quantization language, λ_q a free parameter of the model coupling the position operator $\mathbf{q}$ to the noise vector $\mathbf{w}(t)$, which has zero average and correlations $\mathbb{E}[w_i(s)w_j(s')] = \delta_{ij}\delta(s - s')$ (with i, j labeling the three components of $\mathbf{w}(t)$). It can be proved that, at the level of the master equation, the QMUPL model is a limit version of the CSL model when the number of particles of the system is fixed and the wave function spread is smaller than r_C. In such a case, the coupling constant λ_q is related to the CSL parameters by the relation $\lambda_q = \lambda/2r_C^2$. Therefore, it is reasonable to expect the two models to lead to the same predictions. Moreover, in the QMUPL model the interaction with the noise is linear in the position operator of the system, and therefore it is possible, under only the dipole approximation, to solve exactly the model. The result, when expanded to the first perturbative order in the EM and noise interactions, leads to

$$\frac{d\Gamma_k}{dk} = \frac{\lambda\hbar e^2}{2\pi^2\varepsilon_0 c^3 m_0^2 r_C^2 k} \tag{14}$$

for a free particle. Compared to Eq. (11), a factor 2 of difference is present. As we will see, this factor 2 does not arise from a typo or a silly mistake: it is related to the fact that, for this problem, the correct use of perturbation theory requires some care.

5 Repeating the Perturbative Calculation with the CSL Model

In order to shed light on this discrepancy, during my Master thesis and my PhD under the guidance of Prof. A. Bassi, I repeated both calculations, the one perturbative done in [8, 10] and that exact performed in [12]. I found a result in agreement with that of Bassi and Dürr [12]. The origin of the discrepancy with the previous calculations done by Fu and Adler was related to the calculation of some time integrals in the perturbative terms.

While the discrepancy in the case of the white noise models is just a factor 2, when non-Markovian generalizations of the CSL model are considered it has much more serious consequences. In fact, the emission rate has the form [13][4]:

$$\frac{d\Gamma_k}{dk} = \frac{\lambda\hbar e^2}{4\pi^2\varepsilon_0 c^3 m_0^2 r_C^2 k}\left(\tilde{f}(0) + \tilde{f}(\omega_k)\right). \tag{15}$$

The second term in Eq. (15) is the expected contribution, in agreement with the results of [10], and it is physically meaningful: given a noise with generic spectrum $\tilde{f}(\omega)$, only the frequency of the spectrum resonant with that of the emitted photon $\omega = \omega_k$ plays a relevant role. On the other hand, the first term proportional to $\tilde{f}(0)$ is

[4]Note that in the white noise limit, the noise spectrum becomes $\tilde{f}(\omega) = 1$ for all ω, which is the reason why we get the factor 2.

problematic: the emission depends on the spectrum calculated in zero, implying that even a very weak noise may induce the emission of photons with very large energies. This contribution is clearly nonphysical and a deeper investigation was required to understand its origin.

6 An Exact Analysis with the Non-Markovian QMUPL

In order to shed light on the issue, and in particular on the origin of the nonphysical contribution proportional to $\tilde{f}(0)$, we analyzed the radiation emission using the non-Markovian extension of the QMUPL model. The advantage of this approach is that, like in the work [12] where the white noise QMUPL model was studied, all calculations can be performed exactly, apart for the dipole approximation. The calculation was performed for a free particle and a harmonic oscillator, and the main conclusions were the following [14]:

1. For a free particle there is always a nonphysical contribution. However, if one starts from a bounded particle in a harmonic potential with frequency ω_0, performs all the calculations, and only at the end considers the limit $\omega_0 \to 0$, irrespective of how weak the bounding potential was taken the final result does not contain any term proportional to $\tilde{f}(0)$.
2. For a bounded particle, it is fundamental *not* to treat the EM interaction to the lowest perturbative order. Indeed, when the analysis is done keeping into account the EM exactly, exponential damping terms appear which are not present when calculations are performed to the lowest perturbative order. These damping terms are responsible for suppressing the nonphysical contributions.

7 Getting the Correct Result in the CSL Model

Since the analysis done in [14], discussed in the previous section, clarified that is fundamental to have a bounded particle and to not treat the EM interaction perturbatively, the calculation was repeated for the CSL model fulfilling these conditions [15]. More precisely, we worked under the dipole approximation and treated only the noise perturbatively, while the EM radiation was treated exactly. The calculation, long but straightforward, confirmed all the results found in [14] for the QMUPL model. In particular, when at the end of the calculation the free particle limit ($\omega_0 \to 0$) and lowest order EM interaction are considered $\beta \to 0$, one gets the result:

$$\frac{d\Gamma_k}{dk} = \frac{\lambda \hbar e^2}{4\pi^2 \varepsilon_0 c^3 m_0^2 r_C^2 k} \tilde{f}(\omega_k), \tag{16}$$

with no presence of the nonphysical contribution, and that, in the white noise limit ($\tilde{f}(\omega) \to 1$), it reduces to the result found by Fu in Eq. (11).

However, the approach used in [15], has a fundamental limitation: the calculation is based on moving to an interaction picture where the operators evolution given by the Hamiltonian H in Eq. (3) is solved exactly. Even using the dipole approximation, the evolution of the operators can be solved exactly only for quadratic Hamiltonians, i.e. the free particle or the harmonic oscillator. For more general systems, H is not quadratic anymore, the Heisenberg equations of motion cannot be solved exactly making this approach unfeasible.

This difficulty was overcame in [16], where a more refined perturbative method was applied. The main idea is to use a perturbative approach where the radiative corrections to the particle propagators are considered. In terms of Feynman diagrams, this corresponds to account for the corrections to the propagator due to the emission and re-absorption of a virtual photon. As discussed also in the literature [17], this implies a Lamb shift which induces an exponential decay in time of the propagator. This damping suppresses the terms responsible for the nonphysical contributions proportional to $\tilde{f}(0)$. When the system considered is an harmonic oscillator, the decaying exponents coincide with those found in [14, 15], confirming that this is the appropriate way of introducing these damping in perturbative calculations.

After a long calculation, in [16] it was proved that the emission rate for a system interacting with a generic family of noises N_ℓ is given by:

$$\frac{d\Gamma}{dk} = \sum_{\mu} \int d\Omega_k \left(\frac{\gamma}{\hbar^2}\right) \sum_f \sum_\ell \left| \sum_n \frac{\langle f\,|R_k|\,n\rangle\,\langle n\,|N_\ell|\,i\rangle}{\left[i\left(\Delta_{fn} + \omega_k\right) - \Gamma_n\right]} - \frac{\langle f\,|N_\ell|\,n\rangle\,\langle n\,|R_k|\,i\rangle}{\left[i\left(\Delta_{ni} + \omega_k\right) + \Gamma_n\right]} \right|^2 \times \tilde{f}\left(\Delta_{fi} + \omega_k\right). \tag{17}$$

where the matrix elements with R_k are due to the interaction with the EM field, $\Delta_{ni} = (E_n - E_i)/\hbar$ with E_n energy eigenstates of the Hamiltonian of the system and Γ_n the corresponding damping.

In the case of the CSL model, the discrete index ℓ is replaced by the continuous parameter $\mathbf{x}$ labeling each space point, and the noise operator (using the first quantization language for a particle with position operator $\mathbf{q}$) by:

$$N_\ell \longrightarrow \frac{m}{m_0} \frac{1}{(\sqrt{2\pi} r_C)^3} e^{-\frac{(\mathbf{q}-\mathbf{x})^2}{2r_C^2}}. \tag{18}$$

Then, for a free particle, Eq. (17) reduces to Eq. (16).

8 Conclusions and Perspectives

We discussed the phenomenon of the collapse-induced radiation emission in the CSL and the QMUPL models. The calculation for computing the radiation emission rate presents some non-trivial features when the problem is studied using perturbation theory: in particular, when the EM interaction is treated to the lowest perturbative order, nonphysical contributions appear. After clarifying the origin of these non-physical terms, it was showed in [16] how to properly use perturbation techniques in order to avoid them. The final result reported in Eq. (17) is quite general and it will be the starting point for future analysis. We are currently collaborating with the group of Prof. C. Curceanu on performing a dedicated experiment at the Gran Sasso Laboratories focused on further improving the bounds on the CSL parameters by studying the emission of radiation from Germanium.

References

1. G. C. Ghirardi, *Un'occhiata alle carte di Dio*, Il Saggiatore, Milano (1997).
2. A. Bassi and G. C. Ghirardi, Phys. Rept. **379**, 257 (2003).
3. A. Bassi, K. Lochan, S. Satin, T. P. Singh and H. Ulbricht, Rev. Mod. Phys. **85**, 471–527 (2013).
4. G. C. Ghirardi, A. Rimini and T. Weber, Phys. Rev. D **34**, 470 (1986).
5. G. C. Ghirardi, P. Pearle and A. Rimini, Phys. Rev. A **42**, 78 (1990).
6. L. Diósi, Phys. Rev. A **40**, 1165 (1989).
7. L. Diósi, Phys. Rev. A **42**, 5086 (1990).
8. Q. Fu, Phys. Rev. A **56**, 1806 (1997).
9. S. L. Adler and A. Bassi J. Phys. A: Math. Theor. **40**, 15083–98 (2007).
10. S. L. Adler and F.M. Ramazanoglu, J. Phys. A **40**, 13395 (2007).
11. K. Piscicchia et al., Entropy **19**, 319 (2017).
12. A. Bassi and D. Dürr, J. Phys. A **42**, 485302 (2009).
13. S. L. Adler, A. Bassi, and S. Donadi, J. Phys. A **46**, 245304 (2013).
14. A. Bassi and S. Donadi, Phys. Lett. A **378**, 761 (2014).
15. S. Donadi, D.-A. Deckert, and A. Bassi, Ann. Phys. (N.Y.) **340**, 70 (2014).
16. S. Donadi and A. Bassi, J. Phys. A **48**, 035305 (2015).
17. J. J. Sakurai, *Advanced Quantum Mechanics*, Hardcover, (1967).

Relativistic Quantum Theory

Jürg Fröhlich

Eine neue wissenschaftliche Wahrheit pflegt sich nicht in der Weise durchzusetzen, dass ihre Gegner überzeugt werden und sich als belehrt erklären, sondern vielmehr dadurch, dass die Gegner allmählich aussterben und dass die heranwachsende Generation von vornherein mit der Wahrheit vetraut ist.

Max Planck

Abstract The purpose of this paper is to sketch an approach towards a reconciliation of quantum theory with relativity theory. It will actually be argued that these two theories ultimately rely on one another. A general operator-algebraic framework for relativistic quantum theory is outlined. Some concepts of space-time structure are translated into algebra. Following deep results of Buchholz et al., the key role of massless modes, photons and gravitons, and of Huygens' Principle in a relativistic quantum theory well suited to describe "events" and "measurements" is highlighted. In summary, a relativistic version of the "*ETH Approach*" to quantum mechanics is described.

1 Topics to Be Addressed

Anybody who attempts to work on the foundations—or "interpretation"—of quantum theory realizes quickly that this field is in a state of utmost confusion. Whether authorities in this matter or not, *Richard Feynman* once said: "If someone tells you they understand quantum mechanics then all you've learned is that you've met a liar"; and *Sean Carroll*, of the California Institute of Technology, in a popular article

J. Fröhlich (✉)
Theoretical Physics, ETH Zurich, Wolfgang-Pauli-Str. 27, 8093 Zurich, Switzerland
e-mail: juerg@phys.ethz.ch

V. Allori et al. (eds.), *Do Wave Functions Jump?*, Fundamental Theories of Physics 198,
https://doi.org/10.1007/978-3-030-46777-7_19

that appeared in the 'New York Times' [1], writes: "... quantum mechanics has a reputation for being especially mysterious. What's surprising is that physicists seem to be O.K. with not understanding the most important theory they have. ... Physicists don't understand their own theory any better than a typical smartphone user understands what's going on inside the device. ... The whole thing is preposterous. Why are observations special? What counts as an "observation", anyway? When exactly does it happen? Does it need to be performed by a person? Is *consciousness* somehow involved in the basic rules of reality? Together these questions are known as the "measurement problem" of quantum theory. ..."—Well, obviously a text like this leaves the reader in a state of bewilderment and/or anger! In the same article *Caroll* also writes: "You would naturally think, then, that understanding quantum mechanics would be the absolute highest priority among physicists worldwide. ... Physicists, you might imagine, would stop at nothing until they truly understood quantum mechanics."

Quite some time (perhaps thirty years) ago, I arrived at a conclusion similar to the one *Caroll* reached in the last two sentences quoted above. In 2012, when I retired from my position at ETH and did not have to make a career, anymore, I started to consider it to be one of my obligations to help removing some of the confusion surrounding the foundations of quantum mechanics. I do not have any illusions about the chances of success in pursuing this goal,[1] not because it is impossible to understand quantum mechanics—I actually think it **is possible**—but chiefly because people have so many prejudices about it.

Here is my credo in this endeavor:

- Talking of the "interpretation" of a physical theory presupposes implicitly that the theory has reached its final form, but that it is not completely clear, yet, what it tells us about natural phenomena. Otherwise, we had better speak of the "foundations" of the theory. Quantum Mechanics has apparently not reached its final form, yet. Thus, it is not really just a matter of interpreting it, but of completing its foundations.
- The only form of "interpretion" of a physical theory that I find legitimate and useful is to delineate approximately the ensemble of natural phenomena the theory is supposed to describe and to construct something resembling a "structure-preserving map" from a subset of mathematical symbols used in the theory that are supposed to represent physical quantities to concrete physical objects and phenomena (or events) to be described by the theory. Once these items are clarified the theory is supposed to provide its own "interpretation". (A good example is Maxwell's electrodynamics, augmented by the special theory of relativity.)
- The ontology a physical theory is supposed to capture lies in *sequences of events, sometimes called "histories"*, which form the objects of series of observations

[1] A recent paper of mine on the foundations of quantum mechanics triggered the following comment from a "colleague": *"Hi, again and again. How many time will you recycle your papers? Cannot see (you?) that no one is interested in your obscure thinking. Adding 'ETH' will not help. You are old and essentially useless. Go fishing. Best, A."*.

extending over possibly long stretches of time and which the theory is supposed to describe.
- In discussing a physical theory and mathematical challenges it raises it is useful to introduce clear concepts and basic principles to start from and then use precise and—if necessary—quite sophisticated mathematical tools to formulate the theory and to cope with those challenges.
- To emphasize this last point very explicitly, I am against denigrating mathematical precision and ignoring or neglecting precise mathematical tools in the search for physical theories and in attempts to understand them, derive consequences from them and apply them to solve concrete problems.

In this paper I will sketch some ideas about a formulation of **local relativistic quantum theory** designed to describe "events" and, ultimately, to solve the "measurement problem" alluded to above. (In doing this I try to follow the credo formulated above.) I will specifically address the following topics:

1. Why is it fundamentally impossible to use a physical theory to predict the future?—Sect. 2.
2. Why is quantum theory intrinsically probabilistic?—Sect. 2.
3. How are "locality" and "Einstein causality" expressed in relativistic quantum theory; what is their meaning?—Sect. 3.
4. What are "events" in quantum theory—Sect. 4—and how does one describe their recording? What is meant by "measuring a physical quantity"?—Sect. 5.
5. How do *states of physical systems* evolve in (space-)time, according to quantum theory? What is the probabilistic law governing their evolution?—Sect. 4.
6. How does quantum theory distinguish between past and future; how does it talk about space-time? Could it be that a consistent "Quantum Theory of Events" must necessarily be relativistic and involve massless modes? Could it be that such a quantum theory could explain why space-time is even-dimensional and that it might incorporate gravitation as an "emergent phenomenon"?—Sect. 6.

I wish to mention that various ideas related to ones elaborated on in [2, 3] and in this paper have been described in [4, 5]. In particular, many years ago, the late *Rudolf Haag* has emphasized the importance of introducing a clear notion of "events" in quantum theory and to elucidate their role.

This paper is dedicated to the memory of *Gian Carlo Ghirardi*. My approach to the foundations of quantum mechanics (dubbed *"ETH Approach"*) shares some general features with *GRW* [6]; in particular, an important role is played by "state collapse". I wish to thank *Detlef Dürr* for having invited me to present my ideas in this book.

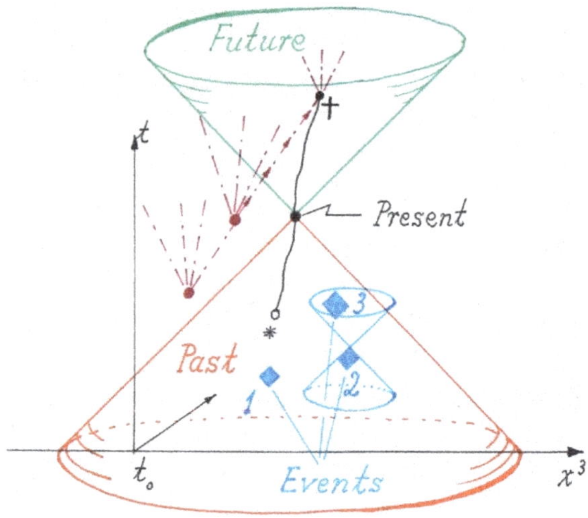

Fig. 1 The "observer" sits at "Present" and is unaware of the dangers lurking from outside his past light-cone (denoted *"Past"*). He might get killed at †, a space-time point in his future light-cone (denoted *"Future"*). Events are numbered in the figure; events 1 and 2 are space-like separated, event 3 is in the future of event 2

2 Why Are We Not Able to Predict the Future by Using Our Physical Theories, and Why Is Quantum Theory Intrinsically Probabilistic?

Imagine that the space-time of our Universe has an event horizon that hides what may happen in causally disconnected regions of space-time. Figure 1, below, illustrates the claim that, for fundamental reasons, observers are then unable to use relativistic theories to fully predict their future; for, **never** do they have access to complete knowledge of the initial conditions of the Universe that would be necessary (but not necessarily sufficient) to predict the future.[2] This argument applies to both, classical *and* quantum theories. But quantum theories have an additional feature that makes it impossible to use them to predict the future precisely: *They are fundamentally probabilistic*.

Figure 1 is supposed to illustrate, furthermore, that the "*Past*" consists of a "History of Events" or "Facts", while the "*Future*" consists of an ensemble of "Potentialities". In a proper formulation of Quantum Mechanics this dichotomy should be retained! In this paper we will try to find out how to implement it in relativistic quantum theory.

Let S be an *"isolated physical system"* to be described by a model of relativistic quantum theory.—*Note:* An isolated system has the property that, over some period of time, its evolution does not depend on anything happening in its complement, i.e., in the rest of the Universe, in the sense that, during a certain period of time, the *Heisenberg-picture dynamics* of physical quantities characteristic of S is, *for all practical purposes*, **independent** of the degrees of freedom in the complement of S,

[2]The same is true if there exist waves propagating at the speed of light along *surfaces* of light-cones.

(a consequence of cluster properties). It should be noted, however, that the state of S can be entangled with the state of its complement!.—

The concept of an *isolated* physical system is important in quantum mechanics, because, *only for such systems*, we know how to describe the time evolution of operators representing physical quantities in the Heisenberg picture (in terms of conjugation of those operators with the unitary propagator of the system). In order to describe the quantum dynamics of an isolated physical system S, we will allways start from the Heisenberg-picture dynamics of "observables" (i.e., of self-adjoint operators representing physical quantities) referring to S. The dynamics of *states* of S is considerably more subtle to understand and is, in a sense, at the *core of our considerations* in this paper—as it has been in the work of Ghirardi, Rimini and Weber.

In this paper we use (for simplicity) the following pedestrian formulation of the quantum mechanics of an isolated physical system S in the Heisenberg picture: States of S are given by density matrices, Ω, acting on a separable Hilbert space, $\mathcal{H}$, of "pure state vectors" of S. Let $\hat{X}$ be a physical quantity of S, and let $X(t) = X(t)^*$ be the self-adjoint linear operator on $\mathcal{H}$ representing $\hat{X}$ at time t. Then the operators $X(t)$ and $X(t')$ representing $\hat{X}$ at two *different* times t and t', respectively, are unitarily conjugated to one another:

$$X(t) = U(t', t)\, X(t')U(t, t'), \tag{1}$$

where, for each pair of times t, t', $U(t, t')$ is the propagator (from t' to t) of the system S, which is a unitary operator acting on $\mathcal{H}$, and $\{U(t, t')\}_{t,t' \in \mathbb{R}}$ satisfy

$$U(t, t') \cdot U(t', t'') = U(t, t''),\ \forall t, t', t'', \quad U(t, t) = \mathbf{1}\,,\ \forall t\,.$$

It is often said that, in the Heisenberg picture, states of S are *independent* of time; and that the Heisenberg picture is equivalent to the Schrödinger picture, where physical quantities are time-independent, but states evolve according to the propagator $U(t, t')$, solving a *deterministic* Schrödinger equation. Even if quantum mechanics were put under the auspices of the so-called "Copenhagen interpretation", this is, of course, nonsense, as has been amply demonstrated on many examples; (see [8, 10, 11], and refs. given there)! For, whenever a "measurement" is made, at some time t, say—we will later speak, more accurately, of an **"event"** happening at approximately time t—the deterministic unitary evolution of the state of S in the Schrödinger picture is **interrupted** at this time, and the state "jumps", or "collapses" into an eigenspace of the "observable" that is measured—more accurately: the state jumps into the image of an orthogonal projection representing the "event" that actually happens at time t, with jumping probabilities as given by *Born's Rule*; (see also [3, 4]). Expressed in the Heisenberg picture, one can say that, while operators representing physical quantities referring to an isolated physical system S evolve in time according to Eq. (1), the *state* of S changes *randomly* whenever an "event" happens; it thus exhibits a *non-trivial, stochastic evolution* in time, a kind of *stochastic branching process* described in [2,

3, 12, 13] and in Sect. 4 of this paper. In order to avoid paradoxes [7–9], it is crucial to assume that the occurrence of an event (for example, the successful completion of a measurement) has an **objective** meaning independent of the "observer"—and independent of whether an "observer" is actually present or not.

One should think that, by now, these things are exceedingly well-known and appreciated, and hence I won't dwell on them any further.—It might be added, however, that, in *Bohmian mechanics*, randomness enters in a way that differs from the one in other formulations of quantum mechanics: Randomness is due, in Bohmian mechanics, to incomplete knowledge of initial conditions; see [14].[3]

3 The Meaning of "Locality" or "Einstein Causality" in Relativistic Quantum Theory

In this section, I sketch ideas on "locality" or "Einstein causality". For, there appears to exist a certain amount of confusion concerning the question in which sense quantum mechanics is *"non-local"* and in which sense it is perfectly *"local"*. Let us consider an isolated system, S, consisting of two spin-$\frac{1}{2}$ particles, p and p', and of equipment serving to measure components of their spins along two directions given by unit vectors $\vec{n}$ and $\vec{n}'$, respectively. We imagine that, after preparation of the initial state, Ω, of S, particle p propagates into a cone, C, opening in the direction of the negative x-axis, while p' propagates into a cone, C', opening in the direction of the positive x-axis, with only tiny probabilities for sojourn outside C and C', respectively. Let us assume that the measurement of the spin of p takes place inside a region $B \subset C$ in an interval $[t_1, t_2]$ of times, while the measurement of the spin of p' takes place in a region $B' \subset C'$ within a time-interval $[t_1', t_2']$, and let us imagine that the space-time regions $B \times [t_1, t_2]$ and $B' \times [t_1', t_2']$ are *space-like separated.* The results of the two measurements are described by two orthogonal projection operators, $\Pi^{p}_{\vec{n},\sigma}$, $\sigma = \pm$, and $\Pi^{p'}_{\vec{n}',\sigma'}$, $\sigma' = \pm$, where "$\sigma = +$" means that the spin of p is aligned with $\vec{n}$ after the measurement has been completed, while "$\sigma = -$" means that the spin of p is anti-parallel to $\vec{n}$ after its measurement, and similarly for p'. The operators $\Pi^{p}_{\vec{n},\sigma}$, $\sigma = \pm$, have the following properties:

$$\Pi^{p}_{\vec{n},+} \cdot \Pi^{p}_{\vec{n},-} = 0, \;\; \Pi^{p}_{\vec{n},+} + \Pi^{p}_{\vec{n},-} = \mathbf{1}\,, \tag{2}$$

and similarly for the operators $\Pi^{p'}_{\vec{n}',\sigma'}$, $\sigma' = \pm$. Moreover, the operators $\Pi^{p}_{\vec{n},\sigma}$ and $\Pi^{p'}_{\vec{n}',\sigma'}$ are localized in *space-like separated* regions, $B \times [t_1, t_2]$ and $B' \times [t_1', t_2']$, respectively, of space-time, for all choices of σ and of σ'. We would like to make an educated guess of the state used by a localized observer, $\mathcal{O}$, to predict his future

[3]The Bohmian point of view cannot be discussed any further in this paper. Suffice it to remark that Bohmian Mechanics is **not** equivalent to the formulation of Quantum Theory proposed in this paper and in [3].

if $\mathcal{O}$ has the property that the past light-cones of all points inside $\mathcal{O}$ contain both regions, $B \times [t_1, t_2]$ and $B' \times [t_1', t_2']$. The answer to the question which of the two spin measurements was initiated or completed *first* then obviously depends on the past "world-tube" of the observer $\mathcal{O}$. This is because $B \times [t_1, t_2]$ and $B' \times [t_1', t_2']$ are space-like separated. Let us suppose that, for an observer $\mathcal{O}$, the spin of p was measured first, that the state of S before any of these measurements were carried out was given by a density matirx Ω, and that between the preparation of the state Ω of S and further observations by $\mathcal{O}$ *only* the measurements of the spins of p and of p' were made. According to the standard "projection postulate" (of the Copenhagen interpretation), the state used by $\mathcal{O}$ to predict future measurement outcomes is then given by

$$\Omega_{\mathcal{O}} = [\mathcal{N}_{(\vec{n},\sigma),(\vec{n}',\sigma')}]^{-1}\, \Pi^{p'}_{\vec{n}',\sigma'} \cdot \Pi^{p}_{\vec{n},\sigma}\, \Omega\, \Pi^{p}_{\vec{n},\sigma} \cdot \Pi^{p'}_{\vec{n}',\sigma'}\,, \tag{3}$$

where $\mathcal{N}_{(\vec{n},\sigma),(\vec{n}',\sigma')} := \mathrm{tr}\Big(\Pi^{p'}_{\vec{n}',\sigma'} \cdot \Pi^{p}_{\vec{n},\sigma}\, \Omega\, \Pi^{p}_{\vec{n},\sigma} \cdot \Pi^{p'}_{\vec{n}',\sigma'}\Big)$ is a normalization factor. Imagine now that $\mathcal{O}'$ is an observer localized in the *same* space-time region as $\mathcal{O}$, but for whom the spin of p' is measured *before* the spin of p. He then proposes to use the state $\Omega_{\mathcal{O}'}$ given by a formula arising form (3) by exchanging the order of $\Pi^{p}_{\vec{n},\sigma}$ and $\Pi^{p'}_{\vec{n}',\sigma'}$. We want to impose the requirement that *the predictions made by* $\mathcal{O}$ *and* $\mathcal{O}'$ *concerning future measurements* (i.e., ones localized in their common future light-cone) *must be compatible*. This implies that the two states $\Omega_{\mathcal{O}}$ and $\Omega_{\mathcal{O}'}$ must agree on the algebra of all "observables" potentially measureable in the future of $\mathcal{O}$ = future of $\mathcal{O}'$. This would be guaranteed if (but does **not** imply that)

$$\Pi^{p'}_{\vec{n}',\sigma'} \cdot \Pi^{p}_{\vec{n},\sigma} = \Pi^{p}_{\vec{n},\sigma} \cdot \Pi^{p'}_{\vec{n}',\sigma'}\,, \tag{4}$$

for arbitrary choices of $(\vec{n}, \sigma)$ and $(\vec{n}', \sigma')$, assuming, as stated above, that the localization regions $B \times [t_1, t_2]$ and $B' \times [t_1', t_2']$ are space-like separated. Equation (4) is what is called *"locality"* or *"Einstein causality"* in relativistic quantum field theory. This is a sufficient (but not necessary) condition to eliminate ambiguities in the predictions of possible future measurement outcomes made by different observers that are due to the impossibility of unambiguously ordering measurements according to the times at which they are initiated (or completed). But Eq. (4) does **not** imply that quantum mechanics is "local" in the following sense: Consider the state

$$\Omega_{(\vec{n},\sigma)} := [\mathcal{N}_{(\vec{n},\sigma)}]^{-1} \Pi^{p}_{\vec{n},\sigma}\, \Omega\, \Pi^{p}_{\vec{n},\sigma}\,,$$

where $\mathcal{N}_{(\vec{n},\sigma)}$ is a normalization factor chosen such that $\mathrm{tr}(\Omega_{(\vec{n},\sigma)}) = 1$. Let A be an "observable" localized in a space-time region space-like separated from $B \times [t_1, t_2]$; (for example $A = \Pi^{p'}_{\vec{n}',\sigma'}$). One might expect that

$$\mathrm{tr}\big(\Omega\, A\big) = \mathrm{tr}\big(\Omega_{(\vec{n},\sigma)}\, A\big)\,,$$

for any operator A with these properties. But, of course, this equality does **not** hold! This fact is what people call the "non-locality" of quantum theory. In quantum field theory, this kind of "non-locality" is neatly reflected in the Reeh-Schlieder theorem [15]. It results from entanglement.

One major purpose of this paper is to render the "projection postulate" (or "collapse postulate"—see Eq. (3)) more precise, to explain its origin and to find out *under what conditions it is applicable*. In contrast to the ideas described in [6], we will not invoke any mechanism extraneous to quantum mechanics that produces "state collapse".

4 Relativistic Quantum Theory, and the Notion of "events"

In this section we propose an algebraic definition of local relativistic quantum theory and then introduce a precise notion of "events". We require some rudimentary knowledge of the theory of operator algebras. In particular, the reader might profit from knowing what a C^*—and what a von Neumann algebra is and what, for example, the Gel'fand-Naimark-Segal (GNS) construction is. What will be used from the theory of operator algebras, in this paper, can be learned in a few hours! A useful reference may be [16].

For the time being, we will consider space-time, $\mathcal{M}$, to be given; but we do not equip $\mathcal{M}$ with a Lorentzian metric. Later, we will try to clarify how properties of algebras of operators representing localized potentialities equip $\mathcal{M}$ with a causal structure. But to start with, we assume $\mathcal{M}$ to be given by Minkowski space, $\mathbb{M}^d$, with $d = 4$.

In relativistic quantum theory, all operators representing physical quantities characteristic of an isolated physical system S can be localized in some space-time regions. Given a region $\mathcal{O} \subset \mathcal{M}$, we denote by $\mathcal{A}(\mathcal{O})$ the algebra generated by all bounded operators localized in $\mathcal{O}$ that represent physical quantities. The family $\{\mathcal{A}(\mathcal{O})\}_{\mathcal{O}\subset\mathcal{M}}$ is called a "net of local algebras". For an introduction to these concepts and to algebraic quantum field theory the reader is advised to consult [17]. In the following considerations, the regions $\mathcal{O}$ are usually taken to be forward or backward light-cones with apex in an arbitrary space-time point $P \in \mathcal{M}$.

A general formulation of local relativistic quantum theory:

We consider an isolated physical system S to be described with the help of a model of local relativistic quantum theory.

Definition 1 By $\mathcal{F}_P$ we denote the *algebra generated by all operators representing physical quantities referring to S (such as potential events) localized in the "future" of the space-time point P, while $\mathcal{P}_P$ denotes the algebra generated by all operators representing physical quantities localized in the "past" of P. □

We assume that all the algebras $\mathcal{F}_P$ are contained in a C^*-algebra $\mathcal{E}$, and

$$\mathcal{E} = \overline{\bigvee_{P \in \mathcal{M}} \mathcal{F}_P}\,, \tag{5}$$

where the closure on the right side is taken in the operator norm of $\mathcal{E}$. We assume that all these algebras are represented on a common separable Hilbert space $\mathcal{H}$ and that all *"states of physical interest"* of S can be identified with density matrices (non-negative trace-class operators normalized to have trace $= 1$) acting on $\mathcal{H}$.[4] In our notation, we will not distinguish between an abstract element of the algebra $\mathcal{E}$ and the linear operator on $\mathcal{H}$ representing it.

Definition 2 We define $\mathcal{E}_P$ to be the von Neumann algebra obtained by closure of the algebra $\mathcal{F}_P$ in the weak operator topology of the algebra, $B(\mathcal{H})$, of all bounded operators on $\mathcal{H}$. □

If S is a physical system in a state of *finite* energy describing only excitations of strictly positive rest mass then

$$\mathcal{E}_P \simeq B(\mathcal{H})\,, \text{ for any point } P \in \mathcal{M}\,. \tag{6}$$

It is expected that this equality always holds in a space-time of *odd* dimension, *even* if massless particles are present. This is because Huygens' Principle does not hold in space-times of odd dimension. (It also does not hold in certain even-dimensional space-times with non-vanishing curvature. But that's another story, which, for reasons that I will not explain in any detail, is not expected to invalidate the following considerations.) The property expressed in Eq. (6) is one most people sub-consciously consider to be always valid. But this is actually *not* the case! (If it were we would probably be unable to introduce a reasonable notion of "events" in quantum theory, and we would never solve the "measurement problem".)

If there exist massless particles, in particular photons and/or gravitons and Dark-Energy modes, and if Huygens' Principle holds in an appropriate sense ($\mathcal{M}$ even-dimensional, specifically $\mathcal{M} = \mathbb{M}^4$),[5] the algebra $\mathcal{E}_P$ tends to have an infinite-dimensional commutant, $\mathcal{E}'_P$. (The commutant, $\mathfrak{M}'$, of an algebra $\mathfrak{M}$ contained in $B(\mathcal{H})$ is the algebra of all bounded operators on $\mathcal{H}$ commuting with *all* operators in $\mathfrak{M}$.) More specifically, within an algebraic framework of local relativisitic quantum field theory over four-dimensional Minkowski space-time, *Detlev Buchholz* has shown [18] that, in the presence of massless particles, $\mathcal{E}'_{P_t} \cap \mathcal{E}_{P_{t_0}}$ is an *infinite-dimensional, non-commutative* algebra, whenever P_{t_0} is a space-time point in the *past* of the space-time point P_t, as indicated in Fig. 2.

In his proof, *Huygens' Principle* is exploited in the form that asymptotic out-fields creating on-shell massless particles escaping to infinity do not propagate into

[4] It is sometimes advantageous to formulate this assumption in a more abstract, algebraic way involving, among other ingredients, the GNS-construction; see, e.g., [17].

[5] Or in the presence of blackholes in space-time.

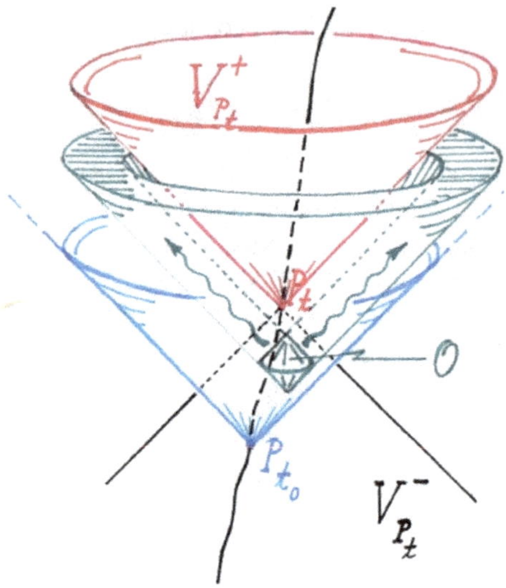

Fig. 2 The black line is the world-line of an "observer" who, at time t, is localized near P_t. Operators representing physical quantities potentially observable by the "observer" in the future of P_t are localized inside the forward light-cone $V^+_{P_t}$. They generate the algebra $\mathcal{E}_{P_t}$. Asymptotic out-field operators describing the emission of (on-shell) photons or gravitons in the region $\mathcal{O}$ propagate along the light-cones contained in $V^+_{P_{t_0}}$ but *not* contained in $V^+_{P_t}$

the *interior* of forward light-cones contained in the future of the space-time region (denoted by $\mathcal{O}$ in Fig. 2) where they are localized, but propagate along the *surface* of forward light-cones with apices in $\mathcal{O}$. Such asymptotic out-fields are then shown to commute with all operators in the algebra $\mathcal{E}_{P_t}$.

One expects that, if space-time is even-dimensional and in the presence of massless particles, the algebras $\mathcal{E}_P$ have the property that all non-zero orthogonal projections belonging to $\mathcal{E}_P$ have an infinite-dimensional range. This implies that there do **not** exist any normal *pure* states on these algebras. Furthermore, they are expected to be isomorphic to a certain "universal" von Neumann algebra, $\mathfrak{R}$,[6] i.e., $\mathcal{E}_P \simeq \mathfrak{R}$, $\forall\, P \in \mathcal{M}$.

We now use these insights to extract a **general algebraic formulation** of *local relativistic quantum theory* compatible with the appearance of "events" and promising a solution of the "measurement problem". We assume that space-time $\mathcal{M}$ is a topological space and that, with every point $P \in \mathcal{M}$, one can associate a von Neumann algebras, $\mathcal{E}_P$, the "algebra of potential events that might possibly happen in the future of P", with the property that $\mathcal{E}_P$ is contained in a C^*-algebra $\mathcal{E}$, for all $P \in \mathcal{M}$.

The family of algebras $\{\mathcal{E}_P\}_{P \in \mathcal{M}}$ equips space-time $\mathcal{M}$ with the following *causal structure*:

[6] $\mathfrak{R}$ is expected to be a von Neumann algebra of type III_1.

Definition 3 A space-time point P' is in the future of a space-time point P, written as $P' \succ P$, (or, equivalently, P is in the past of P', written as $P \prec P'$) iff

$$\boxed{\mathcal{E}_{P'} \subsetneqq \mathcal{E}_P\,, \quad \mathcal{E}'_{P'} \cap \mathcal{E}_P \text{ is an } \infty - \text{dim. non-commutative algebra}} \tag{7}$$

□

Equation (7) expresses what I call the

"Principle of Diminishing Potentialities" (PDP)

This principle is actually a **theorem** in an axiomatic formulation of quantum electrodynamics over four-dimensional Minkowski space proposed by D. Buchholz and the late J. Roberts [20].

Henceforth, the *Principle of Diminishing Potentialities will always be* ***assumed*** *to hold*; and, within our formulation of relativistic quantum theories, (a model of) an *isolated physical system* S is defined by specifying the following data:

$$S = \{\mathcal{M}, \mathcal{E}, \mathcal{H}, \{\mathcal{E}_P\}_{P\in\mathcal{M}} \text{ satisfying } PDP\}, \tag{8}$$

where $\mathcal{M}$ is a model of space-time, $\mathcal{E}$ is a C^*-algebra represented on a Hilbert space $\mathcal{H}$, and $\{\mathcal{E}_P\}_{P\in\mathcal{M}}$ is a family of von Neumann algebras satisfying the "Principle of Diminishing Potentialities" introduced in Eq. (7).

Definition 4 If a space-time point P' is neither in the future of a space-time point P nor in the past of P we say that P and P' are *space-like separated*, written as $P \times P'$. □

Let Σ be a space-like subset of $\mathcal{M}$. If $\mathcal{M} = \mathbb{M}^4$ we imagine that Σ is a subset of a space-like hypersurface of co-dimension 1 in $\mathcal{M}$. Since all the algebras $\mathcal{E}_P$, $P \in \mathcal{M}$, are assumed to be contained in the C^*-algebra $\mathcal{E}$, the following definition is meaningful:

$$\mathcal{E}_\Sigma := \overline{\bigvee_{P\in\Sigma} \mathcal{E}_P}\,, \tag{9}$$

where the closure is taken in the weak topology of $B(\mathcal{H})$. A state, ω_Σ, on the algebra $\mathcal{E}_\Sigma$ is a normalized, positive linear functional on $\mathcal{E}_\Sigma$.

Remark: At this point we should comment on the question of what the operational meaning of a "state" of an isolated system S is, and how one can *prepare* S in a specific state. Obviously these are important questions, which, however, cannot be discussed here; but see [21].

Definition 5 Let $\mathfrak{M}$ be a von Neumann algebra, and let ω be a normal state on $\mathfrak{M}$. For an operator $X \in \mathfrak{M}$, we define $ad_X(\omega)$ to be the linear functional on $\mathfrak{M}$ defined by

$$ad_X(\omega)(Y) := \omega([Y, X]), \quad \forall\, Y \in \mathfrak{M}.$$

We define the *centralizer*, $\mathcal{C}_\omega(\mathfrak{M})$, of the state ω by

$$\mathcal{C}_\omega(\mathfrak{M}) := \{X \,|\, X \in \mathfrak{M}, ad_X(\omega) = 0\}. \tag{10}$$

It is easy to verify that $\mathcal{C}_\omega(\mathfrak{M})$ is a (von Neumann) subalgebra of $\mathfrak{M}$, and that ω is a normalized trace on $\mathcal{C}_\omega(\mathfrak{M})$. (This property implies that centralizers are completely classified!)
Given an algebra $\mathfrak{N}$, the *center*, $\mathcal{Z}(\mathfrak{N})$, is the abelian subalgebra of $\mathfrak{N}$ consisting of all operators in $\mathfrak{N}$ commuting with all other operators in $\mathfrak{N}$. We set

$$\mathcal{Z}_\omega(\mathfrak{M}) := \mathcal{Z}(\mathcal{C}_\omega(\mathfrak{M})) \tag{11}$$

□

Motivation underlying the following notions and definitions is provided in [2, 3, 13].

Definition 6 Given a point $P \in \mathcal{M}$, a *potential event* in the future of P is a family, $\{\pi_\xi \,|\, \xi \in \mathfrak{X}\}$, ($\mathfrak{X}$ a countable set of indices[7]), of orthogonal projections belonging to $\mathcal{E}_P$ with the properties

$$\pi_\xi \cdot \pi_\eta = \delta_{\xi\eta}\pi_\xi, \ \forall\, \xi, \eta \in \mathfrak{X}, \quad \sum_{\xi \in \mathfrak{X}} \pi_\xi = \mathbf{1}. \tag{12}$$

It is expected that events usually have a *finite duration*. This would imply that operators $\{\pi_\xi \,|\xi \in \mathfrak{X}\}$ representing a potential event in the future of the point P would be localized in a *compact* region of space-time contained in the future of P (the future light-cone with apex in P). □

Definition 7 Given a state ω_P on the algebra $\mathcal{E}_P$, we say that an *event happens in the future of the space-time point* P iff the algebra

$$\mathcal{Z}_{\omega_P} := \mathcal{Z}\big(\mathcal{C}_{\omega_P}(\mathcal{E}_P)\big)$$

is generated by the projections $\{\pi_\xi \,|\, \xi \in \mathfrak{X}\} \subset \mathcal{Z}_{\omega_P} \subset \mathcal{E}_P$ of a potential event in the future of P with the properties that the cardinality of $\mathfrak{X}$ is at least 2 and that there exist projections $\pi_{\xi_1}, \ldots, \pi_{\xi_n}$, such that

$$\omega_p(\pi_{\xi_j}) > 0, \qquad \forall j = 1, \ldots, n, \quad \text{for some } n \geq 2. \tag{13}$$

(The quantity $\omega(\pi_\xi)$ will turn out to be the *Born probability* for π_ξ to occur in the future of P.) □

---

[7] Here it is assumed that potential events can be identified with the spectral projections of self-adjoint operators with *discrete* spectrum ($\simeq \mathfrak{X}$); more generally, one could identify potential events with spectral projections of families (abelian algebras) of commuting self-adjoint operators that may have continuous spectrum.

Let ω_P be the state of S on the algebra $\mathcal{E}_P$. It is easy to see that if an event described by the family $\{\pi_\xi \mid \xi \in \mathfrak{X}\} \subset \mathcal{Z}_{\omega_P}$ of projections happens in the future of the point P then

$$\omega_P(X) = \sum_{\xi \in \mathfrak{X}} \omega\big(\pi_\xi \, X \, \pi_\xi\big), \quad \forall X \in \mathcal{E}_P\,, \tag{14}$$

i.e., the state ω_P on the algebra $\mathcal{E}_P$ is a *mixture* of the states

$$\omega_{P,\xi} := \big[\omega_P(\pi_\xi)\big]^{-1} \omega\big(\pi_\xi(\cdot)\pi_\xi\big) \tag{15}$$

labelled by the points $\xi \in \mathfrak{X}$.

The following is a crucial axiom.

Axiom 1 ("State-collapse" postulate)*: If an event happens in the future of a point $P \in \mathcal{M}$, in the sense of Definition 7, then the state to be used to make predictions of further events possibly happening in the future of P is given by ω_{P,ξ_*}, for some $\xi_* \in \mathfrak{X}$ with $\omega_P(\pi_{\xi_*}) > 0$, where ω_{P,ξ_*}, $\xi_* \in \mathfrak{X}$, is defined in Eq.* (15).

The probability that ω_{P,ξ_*} is selected among the states $\{\omega_{P,\xi} \mid \xi \in \mathfrak{X}\}$ is given by Born's Rule, namely it is given by $\omega_P(\pi_{\xi_*})$. The projection π_{ξ_*} is called the "actual event" happening in the future of P. □

Next, we consider two points, P and P', in a subset Σ of $\mathcal{M}$, with $P \times P'$, (i.e., P and P' are space-like separated), We assume that the state ω_Σ defined in Eq. (9) is given, so that the states $\omega_P = \omega_\Sigma|_{\mathcal{E}_P}$ and $\omega_{P'} = \omega_\Sigma|_{\mathcal{E}'_P}$ are known, too. We suppose that, given ω_Σ, events happen in the future of P and of P'. Let $\mathcal{Z}_{\omega_P}$ denote the center of the centralizer of the state ω_P on the algebra $\mathcal{E}_P$, which describes the event $\{\pi_\xi^P | \xi \in \mathfrak{X}^P\}$ happening in the future of P, and let $\mathcal{Z}_{\omega_{P'}}$ be the algebra describing the event happening in the future of the point P'. We require the following axiom.

Axiom 2 (Events in the future of space-like separated points commute)*: Let $P \times P'$. Then all operators in $\mathcal{Z}_{\omega_P}$ commute with all operators in $\mathcal{Z}_{\omega_{P'}}$. In particular,*

$$\big[\pi_\xi^P, \pi_\eta^{P'}\big] = 0, \quad \forall\, \xi \in \mathfrak{X}^P \text{ and all } \eta \in \mathfrak{X}^{P'}. \qquad \square$$

This axiom may be one reflection of what people sometimes interpret as the fundamental **non-locality** of quantum theory: Projection operators representing events in the future of two *space-like separated* points P and P' in space-time are **constrained** to commute with each other! Actually, this implies what in quantum field theory is understood to express **locality** or Einstein causality.

Next, we assume that some slice, $\mathfrak{F}$, in space-time $\mathcal{M}$ is foliated by space-like hypersurfaces, Σ_τ: $\mathfrak{F} := \{\Sigma_\tau | \tau \in [0, 1]\}$, where τ is a time coordinate in the space-time region filled by $\mathfrak{F}$. Let P be an arbitrary space-time point in the leaf Σ_1, and let the "recent past" of P, $V_P^-(\mathfrak{F})$, consist of all points in $\bigcup_{\tau<1} \Sigma_\tau$ that are in the *past* of P, in the sense specified in Definition 3, above. The task we propose to tackle is the following one: We suppose that we know the state ω_{Σ_0} on the algebra $\mathcal{E}_{\Sigma_0}$, (see Eq. (9)). Assuming that Axioms 1 and 2 hold, we propose to determine the state ω_P on $\mathcal{E}_P$, for the given point $P \in \Sigma_1$. Let $\{P_\iota | \iota \in \mathfrak{I}(\mathfrak{F})\}$ denote the subset of points in $V_P^-(\mathfrak{F})$ in whose future events happen (see Definition 7), and let

$$\{\pi_{\xi_\iota}^{P_\iota} | \iota \in \mathfrak{I}(\mathfrak{F})\} \subset \mathcal{E}_{\Sigma_0}$$

be the *actual events* (see Axiom 1) that happen in the future of the points P_ι, $\iota \in \mathfrak{I}(\mathfrak{F})$; (here $\mathfrak{I}(\mathfrak{F})$ is a set of indices labelling the points in $V_P^-(\mathfrak{F})$ in whose future events happen; it is here assumed to be countable). We define a so-called *"History Operator"*

$$H\big(V_P^-(\mathfrak{F})\big) := \vec{\Pi}_{\iota \in \mathfrak{I}(\mathfrak{F})}\, \pi_{\xi_\iota}^{P_\iota}\,, \tag{16}$$

where the ordering in the product $\vec{\Pi}$ is such that a factor $\pi_{\xi_\kappa}^{P_\kappa}$ corresponding to a point P_κ stands to the right of a factor $\pi_{\xi_\iota}^{P_\iota}$ corresponding to a point P_ι iff $P_\kappa \prec P_\iota$, (i.e., if P_κ is in the past of P_ι). But if $P_\iota \times P_\kappa$, i.e., if P_ι and P_κ are space-like separated the order of the two factors is *irrelevant—thanks to* Axiom 2!

The state on the algebra $\mathcal{E}_P$ relevant to make predictions about events happening in the future of P, in the sense of Definition 7, is then given by

$$\omega_P(X) \equiv \omega_P^{\mathfrak{F}}(X) = \big[\mathcal{N}_P^{\mathfrak{F}}\big]^{-1} \omega_{\Sigma_0}\big(H(V_P^-(\mathfrak{F}))^*\, X\, H(V_P^-(\mathfrak{F}))\big)\,, \quad X \in \mathcal{E}_P\,, \tag{17}$$

where the normalization factor $\mathcal{N}_P^{\mathfrak{F}}$ is given by

$$\mathcal{N}_P^{\mathfrak{F}} = \omega_{\Sigma_0}\big(H(V_P^-(\mathfrak{F}))^* \cdot H(V_P^-(\mathfrak{F}))\big)\,. \tag{18}$$

We recall that, according to Definition 7, an event happens in the future of a point $P \in \Sigma_1$ iff the center, $\mathcal{Z}_{\omega_P}$, of the centralizer of the state ω_P on the algebra $\mathcal{E}_P$, defined in (17), contains at least two disjoint orthogonal projections of strictly positive probability, as given by *Born's Rule*; (see Axiom 1).

The quantities $\mathcal{N}_P^{\mathfrak{F}}$ can be used to equip the *tree-like* space (the so-called "non-commutative spectrum" of S) of all possible histories of events in the future of points belonging to the foliation $\mathfrak{F}$ with a *probability measure*; see [3].

The ideas and results discussed here are illustrated in Fig. 3, above.

To conclude this discussion, in the approach to relativistic quantum theory presented in this paper (called "*ETH Approach*"), the **evolution** (along the foliation $\mathfrak{F}$) of the **state** of an isolated physical system S, given the initial state ω_{Σ_0} on the algebra $\mathcal{E}_{\Sigma_0}$ defined in Eq. (9),[8] can be viewed as a *generalized stochastic branching process*, whose state space is what I have called the "non-commutative spectrum"

[8]and assuming the axiom of choice.

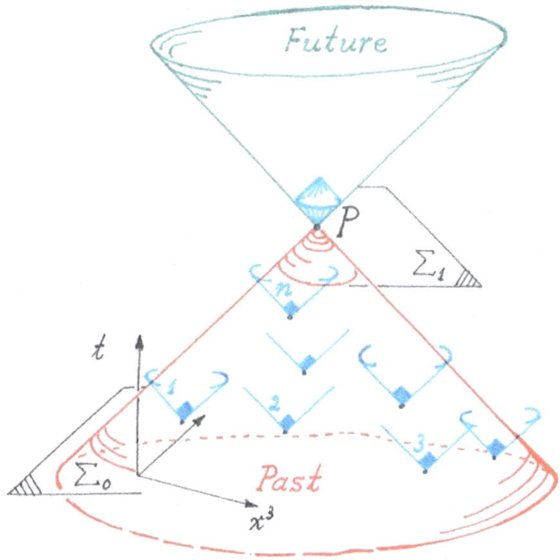

Fig. 3 It is tacitly assumed here that all events that happened in the past of the point P have a strictly finite duration. They are marked by small "diamonds" and are numbered from 1 to n. Notice that 1×2 and $2 \prec n$

of the system S, (see [3], and Eq. (27), Sect. 6, for a definition), and with *branching rules* derived from Definition 7, Axioms 1 and 2 and Eqs. (16)–(18).[9]

Mathematical details can be made precise if space-time is discretized. Additional information can be found in [3, 23, 24].

5 Monitoring Events by Measuring Physical Quantities

Let $S = \{\mathcal{M}, \mathcal{E}, \mathcal{H}, \{\mathcal{E}_P\}_{P \in \mathcal{M}}$ satisfying $PDP\}$ be the data defining an isolated physical system, with the properties specified in Sect. 4, Eq. (8), and assumed to satisfy Axioms 1 and 2. In Sect. 4, we have introduced a precise notion of **"events"** featured by S. In this section, we propose to explain how events can be recorded/monitored by measuring physical quantities referring to S.

For the purposes of the present exposition it is convenient to define a *"physical quantity"* to be an abstract self-adjoint linear operator $\hat{X}$ with the property that, for every point $P \in \mathcal{M}$, there exists a concrete self-adjoint linear operator $X(P) \in \mathcal{E}_P$ acting on the Hilbert space $\mathcal{H}$ of S and representing the quantity $\hat{X}$; (see [3] for a somewhat more general and abstract notion of physical quantities).

Remark: If space-time $\mathcal{M}$ is given by Minkowski space $\mathbb{M}^4$ the operator $X(P)$ is conjugated to the operator $X(P')$ by a unitary operator on the Hilbert space $\mathcal{H}$

[9]This picture has reminded my former student *P.-F. Rodriguez* of the following sentence from the short story *"The Garden of Forking Paths"*, by *Jorge Luis Borges*: "I leave to several futures (not to all) my garden of forking paths".

representing the space-time translation from P to P'. But on general space-times a simple relation between $X(P)$ and $X(P')$ may not exist.

We define

$$\mathcal{O}_S := \left\{ \hat{X}_\iota = \hat{X}_\iota^* \mid \iota \in \mathfrak{I}(S) \right\} \tag{19}$$

to be a list of all physical quantities available, at present, to characterize properties of S for which there exists a prescription of how they can be measured.[10] The list $\mathcal{O}_S$ is not intrinsic to the theoretical description of the system S; rather it specifies those physical quantities referring to S that, during a given era, can be expected to be measurable in real experiments. In quantum theory, this list is *not* an algebra (unless all operators belonging to $\mathcal{O}_S$ commute with one another), and it is usually not even a real linear vector space. The question to be addressed in the following is what we mean by saying that some quantity $\hat{X} \in \mathcal{O}_S$ is measured in the future of a space-time point P, and how such a measurement can be used to record an event that happens in the future of P.

Suppose that, for some point $P \in \mathcal{M}$, the *center* $\mathcal{Z}_{\omega_P}$ (of the centralizer $\mathcal{C}_{\omega_P}(\mathcal{E}_P) \subset \mathcal{E}_P$ of the state ω_P on the algebra $\mathcal{E}_P$) is non-trivial and is generated by a family $\{\pi_\xi \mid \xi \in \mathfrak{X}\}$ of disjoint orthogonal projections describing an event happening in the future of P. Let ε be a positive number; (it will turn out to be a measure of the "resolution" of the recording of this event in a measurement of a physical quantity $\hat{X} \in \mathcal{O}_S$). We let $\{\pi_1, \ldots, \pi_N\}$ be a finite number of disjoint orthogonal projections contained in $\mathcal{Z}_{\omega_P}$ with the property that

$$\omega_P(\pi_j) \geq \varepsilon, \ \forall\, j = 1, \ldots, N, \qquad \omega_P\left(\mathbf{1} - \sum_{i=1}^{N} \pi_i\right) < \varepsilon\,. \tag{20}$$

The projections $\{\pi_1, \ldots, \pi_N\}$ form the basis of an N-dimensional vector space, $\mathcal{V}_{\omega_P}^{(\varepsilon)}$, equipped with a (positive-definite) scalar product, $\langle \cdot, \cdot \rangle$, given by

$$\langle \pi_i, \pi_j \rangle := \omega_P(\pi_i \cdot \pi_j) = \omega_P(\pi_i)\,\delta_{ij} \geq \varepsilon\,\delta_{ij}, \quad \text{for } i, j = 1, \ldots, N\,. \tag{21}$$

Every vector $Z \in \mathcal{V}_{\omega_P}^{(\varepsilon)}$ can be represented as a linear combination,

$$Z = \sum_{j=1}^{N} z_j \pi_j \in \mathcal{Z}_{\omega_P}\,, \text{ for complex numbers } z_1, \ldots, z_N\,. \tag{22}$$

We can thus identify $\mathcal{V}_{\omega_P}^{(\varepsilon)}$ with an N-dimensional subspace, actually an N-dimensional *subalgebra* of $\mathcal{Z}_{\omega_P}$.

Let $\mathcal{H}_{\omega_P}$ be the Hilbert space and Ω_P the cyclic vector in $\mathcal{H}_{\omega_P}$ obtained by applying the Gel'fand-Naimark-Segal (GNS) construction to the pair $(\mathcal{E}_P, \omega_P)$; (see. e.g., [16]). There is a bijection between the vector space $\mathcal{V}_{\omega_P}^{(\varepsilon)}$ and the subspace $\mathcal{W}_{\omega_P}^{(\varepsilon)} \subset \mathcal{H}_{\omega_P}$ spanned by the vectors

[10]For simplicity, we assume that all operators in $\mathcal{O}_S$ have discrete spectrum.

$$\{Z\,\Omega_P \,|\, Z \in \mathcal{V}^{(\varepsilon)}_{\omega_P}\}.$$

By $Q^{(\varepsilon)}$ we denote the orthogonal projection onto $\mathcal{W}^{(\varepsilon)}_{\omega_P}$.

Let $\hat{X} \in \mathcal{O}_S$ be a physical quantitiy characteristic of S, and let $X(P) \in \mathcal{E}_P$ denote the self-adjoint operator representing $\hat{X}$. We consider the spectral decomposition of $X(P)$:

$$X(P) = \sum_{k=1}^{M} x_k \,\Pi_k(P)\,, \tag{23}$$

where the operators $\Pi_k(P) \in \mathcal{E}_P, k = 1, \ldots, M \leq \infty$, are the spectral projections of $X(P)$, with

$$\Pi_k(P) = \Pi_k(P)^*\,,\ \Pi_j(P)\cdot\Pi_k(P) = \delta_{jk}\,\Pi_j(P),\ \forall j, k\,,\ \sum_{k=1}^{M} \Pi_k(P) = \mathbf{1},$$

and $x_1, \ldots, x_M$ are the eigenvalues of $X(P)$ (= eigenvalues of $\hat{X}$), ordered in such a way that the sequence $\big(\omega_P(\Pi_k(P))\big)_{k=1}^{M}$ is *decreasing*. Let $L \leq M$ be such that

$$\omega_P\big(\mathbf{1} - \sum_{k=1}^{L} \Pi_k(P)\big) < \varepsilon\,.$$

Given an operator $A \in \mathcal{E}_P$, we denote by $\epsilon_{\omega_P}(A)$ the *unique* operator in the algebra $\mathcal{V}^{(\varepsilon)}_{\omega_P} \subset \mathcal{Z}_{\omega_P}$ given by

$$Q^{(\varepsilon)} A\Omega_P =: \epsilon_{\omega_P}(A)\Omega_P\,, \qquad \epsilon_{\omega_P}(A) \in \mathcal{V}^{(\varepsilon)}_{\omega_P}\,. \tag{24}$$

The map

$$\epsilon_{\omega_P} : \mathcal{E}_P \to \mathcal{V}^{(\varepsilon)}_{\omega_P}$$

is called a *"conditional expectation"*; (see [25] for a systematic theory). Claiming that a measurement of the physical quantity $\hat{X}$ can be expected to be possible and to record the event $\{\pi_\xi \,|\, \xi \in \mathfrak{X}\}$ generating $\mathcal{Z}_{\omega_P}$ with a resolution of order ε relies on the validity of the following

Basic Assumption:

$$\|\Pi_k(P) - \epsilon_{\omega_P}\big(\Pi_k(P)\big)\| < \varepsilon, \quad \forall k = 1, \ldots, L\,. \tag{25}$$

It is not hard to verify (but see [3], Eqs. (22), (23), for a proof) that this Assumption implies that

$$\omega_P(A) = \sum_{k=1}^{L} \omega_P\big(\Pi_k(P)\, A\, \Pi_k(P)\big) + \mathcal{O}\big(L\,\varepsilon\, \|A\|\big)\,, \qquad \forall\, A \in \mathcal{E}_P\,, \tag{26}$$

i.e., the state ω_P is an **incoherent** superposition of eigenstates of the operator $X(P)$, up to an error of order ε. In this very precise sense, one can say that Assumption (25) implies that there is an approximate measurement of the physical quantity $\hat{X}$ in the future of the point P.

Using a simple lemma (see [22], Lemma 8 and Appendix C), one can show that if ε is sufficiently small Assumption (25) implies that there are orthogonal projections $\pi_k(\hat{X}) \in \mathcal{Z}_{\omega_P}$ with the property that

$$\|\Pi_k(P) - \pi_k(\hat{X})\| < \mathcal{O}(\varepsilon),$$

and

$$\omega_P(A) = \sum_{k=1}^{L} \omega_P\big(\pi_k(\hat{X})\, A\, \pi_k(\hat{X})\big) + \mathcal{O}\big(L\,\varepsilon\, \|A\|\big)\,, \qquad \forall\, A \in \mathcal{E}_P\,.$$

In this precise sense, if $L \geq 2$ a measurement of the quantity $\hat{X}$ in the future of P yields non-trivial information about the **event** described by $\mathcal{Z}_{\omega_P}$ happening in the future of P. If $L = N$ the projections $\{\pi_k(\hat{X}) | k = 1, \ldots, L\}$ must coincide with the projections $\{\pi_j | j = 1, \ldots, N\}$ introduced right before (20), provided $\varepsilon \ll 1$ is sufficiently small. In this case, a measurement of $\hat{X}$ yields very precise information about the event happening in the future of P.

For further discussion of these matters see [3], (Sect. 3, V.).

6 Conclusions and Outlook

In this last section, some scattered remarks and speculations that grow out of the results sketched in Sects. 4 and 5 are presented.

1. In our attempt to cast *local relativistic quantum theory* in a form compatible with the manifestation of what we have defined to be *"events"* and with a solution of the *"measurement problem"*, the *"Principle of Diminishing Potentialities" (PDP)*, (see Definition 3, Sect. 4, Eq. (7), and [3]), plays a fundamental role. We have seen that if space-time is *even-dimensional* (e.g., $\mathcal{M} = \mathbb{M}^4$) and if there exist massless particles—photons, gravitons and, possibly, Dark-Energy modes—satisfying some form of Huygens' Prinicple, (see [18]), then (PDP) holds. One may argue that (PDP) also holds in space-times containing blackholes. From a very general point of view, it appears that a quantum theory satisfying (PDP) is necessarily "relativistic", and the dimension of its space-time must be even.
2. In Definitions 3 and 4 of Sect. 4, we have seen that there is a purely algebraic way to equip space-time $\mathcal{M}$ with a *causal structure*: A space-time point P is in the

past of a space-time point P' (written as $P \prec P'$) iff

$$\mathcal{E}_{P'} \subsetneqq \mathcal{E}_P,$$

and the relative commutant, $\mathcal{E}'_{P'} \cap \mathcal{E}_P$, of the algebra $\mathcal{E}_{P'}$ in $\mathcal{E}_P$ is a non-commutative algebra. Two points P and P' are space-like separated (written as $P \times P'$) iff P is not in the past of P' and P' is not in the past of P. It would be desirable to further elucidate the relationship of the algebras $\mathcal{E}_P$ and $\mathcal{E}_{P'}$ in case the points P and P' are space-like separated.

Ultimately, we would like to **reconstruct** space-time from purely algebraic data concerning a family (or families) of operator algebras equipped with certain relations, in particular inclusions and statements about relative commutants, given a state on these algebras. A (presumably not entirely successful) attempt in this direction has been made in [26].

3. In the formalism described in Sect. 4, **"events"** are localized in the future of certain space-time points, P; in the sense that they are described in terms of the abelian algebras $\mathcal{Z}_{\omega_P} \subset \mathcal{E}_P$, where, for a given point P, $\mathcal{Z}_{\omega_P}$ is the center of the centralizer of the state ω_P on the algebra $\mathcal{E}_P$, with $\mathcal{E}_P$ describing all potentialities in the *future* of P. The *actual event* happening in the future of some point P is an orthogonal projection, π_ξ^P, belonging to $\mathcal{Z}_{\omega_P}$, for some point ξ in an index set $\mathfrak{X}^P$, and having a strictly positive probability as predicted by *Born's Rule*. In view of Axiom 2, Sect. 4, it would be important to have a more precise idea about the space-time regions where the operators π_ξ^P, $\xi \in \mathfrak{X}^P$, are localized. This might actually yield information about the **geometry** of space-time and, ultimately, support the view that gravitation is an "emergent" (or "derived") phenomenon.

   To render these remarks a little more precise, we recall that one expects that all the algebras $\mathcal{E}_P$ are isomorphic to a "universal" von Neumann algebra $\mathfrak{N}$. One would like to know more about properties of states, ω, on $\mathfrak{N}$ for which the centers, $\mathcal{Z}_\omega(\mathfrak{N})$, of the centralizers $\mathcal{C}_\omega(\mathfrak{N})$ of ω are non-trivial, in the sense of Definition 7, Sect. 4. In [3],

$$\mathfrak{Z}_S := \bigcup_\omega \mathcal{Z}_\omega(\mathfrak{N}), \tag{27}$$

   where ω ranges over all "states of physical interest", has been dubbed the *"non-commutative spectrum"* of the system S. It is the "state space" of the stochastic branching process defined by Eqs. (16), (17) and (18) of Sect. 4, which describes the *stochastic evolution of states* of S. Unfortunately, we have very little insight into the structure of the non-commutative spectrum $\mathfrak{Z}_S$.

   It would be important to equip the algebra $\mathfrak{N}$ (and hence $\mathcal{E}_P$, for $P \in \mathcal{M}$) with a local structure, (in the sense that $\mathfrak{N}$ is generated by a net of local sub-algebras), and to attempt to show that *events*, i.e., elements of one of the algebras $\mathcal{Z}_\omega(\mathfrak{N})$, with ω a "state of physical interest", *are typically contained in sub-algebras of* $\mathfrak{N}$ *corresponding to what can be considered a "bounded region" of space-time*. This would help to introduce a more precise version of Axiom 2. But this topic, too, remains to be clarified.

4. One would expect that, for initial conditions given by states, ω_{Σ_0}, of S of "physical interest", (see Eq. (9), Sect. 4), the ensemble of events happening in the future of the points belonging to a foliation $\{\Sigma_\tau | \tau \in [0, 1]\}$ of some slab of space-time (see Sect. 4, after Axiom 2) is *countable*, and that these events are localizable in bounded regions of space-time. One would expect, moreover, that the metric extension of a space-time region within which an event can be localized is constrained by *space-time uncertainty relations* of a kind discussed, e.g., in [27]. This ought to be a consequence of time-energy uncertainty relations and of the possibility that blackholes form in the aftermath of energetic events, which, afterwards, would evaporate.
   Alas, I don't know how to even start to derive these expectations from a more precise formalism of local relativistic quantum theory. Yet, the results reviewed in this paper and in [24] suggest that, once we truly understand what is meant by a local relativistic quantum theory of events, we will view **events** as the **basic building blocks** weaving the fabric of space-time and the *relations between events* as determining the **geometry of space-time**.

To conclude, I want to express the hope that the results, problems and speculations reviewed in this paper might challenge colleagues with more technical knowledge and strength than I am able to muster to go further towards the goal of truly understanding the miracles of quantum theory.

Acknowledgements I am very much indebted to my collaborators on matters related to the results sketched in this paper; among them to Philippe Blanchard, Jérémy Faupin, Martin Fraas, and especially to *Baptiste Schubnel*. I also thank Detlev Buchholz, Gian Michele Graf, Klaus Hepp, Sandu Popescu, Renato Renner, but primarily Detlef Dürr and Shelly Goldstein for many helpful and enjoyable discussions and for serving as patient "sounding boards".

References

1. Sean Carroll, in the 'New York Times'
2. Ph. Blanchard, J. Fröhlich and B. Schubnel, *A 'Garden of Forking Paths' – the Quantum Mechanics of Histories of Events*, Nucl. Phys. B**912** (2016), 463–484
3. J. Fröhlich, *A brief review of the ETH Approach to Quantum Mechanics*, to appear in: "Frontiers in Analysis and Probability", N. Anantharaman and A. Nikeghbali (eds.), Springer-Verlag (2020) [arXiv:1905.06603]
4. R. Haag, *Fundamental Irreversibility and the Concept of Events*, Commun. Math. Phys. **132** (1990), 245-251; R. Haag, *Events, Histories, Irreversibility*, in: *Quantum Control and Measurement*, Proc. ISQM, ARL Hitachi, H. Ezawa and Y. Murayama (eds.), North Holland, Amsterdam 1993; Ph. Blanchard and A. Jadczyk, *Event-Enhanced Quantum Theory and Piecewise Deterministic Dynamics*, Annalen der Physik **4** (1995), 583–599
5. Bernard S. Kay and Varqa Abyaneh, *Expectation values, experimental predictions, events and entropy in quantum gravitationally decohered quantum mechanics*, arXiv:0710.0992 (v1), unpublished; Bernard S. Kay, *The Matter-Gravity Entanglement Hypothesis*, Found Phys **48** (2018), 542–557
6. G. C. Ghirardi, A. Rimini, and T. Weber, *Unified dynamics for microscopic and macroscopic systems*, Phys. Rev. D **34** (1986), 470

7. E. P. Wigner, *Remarks on the mind-body question*, in: I. J. Good, "The Scientist Speculates", Heinemann, London 1961
8. L. Hardy, *Quantum mechanics, local realistic theories, and Lorentz-invariant realistic theories*, Phys. Rev. Letters. **68**, (20) (1992), 2981–2984; and *Nonlocality for two particles without inequalities for almost all entangled states*, Phys. Rev. Letters. **71**, (11) (1993), 1665–1668.
9. D. Frauchiger and R. Renner, *Quantum Theory Cannot Consistently Describe the Use of Itself*, Nature Communications **9** (2018), # 3711
10. J. S. Bell, *Speakable and Unspeakable in Quantum Mechanics*, Cambridge University Press, Cambridge (UK) 1987. See also: J. A. Wheeler and W. H. Zurek, *Quantum Theory and Measurement*, Princeton University Press, Princeton NJ, 1983; K. Hepp, *Quantum Theory of Measurement and Macroscopic Observables*, Helv. Phys. Acta **45** (1972), 237-248; H. Primas, *Asymptotically Disjoint Quantum States*, in: *Decoherence: Theoretical, experimental and Conceptual Problems*, pp 161-178, Ph. Blanchard, D. Giulini, E. Joos, C. Kiefer and I.-O. Stamatescu (eds.), Springer-Verlag, Berlin 2000
11. J. Faupin, J. Fröhlich and B. Schubnel, *On the Probabilistic Nature of Quantum Mechanics and the Notion of 'Closed' Systems*, Ann. Henri Poincaré **17** (2016), 689–731
12. J. Fröhlich and B. Schubnel, *Quantum Probability Theory and the Foundations of Quantum Mechanics*, arXiv:1310.1484, in: *The Message of Quantum Science – Attempts Towards a Synthesis*, Ph. Blanchard and J. Fröhlich (eds.), Springer-Verlag, Berlin-Heidelberg-New York 2015
13. J. Fröhlich, *Quantum Theory and Causality*, Talks at the University of Leipzig (2018), TU-Stuttgart (2019), IHES (2019) and at Vietri sul Mare (Italy) (2019) – slides available on 'ResearchGate'.
14. D. Dürr and S. Teufel, *Bohmian Mechanics – The Physics and Mathematics of Quantum Theory*, Springer-Verlag, Berlin-Heidelberg-New York 2009
15. R. Jost, *The General Theory of Quantized Fields*, AMS Publ., Providence R.I., 1965.
16. O. Bratteli and D. W. Robinson, *Operator Algebras and Quantum Statistical Mechanics, Vol. 1, 2nd* edition, Springer-Verlag, Berlin-Heidelberg-New York, 1997.
17. R. Haag, *Local Quantum Physics – Fields, Particles, Algebras*, Springer-Verlag, Berlin-Heidelberg-New York, 1992.
18. D. Buchholz, *Collision Theory for Massless Bosons*, Commun. Math. Phys. **52**, (1977), 147–173; (see Theorem 9)
19. D. Buchholz and S. Doplicher, *Exotic infrared representations of interacting systems*, Ann. Inst. H.Poincaré (Physique théorique) **40**, no. 2 (1984), 175–184
20. D. Buchholz and J. E. Roberts, *New Light on Infrared Problems: Sectors, Statistics, Symmetries and Spectrum*, Commun. Math. Phys. **330** (2014), 935-972
21. J. Fröhlich and B. Schubnel, *The Preparation of States in Quantum Mechanics*, J. Math. Phys. **57** (2016), 042 101
22. J. Fröhlich and B. Schubnel, *Do We Understand Quantum Mechanics – Finally?*, arXiv:1203.3678, in: Proceedings of conference in memory of Erwin Schrödinger, Vienna, January 2011, publi. in 2012
23. B. Schubnel, *Mathematical Results on the Foundations of Quantum Mechanics*, PhD thesis 2014, available at https://doi.org/10.3929/ethz-a-010428944
24. J. Fröhlich, *'ETH' in Quantum Mechanics*, Notes of Lectures; unpublished lectures; (for extended lectures on quantum mechanics at LMU-Munich, see: https://www.theorie.physik.uni-muenchen.de/TMP/)
25. M. Takesaki, *Conditional Expectations in von Neumann Algebras*, J. Funct. Anal. **9** (1972), 306-321; F. Combes, *Poids et Espérances Conditionnelles dans les Algèbres de von Neumann*, Bull. Soc. Math. France **99** (1971), 73–112
26. U. Bannier, *Intrinsic Algebraic Characterization of Space-Time Structure*, Intl. J. of Theor. Physics **33** (1994), 1797–1809
27. S. Doplicher, K. Fredenhagen and J. E. Roberts, *Spacetime quantization induced by classical gravity*, Physics Letters **331** (1994), 39–44

Classically Gravitating Quantum Systems

André Großardt

Abstract Could gravity be of a fundamentally classical nature? Gravitational collapse models support this idea, which is summarised and whose implications for quantum experiments are briefly reviewed. We also discuss more recent suggestions to use spin as an entanglement witness in order to assess the quantumness of the gravitational force between two masses. Finally, an argument is given why the non-linear evolution resulting from such a classical treatment of gravity may not lead to problems with causality if one also invokes some gravity related stochastic wave function collapse.

1 Introduction

We do not have a complete and consistent theory that describes the gravitational influence that a quantum system in a spatial superposition state has on the motion of a test particle. This simple statement sums up the current state of theoretical physics on scales where both quantum effects and gravity become important.

For some, this assertion comes as a surprise, as it is a widespread belief that quantum gravitational effects become relevant only at the Planck scale and no ambiguities exist at the scale of low-energy tabletop experiments. This belief, however, is based on the assumption that gravity must ultimately be described by a quantum theory of which general relativity is an effective field theory. Low energy gravitational physics is then simply modelled by the perturbative quantisation of the metric tensor.

The idea that gravity could be fundamentally different from the familiar quantum field theories for matter fields is supported by collapse models. The mechanisms proposed by Diósi [1] or Adler [2], for instance, consider gravity as the origin of

A. Großardt (✉)
Theoretisch-Physikalisches Institut, Friedrich-Schiller-Universität Jena, Fröbelstieg 1, 07743 Jena, Germany
e-mail: andre.grossardt@uni-jena.de

V. Allori et al. (eds.), *Do Wave Functions Jump?*, Fundamental Theories of Physics 198,
https://doi.org/10.1007/978-3-030-46777-7_20

the stochastic, nonlinear terms appearing in the modified dynamical laws for the collapse. The gravity-related noise in these models, however, must be of classical rather than quantum nature [2]. In conclusion, if gravity causes a dynamical collapse of the wave function then the gravitational field cannot possess the typical properties of quantum matter fields; specifically the gravitational interaction must induce a nonlinear quantum dynamics in order to account for state reduction [3].

In the limit where a quantum system reaches comparably low energies, we know that it can be described by quantum fields (or even the nonrelativistic Schrödinger equation) in a flat spacetime and the system will obey the superposition principle. A system with a large mass, on the other hand, will not exhibit any quantum effects; it will, however, source a spacetime curvature which—due to the nonlinearity of Einstein's field equations—does not obey the superposition principle [4]. If we carefully increase the mass of a quantum system, making sure that its quantum properties remain observable while simultaneously obtaining a detectable gravitational field, then the question arises whether the superposition principle will survive, leading to a gravitational field incompatible with the principles of general relativity, or whether the classical spacetime structure of general relativity will emerge.

In the latter case, we face the question how quantum matter in a non-classical state acts as a source for the gravitational field, and whether this classical spacetime sourced by quantum matter can have observable consequences within the range of feasible experiments.

2 Semiclassical Gravity in the Nonrelativistic Limit

The basic principle of general relativity is that the curvature of spacetime determines how matter is dynamically evolving, and the distribution of matter (or rather the densities and fluxes of energy and momentum) on spacetime determines how spacetime is curved. The latter part is reflected in Einstein's field equations,

$$R_{\mu\nu} - \frac{1}{2} g_{\mu\nu} R = \frac{8\pi G}{c^4} T_{\mu\nu} , \tag{1}$$

where the left-hand side represents the curvature of spacetime and the right-hand side the matter distribution.

Matter, however, is quantised, promoting the right-hand side of this equation to a Hilbert space operator. A consistent model that couples quantised matter to classical spacetime, therefore, must include some prescription how to derive a classical object $T_{\mu\nu}$ that has the correct classical limit from quantum matter. An obvious choice is the expectation value in a given quantum state:

$$T_{\mu\nu} = \langle \Psi \mid \hat{T}_{\mu\nu} \mid \Psi \rangle . \tag{2}$$

Equations (1) with this definition are known as the semiclassical Einstein equations. In the nonrelativistic limit they result in the Poisson equation [5]

$$\nabla^2 \Phi = 4\pi\, G\, m\, |\Psi|^2 \tag{3}$$

for the Newtonian gravitational potential Φ which is sourced by the absolute value squared of the spatial wave function Ψ; i.e. the probability density $|\Psi|^2$ acts like a mass density. This state dependent potential renders the evolution of the quantum state nonlinear, described by the *Schrödinger-Newton equation*

$$\mathrm{i}\hbar \frac{\partial \Psi(t, \mathbf{r})}{\partial t} = \left(-\frac{\hbar^2}{2m} \nabla^2 - G\, m^2 \int \mathrm{d}^3\mathbf{r}' \, \frac{\left|\Psi(t, \mathbf{r}')\right|^2}{|\mathbf{r} - \mathbf{r}'|} \right) \Psi(t, \mathbf{r})\,. \tag{4}$$

A single quantum particle whose dynamics are described by this equation will exhibit a qualitatively different behaviour. Its wave function—which also corresponds to a gravitational mass distribution—will show a tendency to self-attract, leading to an inhibition of the usual quantum mechanical spreading which for very large masses can even surpass the usual spreading and lead to a collapse-like motion towards a narrower, solitonic solution [6, 7]. For atoms, however, the deviation to the standard evolution is negligible and one needs to consider composite, mesoscopic systems to achieve a significant effect.

2.1 Composite Systems

For a system of N particles of equal mass m, the Schrödinger-Newton equation reads [5, 8]

$$\begin{aligned} \mathrm{i}\hbar \frac{\partial \Psi_N(t, \mathbf{r}_1, \ldots, \mathbf{r}_N)}{\partial t} &= \Bigg(-\frac{\hbar^2}{2m} \sum_{i=1}^{N} \nabla_i^2 \\ &- G\, m^2 \sum_{i=1}^{N} \sum_{j=1}^{N} \int \mathrm{d}^3\mathbf{r}_1' \cdots \mathrm{d}^3\mathbf{r}_N' \, \frac{|\Psi_N(t, \mathbf{r}_1', \ldots, \mathbf{r}_N')|^2}{\left|\mathbf{r}_i - \mathbf{r}_j'\right|} \Bigg) \Psi_N(t, \mathbf{r}_1, \ldots, \mathbf{r}_N)\,. \end{aligned} \tag{5}$$

This equation describes both the usual mutual Newtonian gravitational interaction (terms $i \neq j$) and the gravitational self-interaction (terms $i = j$) which is characteristic of the Schrödinger-Newton equation.

Under the assumption that a large number of particles N is held together by (e.g. electromagnetic) internal forces much stronger than the gravitational interaction, a centre of mass equation can be derived [9]:

$$\mathrm{i}\hbar \frac{\partial \psi(t, \mathbf{r})}{\partial t} = \left(-\frac{\hbar^2}{2M} \nabla^2 + \int \mathrm{d}^3\mathbf{r}' \, \left|\psi(t, \mathbf{r}')\right|^2 \, I_\rho(\mathbf{r} - \mathbf{r}') \right) \psi(t, \mathbf{r})\,, \tag{6}$$

where ψ is the centre of mass wave function for the total mass M, $\mathbf{r}$ denotes the centre of mass coordinate, and $I_\rho(\mathbf{d})$ is the mutual gravitational potential between two equal mass distributions given by the mass density ρ and shifted from each other by a distance $\mathbf{d}$,

$$I_\rho(\mathbf{d}) = -G \int \mathrm{d}^3\mathbf{r}\, \mathrm{d}^3\mathbf{r}' \, \frac{\rho(\mathbf{r})\, \rho(\mathbf{r}')}{|\mathbf{r} - \mathbf{r}' + \mathbf{d}|}\,, \tag{7}$$

with $\rho(\mathbf{r})$ being the mass distribution resulting from the internal forces.

For a point-like object whose wave function extends much wider than its mass distribution, such that ρ can be approximated by a delta distribution, I_ρ is simply the Coulomb-like gravitational potential and the dynamics can be modeled by the single particle equation (4). The equation also simplifies in the opposite situation of a very well localised, macroscopic object, i.e. if the size of the object is much larger than the width of the centre-of-mass wave-function. In this case, significant contributions to the Schrödinger evolution will only occur where $\left|\mathbf{r} - \mathbf{r}'\right|$ is small, and Taylor expansion of I_ρ to second order yields the approximate nonlinear Schrödinger equation

$$\mathrm{i}\hbar \frac{\partial \psi(t, \mathbf{r})}{\partial t} = \left(-\frac{\hbar^2}{2M} \nabla^2 + \frac{\mathbf{r} - \langle \mathbf{r} \rangle}{2} \cdot I_\rho''(\mathbf{0})\mathbf{r} - \frac{\mathbf{r}}{2} \cdot \langle I_\rho''(\mathbf{0})\mathbf{r} \rangle + \frac{1}{2} \langle \mathbf{r} \cdot I_\rho''(\mathbf{0})\mathbf{r} \rangle \right) \psi(t, \mathbf{r})\,, \tag{8}$$

where constant terms in the Hamiltonian have been omitted and $I_\rho''(\mathbf{0})$ denotes the Hessian matrix at $\mathbf{d} = \mathbf{0}$ of the mutual gravitational potential defined in Eq. (7).

For a spherically symmetric mass distribution, $\rho(\mathbf{r}) = \rho(r)$, the Hessian matrix is diagonal, $I_\rho''(\mathbf{0}) = k\, \mathbb{1}$, and results in a quadratic nonlinear potential:

$$\mathrm{i}\hbar \frac{\partial \psi(t, \mathbf{r})}{\partial t} = \left(-\frac{\hbar^2}{2M} \nabla^2 + \frac{k}{2} \left(\mathbf{r} - \langle \mathbf{r} \rangle\right)^2 + \frac{k}{2} \left(\langle \mathbf{r}^2 \rangle - \langle \mathbf{r} \rangle^2 \right) \right) \psi(t, \mathbf{r})\,, \tag{9}$$

with $k = GM^2/R^3$ for a solid sphere of radius R [10]. For microgram masses, the centre of mass wave function is usually narrow in comparison to the picometre scale localisation length σ of the atoms of mass m within condensed matter, and one finds $k \sim GMm/\sigma^3$ [11].

Equation (9) can serve as a starting point for experimental tests of the Schrödinger-Newton equation, for instance through a de-phasing between internal and external oscillations of a squeezed Gaussian state of a nanomechanical oscillator [12]. Closer analysis suggests that effects would become observable for particle masses of the order of nanograms, if such particles can be trapped at a frequency of below 10 Hz with a quality factor of at least 10^6. Alternative experimental proposals [13] have similar requirements. Although these parameters are technologically challenging and, to date, ground state cooling has not been achieved for such systems, there is no fundamental reason why an experimental test should not be possible within the next decade.

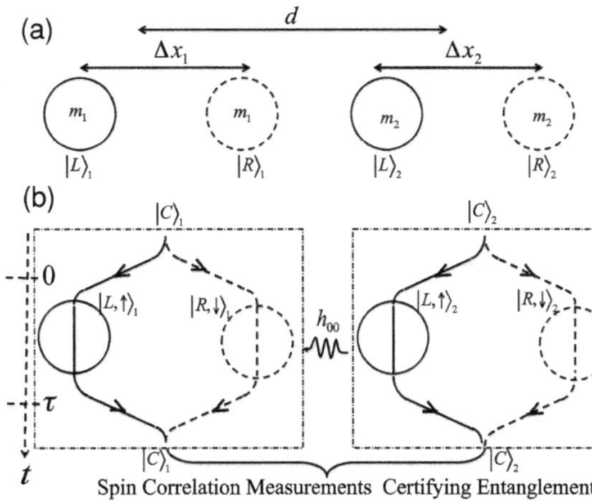

Fig. 1 Schematic depiction of an experiment in which two Stern-Gerlach devices prepare spatial superposition states, the adjacent paths of which interact gravitationally. From [14]

3 Entanglement Generation Through Newtonian Gravity

The classical gravitational interaction described in the previous section will not result in an entanglement of two distant masses. This is different from the expectation one has for a fully quantum theory of gravity and due to the fact that in the limit of two well localised particles (i. e. narrow wave functions) with position operators $\hat{\mathbf{r}}_1$ and $\hat{\mathbf{r}}_2$, the mutual part of the potential in the two-particle version of Eq. (5) takes the form

$$\hat{V}_{12}^{\text{semicl.}} = -\frac{Gm^2}{\left|\hat{\mathbf{r}}_1 - \langle\hat{\mathbf{r}}_2\rangle\right|} - \frac{Gm^2}{\left|\langle\hat{\mathbf{r}}_1\rangle - \hat{\mathbf{r}}_2\right|} . \tag{10}$$

Quantised gravity, in contrast, if treated in analogy to electrodynamics as a quantisation of linear perturbations of the metric tensor around the flat Minkowski metric, yields a Coulomb-like potential in terms of the position operators:

$$\hat{V}_{12}^{\text{quantum}} = -\frac{Gm^2}{\left|\hat{\mathbf{r}}_1 - \hat{\mathbf{r}}_2\right|} . \tag{11}$$

A concrete proposal to use spin correlations as a witness of the entanglement generated in the latter case has been put forward [14]. In this scheme, depicted in Fig. 1, two particles (masses m_1 and m_2) are sent through adjacent Stern-Gerlach devices and split in a "left" and "right" path each, resulting in an initial state

$$\begin{aligned} \mid\Psi_0\rangle &= (\mid L,\uparrow\rangle_1 + \mid R,\downarrow\rangle_1) \otimes (\mid L,\uparrow\rangle_2 + \mid R,\downarrow\rangle_2) \\ &= \mid L,\uparrow\rangle_1 \mid L,\uparrow\rangle_2 + \mid L,\uparrow\rangle_1 \mid R,\downarrow\rangle_2 + \mid R,\downarrow\rangle_1 \mid L,\uparrow\rangle_2 + \mid R,\downarrow\rangle_1 \mid R,\downarrow\rangle_2 . \end{aligned} \tag{12}$$

Assuming that the splitting (Δx_1, Δx_2 in the figure) is large compared to the distance δx between $|R,\downarrow\rangle_1$ and $|L,\uparrow\rangle_2$, only the gravitational interaction between those two parts of the state are significant. In a lowest order WKB approximation, the potential results in a phase shift

$$\phi \approx \frac{G\,m_1\,m_2\,\tau}{\hbar\,\delta x} \tag{13}$$

and the state at time $t = \tau$ will have evolved to the non-separable

$$|\Psi_\tau^{\text{quantum}}\rangle = |L,\uparrow\rangle_1\,|L,\uparrow\rangle_2 + |L,\uparrow\rangle_1\,|R,\downarrow\rangle_2 + \mathrm{e}^{\mathrm{i}\phi}\,|R,\downarrow\rangle_1\,|L,\uparrow\rangle_2 + |R,\downarrow\rangle_1\,|R,\downarrow\rangle_2\,. \tag{14}$$

In semiclassical gravity, on the other hand, both particles will sense a gravitational potential generated by the full superposition state of the respective other particle, i. e. as if there was half the mass in the "left" and half the mass in the "right" position. Every term in the state that contains $|R,\downarrow\rangle_1$ will therefore acquire a phase $\phi/2$ *regardless* of the state of particle 2, and every term that contains $|L,\uparrow\rangle_2$ will acquire a phase $\phi/2$ *regardless* of the state of particle 1. The final state is the still separable

$$|\Psi_\tau^{\text{semicl.}}\rangle = \left(|L,\uparrow\rangle_1 + \mathrm{e}^{\mathrm{i}\phi/2}\,|R,\downarrow\rangle_1\right) \otimes \left(\mathrm{e}^{\mathrm{i}\phi/2}\,|L,\uparrow\rangle_2 + |R,\downarrow\rangle_2\right)\,. \tag{15}$$

After refocusing the two paths, the spin part of the wave function will be

$$|\psi_{\text{spin}}^{\text{quantum}}\rangle = \frac{1}{2}\,|\uparrow\uparrow\rangle + \frac{1}{2}\,|\uparrow\downarrow\rangle + \frac{\mathrm{e}^{\mathrm{i}\phi}}{2}\,|\downarrow\uparrow\rangle + \frac{1}{2}\,|\downarrow\downarrow\rangle \tag{16}$$

or

$$|\psi_{\text{spin}}^{\text{semicl.}}\rangle = \frac{\mathrm{e}^{\mathrm{i}\phi/2}}{2}\,|\uparrow\uparrow\rangle + \frac{1}{2}\,|\uparrow\downarrow\rangle + \frac{\mathrm{e}^{\mathrm{i}\phi}}{2}\,|\downarrow\uparrow\rangle + \frac{\mathrm{e}^{\mathrm{i}\phi/2}}{2}\,|\downarrow\downarrow\rangle \tag{17}$$

for quantised and semiclassical gravity, respectively.

In order to witness the entanglement present in the former state, the authors of [14] suggest to measure spin correlations in two complementary bases and to estimate the value of

$$\mathcal{W} = \left|\langle\sigma_x^{(1)} \otimes \sigma_z^{(2)}\rangle + \langle\sigma_y^{(1)} \otimes \sigma_y^{(2)}\rangle\right|\,. \tag{18}$$

From the states (16) and (17) one calculates

$$\langle\sigma_x^{(1)} \otimes \sigma_z^{(2)}\rangle^{\text{quantum}} = \frac{1}{2}\left(\mathrm{e}^{\mathrm{i}\phi} - 1\right)\,, \qquad \langle\sigma_y^{(1)} \otimes \sigma_y^{(2)}\rangle^{\text{quantum}} = \frac{1}{2}\left(\mathrm{e}^{\mathrm{i}\phi} - 1\right)\,, \tag{19}$$

$$\langle\sigma_x^{(1)} \otimes \sigma_z^{(2)}\rangle^{\text{semicl.}} = \frac{\mathrm{e}^{\mathrm{i}\phi/2}}{2}\left(\mathrm{e}^{\mathrm{i}\phi} - 1\right)\,, \qquad \langle\sigma_y^{(1)} \otimes \sigma_y^{(2)}\rangle^{\text{semicl.}} = 0\,, \tag{20}$$

and therefore finds that in the case of the quantum potential (11)

$$0 \le \mathcal{W} = \left| e^{i\phi} - 1 \right| \le 2 \tag{21}$$

whereas for the semiclassical potential (10)

$$0 \le \mathcal{W} = \frac{1}{2} \left| e^{i\phi} - 1 \right| \le 1 . \tag{22}$$

Hence, if the value of $\mathcal{W}$ is experimentally confirmed to exceed unity, the semiclassical potential can be ruled out.

For a practical realisation of this idea, Bose et al. [14] suggest a mass of 10^{-14} kg, e. g. microdiamonds, with separations of a few hundred micrometres. Unlike the experiments mentioned in the previous section, where one considers centre-of-mass wave-functions that are confined well within the extent of the particle, superpositions must be large compared to the particle size for this type of test; therefore, feasible masses are significantly lower. The closest possible distance between the two particles is limited by Casimir-Polder forces which need to be smaller than the gravitational forces. As a result of the relatively small masses and large distances, gravitational effects are much weaker than in the schemes considered in the previous section.

4 Is Semiclassical Gravity Causally Consistent?

A nonlinearity in the Schrödinger equation, such as the nonlinearity in Eq. (4), can in principle be exploited to send faster-than-light signals [15]. Take, for instance, two entangled spin-$\frac{1}{2}$ particles in the superposition state

$$\frac{1}{\sqrt{2}} \left(|\uparrow\downarrow\rangle + |\downarrow\uparrow\rangle \right) = \frac{1}{\sqrt{2}} \left(|{+}{+}\rangle - |{-}{-}\rangle \right) , \tag{23}$$

where we denote the eigenstates of the σ_z Pauli matrix as $|\uparrow\rangle$, $|\downarrow\rangle$, and the eigenstates of σ_x as

$$|\pm\rangle = \frac{1}{\sqrt{2}} \left(|\uparrow\rangle \pm |\downarrow\rangle \right) . \tag{24}$$

We use the shorthand notation $|\uparrow\downarrow\rangle = |\uparrow\rangle_A \otimes |\downarrow\rangle_B$ and so forth for the two-particle states.

Now we could choose to perform a measurement on particle B in the σ_z basis; then particle A would be in a mixed state, obtained by tracing over the two possible measurement outcomes $|\uparrow\rangle_B$ and $|\downarrow\rangle_B$. Alternatively, the measurement could be performed in the σ_x basis, resulting in tracing over the possible outcomes $|+\rangle_B$ and $|-\rangle_B$ to obtain the reduced state for particle A. However, either case would result in the same density matrix

$$\hat{\rho}_A = \frac{1}{2} |\uparrow\rangle\langle\uparrow| + \frac{1}{2} |\downarrow\rangle\langle\downarrow| = \frac{1}{2} |+\rangle\langle+| + \frac{1}{2} |-\rangle\langle-| , \tag{25}$$

belonging to equivalent mixtures. In standard quantum mechanics, the outcome of any measurement on particle A can be derived from this density matrix and, therefore, does not depend on the choice of basis for the measurement on particle B. The linear evolution of the pure states $|\uparrow\rangle$, $|\downarrow\rangle$, $|+\rangle$, and $|-\rangle$ guarantees that the equality in Eq. (25) holds also at any later time.

This does not apply to the nonlinear Schrödinger-Newton equation (4): Assume that the spin of particle A becomes entangled with its position, e. g. in a magnetic field gradient along the z-axis. We do not write the state of particle B explicitly, which only plays the role of "collapsing" particle A into one of the eigenstates of the σ_z and σ_x bases, respectively. In the "classical" states $|\uparrow\rangle_A$ and $|\downarrow\rangle_A$, the position of particle A will simply evolve as $z_\uparrow(t)$ and $z_\downarrow(t)$ without any self-gravitational influence:

$$|\uparrow\rangle_A \;\to\; |\uparrow\rangle_A \otimes |z_\uparrow(t)\rangle_A\,, \qquad |\downarrow\rangle_A \;\to\; |\downarrow\rangle_A \otimes |z_\downarrow(t)\rangle_A\,. \tag{26}$$

In the superposition states $|+\rangle$ and $|-\rangle$, however, each part of the superposition will be affected by the gravitational potential generated by the respective other part of the superposition. This implies that the evolution is

$$|\pm\rangle_A = \frac{1}{\sqrt{2}}\left(|\uparrow\rangle_A \pm |\downarrow\rangle_A\right) \;\to\; \frac{1}{\sqrt{2}}\left(|\uparrow\rangle_A \otimes |\tilde{z}_\uparrow(t)\rangle_A \pm |\downarrow\rangle_A \otimes |\tilde{z}_\downarrow(t)\rangle_A\right), \tag{27}$$

with

$$\tilde{z}_{\uparrow\downarrow}(t) \approx z_{\uparrow\downarrow}(t) \pm \frac{G\,m}{2}\int_0^t \mathrm{d}t' \int_0^{t'} \mathrm{d}t'' \left|z_\uparrow(t'') - z_\downarrow(t'')\right|^{-2}. \tag{28}$$

Because of this nonlinear, self-gravitational effect, the equality in Eq. (25) no longer holds. In order to know which density matrix correctly captures the statistics of measurements on particle A, it now seems necessary to know the choice of basis made for particle B and, more dramatically, measurement outcomes for particle A seem to be usable to determine the choice of basis for B *regardless of the distance* between particles A and B—and send a binarily encoded message.

The conclusion that is usually drawn from this is that only such dynamical laws for the quantum states can be allowed under which the density matrix evolves linearly and the equality (25) holds at all times. This is the justification why collapse models consider only stochastic nonlinearities that result in a Lindblad structure for the master equation.

This might, however, be a too strict constraint on the possible evolution laws: For a practical realisation of a faster-than-light signal through the communication channel outlined above, consider a superposition with approximately constant separation Δz. Let λ be the spatial resolution with which the particle (its centre of mass position, to be precise) can be detected. If the spatial wave function of the particles can be approximated by two well-defined delta peaks, the self-gravitational effect results in a constant Newtonian gravitational acceleration $\ddot{z} \approx Gm/(2\Delta z^2)$.

We must require $\lambda < \Delta z$ in order to resolve at least the initial separation of states. For an order of magnitude estimation, it is then sufficient to assume that Δz does not vary too much over the time t_D required to achieve a separation of λ. This time can then be estimated as

$$t_D \approx 2\Delta z \sqrt{\frac{\lambda}{G\,m}} \,. \tag{29}$$

The resolution λ is limited by the uncertainty principle, $\delta z\,\delta p_z \approx \hbar$, which implies that

$$\lambda > \delta z + \frac{t_D}{m}\,\delta p_z > \sqrt{\frac{\hbar\,t_D}{m}} > \left(\frac{\hbar^2\,\Delta z^2\,\lambda}{G\,m^3}\right)^{1/4} > \left(\frac{\hbar^2\,\lambda^3}{G\,m^3}\right)^{1/4} , \tag{30}$$

where we use that $\lambda < \Delta z$, and therefore

$$\lambda > \frac{\hbar^2}{G\,m^3} \,. \tag{31}$$

Of course, both a small λ and short time t_D can be achieved if one allows for the mass to become sufficiently large.

However, if gravity is responsible for the collapse of the wave function, a too large mass can result in a collapse *before* there is any chance for the self-gravitational interaction to lead to a significant change in position. Lacking a concrete model for gravity induced collapse, we may resort to the ideas by Diósi and Penrose according to which massive superpositions are reduced with a rate proportional to their gravitational self-energy, i. e. with collapse time

$$t_C \approx \frac{\hbar\,R_0}{G\,m^2} \,. \tag{32}$$

R_0 is a free parameter which is generally associated with a spatial coarse graining of the mass density distribution. The condition that there be no collapse before the separation time t_D has passed implies

$$t_C^2 \approx \frac{\hbar^2\,R_0^2}{G^2\,m^4} > t_D^2 > \frac{\lambda^3}{Gm} \qquad \Leftrightarrow \qquad \frac{\hbar^2}{G\,m^3} > \frac{\lambda^3}{R_0^2} \,. \tag{33}$$

Equations (31) and (33) together can only be satisfied if $R_0 > \lambda$.

In order to effectively make use of the self-gravitational nonlinearity for sending a faster-than-light signal, the spatial resolution must be *better* than the spatial coarse graining length R_0. Whether this is possible, even in principle, in all (or any) models for gravitational state reduction is at least questionable. It seems plausible that an (unkown) mechanism through which gravity collapses the wave function could at the same time prevent any possibility to exploit the nonlinear semiclassical gravitational interaction to violate causality.

Ultimately, the very presence of the dynamical collapse could render the argument against deterministic nonlinearities in the Schrödinger equation invalid, and collapse models may include deterministic nonlinear terms alongside the stochastic ones without permitting faster-than-light signals. Specifically in the case of gravitational collapse, the deterministic nonlinearities present in the Schrödinger-Newton equation (4) have a very appealing physical motivation. Taking them seriously as a possible modification of quantum mechanics could turn out to be rewarding.

References

1. L. Diósi. Models for universal reduction of macroscopic quantum fluctuations. *Physical Review A*, 40 (3): 1165–1174, 1989. https://doi.org/10.1103/PhysRevA.40.1165.
2. S. L. Adler. Gravitation and the noise needed in objective reduction models. In M. Bell and S. Gao, editors, *Quantum Nonlocality and Reality: 50 Years of Bell's Theorem*. Cambridge University Press, Cambridge, 2016.
3. A. Bassi and G. Ghirardi. A general argument against the universal validity of the superposition principle. *Physics Letters A*, 275 (5-6): 373–381, 2000. https://doi.org/10.1016/S0375-9601(00)00612-5.
4. R. Penrose. On gravity's role in quantum state reduction. *General Relativity and Gravitation*, 28 (5): 581–600, 1996. https://doi.org/10.1007/BF02105068.
5. M. Bahrami, A. Großardt, S. Donadi, and A. Bassi. The Schrödinger-Newton equation and its foundations. *New Journal of Physics*, 16: 115007, 2014. https://doi.org/10.1088/1367-2630/16/11/115007.
6. D. Giulini and A. Großardt. Gravitationally induced inhibitions of dispersion according to the Schrödinger-Newton equation. *Classical and Quantum Gravity*, 28 (19): 195026, 2011. https://doi.org/10.1088/0264-9381/28/19/195026.
7. J. R. v. Meter. Schrödinger-Newton 'collapse' of the wavefunction. *Classical and Quantum Gravity*, 28 (21): 215013, 2011. https://doi.org/10.1088/0264-9381/28/21/215013.
8. L. Diósi. Gravitation and quantum-mechanical localization of macro-objects. *Physics Letters A*, 105 (4-5): 199–202, 1984. https://doi.org/10.1016/0375-9601(84)90397-9.
9. D. Giulini and A. Großardt. Centre-of-mass motion in multi-particle Schrödinger-Newton dynamics. *New Journal of Physics*, 16: 075005, 2014. https://doi.org/10.1088/1367-2630/16/7/075005
10. H. Iwe. Coulomb Potentials Between Spherical Heavy Ions. *Zeitschrift für Physik*, 304 (4): 347–361, 1982. https://doi.org/10.1007/BF01421517.
11. A. Großardt, J. Bateman, H. Ulbricht, and A. Bassi. Effects of Newtonian gravitational self-interaction in harmonically trapped quantum systems. *Scientific Reports*, 6: 30840, 2016. https://doi.org/10.1038/srep30840.
12. H. Yang, H. Miao, D.-S. Lee, B. Helou, and Y. Chen. Macroscopic Quantum Mechanics in a Classical Spacetime. *Physical Review Letters*, 110 (17): 170401, 2013. https://doi.org/10.1103/PhysRevLett.110.170401.
13. A. Großardt, J. Bateman, H. Ulbricht, and A. Bassi. Optomechanical test of the Schrödinger-Newton equation. *Physical Review D*, 93: 096003, 2016b. https://doi.org/10.1103/PhysRevD.93.096003.
14. S. Bose, A. Mazumdar, G. W. Morley, et al. Spin Entanglement Witness for Quantum Gravity. *Physical Review Letters*, 119: 240401, 2017. https://doi.org/10.1103/PhysRevLett.119.240401.
15. N. Gisin. Stochastic quantum dynamics and relativity. *Helvetica Physica Acta*, 62 (4): 363–371, 1989. https://doi.org/10.5169/seals-116034.

Collapse Models and Cosmology

Jérôme Martin and Vincent Vennin

Abstract Attempts to apply quantum collapse theories to Cosmology and cosmic inflation are reviewed. These attempts are motivated by the fact that the theory of cosmological perturbations of quantum-mechanical origin suffers from the single outcome problem, which is a modern incarnation of the quantum measurement problem, and that collapse models can provide a solution to these issues. Since inflationary predictions can be very accurately tested by cosmological data, this also leads to constraints on collapse models. These constraints are derived in the case of Continuous Spontaneous Localization (CSL) and are shown to be of unprecedented efficiency.

1 Introduction

Quantum Mechanics finds itself in a somehow paradoxical situation. On one hand, it is an extremely efficient and well-tested theory whose experimental successes are impressive and unquestioned. On the other hand, understanding and interpreting the formalism on which it rests is still a matter of debates. This on-going discussion has led to a variety of points of view ranging from challenging that there is an actual problem, to developing different ways of understanding the theory or, in other words, different "interpretations" [1].

J. Martin (✉) · V. Vennin
d'Astrophysique de Paris, UMR 7095-CNRS, Université Pierre et Marie Curie,
98bis boulevard Arago, 75014 Paris, France
e-mail: jmartin@iap.fr

V. Vennin
e-mail: vincent.vennin@apc.in2p3.fr

V. Vennin
Laboratoire Astroparticule et Cosmologie, Université Denis Diderot Paris 7,
75013 Paris, France

V. Allori et al. (eds.), *Do Wave Functions Jump?*, Fundamental Theories of Physics 198,
https://doi.org/10.1007/978-3-030-46777-7_21

Giancarlo Ghirardi, to whom this book and chapter are dedicated, has made fundamental contributions to this question. In fact, the approach proposed by Ghirardi (together with his collaborators, Rimini and Weber and, independently, Pearle), the so-called collapse models [2–5], unlike the other interpretations, goes beyond simply advocating for a different scheme to capture the meaning of the Quantum Mechanics formalism. It is actually an alternative to Quantum Mechanics and, as such, it should not be considered as an interpretation but rather as another, rival, theory. In some sense, collapse models enlarge Quantum Mechanics, which becomes only one particular theory in a larger parameter space, in the same way that, for instance, General Relativity is only one point in the parameter space of scalar-tensor theories [6]. As a consequence, the great advantage of collapse theories is that they make predictions that are different from those of Quantum Mechanics and that can thus be falsified. This was of course realized from the very beginning by Ghirardi and, nowadays, there exists a long list of experiments aiming at constraining collapse models [1].

These experiments, however, are all performed in the lab. In the present article, it is pointed out that using Quantum Mechanics and/or collapse models in a cosmological context can shed new light on those theories.

One of the most important insights in Cosmology is the realization that galaxies are of quantum-mechanical origin [7]. They are indeed nothing but quantum fluctuations, stretched to very large distances by cosmic expansion during a phase of inflation [8–12] and amplified by gravitational instability. This discovery has clearly far-reaching implications for Cosmology but also for foundational issues in Quantum Mechanics. Indeed, in Cosmology, Quantum Mechanics is pushed to new territories not only in terms of scales (the typical energy, length or time scales relevant for Cosmology are very different from those characterizing lab experiments) but also in terms of concepts: applying Quantum Mechanics to a single system with no exterior, classical, domain is not trivial [13, 14].

Among the first physicists who realized that Cosmology can be an interesting playground for Quantum Mechanics was John Bell, see for instance his article "*Quantum mechanics for cosmologists*" [15]. As Ghirardi recalled and discussed in detail during the colloquium he gave at the Institut d'Astrophysique de Paris (IAP) on March 22nd, 2012, he and John Bell were good friends and enjoyed interacting together. In his talk,[1] Ghirardi mentioned that Bell emphasized the importance of developing a relativistic, Lorentz invariant, version of collapse models which is of course a prerequisite for Cosmology. He also stressed that one important feature of collapse models is that there is "no mention of measurements, observers and so on", a property that is clearly relevant for Cosmology. Therefore, even if Ghirardi never explicitly worked at the interface between Cosmology and Quantum Foundations, he clearly considered this subject as a promising direction of research.

Recently, the collapse models have started to be considered in Cosmology [16–24], in particular in the context of cosmic inflation, with two essential motivations: to avoid conceptual problems related to the absence of an observer in the very early

[1]The slides of his talk can be found at this http://www.iap.fr/vie_scientifique/seminaires/Seminaire_GReCO/2012/presentations/ghirardi.pdf.

universe; and to use the high-accuracy cosmological data constraining inflation as a probe of the free parameters characterizing collapse models [24]. The goal of this paper is to briefly review these recent works. It is organized as follows. In the next section, Sect. 2, we briefly review cosmic inflation and the theory of cosmological perturbations of quantum-mechanical origin. Then, in Sect. 3, we explain why collapse theories can be useful in Cosmology. In Sect. 4, we discuss how these theories can be implemented concretely and, in Sect. 5, we use cosmological observations to put constraints on the parameters characterizing collapse models. Finally, in Sect. 6, we present our conclusions.

2 Cosmic Inflation and Cosmological Perturbations

In Cosmology, the theory of inflation is a description of the physics of the very early universe [8–12]. It is a phase of exponential, accelerated, expansion [meaning that $\ddot{a} > 0$ where $a(t)$ is the scale factor describing how cosmic expansion proceeds and t is the cosmic time] first introduced to fix some undesirable features of the standard model of Cosmology [25]. Since it occurs in the early universe, it is characterized by a very high energy scale, that could be as large as 10^{15} GeV. Soon after inflation was proposed, in the late seventies and early eighties, it was also realized that it provides an efficient mechanism for structure formation. In the present context, "structures" refer to the small inhomogeneities that are the seeds of the Cosmic Microwave Background (CMB) anisotropies and of the galaxies. They can be represented by an inhomogeneous scalar field called the "curvature perturbation" [7, 26], and denoted $\zeta(t, \boldsymbol{x})$. It represents small ripples propagating on top of an homogeneous and isotropic background. The idea is then to promote this scalar field to a quantum scalar field, which thus undergoes unavoidable quantum fluctuations. These quantum fluctuations are then amplified during inflation and, later on in the history of the universe, give rise to galaxies.

This may seem a rather drastic idea, but one can show that all the predictions of this theory are in perfect agreement with astrophysical observations [27–33]. In particular, the statistics of ζ are quasi Gaussian (no deviation from Gaussianity has been detected so far [34]), and can thus be fully characterized in terms of its power spectrum $\mathcal{P}_\zeta(k)$, which is the square of its Fourier amplitude. It represents the "amount" of inhomogeneities at a given scale. It was known as an empirical fact, well before the advent of inflation, that cosmological data are consistent with a primordial scale-invariant power spectrum, that is to say with a function $\mathcal{P}_\zeta(k)$ that is k-independent. But the theoretical origin of this scale-invariance was not known. Inflation definitively gained respectability when it was realized that it leads to this type of power spectrum for the quantum fluctuations mentioned before. Its convincing power is even higher today because, in fact, inflation does not predict an exact scale-invariant power spectrum, but rather an almost scale-invariant power spectrum: if one writes the power spectrum as $\mathcal{P}_\zeta(k) \sim k^{n_\mathrm{S}-1}$, where n_S is the so-called spectral index, exact scale-invariance corresponds to $n_\mathrm{S} = 1$ while inflation leads to $n_s \neq 1$ but $|n_s - 1| \ll 1$. As a consequence, if inflation is correct, then one should observe a small deviation from $n_s = 1$. In 2013, the European Space Agency (ESA) satellite Planck measured the CMB anisotropies with exquisite precision and found [27] $n_s = 0.9603 \pm 0.0073$, thus establishing that, if n_s is indeed close to one, it differs from one at a (5σ) significant level. The most recent release [32, 33], in 2018, has confirmed this measurement with $n_s = 0.9649 \pm 0.0042$. This confirmation of a crucial inflationary prediction has given a strong support to the idea that galaxies are of quantum-mechanical origin.

At the technical level, it is well known that a field in flat space-time can be interpreted as an infinite collection of harmonic oscillators, each oscillator corresponding to a given Fourier mode. Likewise, a scalar field living in a cosmological, curved, space-time can be viewed as an infinite collection of *parametric* oscillators, the fundamental frequency of each oscillator becoming a time-dependent function because of cosmic expansion (for a review, see Ref. [35]). Upon quantization, harmonic oscillators naturally lead to the concept of coherent states while parametric oscillators lead to the concept of squeezed states [36]. In the Heisenberg picture, the curvature perturbation operator can be expanded as

$$\hat{\zeta}(\eta, \boldsymbol{x}) = \frac{1}{(2\pi)^{3/2}} \frac{1}{z(\eta)} \int \frac{\mathrm{d}\boldsymbol{k}}{\sqrt{2k}} \left[\hat{c}_{\boldsymbol{k}}(\eta) e^{i\boldsymbol{k}\cdot\boldsymbol{x}} + \hat{c}_{\boldsymbol{k}}^{\dagger}(\eta) e^{-i\boldsymbol{k}\cdot\boldsymbol{x}} \right], \tag{1}$$

where $\hat{c}_{\boldsymbol{k}}(\eta)$ and $\hat{c}_{\boldsymbol{k}}^{\dagger}(\eta)$ are the annihilation and creation operators satisfying the usual equal-time commutation relations, $[\hat{c}_{\boldsymbol{k}}(\eta), \hat{c}_{\boldsymbol{p}}^{\dagger}(\eta)] = \delta(\boldsymbol{k} - \boldsymbol{p})$, $z(\eta)$ is a function that depends on the scale factor and its derivatives only, and η denotes the conformal time, related to cosmic time via $\mathrm{d}t = a\mathrm{d}\eta$. The dynamics of $\hat{\zeta}(\eta, \boldsymbol{x})$ is controlled by the following Hamiltonian, which is directly obtained from expanding the Einstein-Hilbert action plus the action of a scalar field at second order[2] in perturbation theory [35],

$$\hat{H} = \int_{\mathbb{R}^3} \mathrm{d}^3\boldsymbol{k}\, \hat{H}_{\mathrm{free}}(\boldsymbol{k}) + g(\eta) \int_{\mathbb{R}^3} \mathrm{d}^3\boldsymbol{k}\, H_{\mathrm{int}}(\boldsymbol{k}). \tag{2}$$

In this expression, $g(\eta) = z'/(2z)$ is a time-dependent "coupling constant", and

$$\hat{H}_{\mathrm{free}}(\boldsymbol{k}) = \frac{k}{2} \left(\hat{c}_{\boldsymbol{k}} \hat{c}_{\boldsymbol{k}}^{\dagger} + \hat{c}_{-\boldsymbol{k}}^{\dagger} \hat{c}_{-\boldsymbol{k}} \right), \quad \hat{H}_{\mathrm{int}}(\boldsymbol{k}) = -i \left(\hat{c}_{\boldsymbol{k}} \hat{c}_{-\boldsymbol{k}} - \hat{c}_{-\boldsymbol{k}}^{\dagger} c_{\boldsymbol{k}}^{\dagger} \right). \tag{3}$$

The first term, $\hat{H}_{\mathrm{free}}$, is the Hamiltonian of a collection of harmonic oscillators and the second one, $\hat{H}_{\mathrm{int}}$, represents the interaction of the quantum perturbations with the classical background. If space-time is not dynamical (Minkowski), then $g(\eta) = 0$.

---

[2]This second-order expansion of the action is valid at linear order in perturbation theory, which is known to provide an excellent description of primordial fluctuations, given their small amplitude. This is the order at which the calculation is performed in this work, as in the standard treatment. At higher order, mode coupling effects are expected, which would made the use of the CSL theory technically more challenging (as for the case of standard quantum mechanics) but these effects are clearly suppressed by the amplitude of perturbations, hence they cannot change our conclusions.

In the inflationary paradigm, a crucial assumption, without which the theory would not be empirically successful, is that the initial state of the system is the so-called "Bunch-Davies" or "adiabatic" vacuum state [37], which can be written as

$$|0\rangle = \bigotimes_{\boldsymbol{k}} |0_{\boldsymbol{k}}\rangle, \tag{4}$$

with $\hat{c}_{\boldsymbol{k}}(\eta_{\rm ini})|0_{\boldsymbol{k}}\rangle = 0$, $\eta_{\rm ini}$ being the conformal time at which the initial state is chosen. The time evolution of the curvature perturbation $\hat{\zeta}(\eta, \boldsymbol{x})$ is then given by the Heisenberg equation $\mathrm{d}\hat{c}_{\boldsymbol{k}}/\mathrm{d}\eta = -i[\hat{c}_{\boldsymbol{k}}, \hat{H}]$. This equation can be solved by means of a Bogoliubov transformation, $\hat{c}_{\boldsymbol{k}}(\eta) = u_k(\eta)\hat{c}_{\boldsymbol{k}}(\eta_{\rm ini}) + v_k(\eta)\hat{c}^{\dagger}_{-\boldsymbol{k}}(\eta_{\rm ini})$, where the functions $u_k(\eta)$ and $v_k(\eta)$ obey

$$i\frac{\mathrm{d}u_k}{\mathrm{d}\eta} = ku_k(\eta) + i\frac{z'}{z}v_k^*(\eta), \quad i\frac{\mathrm{d}v_k}{\mathrm{d}\eta} = kv_k(\eta) + i\frac{z'}{z}u_k^*(\eta). \tag{5}$$

These functions must satisfy $|u_k(\eta)|^2 - |v_k(\eta)|^2 = 1$ in order for the commutation relation between $\hat{c}_{\boldsymbol{k}}$ and $\hat{c}^{\dagger}_{\boldsymbol{p}}$ to be satisfied. If one introduces the Bogoliubov transformation into the expression (1) for the curvature operator, one obtains

$$\hat{\zeta}(\eta, \boldsymbol{x}) = \frac{1}{(2\pi)^{3/2}}\frac{1}{z(\eta)}\int\frac{\mathrm{d}\boldsymbol{k}}{\sqrt{2k}}\left[(u_k + v_k^*)(\eta)\hat{c}_{\boldsymbol{k}}(\eta_{\rm ini})e^{i\boldsymbol{k}\cdot\boldsymbol{x}} + (u_k^* + v_k)(\eta)\hat{c}^{\dagger}_{\boldsymbol{k}}(\eta_{\rm ini})e^{-i\boldsymbol{k}\cdot\boldsymbol{x}}\right]. \tag{6}$$

From Eqs. (5), it is easy to establish that the quantity $u_k + v_k^*$ obeys the equation $(u_k + v_k^*)'' + \omega^2(u_k + v_k^*) = 0$ with $\omega^2 = k^2 - z''/z$. This is the equation of a parametric oscillator, namely a harmonic oscillator with time-dependent fundamental frequency, and, here, this time dependence is entirely controlled by the dynamics of the underlying background space-time. Let us notice that the initial conditions are given by $u_k(\eta_{\rm ini}) = 1$ and $v_k(\eta_{\rm ini}) = 0$, which implies that $(u_k + v_k^*)(\eta_{\rm ini}) = 1$. Having solved the time evolution of the system, one can then calculate the two-point correlation function of the curvature perturbation. It needs to be evaluated in the state $|0\rangle$ since, in the Heisenberg picture, states do not evolve in time, and one has

$$\left\langle 0\left|\zeta^2(\eta, \boldsymbol{x})\right|0\right\rangle \equiv \int_0^{+\infty}\frac{\mathrm{d}k}{k}\mathcal{P}_\zeta(k) = \int_0^{\infty}\frac{\mathrm{d}k}{k}k^2\left|\frac{u_k + v_k^*}{z}\right|^2. \tag{7}$$

This shows how the power spectrum $\mathcal{P}_\zeta(k)$ mentioned above can be determined explicitly once the differential equation for $u_k + v_k^*$ has been solved. Notice that it is, a priori, a function of time. However, on large scales, $u_k + v_k^* \propto z$, and this time dependence disappears.

Let us now describe the same phenomenon but in the Schrödinger picture. We first notice that the Bogoliubov transformation introduced above can be written

$$\hat{c}_{\boldsymbol{k}}(\eta) = \hat{R}_k^{\dagger}\hat{S}_k^{\dagger}\hat{c}_{\boldsymbol{k}}(\eta_{\rm ini})\hat{S}_k\hat{R}_k, \tag{8}$$

where the operators $\hat{R}_k$ and $\hat{S}_k$, called the rotation and squeezing operators respectively, are defined by $\hat{R}_k = e^{\hat{D}_k}$ and $\hat{S}_k = e^{\hat{B}_k}$, with

$$\begin{aligned}\hat{B}_{\boldsymbol{k}} &= r_k e^{-2i\varphi_k} \hat{c}_{-\boldsymbol{k}}(\eta_{\rm ini})\hat{c}_{\boldsymbol{k}}(\eta_{\rm ini}) - r_k e^{2i\varphi_k} \hat{c}^{\dagger}_{-\boldsymbol{k}}(\eta_{\rm ini})\hat{c}^{\dagger}_{\boldsymbol{k}}(\eta_{\rm ini}),\\ \hat{D}_{\boldsymbol{k}} &= -i\theta_{k,1}\hat{c}^{\dagger}_{\boldsymbol{k}}(\eta_{\rm ini})\hat{c}_{\boldsymbol{k}}(\eta_{\rm ini}) - i\theta_{k,2}\hat{c}^{\dagger}_{-\boldsymbol{k}}(\eta_{\rm ini})\hat{c}_{-\boldsymbol{k}}(\eta_{\rm ini}).\end{aligned} \tag{9}$$

They are expressed in terms of the squeezing parameter $r_k(\eta)$, the squeezing angle $\varphi_k(\eta)$ and the rotation angle $\theta_k(\eta) \equiv \theta_{k,1}(\eta) = \theta_{k,2}(\eta)$, which are related to the functions $u_k(\eta)$ and $v_k(\eta)$ via $u_k(\eta) = e^{-i\theta_k}\cosh r_k$ and $v_k(\eta) = -ie^{i\theta_k+2i\varphi_k}\sinh r_k$. In the Schrödinger picture, the state evolves with time into a two-mode squeezed state [38]

$$|0\rangle \rightarrow |\Psi_{\rm 2sq}\rangle = \bigotimes_{\boldsymbol{k}} \hat{S}_k\hat{R}_k|0_{\boldsymbol{k}}, 0_{-\boldsymbol{k}}\rangle = \bigotimes_{\boldsymbol{k}} \frac{1}{\cosh r_k(\eta)} \sum_{n=0}^{\infty} e^{-2in\varphi_k(\eta)} \tanh^n r_k(\eta)|n_{\boldsymbol{k}}, n_{-\boldsymbol{k}}\rangle, \tag{10}$$

where $|n_{\boldsymbol{k}}\rangle$ is an eigenvector of the particle number operator in the mode $\boldsymbol{k}$. In Cosmology, the value of the squeezing parameter, for the modes $\boldsymbol{k}$ probed in the CMB, is $r_k \simeq 10^2$ towards the end of inflation, which is much larger than what can be achieved in the lab. Moreover, this state is, as apparent on the previous expression, entangled. It is therefore reasonable to conclude that the quantum state $|\Psi_{\rm 2sq}\rangle$ is a highly non-classical state.

The above squeezed state can also be written in terms of a wave-functional, which usually corresponds to writing the state in the "position" basis. This, however, is not as straightforward as it might seem in the present context. Indeed, the curvature perturbation and its conjugate momentum are related to the creation and annihilation operators through

$$z(\eta)\hat{\zeta}_{\boldsymbol{k}} = \frac{1}{\sqrt{2k}}\left(\hat{c}_{\boldsymbol{k}} + \hat{c}^{\dagger}_{-\boldsymbol{k}}\right), \quad z(\eta)\hat{\zeta}'_{\boldsymbol{k}} = -i\sqrt{\frac{k}{2}}\left(\hat{c}_{\boldsymbol{k}} - \hat{c}^{\dagger}_{-\boldsymbol{k}}\right). \tag{11}$$

We notice that the curvature perturbation and its conjugate momentum are not Hermitian operators since the above relations imply that $\hat{\zeta}^{\dagger}_{\boldsymbol{k}} = \hat{\zeta}_{-\boldsymbol{k}}$, which simply translates the fact that the curvature perturbation is a real field. As a consequence, $\hat{\zeta}_{\boldsymbol{k}}$ cannot play the role of the position operator. Moreover, these expressions mix creation and annihilation operators of momentum $\boldsymbol{k}$ and $-\boldsymbol{k}$, while it seems more natural to define a position operator for each mode $\boldsymbol{k}$. This, however, can be done if one introduces the operators $\hat{q}_{\boldsymbol{k}}$ and $\hat{\pi}_{\boldsymbol{k}}$ defined by [39]

$$z(\eta)\hat{\zeta}_{\boldsymbol{k}} = \frac{1}{2}\left[\hat{q}_{\boldsymbol{k}} + \hat{q}_{-\boldsymbol{k}} + \frac{i}{k}\left(\hat{\pi}_{\boldsymbol{k}} - \hat{\pi}_{-\boldsymbol{k}}\right)\right], \quad z(\eta)\hat{\zeta}'_{\boldsymbol{k}} = \frac{1}{2i}\left[k\left(\hat{q}_{\boldsymbol{k}} - \hat{q}_{-\boldsymbol{k}}\right) + i\left(\hat{\pi}_{\boldsymbol{k}} + \hat{\pi}_{-\boldsymbol{k}}\right)\right]. \tag{12}$$

From those relations, it is easy to establish that

$$\hat{q}_{\boldsymbol{k}} = \frac{1}{\sqrt{2k}}\left(\hat{c}_{\boldsymbol{k}} + \hat{c}_{\boldsymbol{k}}^{\dagger}\right), \quad \hat{\pi}_{\boldsymbol{k}} = -i\sqrt{\frac{k}{2}}\left(\hat{c}_{\boldsymbol{k}} - \hat{c}_{\boldsymbol{k}}^{\dagger}\right), \tag{13}$$

so that $\hat{q}_{\boldsymbol{k}}$ and $\hat{\pi}_{\boldsymbol{k}}$ involve only creation and annihilation operators for a fixed mode $\boldsymbol{k}$. It is also easy to check that $[\hat{q}_{\boldsymbol{k}}, \hat{\pi}_{\boldsymbol{k}}] = i$, such that $\hat{q}_{\boldsymbol{k}}$ and $\hat{\pi}_{\boldsymbol{k}}$ are the proper generalization of "position" and "momentum" for field theory. Then, it follows that the total wave-functional of the system can be written as a product of wave-functions for each mode, namely $\Psi_{2\mathrm{sq}}[\eta; q] = \prod_{\boldsymbol{k}} \Psi_{\boldsymbol{k}}(q_{\boldsymbol{k}}, q_{-\boldsymbol{k}})$, with

$$\Psi_{\boldsymbol{k}}\left(q_{\boldsymbol{k}}, q_{-\boldsymbol{k}}\right) = \langle q_{\boldsymbol{k}}, q_{-\boldsymbol{k}}|\Psi_{\boldsymbol{k}}\rangle = \frac{e^{A(r_k,\varphi_k)(q_{\boldsymbol{k}}^2+q_{-\boldsymbol{k}}^2)-B(r_k,\varphi_k)q_{\boldsymbol{k}}q_{-\boldsymbol{k}}}}{\cosh r_k\sqrt{\pi}\sqrt{1-e^{-4i\varphi_k}\tanh^2 r_k}}, \tag{14}$$

where the functions $A(r_k, \varphi_k)$ and $B(r_k, \varphi_k)$ are defined by

$$A\left(r_k, \varphi_k\right) = \frac{e^{-4i\varphi_k}\tanh^2 r_k + 1}{2(e^{-4i\varphi_k}\tanh^2 r_k - 1)}, \quad B\left(r_k, \varphi_k\right) = \frac{2e^{-2i\varphi_k}\tanh r_k}{e^{-4i\varphi_k}\tanh^2 r_k - 1}. \tag{15}$$

Initially $r_k = 0$, so $A = -1/2$ and $B = 0$, and $\Psi_{\boldsymbol{k}}(q_{\boldsymbol{k}}, q_{-\boldsymbol{k}}) \propto e^{-q_{\boldsymbol{k}}^2/2}e^{-q_{-\boldsymbol{k}}^2/2}$. Each mode $\boldsymbol{k}$ and $-\boldsymbol{k}$ is decoupled and placed in their ground state (namely, the Bunch-Davies vacuum mentioned above). Then, the state evolves, r_k becomes non-vanishing and $\Psi_{\boldsymbol{k}}(q_{\boldsymbol{k}}, q_{-\boldsymbol{k}})$ can no longer be written as a product $\Psi(q_{\boldsymbol{k}})\Psi(q_{-\boldsymbol{k}})$. This is of course another manifestation of the fact that the state becomes entangled.

The wave-functional $\Psi_{2\mathrm{sq}}$ can also be written in the basis $|\zeta_{\boldsymbol{k}}^{\mathrm{R}}, \zeta_{\boldsymbol{k}}^{\mathrm{I}}\rangle$, where one defines $\hat{\zeta}_{\boldsymbol{k}} \equiv (\hat{\zeta}_{\boldsymbol{k}}^{\mathrm{R}} + i\hat{\zeta}_{\boldsymbol{k}}^{\mathrm{I}})/\sqrt{2}$, which implies that

$$z\hat{\zeta}_{\boldsymbol{k}}^{\mathrm{R}} = \frac{1}{\sqrt{2}}\left(\hat{q}_{\boldsymbol{k}} + \hat{q}_{-\boldsymbol{k}}\right), \quad z\hat{\zeta}_{\boldsymbol{k}}^{\mathrm{I}} = \frac{1}{k\sqrt{2}}\left(\hat{\pi}_{\boldsymbol{k}} - \hat{\pi}_{-\boldsymbol{k}}\right). \tag{16}$$

In that case, $\Psi_{2\mathrm{sq}}[\eta, \zeta] = \prod_{\boldsymbol{k}} \Psi_{\boldsymbol{k}}(\zeta_{\boldsymbol{k}}^{\mathrm{R}})\Psi_{\boldsymbol{k}}(\zeta_{\boldsymbol{k}}^{\mathrm{I}})$, where the individual wave-functions can be expressed as $\Psi_{\boldsymbol{k}}(\zeta_{\boldsymbol{k}}^s) \equiv \Psi_{\boldsymbol{k}}^s = N_{\boldsymbol{k}}e^{-\Omega_{\boldsymbol{k}}(a\zeta_{\boldsymbol{k}}^s)^2}$, where $|N_{\boldsymbol{k}}| = (2\Re\mathrm{e}\,\Omega_{\boldsymbol{k}}/\pi)^{1/4}$ and $s = \mathrm{R}, \mathrm{I}$. The behavior of $\Omega_{\boldsymbol{k}}(\eta)$ is determined by the Schrödinger equation, which leads to $\Omega_{\boldsymbol{k}}' = -2i\Omega_{\boldsymbol{k}}^2 + i\omega^2(k, \eta)/2$, where we remind that $\omega^2(k, \eta)$ is the time-dependent fundamental frequency of each oscillator. Several remarks are in order at this point. First, the wave-functional $\Psi_{2\mathrm{sq}}[\eta, \zeta]$ can be obtained from $\Psi_{2\mathrm{sq}}[\eta, q]$ by canonical transformation [35, 40]. Second, finding the time dependence of the function $\Omega_{\boldsymbol{k}}(\eta)$ is clearly equivalent to solving the equation of motion (5). Third, given the previous considerations about entanglement, it may seem surprising that $\Psi_{\boldsymbol{k}}(\zeta_{\boldsymbol{k}}^{\mathrm{R}}, \zeta_{\boldsymbol{k}}^{\mathrm{I}})$ can be written in a separable form, as a product of $\Psi_{\boldsymbol{k}}(\zeta_{\boldsymbol{k}}^{\mathrm{R}})$ and $\Psi_{\boldsymbol{k}}(\zeta_{\boldsymbol{k}}^{\mathrm{I}})$. But, in fact, entanglement depends on how a system is divided into two bipartite sub-systems. This is confirmed by a calculation of the quantum discord which may be vanishing for a partition and non-vanishing for another [39]. Finally, in the wave-functional approach, the two-point correlation function that was calculated in Eq. (7) in the Heisenberg picture can be obtained with the following formula

$$\langle 0 | \zeta^2(\eta, \boldsymbol{x}) | 0 \rangle = \int \prod_{\boldsymbol{k}} \mathrm{d}\zeta_{\boldsymbol{k}}^{\mathrm{R}}\, \mathrm{d}\zeta_{\boldsymbol{k}}^{\mathrm{I}}\, \Psi_{\boldsymbol{k}}^*(\zeta_{\boldsymbol{k}}^{\mathrm{R}}, \zeta_{\boldsymbol{k}}^{\mathrm{I}})\, \zeta^2(\eta, \boldsymbol{x})\; \Psi_{\boldsymbol{k}}(\zeta_{\boldsymbol{k}}^{\mathrm{R}}, \zeta_{\boldsymbol{k}}^{\mathrm{I}}). \tag{17}$$

This leads to the power spectrum

$$\mathcal{P}_\zeta(k) = \frac{k^3}{2\pi^2} \frac{1}{4\Re\mathrm{e}\, \Omega_k}, \tag{18}$$

which can be checked to match the one obtained in Eq. (7).

Having explained how the theory of quantum-mechanical inflationary perturbations can be used to calculate the power spectrum $\mathcal{P}_\zeta(k)$ of the fluctuations, let us now briefly describe how this power spectrum can be related to astrophysical observations. In modern Cosmology, there exist many different observables that probe various properties of the universe. Among the most important ones is clearly the CMB temperature anisotropy mentioned before. It is the earliest probe, that is to say the closest to the inflationary epoch, that we have at our disposal. The CMB radiation is a relic thermal radiation emitted in the early universe at a redshift of $z_{\mathrm{lss}} \simeq 1100$. Since the early universe is extremely homogeneous and isotropic, the temperature of this radiation (namely $\sim 2.7\,\mathrm{K}$) is almost independent of the direction towards which we observe it. In fact, the early universe is not exactly homogeneous and isotropic, precisely because of the presence of the curvature perturbations discussed before. They manifest themselves by tiny variations of the CMB temperature, at the level $\delta T/T \simeq 10^{-5}$. The CMB anisotropy is thus the earliest observational evidence of curvature perturbations. More explicitly, the Sachs-Wolfe effect [41] relates the curvature perturbation $\hat{\zeta}_{\boldsymbol{k}}$ to the temperature anisotropy $\widehat{\delta T/T}$ through the following formula

$$\frac{\widehat{\delta T}}{T}(\boldsymbol{e}) = \int \frac{\mathrm{d}\boldsymbol{k}}{(2\pi)^{3/2}} \left[F(\boldsymbol{k}) + i\boldsymbol{k}\cdot\boldsymbol{e}\, G(\boldsymbol{k})\right] \hat{\zeta}_{\boldsymbol{k}}(\eta_{\mathrm{end}}) e^{-i\boldsymbol{k}\cdot\boldsymbol{e}(\eta_{\mathrm{lss}}-\eta_0)+i\boldsymbol{k}\cdot\boldsymbol{x}_0}, \tag{19}$$

where $\boldsymbol{e}$ is a unit vector that indicates the direction on the celestial sphere towards which the observation is performed. The conformal times η_{lss} and η_0 are the last scattering surface (lss) and present day (0) conformal times, respectively. The vector $\boldsymbol{x}_0$ represents the Earth's location. The quantities $F(\boldsymbol{k})$ and $G(\boldsymbol{k})$ are the so-called form factors, which encode the evolution of the perturbations after they have crossed in the Hubble radius after inflation. In practice, the temperature anisotropy given by Eq. (19) can be Fourier expanded in terms of the spherical harmonics $Y_{\ell m}$, namely

$$\frac{\widehat{\delta T}}{T}(\boldsymbol{e}) = \sum_{\ell=2}^{+\infty} \sum_{\ell=-m}^{\ell=m} \hat{a}_{\ell m} Y_{\ell m}(\boldsymbol{e}). \tag{20}$$

Using the completeness of the spherical harmonics basis and Eq. (19), it is easy to establish that, on large scales, namely in the limit $F(\boldsymbol{k}) \to 1$ and $G(\boldsymbol{k}) \to 0$, one has

$$\hat{a}_{\ell m} = \frac{4\pi}{(2\pi)^{3/2}} e^{i\pi\ell/2} \int_{\mathbb{R}^3} \mathrm{d}\boldsymbol{k}\, \hat{\zeta}_{\boldsymbol{k}}(\eta_{\mathrm{lss}})\, j_\ell[k(\eta_{\mathrm{lss}} - \eta_0)]\, Y^*_{\ell m}(\boldsymbol{k}), \tag{21}$$

where j_ℓ is a spherical Bessel function. A CMB map is nothing but a collection of numbers $a_{\ell m}$. The statistical properties of a map is characterized by its powers spectrum, which can be written as

$$\left\langle 0 \left| \widehat{\frac{\delta T}{T}}(\boldsymbol{e}_1) \widehat{\frac{\delta T}{T}}(\boldsymbol{e}_2) \right| 0 \right\rangle = \sum_{\ell=2}^{+\infty} \frac{2\ell+1}{4\pi} C_\ell P_\ell(\cos\delta), \tag{22}$$

where P_ℓ is a Legendre polynomial and δ the angle between the direction $\boldsymbol{e}_1$ and $\boldsymbol{e}_2$. The coefficients C_ℓ are the so-called multipole moments and are related to the $\hat{a}_{\ell m}$ by $\langle 0|\hat{a}_{\ell m}\hat{a}^\dagger_{\ell' m'}|0\rangle = C_\ell \delta_{\ell\ell'}\delta_{mm'}$. From Eq. (21), one can also write

$$C_\ell = \int_0^{+\infty} \frac{\mathrm{d}k}{k} \mathcal{P}_\zeta(k)\, j^2_\ell[k(\eta_{\mathrm{lss}} - \eta_0)], \tag{23}$$

thus establishing the relation between the power spectrum $\mathcal{P}_\zeta$ and a CMB map. Let us emphasize again that this relation is in fact oversimplified since it is obtained in the large-scale limit. In order to be realistic, one should take into account the behavior of the perturbations once they re-enter the Hubble radius after inflation which, technically, implies to consider the full form factors $F(\boldsymbol{k})$ and $G(\boldsymbol{k})$. This is a non-trivial task, which requires numerical calculations. It leads to a modulation of the signal and to the appearance of oscillations or peaks in the multipole moments, the so-called Doppler or acoustic peaks.

3 Motivations

The previous framework is usually viewed as very efficient. In particular, the multipole moments (23) calculated with the inflationary power spectrum fit very well the CMB maps obtained by the Planck satellite. Why, then, is the theory of quantum perturbations still considered by some as unsatisfactory or incomplete? The main reason is related to foundational issues in Quantum Mechanics, more precisely to the so-called measurement problem. In the context of inflation, this discussion is especially subtle and, hence, interesting for the following reasons.

On one hand, the inflationary perturbations are placed in a Gaussian state, which means that the corresponding Wigner function is also a Gaussian and, therefore, is positive-definite [42]. The Wigner function can thus be used and interpreted as a classical stochastic distribution [39, 43, 44], in the sense that any two-point Hermitian correlation function can always be reproduced with this Gaussian classical stochastic distribution [39]. This is also the case for any higher-order correlation function involving position only, in particular, any function of the curvature perturbation. It is sometimes argued that these properties require large quantum squeezing but, in fact, a large value of r is needed only for those higher correlation functions mixing position and momentum (which are, in any case, not observable since they involve the momentum, that is to say the decaying mode of the perturbations [39]). Nevertheless, the fact that all observable correlation functions can be reproduced by stochastic averages is often interpreted as the signature that a quantum-to-classical transition has taken place.

On the other hand, we have argued before that the perturbations are very "quantum". They are placed in a very strongly squeezed state, which is a highly entangled state. Indeed, in the limit of infinite squeezing, a squeezed state tends to an Einstein Podolski Rosen state, which was used in the EPR argument to discuss the "weird" (namely non-classical) features of Quantum Mechanics. It is hard to think about a system that would be more "quantum" than this one! As a consequence, the statement that the system has become classical should, at least, require some clarification. In fact, characterizing the system as "classical" because some correlation functions can be mimicked with a stochastic Gaussian process suffers from a number of problems. First, even in the large-squeezing limit, there are so-called "improper operators", for which the Weyl transform takes some values outside the spectrum of the operator. The measurement of these operators can never be described with a classical stochastic distribution [45]. This, for instance, leads to the possibility to violate Bell inequalities even if the Wigner function always remains positive, a property which clearly signals departure from classicality [46–48]. In fact, the question of whether Bell's inequality can be violated in a situation where the Wigner function is positive-definite has been a concern for a long time and was discussed by John Bell himself [49]. The corresponding history, told in Ref. [50], is a chapter of the history of Quantum Mechanics and is associated to the difficulties to define a classical limit. Second, there is the definite outcome question. With the theory of decoherence [51, 52], it is possible to understand why we never observe a superposition of states corresponding to macroscopic configurations but this is not sufficient to explain why a specific state is singled out in the measurement process. In some sense, with the help of quantum decoherence, the quantum measurement problem has been reduced to the definite outcome problem, which is at the core of the foundational issues of Quantum Mechanics. In a cosmological context, let us mention that decoherence has been studied and it has been suggested that it is likely to be at play during inflation [53–55]. But the definite outcome problem is still there and is neither solved by decoherence (as already mentioned), nor by the emergence of "classical" stochastic properties as described above.

In fact, one could even argue that this question, in the context of inflation and Cosmology, is worst than in the lab for the following reasons. We have seen that the operators $\widehat{\delta T}/T(\boldsymbol{e})$ (one for each direction $\boldsymbol{e}$) are observable quantities. Since a measurement of these observables has been performed by the COBE, WMAP and Planck satellites, according to the basic postulates of Quantum Mechanics, the system must be placed in one of the eigenstates of $\widehat{\delta T}/T(\boldsymbol{e})$, that we denote $|\ \rangle_{\mathrm{Planck}}(\boldsymbol{e})$, and that satisfies

$\widehat{\frac{\delta T}{T}}(e)|$  $\rangle_{\rm Planck}(e) = \frac{\delta T}{T}(e)|$  $\rangle_{\rm Planck}(e).$

However, the state $|\Psi_{2\,\rm sq}\rangle$ [recall that this state is defined in Eq. (10)] is not an eigenstate of the temperature anisotropy operator. This can be established with a direct and explicit calculation, but a physically more intuitive method is based on the concept of symmetry [56]. In order to simplify the discussion, let us first use the fact that the curvature perturbation can be viewed as a massless scalar field living in a Friedmann-Lemaître-Robertson-Walker (FLRW) universe with an action given by $S = -1/2 \int \mathrm{d}^4\boldsymbol{x}\, \sqrt{-g}\, g^{\mu\nu}\, \partial_\mu\zeta\, \partial_\nu\zeta$. Then, let us define the 4-momentum operator by

$$\hat{P}_\mu = -\int \mathrm{d}^3\boldsymbol{x}\, \sqrt{^{(3)}g}\, \hat{T}^0{}_\mu, \tag{24}$$

where $\hat{T}_{\mu\nu}$ is the stress energy tensor that can be calculated from the action given above, $\hat{T}_{\mu\nu} = \partial_\mu\hat{\zeta}\partial_\nu\hat{\zeta} - g_{\mu\nu}g^{\alpha\beta}\partial_\alpha\hat{\zeta}\partial_\beta\hat{\zeta}/2$ and $^{(3)}g$ the determinant of the three-dimensional spatial metric. In cosmic time, one can check that $\hat{P}_0$ exactly corresponds to the generator of the time evolution of the system, namely the Hamiltonian. On the other hand, the generator of the space translation along x_i is given by $\hat{P}_i = a \int \mathrm{d}^3\boldsymbol{x}\, \dot{\hat{\zeta}}\, \partial_i\hat{\zeta}$. Expressed in terms of creation and annihilation operators, one obtains $\hat{P}_i \propto \int \mathrm{d}\boldsymbol{k}\, k_i \hat{c}^\dagger_{\boldsymbol{k}}\hat{c}_{\boldsymbol{k}}$. It follows immediately from this expression that $\hat{P}_i|0\rangle = 0$ and the same conclusion would be obtained by applying the generator of rotations (angular momentum operator). This expresses the fact that the vacuum state is homogeneous and isotropic, i.e. it possesses the symmetries of the FLRW background. Moreover, one has $[\hat{H}_{\rm free}, \hat{P}_i] = 0$ and $[\hat{H}_{\rm int}, \hat{P}_i] = 0$, hence $[\hat{H}, \hat{P}_i] = 0$, which implies that the homogeneity and isotropy of the state is preserved during cosmic expansion. As a result, one has $\hat{P}_i|\Psi_{2\,\rm sq}\rangle = 0$, and $|\Psi_{2\,\rm sq}\rangle$ still represents a universe without any structure. Since $\hat{P}_i|$  $\rangle_{\rm Planck}(e)$, the transition between the two-mode squeezed state (10) and a state corresponding to a specific outcome for CMB anisotropies, namely

$|\Psi_{2\,\rm sq}\rangle = \sum$  $c($ $)|$  $\rangle \to |$  $\rangle_{\rm Planck}(e),$

cannot be generated by the Schrödinger equation. This is a concrete manifestation of the measurement and single outcome problems of Quantum Mechanics, which appear much more serious in a cosmological context than in standard lab situations, since the transition (26) seems to have taken place in the absence of any observer.

This leads to a first motivation for considering collapse models in Cosmology. In this class of theories, the collapse of the wave-function is a dynamical process controlled by a modified Schrödinger equation, which does not rely on having an observer. Another motivation is related to the fact that collapse models are falsifiable. Indeed, since they are based on a modified Schrödinger equation, they imply different predictions than standard Quantum Mechanics. Given that the inflationary predictions can be accurately tested with astrophysical data, one can then use them in order to test Quantum Mechanics and collapse models in physical regimes that are completely different from those usually probed in the lab. This also shows that solving the quantum measurement problem can have concrete implications for comparing the inflationary paradigm with the data. Therefore, the question of how a particular realization is produced is not of academic interest only, since it may also alter the properties of the possible realizations themselves.

4 Inflation and Collapse

There is no unique collapse model but different versions that come in different flavors. They are, however, all based on a modified Schrödinger equation that, for a non-relativistic system, reads [4]

$$\begin{aligned} \mathrm{d}\Psi(t,\boldsymbol{x}) = \Bigg[& -i\hat{H}\mathrm{d}t + \frac{\sqrt{\gamma}}{m_0}\sum_i\left(\hat{C}_i - \left\langle\Psi\left|\hat{C}_i\right|\Psi\right\rangle\right)\mathrm{d}W_i(t) \\ & - \frac{\gamma}{2m_0^2}\sum_i\left(\hat{C}_i - \left\langle\Psi\left|\hat{C}_i\right|\Psi\right\rangle\right)^2\mathrm{d}t\Bigg]\Psi(t,\boldsymbol{x}), \end{aligned} \tag{25}$$

where $\hat{H}$ is the Hamiltonian of the system and $\hat{\boldsymbol{C}}$ a collapse operator to be chosen (with three components denoted $\hat{C}_i$, $i = x, y, z$). The parameter γ is a new fundamental constant the dimension of which depends on the choice of $\hat{\boldsymbol{C}}$, and m_0 is a reference mass usually taken to be the mass of a nucleon. Finally, $\mathrm{d}W_i(t)$ is a stochastic noise with $\mathbb{E}[\mathrm{d}W_i(t)\mathrm{d}W_j(t')] = \delta_{ij}\delta(t-t')$ where $\mathbb{E}[.]$ denotes the stochastic average. Notice that the above equation is not sufficient to define the CSL model because we have not yet specified what the collapse operator is.

Then, let us consider a field $\hat{\zeta}(t,\boldsymbol{x})$ and here, of course, we have in mind curvature perturbation. Quantum mechanically, it is described by a wave-functional $\Psi[\zeta(\boldsymbol{x})]$ and we need to know which form the general dynamical collapse equation (25) takes in this case. A first question that immediately arises is that the above equation (25) is, in principle, valid in the non-relativistic regime only while one needs to go beyond since we want to apply collapse models to Cosmology and Field Theory. Attempts to develop a relativistic version of the collapse models are being carried out, see e.g. Refs. [4, 57–59] but they are not completed yet. Therefore, either one stops at this stage and waits for a fully satisfactory relativistic version to come, or one proceeds using reasonable assumptions, at the price of being maybe on shaky grounds. Here, we use collapse theories in Cosmology where there is a natural notion of time (the Hubble flow). Technically, this often means that the relativistic equations describing a phenomenon are well-approximated by the corresponding non-relativistic equations only modified by the appearance of the scale factor at some places. The prototypical example of such an approach is "Newtonian Cosmology" for which the laws that describe the time evolution of an expanding homogeneous and isotropic universe can be deduced from Newtonian dynamics and gravitation. Although the derivation is not strictly self-consistent it nevertheless provides some intuitive insights and represents a valuable first step. In some sense, here, we follow the same logic and, therefore, we will simply postulate that Eq. (25) can also be used in this context where the Hamiltonian of the system is simply the Hamiltonian (2) that is obtained from the theory of relativistic cosmological perturbations.

In order to see what this implies in practice, it is convenient to view space-like sections as an infinite grid of discrete points. In this case, the functional can be interpreted as an ordinary function of an infinite number of variables v_i, $\Psi(\cdots, v_i, v_j, \cdots)$, where $v_i \equiv v(\boldsymbol{x}_i)$ is the value of the field at each point of the grid. Therefore, instead of dealing with a three-dimensional index i as before, we now deal with an infinite-dimensional one. As a consequence, we can write an equation similar to Eq. (25) for $\Psi(v_i)$ where, now, the operators $\hat{H}$ and $\hat{\boldsymbol{C}}$ are functions of the "position" $\hat{v}_i$ and "momentum" $\hat{p}_i = -i\partial/\partial v_i$. Then, taking the continuous limit, "$\sum_i \to \int \mathrm{d}\boldsymbol{x}_\mathrm{p}$", we arrive at

$$\begin{aligned}\mathrm{d}\left|\Psi[\zeta(\boldsymbol{x}_\mathrm{p})]\right\rangle = \Bigg\{ & -i\hat{H}\mathrm{d}t + \frac{\sqrt{\gamma}}{m_0}\int \mathrm{d}\boldsymbol{x}_\mathrm{p}\left[\hat{C}(\boldsymbol{x}_\mathrm{p}) - \left\langle\hat{C}(\boldsymbol{x}_\mathrm{p})\right\rangle\right]\mathrm{d}W_t(\boldsymbol{x}_\mathrm{p}) \\ & - \frac{\gamma}{2m_0^2}\int \mathrm{d}\boldsymbol{x}_\mathrm{p}\left[\hat{C}(\boldsymbol{x}_\mathrm{p}) - \left\langle\hat{C}(\boldsymbol{x}_\mathrm{p})\right\rangle\right]^2\mathrm{d}t\Bigg\}\left|\Psi[\zeta(\boldsymbol{x}_\mathrm{p})]\right\rangle. \qquad (26)\end{aligned}$$

The quantity $\mathrm{d}W_t(\boldsymbol{x}_\mathrm{p})$ is still a stochastic noise but we now have one for each point in space. A fundamental aspect of the theory is to specify this noise, and each possibility corresponds to a different version of the theory. A priori, as already mentioned, the noise can be white or colored but, so far in the context of Cosmology, only white noises have been considered. They satisfy $\mathbb{E}[\mathrm{d}W_t(\boldsymbol{x}_\mathrm{p})\mathrm{d}W_{t'}(\boldsymbol{x}'_\mathrm{p})] = \delta(\boldsymbol{x}_\mathrm{p} - \boldsymbol{x}'_\mathrm{p})\delta(t - t')$. Let us also notice that $\boldsymbol{x}_\mathrm{p}$ denotes the physical coordinate, as opposed to the comoving one $\boldsymbol{x}$ ($\boldsymbol{x}_\mathrm{p} = a\boldsymbol{x}$) usually employed in Cosmology, and in terms of which Eq. (26) takes the form [24]

$$\begin{aligned}\mathrm{d}\left|\Psi[\zeta(\boldsymbol{x})]\right\rangle = \Bigg\{ & -i\hat{H}\mathrm{d}t + \frac{1}{m_0}\sqrt{\frac{\gamma}{a^3}}\int \mathrm{d}\boldsymbol{x}\, a^3\left[\hat{C}(\boldsymbol{x}) - \left\langle\hat{C}(\boldsymbol{x})\right\rangle\right]\mathrm{d}W_t(\boldsymbol{x}) \\ & - \frac{\gamma}{2m_0^2}\int \mathrm{d}\boldsymbol{x}\, a^3\left[\hat{C}(\boldsymbol{x}) - \left\langle\hat{C}(\boldsymbol{x})\right\rangle\right]^2\mathrm{d}t\Bigg\}\left|\Psi[\zeta(\boldsymbol{x})]\right\rangle, \qquad (27)\end{aligned}$$

where $\mathrm{d}W_t(\boldsymbol{x}_\mathrm{p}) = a^{-3/2}\mathrm{d}W_t(\boldsymbol{x})$ so that $\mathrm{d}W_t(\boldsymbol{x})$ is still white, namely $\mathbb{E}\left[\mathrm{d}W_t(\boldsymbol{x})\mathrm{d}W_{t'}(\boldsymbol{x}')\right] = \delta(\boldsymbol{x} - \boldsymbol{x}')\delta(t - t')\mathrm{d}t^2$. We emphasize that the above stochastic equation is the usual CSL equation: it is just written down in a situation where the number of variables becomes infinite.

Of course, we are not forced to describe the field $\hat{\zeta}(\boldsymbol{x})$ in real space and we can also write it in Fourier space. In that case, the wave-functional becomes a function of all Fourier components of the field, $\Psi(\cdots, \zeta_{\boldsymbol{k}}, \zeta_{\boldsymbol{k}'}, \cdots)$, that is to say we deal, again, with the same situation as described by Eq. (25) but, now, with a continuous index $\boldsymbol{k}$ instead of $i = x, y, z$. The advantage of this approach is that, because we work in the framework of linear perturbations theory, one can write the wave-function as $\Psi(\cdots, \zeta_{\boldsymbol{k}}, \zeta_{\boldsymbol{k}'}, \cdots) = \prod_{\boldsymbol{k}} \Psi_{\boldsymbol{k}}^{\mathrm{R}} \Psi_{\boldsymbol{k}}^{\mathrm{I}}$. As explained before, we have used the notation $s = \mathrm{R}, \mathrm{I}$ so that $\Psi_{\boldsymbol{k}}^s \equiv \Psi(\zeta_{\boldsymbol{k}}^s)$. This is the great advantage of going to Fourier space compared to real space: it drastically simplifies the wave-function. One may, however, wonder whether the non-linearities necessarily present in the theory (recall that the new terms in the Schrödinger equation are necessarily stochastic and non-linear) could bring to naught the technical convenience of using the Fourier transform. Usually, only when a theory is linear, the Fourier modes evolve independently (no mode coupling) and it is useful to go to Fourier space. This corresponds to a situation where the Hamiltonian is quadratic. A point, which is usually not very well appreciated, is that this does not necessarily imply the absence of interactions. It is true that, in field theory, interactions are associated with non-quadratic terms in the action but one exception is the interaction of a quantum field with a classical source. In this case, the action remains quadratic but the fundamental frequency of the system acquires a time dependence given by the source. This is typically the case for the Schwinger effect [35, 60] but also for Cosmology. In this last situation, the source is just the dynamics of the background space-time itself. In the following, we restrict ourselves to quadratic Hamiltonians since this is sufficient to describe cosmological perturbations during inflation (of course, if one wants to calculate higher-order statistics, such as Non-Gaussianities, then non-linear terms in the Hamiltonian must be taken into account).

However, in the present situation, even if one restricts oneself to quadratic Hamiltonians, one also has the extra non-linear and stochastic terms in the modified Schrödinger equation and, as noticed above, there is the concern that they could be responsible for the appearance of mode couplings. Fortunately, this is not the case. Indeed, if one recalls that the Hamiltonian of the system reads $\hat{H} = \int_{\mathbb{R}^{3+}} \mathrm{d}\boldsymbol{k} \sum_{s=\mathrm{R},\mathrm{I}} \hat{H}_{\boldsymbol{k}}^s$ and if one introduces the Fourier transform of the collapse operator, $\hat{C}(\boldsymbol{x}) = (2\pi)^{-3/2} \int \mathrm{d}\boldsymbol{k}\, \hat{C}(\boldsymbol{k}) e^{-i\boldsymbol{k}\cdot\boldsymbol{x}}$ (and a similar formula for the noise), then straightforward calculations lead to [24]

$$
\mathrm{d}\left|\Psi_{\boldsymbol{k}}^s\right\rangle = \left\{ -i \hat{H}_{\boldsymbol{k}}^s \mathrm{d}t + \frac{\sqrt{\gamma a^3}}{m_0} \left[\hat{C}^s(\boldsymbol{k}) - \left\langle \hat{C}^s(\boldsymbol{k})\right\rangle\right] \mathrm{d}W_t^s(\boldsymbol{k}) \right.
\left. - \frac{\gamma a^3}{2 m_0^2} \left[\hat{C}^s(\boldsymbol{k}) - \left\langle \hat{C}^s(\boldsymbol{k})\right\rangle\right]^2 \mathrm{d}t \right\} \left|\Psi_{\boldsymbol{k}}^s\right\rangle . \tag{28}
$$

We see that we can write a CSL equation for each Fourier mode. In other words, it seems that the presence of the extra stochastic and non-linear terms does not destroy the property that the modes still evolve separately [24]. In order to better understand the origin of this property, let us come back to Eq. (25). Let us assume that we are in the particular situation where $\hat{H} = H(\hat{\boldsymbol{x}}, \hat{\boldsymbol{p}}) = H_1(\hat{x}_1, \hat{p}_1) + H_2(\hat{x}_2, \hat{p}_2) + H_3(\hat{x}_3, \hat{p}_3)$ and $\hat{C}_i = C_i(\hat{\boldsymbol{x}}, \hat{\boldsymbol{p}}) = C_i(\hat{x}_i, \hat{p}_i)$, namely the component $\hat{C}_i$ only depends on $\hat{x}_i$ and $\hat{p}_i$ [in other words, we do not have, for instance, $\hat{C}_x = C_x(\hat{y}, \hat{p}_y)$]. Then writing $\Psi = \prod_i \Psi_i(x_i)$, it is easy to show that

$$\mathrm{d}\Psi_i = \left[-i\hat{H}_i \mathrm{d}t + \frac{\sqrt{\gamma}}{m_0} \left(\hat{C}_i - \left\langle \Psi_i \left| \hat{C}_i \right| \Psi_i \right\rangle \right) \mathrm{d}W_i - \frac{\gamma}{2m_0^2} \left(\hat{C}_i - \left\langle \Psi_i \left| \hat{C}_i \right| \Psi_i \right\rangle \right)^2 \mathrm{d}t \right] \Psi_i, \tag{29}$$

where we have used the fact that

$$\left\langle \Psi \left| \hat{C}_i \right| \Psi \right\rangle = \left\langle \prod_j \Psi_j \left| \hat{C}_i \right| \prod_k \Psi_k \right\rangle = \left\langle \prod_{j \neq i} \Psi_j \middle| \prod_{k \neq i} \Psi_k \right\rangle \left\langle \Psi_i \left| \hat{C}_i \right| \Psi_i \right\rangle = \left\langle \Psi_i \left| \hat{C}_i \right| \Psi_i \right\rangle. \tag{30}$$

We see that we can write an independent equation for each Ψ_i. In inflationary perturbations theory, the two properties needed to obtain this independent equation are also satisfied, namely the Hamiltonian is a sum of the Hamiltonians for each Fourier mode and $\hat{C}^s(\boldsymbol{k})$ only depends on $\boldsymbol{k}$ and not on other modes. This is the reason why one can obtain an equation (28) for each Fourier mode.

Then comes the choice of the collapse operator $\hat{C}(\boldsymbol{x}_\mathrm{p})$. Many different possibilities have been discussed in the literature and each of them correspond to a different version of the theory. In the context of standard Quantum Mechanics, if $\hat{C}(\boldsymbol{x}_\mathrm{p})$ is the position operator, then we have Quantum Mechanics with Universal Position Localization (QMUPL) while if $\hat{C}(\boldsymbol{x}_\mathrm{p})$ is the mass density operator, we deal with the Continuous Spontaneous Localization (CSL) model [4]. In the context of Field Theory and Cosmology, two choices have been studied. The first one corresponds to $\hat{C}^s(\boldsymbol{k}) \propto a^p \hat{\zeta}^s_{\boldsymbol{k}}$, where p is a free parameter. Since, in some sense, field amplitude plays the role of position, this case represents the field-theoretic version of QMUPL. Except for p, this version is characterized by one parameter, γ. The other possibility is CSL, which relies on coarse-graining the mass density over the distance r_c. This corresponds to

$$\hat{C}(\boldsymbol{x}) = \left(\frac{a}{r_\mathrm{c}} \right)^3 \frac{1}{(2\pi)^{3/2}} \int \mathrm{d}\boldsymbol{y}\, \hat{\delta}_\mathrm{g}(\boldsymbol{x} + \boldsymbol{y}) e^{-\frac{|\boldsymbol{y}|^2 a^2}{2r_\mathrm{c}^2}}, \tag{31}$$

where $\hat{\delta}_\mathrm{g}$ is the energy density contrast relative to a "Newtonian" time slicing (see the beginning of the next section for a more complete discussion). At this point, we meet again the problem that a fully relativistic and covariant collapse model is not available. Indeed, the definition of energy density is not unique in General Relativity and an infinite number of other choices could have been contemplated, by considering the energy density contrast relative to other slicings [24]. Without additional criterions, there is presently no mean to decide which version makes more sense. However, what can be done is to constrain these different versions with CMB data. In fact, and we come back to this question in the next section, Sect. 5, we can show that the situation is not as problematic as it may seem and that (almost) all possible choices lead to the same result. In this sense, the results obtained in the following are rather generic.

Once the collapse operator and the noise have been chosen, Eq. (28) is entirely specified and the next step is then to solve it. The solution is given by a wave-function evolving stochastically in Hilbert space. As discussed above, the initial conditions are Gaussian and the Hamiltonian being quadratic, the Gaussian character of the wave-function is preserved in time. Therefore, without loss of generality, one can write the most general stochastic wave-function as

$$\Psi_{\boldsymbol{k}}^{s}\left(\zeta_{\boldsymbol{k}}^{s}\right) = \left|N_{\boldsymbol{k}}\left(\eta\right)\right| \exp\Big\{-\Re\mathrm{e}\,\Omega_{\boldsymbol{k}}\left(\eta\right) z^{2}\left[\zeta_{\boldsymbol{k}}^{s} - \bar{\zeta}_{\boldsymbol{k}}^{s}\left(\eta\right)\right]^{2} + i\sigma_{\boldsymbol{k}}^{s}(\eta) + iz\chi_{\boldsymbol{k}}^{s}(\eta)\zeta_{\boldsymbol{k}}^{s} - iz^{2}\Im\mathrm{m}\,\Omega_{\boldsymbol{k}}(\eta)\left(\zeta_{\boldsymbol{k}}^{s}\right)^{2}\Big\}, \tag{32}$$

where the free functions $\Omega_{\boldsymbol{k}}(\eta)$, $\bar{\zeta}_{\boldsymbol{k}}^{s}(\eta)$, $\sigma_{\boldsymbol{k}}^{s}(\eta)$ and $\chi_{\boldsymbol{k}}^{s}(\eta)$ are (a priori) stochastic quantities.

Let us now discuss how collapse models can be, in the context of Cosmology, related to observations. This needs to be carefully studied since we now have two ways to calculate averages, the quantum average and the stochastic average. For instance, the quantum average of a given observable $\mathcal{O}(\hat{\zeta}_{\boldsymbol{k}}^{s})$, $\langle\mathcal{O}(\hat{\zeta}_{\boldsymbol{k}}^{s})\rangle \equiv \int |\Psi_{\boldsymbol{k}}^{s}|^{2}\mathcal{O}(\zeta_{\boldsymbol{k}}^{s})\mathrm{d}\zeta_{\boldsymbol{k}}^{s}$, which, in the standard context, would be a number is, here, a stochastic quantity. So only $\mathbb{E}[\langle\mathcal{O}(\hat{\zeta}_{\boldsymbol{k}}^{s})\rangle] = \int \mathbb{E}[|\Psi_{\boldsymbol{k}}^{s}|^{2}]\mathcal{O}(\zeta_{\boldsymbol{k}}^{s})\mathrm{d}\zeta_{\boldsymbol{k}}^{s}$ is a number. The quantity

$$|\Psi_{\boldsymbol{k}}^{s}(\zeta_{\boldsymbol{k}}^{s})|^{2} = z\sqrt{\frac{2\Re\mathrm{e}\,\Omega_{\boldsymbol{k}}}{\pi}}\exp\left[-2z^{2}\Re\mathrm{e}\,\Omega_{\boldsymbol{k}}\left(\zeta_{\boldsymbol{k}}^{s} - \bar{\zeta}_{\boldsymbol{k}}^{s}\right)^{2}\right], \tag{33}$$

which is centered at $\bar{\zeta}_{\boldsymbol{k}}^{s}$ and has width $(4z^{2}\Re\mathrm{e}\,\Omega_{\boldsymbol{k}})^{-1}$, describes a Gaussian wave-packet whose mean and variance evolve stochastically (in fact, in the particular case considered here, it turns out that the variance is a deterministic quantity and that only the mean is stochastic). Therefore, for a specific realization, one expects, as time passes, that $|\Psi_{\boldsymbol{k}}^{s}(\zeta_{\boldsymbol{k}}^{s})|^{2}$ stochastically shifts its position $\bar{\zeta}_{\boldsymbol{k}}^{s}(\eta)$ while its width decreases until $\bar{\zeta}_{\boldsymbol{k}}^{s}$ settles down to a particular position $\bar{\zeta}_{\boldsymbol{k}}^{s}(\eta_{\mathrm{coll}})$, with an (almost) vanishing width. In this way, the macro-objectification problem of Quantum Mechanics is solved and a single outcome has been produced. The interest of this approach for Cosmology is that it does so without invoking the presence of an observer, and only thanks to the modified dynamics of the wave-function. If one then considers another realization, a qualitatively similar behavior is observed but, of course, the final value $\bar{\zeta}_{\boldsymbol{k}}^{s}(\eta_{\mathrm{coll}})$ (in fact the whole trajectory) needs not be the same. If we repeat many times the same experiment and have at our disposal many realizations, one can then calculate, say, $\mathbb{E}[\langle\hat{\zeta}_{\boldsymbol{k}}^{s}\rangle] = \mathbb{E}[\bar{\zeta}_{\boldsymbol{k}}^{s}]$ or $\mathbb{E}[\langle\hat{\zeta}_{\boldsymbol{k}}^{s}\rangle^{2}] = \mathbb{E}[\bar{\zeta}_{\boldsymbol{k}}^{s\,2}]$. This allows us to calculate the dispersion of $\bar{\zeta}_{\boldsymbol{k}}^{s}$ according to

$$\mathcal{P}_{\zeta}(k) = \frac{k^{3}}{2\pi^{2}}\left\{\mathbb{E}\left[\bar{\zeta}_{\boldsymbol{k}}^{s\,2}\right] - \mathbb{E}^{2}\left[\bar{\zeta}_{\boldsymbol{k}}^{s}\right]\right\}, \tag{34}$$

which makes the connection with the previous considerations.

In fact, in Cosmology, a legitimate question is why the above-defined dispersion $\mathcal{P}_\zeta$ is equivalent to (or, even, has something to do with) the power spectrum of curvature perturbations. Indeed, in order to give an operational meaning to the above quantity, one needs to have access to a large number of realizations. This is necessary if one wants to identify the mathematical object $\mathbb{E}[.]$ with the relative frequency of occurrence. Clearly, in Cosmology, we deal with only one realization (one universe) and there is no way to repeat the experiment. In fact, this question is by no mean an issue only for the collapse models since, even in the standard approach, the predictions are expressed in terms of ensemble averages.

Here, the key idea, admittedly not always explicitly stated in the inflationary literature, is the use of an ergodic-like principle, which consists in identifying ensemble averages with spatial averages [61]. A very schematic description of this procedure is as follows. For a given Fourier mode $\boldsymbol{k}$, one can divide the celestial sphere into different patches, and construct an estimate of the amplitude of the curvature perturbation at this Fourier mode in each patch. Interpreting each patch as a different realization, one can then calculate the ensemble average of these "measurements", which is thus nothing but a spatial average. In this sense, "repeating the experiment" is replaced with "looking at different regions on the sky". Obviously, to be able to evaluate the Fourier mode $\boldsymbol{k}$ in a certain patch, the size of the patch has to be larger than the wavelength associated to $\boldsymbol{k}$. However, the celestial sphere being compact, only a finite number of patches with a certain minimum size can be drawn on it. This is why the ensemble average can be calculated only over a finite number of "realizations", and the larger the wavelength (i.e. the smaller k) is, the larger the patches need to be, hence the fewer "realizations" are available. This introduces an unavoidable error which is called the "cosmic variance" in the Cosmology literature, see Ref. [61] for more details.

5 Comparison with Observations

In this section, we briefly discuss the observational status of collapse models in Cosmology. As already mentioned, only few cases have been investigated so far: QMUPL and CSL, both with a white noise and using a naive generalization of non-relativistic collapse models to field theory. A discussion of QMUPL in Cosmology can be found in Refs. [19, 62] and, here, we focus on CSL since this is the model that has drawn the most attention [24].

The CSL theory consists in assuming that the collapse operator is mass or energy density. In a cosmological context, as already briefly mentioned in the previous section, this corresponds to $\hat{C} = \rho + \widehat{\delta\rho}$, where ρ is the energy density stored in the inflaton field and $\hat{\delta} \equiv \widehat{\delta\rho}/\rho$ is the density contrast. In fact, only the density contrast will be playing a role in what follows because, in inflationary perturbations theory, ρ is a classical quantity and, therefore, cancels out in the modified Schrödinger equation. In General Relativity, however, as already mentioned, there is no unique definition for δ. Nevertheless, see Ref. [24], what matters is in fact the scale dependence of δ, in particular its behavior on large scales. Conveniently, one can show that, for all reasonable choices, all the δ's behave similarly (namely, in the same way as the Newtonian density contrast "δ_{g}") except for one particular case, the so-called "δ_{m}" density contrast. Therefore, even if the choice of δ is ambiguous, the final result turns out to be (almost) independent of this choice.

Once the collapse operator has been chosen, one can solve the modified Schrödinger equation and calculate the CSL inflationary power spectrum along the lines explained in the previous sections. This power spectrum depends on the two CSL parameters γ and r_{c}. Quite intuitively, one finds that the extra CSL terms operate only if the physical wavelength of a Fourier mode is larger than the localization scale r_{c}. In an expanding universe, physical wavelengths increase with time, so this implies that for any given wavenumber $\boldsymbol{k}$, there is a time before which its physical wavelength is smaller than r_{c}, hence the CSL corrections are absent. This is a crucial feature since it guarantees that the usual way of setting initial conditions in the Bunch-Davies vacuum, which is a very important aspect of the inflationary paradigm, is still available.

When the physical wavelength of a Fourier mode becomes larger than r_{c}, the CSL terms become important and collapse occurs. This generates the power spectrum [24]

$$\mathcal{P}_{\zeta}(k) = \frac{k^3}{2\pi^2}\frac{1}{4\Re\mathrm{e}\,\Omega_{\boldsymbol{k}}|_{\gamma=0}}\left[1+\mathcal{O}(1)\frac{\gamma}{m_0^2}\rho\,\epsilon_1\left(\frac{r_{\mathrm{c}}}{\ell_{\mathrm{H}}}\right)^{\mathfrak{a}}_{\mathrm{end}}\left(\frac{k}{aH}\right)^{\mathfrak{b}}_{\mathrm{end}}-\frac{\Re\mathrm{e}\,\Omega_{\boldsymbol{k}}|_{\gamma=0}}{\Re\mathrm{e}\,\Omega_{\boldsymbol{k}}}\right]. \tag{35}$$

In the limit where $\gamma = 0$, one checks that the power spectrum vanishes, since no perturbation is being produced, in agreement with the discussion presented in Sect. 4. Let us also recall that the "standard" result, obtained in the Copenhagen interpretation, is given by Eq. (18), which matches the prefactor in Eq. (35), and that $\Re\mathrm{e}\,\Omega_{\boldsymbol{k}}$ is proportional to the inverse variance of the wave-packet. If γ is sufficiently large so that the collapse occurs, the width of the wave-function is much smaller than what it would be in the unmodified theory, hence the third term in the square brackets of Eq. (35) can be neglected when compared to the first term. In that case, the power spectrum takes the form of the standard result, plus a correction proportional to γ. This CSL correction is also proportional to $\rho\epsilon_1$, where ϵ_1 is the first slow-roll parameter and ρ the energy density at the end of inflation. Let us recall that, during inflation, ρ is quasi constant and can be as large as

$$\rho \sim 10^{80}\mathrm{g}\times\mathrm{cm}^{-3}. \tag{36}$$

We see here why Cosmology is a natural place to probe collapse theories: it tests them in regimes that are completely different, in terms of energy, time or length scales, than those relevant in the lab. Since the amplitude of the CSL new terms are controlled by the energy density, it makes sense to constrain them in physical conditions where ρ is as large as possible. This is why, for instance, the CSL mechanism was also applied

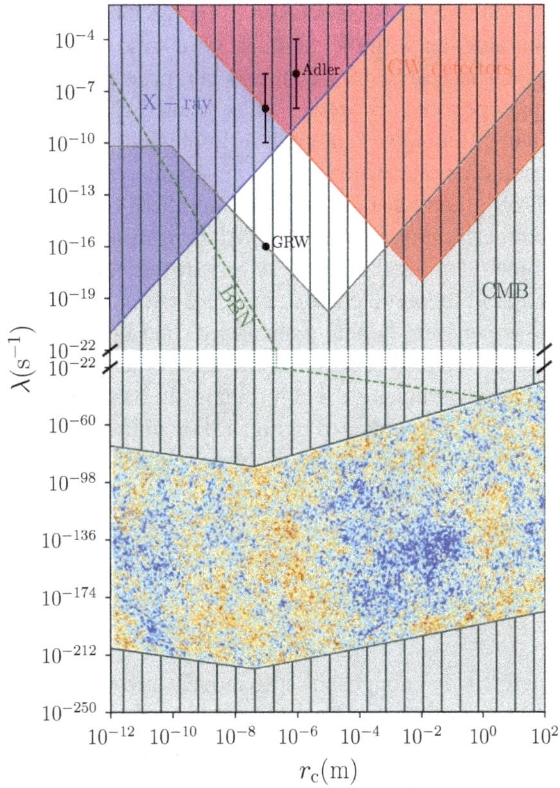

Fig. 1 Observational constraints on the two parameters r_c and λ of the CSL model obtained in Ref. [24]. The white region is allowed by laboratory experiments while the "CMB map" region is allowed by CMB measurements. The green dashed line stands for the upper bound on λ if inflation proceeds at the Big-Bang Nucleosynthesis (BBN) scale

to neutron stars in Ref. [63]. Primordial Cosmology is a situation where ρ is even larger and, therefore, one can expect it to be even more appropriate when it comes to establishing constraints on CSL.

The second crucial piece of information that comes from Eq. (35) is that the CSL corrections are not scale invariant. Their scale dependence is $\propto k^{\mathfrak{b}}$ where $\mathfrak{b} = -1$ if the scale r_c is crossed out during inflation and $\mathfrak{b} = -10$ if r_c is crossed out during the subsequent radiation dominated era. In this last case, there is an additional factor $\propto (r_c/\ell_H)^{\mathfrak{a}}$, where ℓ_H is the Hubble radius at the end of inflation, with $\mathfrak{a} = -9$ (if r_c is crossed out during inflation, this term is not present and $\mathfrak{a} = 0$). In other words, detectable CSL corrections would be strongly incompatible with CMB measurements. Since we have seen that they are typically very large, we expect the constraints that can be inferred from them to be very efficient.

These constraints are represented in Fig. 1 in the space (r_c, λ) where $\lambda = \gamma/(8\pi^{3/2}r_c^3)$. In this plot, the white region corresponds to the parameter space allowed by lab experiments while the "CMB map" region corresponds to parameter space allowed by CMB measurements. Evidently, the most striking feature of the plot is that the two regions do not overlap. Taken at face value, this implies that CSL is ruled out! However, this conclusion should be toned down. First, we should notice that if the collapse operator is taken to be $\delta_{\rm m}$, then the CMB constraints are no longer in contradiction with the lab ones. Of course, in some sense, $\delta_{\rm m}$ is "of measure zero" in the space of density contrasts but, nevertheless, this shows that one can find collapse operators for which CSL is rescued. Second, one has to remember that we used a naive (too naive?) method to implement the collapse mechanism in field theory. It could be that, when a truly covariant version of collapse models is available [4, 57–59], the final result will be modified. For instance, the constraints on the CSL parameters coming from the CMB constraints on one hand, and from lab experiments on the other hand, operate at very different energy scales. One could imagine that, in a field-theoretic context, the CSL parameters run with the energy scale at which the experiment is being performed, and that one cannot simply compare the constraints obtained at different energies. Finally, we used a white noise in the modified Schrödinger equation and it remains to be seen if using a colored noise can modify the constraints obtained in Fig. 1. For all these reasons, it is necessary to be cautious and testing the robustness of the conclusions obtained here will certainly be a major goal in the future.

6 Conclusions

Interestingly enough, collapse models advocated by Giancarlo Ghirardi (and others) and cosmic inflation have almost the same age. Roughly speaking, they were both introduced at the end of the seventies and beginning of the eighties. Nevertheless, until recently, they had never met. In this article, we have described the recent attempts to apply collapse models to inflation. We have argued that there is a good scientific case motivating those attempts. In particular, for collapse models to be interesting and to insure proper localization, the collapse operators must be related to the energy density. As a consequence, the most efficient tests of collapse models will be in physical situations where the energy density is as large as possible. Without any doubt, this is to be found in the early universe. We have shown that, indeed, the high-accuracy data now at our disposal leads to extremely competitive constraints, that anyone interested in collapse theories can no longer ignore. We hope this will cause further investigations to test the robustness of these results.

Finally, after 40 years, collapse theories and cosmic inflation have met and we are convinced that Giancarlo Ghirardi would have been fascinated by the fact that his great insights about Quantum Mechanics can even find applications in Cosmology.

References

1. A. Bassi, K. Lochan, S. Satin, T. P. Singh, and H. Ulbricht, Rev. Mod. Phys. **85**, 471 (2013), 1204.4325.
2. G. C. Ghirardi, A. Rimini, and T. Weber, Phys. Rev. **D34**, 470 (1986).

3. P. M. Pearle, Phys. Rev. **A39**, 2277 (1989).
4. G. C. Ghirardi, P. M. Pearle, and A. Rimini, Phys. Rev. **A42**, 78 (1990).
5. A. Bassi and G. C. Ghirardi, Phys. Rept. **379**, 257 (2003), quant-ph/0302164.
6. G. Esposito-Farese, AIP Conf. Proc. **736**, 35 (2004), gr-qc/0409081.
7. V. F. Mukhanov and G. Chibisov, JETP Lett. **33**, 532 (1981).
8. A. A. Starobinsky, Phys. Lett. **B91**, 99 (1980).
9. A. H. Guth, Phys. Rev. **D23**, 347 (1981).
10. A. D. Linde, Phys.Lett. **B108**, 389 (1982).
11. A. Albrecht and P. J. Steinhardt, Phys. Rev. Lett. **48**, 1220 (1982).
12. A. D. Linde, Phys. Lett. **B129**, 177 (1983).
13. J. von Neumann, *Mathematical foundations of quantum mechanics* (1955).
14. J. B. Hartle (2019), 1901.03933.
15. J. S. Bell and N. D. Mermin, Physics Today **41**, 89 (1988).
16. A. Perez, H. Sahlmann, and D. Sudarsky, Class. Quant. Grav. **23**, 2317 (2006), gr-qc/0508100.
17. P. M. Pearle, in *Foundational Questions Institute Inaugural Workshop (FQXi 2007 Reykjavik, Iceland, July 21-26, 2007* (2007), 0710.0567.
18. K. Lochan, S. Das, and A. Bassi, Phys. Rev. **D86**, 065016 (2012), 1206.4425.
19. J. Martin, V. Vennin, and P. Peter, Phys. Rev. **D86**, 103524 (2012), 1207.2086.
20. P. Cañate, P. Pearle, and D. Sudarsky, Phys. Rev. **D87**, 104024 (2013), 1211.3463.
21. M. P. Piccirilli, G. León, S. J. Landau, M. Benetti, and D. Sudarsky, Int. J. Mod. Phys. **D28**, 1950041 (2018), 1709.06237.
22. G. León, A. Majhi, E. Okon, and D. Sudarsky, Phys. Rev. **D98**, 023512 (2018), 1712.02435.
23. G. León, A. Pujol, S. J. Landau, and M. P. Piccirilli, Phys. Dark Univ. **24**, 100285 (2019), 1902.08696.
24. J. Martin and V. Vennin (2019), 1906.04405.
25. J. Martin (2019a), 1902.05286.
26. H. Kodama and M. Sasaki, Prog. Theor. Phys. Suppl. **78**, 1 (1984).
27. P. Ade et al. (Planck), Astron.Astrophys. **571**, A16 (2014), 1303.5076.
28. J. Martin, C. Ringeval, and V. Vennin, Phys. Dark Univ. **5-6**, 75–235 (2014a), 1303.3787.
29. J. Martin, C. Ringeval, R. Trotta, and V. Vennin, JCAP **1403**, 039 (2014b), 1312.3529.
30. J. Martin, C. Ringeval, and V. Vennin, Phys. Rev. Lett. **114**, 081303 (2015), 1410.7958.
31. J. Martin (2015), 1502.05733.
32. Y. Akrami et al. (Planck) (2018a), 1807.06205.
33. Y. Akrami et al. (Planck) (2018b), 1807.06211.
34. Y. Akrami et al. (Planck) (2019), 1905.05697.
35. J. Martin, Lect. Notes Phys. **738**, 193 (2008), 0704.3540.
36. C. Cohen-Tannoudji, J. Dupont-Roc, and G. Grunberg, *Atom - Photon Interactions: Basic Process and Appilcations* (Wiley-Interscience, 1992).
37. T. Bunch and P. Davies, Proc.Roy.Soc.Lond. **A360**, 117 (1978).
38. L. Grishchuk and Y. Sidorov, Phys. Rev. **D42**, 3413 (1990).
39. J. Martin and V. Vennin, Phys. Rev. **D93**, 023505 (2016a), 1510.04038.
40. J. Grain and V. Vennin (2019), 1910.01916.
41. R. K. Sachs and A. M. Wolfe, Astrophys. J. **147**, 73 (1967), [Gen. Rel. Grav.39,1929(2007)].
42. A. Kenfack and K. Zyczkowski, Journal of Optics B: Quantum and Semiclassical Optics **6**, 396 (2004), quant-ph/0406015.
43. D. Polarski and A. A. Starobinsky, Class. Quant. Grav. **13**, 377 (1996), gr-qc/9504030.
44. A. Albrecht, P. Ferreira, M. Joyce, and T. Prokopec, Phys. Rev. **D50**, 4807 (1994), astro-ph/9303001.
45. M. Revzen, Foundations of Physics **36**, 546 (2006).
46. M. Revzen, P. A. Mello, A. Mann, and L. M. Johansen, A **71**, 022103 (2005), quant-ph/0405100.
47. J. Martin and V. Vennin, Phys. Rev. **A93**, 062117 (2016b), 1605.02944.
48. J. Martin and V. Vennin, Phys. Rev. **D96**, 063501 (2017), 1706.05001.
49. J. S. Bell, Annals of the New York Academy of Sciences **480**, 263 (1986).
50. J. Martin, Universe **5**, 92 (2019b), 1904.00083.

51. W. H. Zurek, Phys. Rev. **D24**, 1516 (1981).
52. M. Schlosshauer, Rev. Mod. Phys. **76**, 1267 (2004), quant-ph/0312059.
53. C. P. Burgess, R. Holman, and D. Hoover, Phys.Rev. **D77**, 063534 (2008), astro-ph/0601646.
54. J. Martin and V. Vennin, JCAP **1805**, 063 (2018a), 1801.09949.
55. J. Martin and V. Vennin, JCAP **1806**, 037 (2018b), 1805.05609.
56. M. Castagnino, S. Fortin, R. Laura, and D. Sudarsky, Found. Phys. **47**, 1387 (2017), 1412.7576.
57. R. Tumulka, Proc. Roy. Soc. Lond. **A462**, 1897 (2006), quant-ph/0508230.
58. D. J. Bedingham, Found. Phys. **41**, 686 (2011), 1003.2774.
59. D. Bedingham, D. Dürr, G. Ghirardi, S. Goldstein, R. Tumulka, and N. Zanghì, Journal of Statistical Physics **154**, 623 (2014), 1111.1425.
60. J. S. Schwinger, Phys. Rev. **82**, 664 (1951), [,116(1951)].
61. L. P. Grishchuk and J. Martin, Phys. Rev. **D56**, 1924 (1997), gr-qc/9702018.
62. S. Das, K. Lochan, S. Sahu, and T. P. Singh, Phys. Rev. **D88**, 085020 (2013), [Erratum: Phys. Rev.D89,no.10,109902(2014)], 1304.5094.
63. A. Tilloy and T. M. Stace, Phys. Rev. Lett. **123**, 080402 (2019), 1901.05477.

Spontaneous Collapse Theories and Cosmology

Daniel Sudarsky

Abstract The account for the emergence of the primordial seeds of structure in the universe as a result of "quantum fluctuations" during the inflationary epoch, is, on one hand, remarkably successful at the empirical level, and, on the other hand, it faces severe conceptual shortcomings tied to the conceptual difficulties that afflict quantum theory in general. In the cosmological context, such problems become exacerbated by the relative simplicity of the form that the questions takes. This, at the same time, makes their investigation rather direct, and the case for novel physics, such as that represented by spontaneous collapse theories, extremely compelling. We will discuss those aspects and argue that the most natural framework for the consideration of the relevant issues in this context is that provided by semi-classical gravity. We will see that such line of research offers a path to deal with the conceptual difficulties alluded. Moreover, in this particular case, it also offers a natural resolution of one of the few instances where predictions of the standard approaches to the subject are in tension with the empirical results, namely that referring to the primordial gravity waves.

[1] We will use "spontaneous collapse" to refer to the kind of state reduction considered in theories that attempt to address the "measurement problem" via a modification of Schrödinger's evolution, and that does not explicitly tie such "reduction" to "measurement situations" or "interactions" with "observers". We will make use of the generic term "collapse" to include both the situations above, as well as possible considerations where the reduction of the quantum state is meant to be triggered by "measurements", "interaction with measuring apparatuses" or observers as in the Copenhagen or Von Neumann's approaches. That distinction will not be made explicitly when the context prevents any possibility of confusion and the wording would become too cumbersome, including discussions involving "collapse operators" and "the collapse rate constant".

[2] That dilution is often taken as characterized by a factor of e^N, with N "the number of e-folds" of inflation (the logarithm of the factor by which the scale factor of the Universe grows during the inflationary regime) usually taken to be at least 60.

D. Sudarsky (✉)
Departamento de Gravitaci6n y Teoría de Campos, Instituto de Ciencias Nucleares, Universidad Nacional Autónoma de México, A. Postal 70-543, 045010 Mexico City, Mexico
e-mail: sudarsky@nucleares.unam.mx

V. Allori et al. (eds.), *Do Wave Functions Jump?*, Fundamental Theories of Physics 198,
https://doi.org/10.1007/978-3-030-46777-7_22

1 Introduction

As clearly anticipated by Bell [1], cosmology provides a particularly clear context where the difficulties of quantum theory, viewed as a fundamental description of nature, are exposed. The issue was also recognized in [2], driving the authors to look for an alternative formulation of quantum theory that was appropriate to deal with cosmology. The focus was on an approach [3] which was soon shown to suffer from serious deficiencies [4, 5]. It is, thus, perhaps not surprising that spontaneous collapse [1] theories are increasingly being considered as playing an important role in such context. Indeed, current approaches to cosmology which rely on the hypothesis of an early inflationary epoch to address various "naturalness" difficulties [6] in the traditional Big Bang model, all have quantum theory playing a central role. In particular, inflation is supposed to smooth out all inhomogeneities and dilute all sorts of matter content present in the pre-inflationary stage.[2] The result is a universe that is devoid of matter and is homogeneous and isotropic to an exceedingly high degree. Therefore, any successful scenario must account for both the repopulation of the universe with matter,[3] and for the emergence of the primordial inhomogeneities that eventually grow to form all cosmic structure we see around us. The latter is where the present cosmological models strongly rely on quantum aspects, more specifically, on the so called "quantum fluctuations" of the vacuum. The usual treatment is, however, plagued with unjustifiable steps based on serious misunderstandings of quantum theory, as we will note shortly. This issue was first discussed in [7], where we argued that something like a spontaneous collapse theory could help in addressing the problems behind the unjustified steps. That took place before we learned that, by that time, the research program on spontaneous collapse theories was well underway with concrete and viable proposals represented by the GRW and CSL theories [8–14], where G.C. Ghirardi left his most unerasable marks, and had already made extraordinary advances [15]. The event illustrates, not only our level of ignorance regarding the field at that time, a condition that still afflicts the great majority of the cosmology community,[4] but also, the compelling force of the spontaneous collapse idea in the cosmological context.[5]

In this paper, we want to make the case that, not only is inflationary cosmology an ideal ground to contemplate the role of spontaneous collapse theories, and, in particular, their interface with gravitation, but also that the actual predictions of inflation can be substantially modified. Actually, the modified predictions appear to be, at the time of the writing of this article, more empirically adequate than the standard ones. This is so, at least, within what we believe to be the most appropriate implementation of the marriage between the description of space-time and quantum theory.

[3]A process known as "reheating". For a discussion see for instance [16].

[4]Although, there are some well known texts that do acknowledge the problem. See for instance [17, 18].

[5]Besides the embarrassment, something positive can be found in our lack of knowledge at the time.

This manuscript is organized as follows: In Sect. 2, we offer a brief review of the basic aspects of inflationary cosmology, including the usual account of the "prediction" regarding the primordial inhomogeneities (for simplicity, many technical details will be just mentioned in passing or completely ignored). In Sect. 3, we present a discussion of the problematic aspects of such accounts, as well as a discussion of the approach we have taken to deal with these and related problems. Section 4 is devoted to a practical implementation of the ideas developed in the previous section to the inflationary cosmological context. This includes the analysis of the emergence of seeds of cosmic structure, and the calculation leading to our prediction of the primordial spectrum of scalar perturbations, as well as a very brief discussion of that corresponding to the primordial gravity waves. As we will see it is in regards to the latter that we find the most dramatic modifications in the theoretical predictions. We finish in Sect. 5 with a short review of what has been accomplished so far, and the aspects of the approach that require further elucidation and development.

We will use the $(-+++)$ signature for the space-time metric, and Wald's conventions for the Riemann tensor [19]. Greek indices will be used to denote space-time coordinates, and latin indices to denote spatial coordinates on suitable identified spatial sections.

2 Cosmological Inflationary Model

The basic idea of inflation is that the standard radiation-dominated Big Bang regime is preceded by a period of accelerated expansion, controlled by something that behaves as a large cosmological constant, but which is later "turned off", as a result of its own dynamics. The specific realization of this idea is based on the introduction of a new scalar field ϕ with a potential V, which acts as the cosmological constant when the field is away from its minimum, usually taken to correspond to $\phi = 0$ (with $V(0) = 0$), and the value the scalar field evolves towards, as the expansion progresses. The theory is specified by the action:

$$S = \int d^4x \sqrt{-g}\{\frac{R}{16\pi G} - (1/2)\nabla^{\mu}\phi\nabla_{\mu}\phi - V(\phi) + \mathcal{L}_{Matt}\} \tag{2.1}$$

where G is Newton's constant, R stands for the Ricci scalar of the space-time metric $g_{\mu\nu}$, indices are raised using the inverse metric $g^{\mu\nu}$, and ∇_{μ} are metric compatible derivative operators (which when acting on scalar fields coincide with the usual coordinate partial derivative operators). $\mathcal{L}_{Matt}$ stands for the Lagrangian of the of matter fields other than ϕ (i.e. say the standard model of particle physics).

Standard variational principle leads to Einstein's equation for the space-time metric,

$$R_{\mu\nu} - (1/2)g_{\mu\nu}R = 8\pi G(T^{(Matt)}_{\mu\nu} + T^{(\phi)}_{\mu\nu}) \tag{2.2}$$

where $R_{\mu\nu}$ stands for the Ricci tensor of the space-time metric, the first term in the RHS represents the energy momentum of ordinary matter, which will be absent during inflation,[6] and the second that of the scalar field which is given by,

$$T^{(\phi)}_{\mu\nu} = \nabla_\mu\phi\nabla_\nu\phi - (1/2)g_{\mu\nu}(\nabla^\rho\phi\nabla_\rho\phi + 2V), \tag{2.3}$$

as well as the Klein -Gordon equation for the scalar field:

$$\nabla^\mu\phi\nabla_\mu\phi = -\frac{\partial V}{\partial\phi}. \tag{2.4}$$

The space-time metric one considers is, up to small perturbations (restricted to the relevant degrees of freedom for the problem at hand, namely the scalar perturbation known as the Newtonian potential ψ and the tensor perturbation h_{ij}), and using a specific gauge,[7] that of a spatially flat Robertson Walker cosmology[8]:

$$ds^2 = a^2(\eta)\{-[1 + 2\psi(\vec{x}, \eta)]d\eta^2 + [(1 - 2\psi(\vec{x}, \eta)\delta_{ij} + h_{ij}(x, \eta)]dx^i dx^j\} \tag{2.5}$$

where a stands for the cosmological scale factor, while the scalar field is expressed as $\phi = \phi_0(\eta) + \delta\phi(\vec{x}, \eta)$. Therefore, here, one is separating the treatment of the homogeneous (or zero) mode, from the modes that exhibit nontrivial spatial dependence.

The background (on top of which the perturbations of interest will be considered) is taken to represent a spatially flat homogeneous and isotropic space-time, and corresponds to setting $\psi = 0$, $h_{ij} = 0$, $\delta\phi = 0$. For such situation, Einstein's equations yield:

$$3\mathcal{H}^2 = 4\pi G(\dot{\phi}_0^2 + 2a^2 V_0), \tag{2.6}$$

$\mathcal{H} \equiv \dot{a}/a$ where "$\cdot$" $= \frac{\partial}{\partial\eta}$, while scalar field equation is:

$$\ddot{\phi}_0(\eta) + 2\dot{\phi}_0(\eta)\mathcal{H} + a^2\frac{\partial V}{\partial\phi} = 0 \tag{2.7}$$

Note that the relation between the standard co-moving time t and the conformal time η we are using is given by $dt/d\eta = a$. A further assumption regarding this

---

[6]In accordance to the view that the dilution caused by even the very early stages of inflation is sufficient to essentially erase all contributions to the energy momentum coming from other sectors.

[7]In working with perturbation theory in a general relativistic context, one invariably encounters ambiguities known as "the gauge freedom", and the cosmological setting is no exception. Fortunately, in this situation there are various approaches to deal with it in a satisfactory manner. One approach works with gauge "invariant variables" [20], and another just fixes the "gauge". We will not discuss this issue further, and will work in the context of a fixed gauge.

[8]The spatially flat Robertson Walker space-time metric corresponds to that in Eq. (2.5) only when setting $\psi(\vec{x}, \eta) \equiv 0$ and $h_{ij}(\vec{x}, \eta) \equiv 0$.

background is that the solution corresponds to a "slow roll" situation characterized by the smallness of the so "called slow parameters", in particular $\epsilon \equiv 1 - \frac{\dot{\mathcal{H}}}{\mathcal{H}^2}$.

Under those conditions, the cosmological expansion is almost exponential, corresponding, in conformal time, to $a(\eta) \approx Ce^{H_I t} = -\frac{1}{\eta H_I}$. For definiteness, we set $a = 1$ at the "present cosmological time", the starting time for inflation $\eta = -\mathcal{T}$ and its end point at $\eta = \eta_0 < 0$ so that the inflationary regime correspond $\eta \in (-\mathcal{T}, \eta_0)$. The inflationary epoch is supposed to be followed by a standard hot Big Bang cosmological development, with radiation and matter dominated epochs.[9]

On top of this background, one considers the perturbations or fluctuations characterized by nontrivial $\psi(\vec{x}, \eta)$, $h_{ij}(\vec{x}, \eta)$, and $\delta\phi(\vec{x}, \eta)$. Until this point, our treatment coincides with the standard one. So, let us start by reviewing and examining the "established lore".

Before proceeding, however, let us clarify that the goal of the kind of study one wants to undertake is to obtain an expression of the "power spectrum" of the various kinds of perturbations occurring in the universe. In particular, we will be considering the scalar density perturbations (and later the so called tensor or primordial wave perturbations). In any event, it is worthwhile clarifying for the reader the usage of these concepts.

The question is conveniently discussed representing the quantity of interest $\chi(\eta, \vec{x})$, which are functions of $(\vec{x}, \eta)$, in terms of their spatial Fourier transform coefficients $\chi_{\vec{k}}(\eta)$:

$$\chi(\vec{x}, \eta) = \frac{1}{(2\pi)^{3/2}} \int d^3k \chi_{\vec{k}}(\eta) e^{i\vec{k}.\vec{x}}. \tag{2.8}$$

For our universe, these variables, and thus the corresponding Fourier coefficients, take some specific values. However, cosmologists often proceed by considering an hypothetical ensemble of possible universes of which ours is a "fair" or "typical" representative. The idea is then to describe the statistical distributions of the relevant coefficients over such imaginary ensemble of universes.

In this way, one defines $\mathcal{P}_\chi(\eta, k)$, the power spectrum of χ, through the expression:

$$\overline{\chi_{\vec{k}}(\eta)\chi^*_{\vec{k}'}(\eta)} = \delta^3(\vec{k} - \vec{k}')\mathcal{P}_\chi(\eta, k), \tag{2.9}$$

where the overline represents the average of the quantity in question over such ensemble of universes. It follows from the above definitions that one might characterize the spectrum by

$$\overline{\chi(\vec{x}, \eta)\chi(\vec{y}, \eta)} = \frac{1}{(2\pi)^3} \int d^3k \mathcal{P}_\chi(\eta, k) e^{i\vec{k}.(\vec{x}-\vec{y})}. \tag{2.10}$$

The empirical analysis then proceeds by considering that our universe should be a typical random element of that ensemble.

[9]This transition is thought to be the result of the "reheating" process [16], where the "energy" stored in the inflaton field is transferred to ordinary components of the hot Big Bang universe.

2.1 The Standard Treatment

As it is customary, we will first focus on the so called scalar perturbations. The first step in this treatment requires dealing with the fact that the perturbed evolution equations link $\psi(\eta, \vec{x})$ and $\delta\phi(\eta, \vec{x})$. This is resolved in this context by passing to a characterization of the situation in terms of the combined variable introduced in [36, 37]:

$$v \equiv a\left(\delta\phi + \frac{\dot{\phi}_0}{\mathcal{H}}\psi\right), \tag{2.11}$$

This new field variable is now subjected to a quantum treatment, which is carried out in the standard manner appropriate for quantum fields in a curved space time background. That is, one constructs a Fock- Hilbert space and writes the field variable in terms of suitable creation and annihilation operators.[10]

$$\hat{v}(\eta, \vec{x}) = \frac{1}{(2\pi)^{3/2}} \int d^3k \left(\hat{a}_{\vec{k}} v_{\vec{k}}(\eta) e^{i\vec{k}.\vec{x}} + \hat{a}^{\dagger}_{\vec{k}} v^{*}_{\vec{k}}(\eta) e^{-i\vec{k}.\vec{x}}\right), \tag{2.12}$$

The specific choice of the mode functions corresponds to the selection of a particular vacuum state $|0\rangle$ (characterized by the requirement that $\hat{a}_{\vec{k}}|0\rangle = 0$). In the situation at hand, the natural choice for the mode functions $v_{\vec{k}}(\eta)$ corresponds to that which mimics the usual choice in Minkowski space-time, in the limit in which $\eta \to -\infty$.[11] This leads to the so called "Bunch Davies" state, but other possibilities are available.

As noted before, and as a result of the exponential expansion, after a short time into the inflationary regime the Universe is taken to be homogeneous and isotropic (H&I), both in the part that could be described at the "classical level", as well as that which is characterized at the quantum level.

A fundamental observation is that this vacuum state (as well as most of the alternatives considered) is a fully homogeneous and isotropic state. This can be readily seen by considering a spatial displacement of the state by $\vec{D}$. This is just given by $e^{i\hat{\vec{P}}.\vec{D}}|0\rangle = |0\rangle$ (where $\hat{\vec{P}}$ is the operator representing generator of spatial displacements in the Hilbert space, namely the "total momentum operator"), as follows immediately from the fact the $\hat{\vec{P}}|0\rangle = 0$, indicating that the state is completely homogeneous. Similar considerations hold regarding isotropy, namely the invariance of the state under rotations.

---

[10]Applying such procedure to gravitational perturbations themselves is rather worrisome. First, because the resulting interacting theory is non-renormalizable (a problem that is now a days deemed as non essential, as follows from the effective field theory point of view). On the other hand, and more importantly, the causal structure of the quantum field theory might deviate from that "true causal structure" dictated by the combination of the metric background together with the perturbations [34, 35].

[11]That is, one sets initial conditions for the mode functions that would correspond to the "positive energy solutions" if used as initial data in Minkowski space-time.

At this point, following the standard approach, one is directed to consider that the relevant quantity to investigate the behavior of these "perturbations", is the so called two point correlation function evaluated in the vacuum state $\langle 0|\hat{v}(\vec{x},\eta)\hat{v}(\vec{y},\eta)|0\rangle$. The argument is that such object represents the "quantum fluctuations".[12] From there, one extracts the so called "power spectrum":

$$\langle 0|\hat{v}(\vec{x},\eta)\hat{v}(\vec{y},\eta)|0\rangle = \frac{1}{(2\pi)^3}\int d^3k e^{i\vec{k}(\vec{x}-\vec{y})}\mathcal{P}_v(k). \tag{2.13}$$

The result is (in the limit of infinitely slow roll conditions) $\mathcal{P}_v(k) \sim k^{-3}$.

Now, the point is that the characterization of the quantity $\delta(\eta,\vec{x}) \equiv \frac{\delta\rho(\eta,\vec{x})}{\bar{\rho}(\eta)}$, where $\bar{\rho}(\eta)$ is the spatial average of the universe's density $\rho(\eta,\vec{x})$, with $\delta\rho(\eta,\vec{x}) \equiv \rho(\eta,\vec{x}) - \bar{\rho}(\eta)$, in terms of the "power spectrum" as

$$\overline{\delta(\vec{x},\eta)\delta(\vec{y},\eta)} = \frac{1}{(2\pi)^3}\int d^3k\mathcal{P}_\delta(\eta,k)e^{i\vec{k}.(\vec{x}-\vec{y})} \tag{2.14}$$

where now the average is over pairs of equally separated points in our universe seems to correspond to a Harrison-Zeldovich (HZ), or scale invariant[13] spectrum namely $\mathcal{P}_\delta(k) \propto k^{-3}$. Indeed, after the inclusion of well understood late time physical effects, such as plasma oscillations, etc. (as well as a slight "tilt" or small deviation from the exact exponent of -3, which depends on the small roll parameter ϵ) that spectrum is in excellent agreement with the observations [21].

The issue is, however, that the "power spectrum" we obtained from quantum considerations is now being taken to characterize the primordial inhomogeneities. These include the seeds of all cosmic structure, and thus are what eventually lead to the generation of galaxies, stars, planets, eventually life, and then creatures like ourselves, capable of wondering about the origin of it all.

This all sounds like a really astonishing account, which is, moreover, so full of profound poetic undertones that is hard to resist. Furthermore, as noted above, the theoretical account seems to fit observations to a remarkable extent, so it is not surprising that the prevailing attitude among cosmologists is "what else can one ask for ?".

---

[12]In my view, a substantial amount of the confusion among practitioners is due, in part, to the unfortunate usage of the word "*fluctuations*" to refer to various rather different notions: *(i)Statistical variations in an otherwise symmetric ensemble, (ii) Spatial variations in a single extended object, which is homogeneous at large scales, and (iii) quantum indeterminacies.* In our case, these are uncertainties or indeterminacies **in** the quantum state, **for** the field and conjugate momentum operators. I.e. an instance of (iii), which is often taken by cosmologists to represent either an instance of (i) or an instance of (ii).

[13]Note that as the LHS of (2.14) is dimensionless, so $\mathcal{P}_\delta(k)$ has to have dimensions of $length^3$ (as k has dimensions of $length^{-1}$. Thus, any kind of power law would require a length scale L_0 so that $\mathcal{P}_\delta = L_0^{3+n}k^n$, unless $n = -3$. This spectrum was favored in early phenomenological cosmology considerations [23, 24].

The reader should already find some discomfort in the fact that all 3 notions of fluctuations seem to be subject to a rather liberal interchange through the usage of the same wording "*power spectrum*" for the characterization of quantities appearing in Eqs. 2.10, 2.13 and 2.14. We will ignore for the most part in this work the distinction between ensemble averages and spatial averages over a single inhomogeneous universe, and focus on the more problematic relationship between the quantum characterization and the former two (for further discussion on these issues see [22]).

One often hears in the cosmology community this issue characterized as the problem of the "quantum to classical transition". This, in my view, is a misnomer. There is presumably nothing that is truly classical at a fundamental level, and, therefore, a classical description can, at best, be one corresponding to some kind of approximated characterization of the situation. Thus, the framing of the issue in that manner is really missing the point. The real question is the transition from a situation of full homogeneity and isotropy to one that is not. Moreover, as the process is taken to be described in quantum mechanical terms, what we need to account for is a transition from a quantum state that is homogeneous and isotropic to one that is not.[14] The degree to which the latter might be suitably described in classical terms is a question of the accuracy of the approximation, and although important, it is, in a sense, secondary to the main issue that concerns us here, and which we will be expanding on below.

3 The Case for a Modified Approach and a Specific Proposal

We have so far hinted at, and will shortly discuss in more detail, a problematic aspect of the presently accepted accounts of the emergence of primordial perturbations during inflation as a result of quantum fluctuations. It is easy to see that this is, of course, a particular instance of a broader problem afflicting quantum theory, namely the so called measurement problem. In this regard, it is convenient to remind ourselves of the result of [25] establishing the logical inconsistency of accepting the following three postulates regarding quantum theory (viewed as a fundamental theory):

(A) The description of an isolated physical system by its quantum state is complete (and thinking about our case, we ought to note that the universe is the epitome of an isolated system). (B) The evolution of such system is always dictated by Schrödinger's equation.[15] (C) Individual concrete experiments lead to definite results.

[14]This is actually only a problem for those accounts of quantum theory, that posit that the characterization of a system by its quantum state is complete, and could be easily addressed by approaches involving theories with (non-local) hidden variables, such as de-Broglie-Bohm type approaches.

[15]We ought to emphasize that for instance, the Copenhagen, interpretation explicitly forsakes this postulate as it includes the clause that the state of a system instantaneously turns into one of the eigen-states of the quantity that is being measured (thereby failing to evolve according to the Schrödinger's equation at that time) in a stochastic manner with probabilities dictated by Born's

The need to forsake (at least) one of the above forces one towards a specific conceptual path depending on the choice one makes. Concretely speaking, forsaking (A) seems to lead naturally to hidden variable theories, such as de-Broglie Bohm or "pilot wave" theory. Forsaking (B), one is naturally led to collapse theories, which for the cosmological case seem to leave no option but those of the spontaneous kind, (as there is clearly no role for conscious observers or measuring devices that might be meaningfully brought to bear to the situation at hand). Finally, forsaking (C) seems to be the starting point of approaches such as the Everettian type of interpretations. These, again, seem quite difficult to be suitably implemented in the context at hand, simply because observers, minds, and such, notions that play an important role in most attempts to characterize the world branching structure in those approaches, can only be accounted for within a universe in which structure has already developed, well before the emergence of the said entities.

We will focus in the present manuscript on the consideration of path (B), although it seems clear that at least path (A) seems to offer quite a reasonable alternative. However, before entering that discussion, we want to make the case that the account described in Sect. 2.1 is not satisfactory at all. That is, despite its phenomenological success, the picture it offers about the historical development of our universe is, frankly speaking, conceptually inadequate.

3.1 Conceptual Difficulties in the Standard Approach

The first problem, as already noted, is that, according to the above framework, the universe was H&I, both at the level of the part that is described in classical terms, the background metric and background scalar field, as well as in those aspects that are treated in quantum mechanical terms, the metric and scalar field perturbations. The latter can be seen in the fact that the quantum state, the so called Bunch Davies vacuum, is invariant under rotations and translations. Actually, this symmetry is itself the expected result from the early stages of inflation.[16]

---

rule. Of course the point is that the theory is rather unclear about what kind of interaction does qualify to be considered a measurement.

[16]The argument is that, even if the situation was not exactly homogeneous and isotropic at the classical level, and the state of the quantum field was not exactly the vacuum, the inflationary process itself would drive relatively a broad set of initial conditions towards precisely such homogeneous and isotropic stage for the space-time metric and the quantum state of all fields towards the vacuum. One might then expect small deviations of order e^{-N} to survive after N e- folds. However such minuscule relics from the initial stages of inflation are deemed just too small to be of any relevance. Note that, if those were relevant at all, they would destroy the predictability of the model. The point is that they would be completely unpredictable in the absence of a fundamental quantum gravity theory including a precise prescription of "initial conditions", or whatever replaces that, if such theory is a timeless theory, such as a canonical version of quantum gravity. We will thus ignore any such possible remnants from the pre-inflationary regime and talk as if the situation is exactly homogeneous and isotropic, where the caveat "*up to possible corrections of order* e^{-70}" is implicitly understood.

The issue is the following conundrum: According to the inflationary characterization of the very early universe, the starting point of the analysis corresponds to a situation that is completely homogeneous and isotropic, while the dynamics controlling the evolution explicitly preserves such symmetries. How is it then that we end up with a situation that does not share those symmetries? Indeed, how is it that we are able to make predictions about those inhomogeneities and anisotropies at all? Multiple attempts have been made [26, 27] to try addressing these questions within orthodox and traditional physical practice, but all those have come up short. For a detailed discussion see [28].

The second conceptual difficulty lies, as we discuss below, in the fact that it is rather unclear what is the theoretical framework one is relying on when proceeding according to the standard treatment. Regarding the matter fields, one might consider that one is working with quantum field theory on curved spacetime (QFT in CS), a subject with a rather well developed formalism (see [59]). This is so even though the scalar field is being separated into the "classical background" and the perturbations. The point is that one might regard the separation of the zero mode of the quantum field $\hat{\phi}_0(\eta)$ from the other $\delta\hat{\phi}(\eta, \vec{x})$ (space-dependent modes) and consider that the quantity $\phi_0(\eta)$ appearing in say Eq. (2.7) actually represents the expectation value of the zero mode $\langle\phi_0(\eta)\rangle$ in some highly excited coherent state, while the vacuum state refers only to the state of the spatially nontrivial modes. Regarding gravitation, however the issue is much more delicate. The fact is that one is certainly not working with a quantum gravity theory, or any approximation thereof, simply because we do not have a developed and workable version of such theory.[17] One is not working with classical gravity either, as there are at least some parts of the space-time metric that are being treated in a quantum language (ψ and h_{ij}). What one is doing is separating the metric into background and perturbations, treating the first part in a classical language, and the second part as a quantum field theory on the background space-time provided by the first. This has several problematic features. First, the separation of the metric into background and perturbations can only be viewed as a matter of convenience, and not as something of a fundamental nature. Actually, by suitable changes of coordinates one can pass from one such particular separation into another. Thus it is rather unclear in what sense such a distinct treatment for both is justified. One might think that what one is doing here is similar to what was done regarding the inflaton field, and that was just described above. However, that is not what is going on, simply because, as we noted, we do not have the full quantum

[17]There are multiple approaches in the attempt to construct a fully satisfactory theory of quantum gravity. These include the most popular such as String Theory and Loop Quantum gravity, both still confronting severe obstacles. There are others which are less well known but not for that less deserving, such as causal dynamical triangulations, causal sets, non-commutative geometry, etc. The point is, nevertheless, that none of those approaches is able, at this time, to both recover in a rigorous way general relativity as an approximation, and truly contend with the full quantum nature of what one can expect to be a fundamental theory of space-time, namely one that can deal, not only with causal structures subject to quantum indeterminism, but can also incorporate notions of space-time that accommodate states of matter in superposition of substantially different energy momentum distributions.

gravity theory that would justify that (i.e. we do not have, for quantum gravity, any theory that might be said to be playing the analogous role as quantum field theory on curved space-time). Moreover, we must note that when considering the treatment of the metric perturbations as a QFT in CS, one is, actually, doing severe violence to that framework from the start. The point is that the construction of QFT on CS has as **a basic postulate** that the quantum fields so constructed must have causal commutation relations (i.e. fields at space-like separated points must commute). In the present context one would be imposing the commutation relations for the fields according to the causal structure of the background space-time, rather than that of the "actual" physical space-time, which would be in part characterized by those quantum objects themselves. These and related issues have been discussed by other authors (see for instance [34, 35]). One might dismiss all those concerns and argue somehow that what one is doing is justified as an approximation, and that would probably be something one would tend to agree with. However it seems clear that a conceptually satisfactory picture would only be at hand if one had a clear idea of "an approximation to what" one is supposed to be considering.

We will be motivated by the quest for a clear explanation, framed within a general theoretical setting, that offers at least plausible answers to reasonable questions naturally arising in the situation at hand.

3.2 Spontaneous Collapse of the Quantum State and Einstein's Semi-classical Equations

First, we note that what seems to be required to address the issue at hand is to consider a physical process occurring in time, explaining the emergence of the seeds of structure. After all, emergence means (in this context): *something that was not there at an early time, is there at a later time.*

We need to explain the breakdown of the symmetry of the initial state. We do not want to have to put the inhomogeneities by hand at the start (as that would remove all the hope of predictability of the inflationary model). The theory we are dealing with does not lead through its standard Hamiltonian evolution to a breakdown of the symmetries we are considering. Therefore, something else is required. Spontaneous collapse theories do naturally contain the elements to achieve what is needed: departure from unitary evolution and stochasticity.

Thus, we will *add* to the standard inflationary accounts of very early cosmology, the spontaneous collapse of the wave function. On the other hand, that is not something that can be done straightforwardly. We need to discuss the obstacles such program faces, and the paths we have taken in the quest to overcome those.

We should start by noting that as the spontaneous collapse theories are described by a modification of the dynamics, concretely the time evolution of the state of quantum systems (in our case quantum fields), we seem to be forced to rely on a classical description of the space-time geometry. However, as we will see, for the sake of conceptual clarity,[18] there are even stronger reasons to proceed in this way.

It is worthwhile emphasizing at this point that the interface between quantum theory and gravitation need not involve the *Planck regime*: Consider, for instance, the issues that would have to be confronted in attempting to describe the space-time associated with a macroscopic body in quantum superposition of states localized in two distant regions. A rather influential work [29] considers such an experiment and claims to show semi-classical GR is simply not viable. The core of the argument is the following dichotomy: (1) If there are no quantum collapses, then semi-classical GR conflicts with their experiment. (2) If there are quantum collapses, then semi-classical GR equations are internally inconsistent. In this last regard, the issue is that a quantum collapse would generally be associated with failure of one side of Einsteins's equation, namely that containing the expectation value of the energy momentum tensor, from being divergence free, while the other side is automatically divergence free as a result of Bianchi's identities (see for instance Eq. (3.1)). This and related issues have been considered by various authors, but there is no clear consensus in the conclusions (see for instance [30–33]).

Thus, if the conclusions of [29] were correct, how could one possibly make sense of our approach? The point is that we might regard semi-classical GR, not as a fundamental theory but just as an *approximated description with limited domain of applicability*. We then consider the present line of research as an attempt to push that domain beyond what is usually viewed as a natural boundary. In the present context, we want to consider spontaneous collapses as the missing element that provides plausible resolution to the basic questions discussed in the previous subsection. Indeed, as it is clear that during the spontaneous collapse the equations can not be valid, we can not hope to consider the approach as fundamental. The proposal is, then, to follow an hydro-dynamical analogy: The Navier-Stokes equations for a fluid can not hold in some situations, for instance when a wave is breaking in the ocean. But they can be taken to hold to a very high approximation before and after that. Thus, we take semi-classical GR equations to hold before and after a spontaneous collapse, but not at the "time" it is occurring. In order for such approach to be fully specified, it must be supplemented by a well defined formalism that includes a recipe of how to join the descriptions "just before" and "just after" the spontaneous collapse.

In order to make a concrete proposal for considering the ideas described above, we need to specify a manner which might sensibly incorporate spontaneous collapse into the context of semiclassical GR. At the formal level we take as starting point a slight modification of what is normally described as semi-classical gravity, the theory of classical gravitation, together with the theory of quantum fields on a curved space-time. We will proceed by relying on the notion of *Semi-classical Self-consistent Configuration* (SSC) introduced in [61].

---

[18]In the pursuit of the goal of conceptual clarity, we will try to avoid perturbation theory from playing the dual role it is often relied upon: That of making the calculations manageable, and at the same time helping hide from explicit view some of the most serious foundational difficulties. We will not be able to avoid the former, but we will make all efforts to prevent the latter.

Definition: The set $\{g_{\mu\nu}(x), \hat{\varphi}(x), \hat{\pi}(x), \mathcal{H}, |\xi\rangle \in \mathcal{H}\}$ represents a SSC iff $\hat{\varphi}(x), \hat{\pi}(x)$ $\mathcal{H}$ corresponds to QFT (that is, $\mathcal{H}$ is a Hilbert space and $\hat{\varphi}(x)$, $\hat{\pi}(x)$ represent the quantum field and canonical conjugate momentum operators, as distributional valued operators acting on it and realizing the canonical commutation relations, and satisfying the corresponding evolution equations) constructed over the space-time with metric $g_{\mu\nu}(x)$, and the state $|\xi\rangle \in \mathcal{H}$ is such that:

$$G_{\mu\nu}[g(x)] = 8\pi G\langle\xi|\hat{T}_{\mu\nu}[g(x), \hat{\varphi}(x), \hat{\pi}(x)]|\xi\rangle \tag{3.1}$$

The scheme is simply the standard QFT construction on a given space-time, except for the requirement that there be a special quantum state taken to be the one corresponding to the physical situation at hand, and such that the above equation holds. That requirement gives the whole scheme the kind of self referential features which occur in the Schrödinger-Newton system [41–44]. One might regard the SSC formalism as the General Relativistic version of the latter. We note that most other states in $\mathcal{H}$ will fail to satisfy Eq. (3.1)[19] and thus would have to be considered as un-physical.[20]

Next, let us consider a spontaneous collapse *a la* GRW, i.e. a sudden transition from a given quantum state $|\xi\rangle$ to another $|\tilde{\xi}\rangle$, as dictated by the theory and the "stochastic choice". That would leave us with something that is no longer a SSC as the new state will fail to satisfy Eq. (3.1).[21] In order to remain as close as possible to such formalism, we will contemplate, instead, a spontaneous jump from one complete SSC to another one. That is, the usual "GRW jumps" must be considered now as generalized jumps of the form one full SSC to another, i.e. the spontaneous transition must now be regarded as $SSC1 \to SSC2$. It is, however, clear that generically the Eq. (3.1) will not hold during the jump itself.

In order for the scheme to be well defined, we must supply matching conditions: for both space-time and for the states in the Hilbert space. This has been studied in some detail in [61] and it involves various delicate issues.[22] All those aspects will

[19]To see this, simply consider a given SSC, and an arbitrary smooth function of compact support f, take $\hat{\phi}(f) \equiv \int \sqrt{-g}d^4xf(x)\hat{\phi}(x)$, and define the new state $|\chi\rangle = \hat{\phi}(f)|\xi\rangle$. It should be quite clear that the expectation value of the energy momentum for the state $|\chi\rangle$ will differ from the corresponding one for $|\xi\rangle$. So if the latter satisfies Eq. (3.1), then the former will not.

[20]In this sense, we are advocating a point of view where there is already, at this stage, a breakdown of the superposition principle, because, even if there are two states $|\xi\rangle$ and $|\tilde{\xi}\rangle$ which happen to satisfy Eq. (3.1), generic superpositions of these states would not. The principle would have to be considered valid within the present formalism only as a certain type of approximation.

[21]Indeed, one of the most important characteristics of, say GRW or CSL theories, is their tendency to increase the localization of states in the sense of suppressing superpositions of states corresponding to rather different mass densities distributions in physical space. As such, it seems clear that a spontaneous collapse would generically imply an important change in the spatio-temporal form of the expectation value of the energy momentum tensor.

[22]One of the requirements is that a reasonable spontaneous collapse be such that, when starting with a state with a reasonably defined renormalized energy momentum tensor(i.e. a so called Hadamard state), the spontaneous collapse dynamics leads to a state with the same characteristics (i.e. another Hadamard state). That issue has been explored in [60]. A second problem is that the Hilbert space of

certainly become even more complex and vexing when considered in the context of a theory involving continuous spontaneous collapse such as CSL.

Next, let us consider how could all this possibly fit with our current views regarding an ultimate theory including quantum gravity. Let us start by recalling that the whole quantum gravity program is still confronted with various outstanding issues and conceptual difficulties. Among those we should mention (i) *The Problem of Time* [45] i.e. the fact that canonical approaches to quantum gravity lead to timeless theories, and (ii) the difficulties concerning the identification of suitable observables [46]. More generally, there is the issue (iii) of how to recover space-time and something resembling general relativity from various of the existing approaches, and specifically those of the canonical type.

Solutions to (i) often rely on the use of a dynamical variable as a physical clock and consider relative probabilities (and wave functions). Following that line seems to lead to something like an approximated version of the Schrödinger equation, but with corrections that violate unitarity (see [47]). Although tantalizing, it is not clear, however, that the specific form that such issue takes in the analysis of that work is of the kind that could lead to a resolution of the questions at hand here.

Regarding (iii) there are many suggestions indicating space-time might be an emergent phenomena (see for instance [48–50]). In that case, it is not clear that g_{ab}, as such, should be "quantized" any more than the heat equation should. Under such circumstances, the classical level of description might be the only setting in which notions of space-time may be talked about meaningfully.[23]

The point is that any talk about space-time concepts, in anything close to the standard sense, implies that one is already working within some classical description, as we simply have no idea of how to think of a quantum space-time. Therefore, even when considering that one might have started from some hypothetical satisfactory and fully workable theory of quantum gravity, by the time we reach the level of discussion where we can talk of space-time in the usual terms, we would have proceeded through a long chain of approximations and simplifications. Under those conditions, it does not seem unnatural to expect that some traces of the full quantum gravity regime might survive and remain relevant at the stage we are dealing with. Moreover, these might, from the perspective of the standard space-time language

the second SSC might not be unitarily equivalent to the first one. Even if they are, the unitary mapping might not be unique, making it difficult to identify the state in the new Hilbert space resulting from the spontaneous collapse theory applied to the state of the SSC previous to the collapse. Initial proposals to deal with this issue were discussed and implemented in simple situations in [61, 62], and further studies are under development [63].

[23]In thinking about the heat equation, it seems clear that, while at macroscopic effective level, heat flow is a concept that can be truly made sense of, when considering the situation at the more fundamental level, of say, many particle quantum mechanics, the notions involved would become, at best, secondary. Raising, for instance, a question regarding the "quantum operator" characterizing heat is unlikely to lead to any meaningful answer. Analogously, it might well be, although we have, of course, no proof one way of the other, that no sensible definition of a quantized space-time metric is truly compatible with whatever the fundamental quantum description of gravity is.

"look like spontaneous collapses".[24] Let us think again of the hydrodynamic analogy. Here, we might consider the situation where, say, foam forms after the breaking of an ocean wave, a situation that presumably poses no great problems (at least in principle) if we go all the way down to the description of what we call the fluid, in terms of molecular dynamics. It clearly would look rather strange if we attempt to incorporate some kind of phenomenological terms describing such effects within the Navier-Stokes formalism.

One might characterize the present approach to the consideration of gravity / quantum interface as an essentially bottom up approach, in contrast with the usual "top-down" approach, where one starts with what is presumably a well defined proposal for the full theory of quantum gravity, and then work towards establishing a connection between the formalism and the empirical world. The strategy adopted here starts by considering theories that are rather well understood, supported by substantial experimental evidence, specifically general relativity and quantum field theory treatment of matter fields (in this curved space-time version). Then we push their range of applicability towards the domain where presumably new physics might be required, seeking, in the process, to obtain clues about the features of the ultimate theory.

4 Practical Treatment Adapted to the Cosmological Setting

Trying to apply the above formalism in any specific concrete situation of interest, having no extraordinarily simplifying features, and many relevant degrees of freedom, is evidently an almost impossible task at the practical level.[25] We will, therefore, work by making several suitable simplifications, including using the simplest inflationary model where the potential[26] is just $V = (1/2)m^2\phi^2$; focusing our attention on the usage of the formalism in the inflationary cosmological setting, while trying to follow its basic rules. Furthermore, instead of constructing a new complete quantum field theory in a curved space-time that results from previous spontaneous collapses, we will keep using a single QFT theory construction, that corresponding to the background space-time. That is, we will consider jumps in the quantum state, and the corresponding changes in the space-time metric. However we will neglect the

[24] At this point, and taking a completely agnostic posture regarding what the fundamental theory of quantum gravity might look like, it is hard to offer anything beyond a simple analogy. Thus, one might want to consider a scientist trying to come to terms with, say, the formation of foam when a ocean wave breaks on the shore, while having no clue about the molecular nature of what he normally describes as a fluid using standard tools of hydrodynamics.

[25] In [61], the treatment was applied to the excitation of a single anisotropic mode in the space-time metric. The formalism was later applied [62] to the case of the consecutive excitation of a second mode, allowing the study of essential aspects of the generation of tensor modes.

[26] This specific potential is usually taken as disfavoured by observations [53]. More recent analysis claim quite generally that convex potentials are excluded at the 95% confidence levels [54]. As noted below, that conclusion does not extend to our approach.

requirement of simultaneously changing the Fock-Hilbert space, which will be kept fixed (i.e. we will keep using the Hilbert space constructed for the background SSC, namely the one corresponding to the situation before the breakdown of homogeneity and isotropy).

This simplified treatment seems justified by the smallness of the so called perturbations of the metric (characterized by the 10^{-5} deviations from isotropy observed in the cosmic microwave radiation), and the corresponding smallness of the modification of the Hilbert space. The second simplification is that, although the zero mode of the scalar field (i.e. the mode corresponding to $\vec{k} = 0$ and thus involving no spatial dependence) must, according to our formalism, be treated quantum-mechanically (as should all matter fields), we will describe it as a classical background in the practical calculations. The point is that the zero mode of the field will be taken, as described in Sect. 3.1, to be in a highly excited (and sharply peaked) state, and take $\phi_0(\eta)$ (of Sect. 2) as corresponding to the expectation value of the zero mode $\langle\hat{\phi}_0(\eta)\rangle$. The quantum treatment of the zero mode was included in the works [61, 62] mentioned above. That analysis indicates that, to the level of approximation one is working with, the final results are the same as those we will describe now. The space dependent modes will, just as in the standard approach, be treated quantum mechanically, and taken to start in the "vacuum state". However, the field in question is now the scalar field $\delta\phi$, rather than the composed field v (involving both, matter and metric perturbations) of Sect. 2.1.

It turns out to be convenient to make a change of variables in field space and work with the re-scaled field $y \equiv a(\eta)\delta\phi(\eta, \vec{x})$. The momentum canonical conjugate to that field in conformal time is $\pi = a\delta\phi'$. These objects are now treated according to the standard methods of quantum field theory on curved space-time [59]. In our case, that space-time background is provided by the metric (2.5) with the perturbations set to 0. Thus, the field operator can be represented by the operator as:

$$\hat{y}(\eta, \vec{x}) = \frac{1}{(2\pi)^{3/2}} \int d^3k \left(\hat{a}_{\vec{k}} y_{\vec{k}}(\eta) e^{i\vec{k}.\vec{x}} + \hat{a}^{\dagger}_{\vec{k}} y^{*}_{\vec{k}}(\eta) e^{-i\vec{k}.\vec{x}}\right), \tag{4.1}$$

with the modes $y_{\vec{k}}(\eta)$ chosen again according to the Bunch Davies prescription, so that the state satisfying $\hat{a}_{\vec{k}}|0\rangle = 0$ is the Bunch Davies vacuum.

The basic scheme of the analysis is now the following: During the early stages of inflation taken to correspond to $\eta = -\mathcal{T}$, the starting point of inflation (see Sect. 2), the state of the field $\hat{y}$ is the Bunch-Davies vacuum, and the space-time is homogeneous and isotropic. In that state, the operators corresponding to the Fourier components of the field and momentum conjugate ($\pi = a\delta\phi'$), $\hat{y}_k$, $\hat{\pi}_k$ are characterized by gaussian wave functions centered at 0 with uncertainties Δy_k and $\Delta\pi_k$. To the extent that the spontaneous collapse dynamics is ignored, and as we are working in the Heisenberg picture, the state of the field does not change with time.

The spontaneous collapse dynamics, treated using the interaction picture, modifies the evolution of the quantum state during the inflationary epoch, resulting in a change of the expectation values of $\hat{y}_k(\eta)$ and $\hat{\pi}_k(\eta)$. We will assume the spontaneous collapse occurs mode by mode,[27] and is described by some version of a spontaneous collapse theory, suitably adapted to the situation at hand (we will be more explicit in this point later on).

It is worth noting that, in this approach, it is clear that our universe would correspond to one specific realization of the stochastic objects or functions occurring in the spontaneous collapse dynamics. Moreover, we will consider that to each field mode $\vec{k}$ corresponds an individual and independent stochastic object. We will shortly see schematically how this is realized within the context of a specific theory.

Let us now consider the scalar metric perturbations $\psi(\eta, \vec{x})$, as well as the quantities of direct observational interest. We will do so in a schematic way ignoring at this point the late time physics, which we take as well understood, and which is usually incorporated by the introduction of so called "*transfer functions*" [51], which modify the results coming directly from the inflationary regime.

The Fourier decomposition of the relevant semi classical Einstein's Equations takes the form (see Eq. (64) of [7] in the limit of exponential expansion):

$$- k^2\psi(\eta)_{\vec{k}} = \frac{4\pi G\phi_0'(\eta)}{a}\langle\hat{\pi}(\vec{k}, \eta)\rangle = c\langle\hat{\pi}(\vec{k}, \eta)\rangle. \tag{4.2}$$

At $(\eta = -\mathcal{T})$ the state is the vacuum, an homogeneous and isotropic state, and as noted, in the absence of the spontaneous collapse part of the dynamics, that will remain the case forever. In particular we would have $\langle\hat{\pi}(\vec{k}, \eta)\rangle = 0$, and, therefore, the space-time would also be completely homogeneous and isotropic. The spontaneous collapse will change that, so that by the end of inflation those expectation values will generically differ from zero. Note that from Eq. (4.2) we can reconstruct the space-time Newtonian potential simply by taking $\psi(\eta, \vec{x}) = \frac{1}{(2\pi)^{3/2}} \int d^3k\psi_{\vec{k}}(\eta)e^{i\vec{k}.\vec{x}}$. Of course we have to keep in mind that, if we are interested in the regimes that are more directly empirically accessible, we must consider the evolution of the physical situation from the end of inflation at $\eta = \eta_0$ through the reheating epoch and standard radiation and matter dominated eras of the hot Big Bang. We will be rather schematic in considering those aspects.

Let us focus on the quantity of main observational interest $\frac{\Delta T(\theta,\varphi)}{\bar{T}}$. It is the relative deviation from the sky mean of the temperature of the CMB, coming from a certain direction in the sky, specified by the angles θ, φ, and corresponding to the point on the intersection of our past light cone with the last scattering surface (i.e. the hypersurface corresponding to the moment the hot cosmic plasma cools sufficiently for hydrogen atoms to form, and photons to effectively decouple) at $(\eta = \eta_D)$. This is related to the fact that photons emanating from that region undergo, besides the

[27] There are two arguments that seem to justify such assumption: on the one hand, GRW and CSL dynamics seem to function in that manner in situations involving multiple degrees of freedom. More importantly, as we are dealing with changes in the state that could be regarded as very small perturbations, it seems clear that the linear level treatments should be accurate enough, and that the kind of correlation generating collapses would only occur in higher order treatments. Indeed, we have seen an indication of that kind of correlation-generation occuring in the second order treatment carried out in [62].

cosmological red shift associated with the universe's expansion, an additional red-shift as they overcome a local gravitational barrier characterized by the value, at the emission event, of Newtonian potential ψ. Including effects of plasma physics, that quantity is then given by:

$$\frac{\Delta T(\theta,\varphi)}{\bar{T}} = (1/3)\psi(\eta_D, R_D, \theta, \varphi) = \sum_{lm} \alpha_{lm} Y_{lm}(\theta,\varphi) \tag{4.3}$$

where the last expression provides the decomposition of the sky map into spherical harmonics and defines the coefficients α_{lm}.

In our approach, we can therefore directly obtain the expression:

$$\frac{\Delta T(\theta,\varphi)}{\bar{T}} = (c/3)\int d^3k e^{i\vec{k}\cdot\vec{x}} \frac{1}{k^2}\langle\hat{\pi}(\vec{k},\eta_D)\rangle, \tag{4.4}$$

where the Newtonian potential is evaluated on the η_D, the conformal time corresponding to the decoupling surface (also known as the surface of last scattering), and at the co-moving radius R_D of that surface intersection with our past light cone at the corresponding direction in the sky. The specific version and implementation of the spontaneous collapse theory, as well as the specific realizations of the stochastic processes involved, characterize the quantity $\langle\hat{\pi}(\vec{k},\eta_D)\rangle$. This, of course, depends also on the part of the evolution that does not directly tied to the spontaneous collapse dynamics (i.e. that tied to the scalar field free Hamiltonian in the given background space-time).

Thus we find,

$$\alpha_{lm} = c\int d^2\Omega Y^*_{lm}(\theta,\varphi)\int d^3k e^{i\vec{k}\cdot\vec{x}} \frac{1}{k^2}\langle\hat{\pi}(\vec{k},\eta)\rangle. \tag{4.5}$$

We note that we can not really extract a direct prediction from this expression, simply because the complex quantities $\langle\hat{\pi}(\vec{k},\eta)\rangle$ are determined by stochastic processes (one for each $\vec{k}$). However, we would obtain an explicit prediction for each angle, if we knew the result of all such stochastic processes. It is worth noting that no analogous to this expression exists in the standard approaches. Those simply do not offer an expression, not even in principle, for this quantity, which is actually that of direct observational interest. The above quantity, in the usual approach, would actually be simply zero.

Following the present approach, we are in a better situation, at least in principle, but how can this approach produce actual predictions? The point is that the Eq. (4.5) shows that the quantity of interest is the sum over a large number (actually an integral) of stochastically determined complex quantities (one for each $\vec{k}$). So the quantity of interest can be thought of as a result of a "random walk" on the complex plane.

As it is usually the case in such situations, one cannot predict the end point of such "random walk", but one can focus on the equivalent to the magnitude of the "total displacement", $|\alpha_{lm}|^2$, and estimate its most likely value, which we denote $|\alpha^{(ML)}_{lm}|^2$.

We will evaluate the latter by identifying it with that corresponding to the ensemble average over the possible realizations of the set of stochastic processes.

Thus, we proceed to compute the ensemble average (represented by an overline) at "late times". The relevant quantity is then:

$$\overline{(\langle\hat{\pi}(\mathbf{k},\eta)\rangle\langle\hat{\pi}(\mathbf{k}',\eta)\rangle^*)} = f(k)\delta(\mathbf{k}-\mathbf{k}'). \tag{4.6}$$

where we have used the fact that the different modes are taken as statistically independent,[28] and that, at the ensemble level, we are dealing with an isotropic system, even though each individual element of the ensemble (and, in particular, the realization that actually corresponds to our universe) is **not** isotropic.

Therefore, we are led to the following estimate,

$$|\alpha_{lm}^{(ML)}|^2 = \overline{|\alpha_{lm}|^2} = (4\pi c/3)^2 \int_0^\infty dk j_l(kR_D)^2 \frac{1}{k^2} f(k). \tag{4.7}$$

Agreement with observations requires $f(k) \sim k$ (which would correspond to a Harrison-Zeldovich scale free power spectrum $\mathcal{P}(k) \sim k^{-3}$). Note that, if this was actually the exact form of the spectrum, we would have that $|\alpha_{lm}^{(ML)}|^2$ would end up being independent of R_D (changing variables to $z \equiv R_D k$, the LHS of (4.7) would take the form $(4\pi c)^2 \int_0^\infty dz j_l(z)^2 \frac{1}{z} = (4\pi c)^2 \frac{\pi}{l(l+1)}$), reflecting the "scale invariance" of this particular spectral shape. This would result in a statistically featureless CMB, which is not what is really observed. The very interesting and famous oscillations that have been the focus of the recent CMB studies, are thought to be the result of features generated by late time physical process acting on top of the primordial flat spectrum emerging from the inflationary regime.[29] The point, however, is that, when on top of the flat spectrum one places the effects of the late time physics, including acoustic plasma oscillations, all of which is encoded in the so called "transfer functions" [51], that we are ignoring (for simplicity), one obtains, after fixing a few free parameters a spectrum that represents a remarkable match to the observations.[30]

A detailed analysis of the problem based on one of the simplest inflationary models (a single scalar field with a simple quadratic potential term), together with a version of CSL adapted to the situation at hand (i.e. one involving quantum fields in a cosmological setting) has been performed in [38].

---

[28] As we noted before, at higher order in perturbation theory, within the approach developed here and based on the SSC formalism, one naturally encounters deviations from statistically independence [62]. An early exploration of the possible consequences of the kind of correlations that naturally emerge include possible modifications in the part of the spectrum that refers to the very large angular scales [72]. See also [71].

[29] There is an additional feature coming from the inflationary regime itself, known as the tilt in the spectrum, which we will be ignoring in the present simplified treatment.

[30] This success after the best fit matching of few parameters, is achieved both, when following the standard accounts [54] for the primordial spectrum, and when following ours [55].

The starting point is the version of the theory described by the following two equations: The modified Schrödinger equation, whose solution is:

$$|\psi, t\rangle = \hat{\tau} e^{-\int_0^t dt' \left[i\hat{H} + \frac{1}{4\lambda}[w(t') - 2\lambda \hat{A}]^2 \right]} |\psi, 0\rangle \tag{4.8}$$

where $\hat{\tau}$ is the time-ordering operator, λ is a parameter of the theory, $\hat{A}$ is a self adjoint operator on the system's Hilbert space (usually referred to as the "collapse-driving-operator" or "collapse-operator" for short), and $w(t)$ is a random classical function of time, of white noise type, with a probability rule given by the equation,

$$PDw(t) \equiv \langle \psi, t | \psi, t \rangle \prod_{t_i=0}^{t} \frac{dw(t_i)}{\sqrt{2\pi\lambda/dt}}. \tag{4.9}$$

The state vector norm evolves dynamically (does *not* equal 1), so expectation values, such as those needed in expressions such as (4.5), must be computed with suitably re-normalized states.

The version of the theory adapted to the cosmological case at hand was based on an equation of the form:

$$|\psi, t\rangle = \hat{\tau} e^{-i\int_{-\mathcal{T}}^{\eta} d\eta' \hat{H} - \frac{1}{4\lambda}\int_{-\mathcal{T}}^{\eta} d\eta' \int d\mathbf{x}[w(\mathbf{x},\eta') - 2\lambda \tilde{y}(\mathbf{x})]^2} |\psi, -\mathcal{T}\rangle. \tag{4.10}$$

where the operator playing the roll of the collapse operator[31] $\hat{A}$ of Eq. (4.8) was taken to be linear in the field $\hat{y}(\eta, \vec{x})$ (the quantum field operator corresponding to the re-scaled field $y = a(\eta)\delta\phi$). The result of the analysis indicated that the choice for a collapse operator that leads to results compatible with the scale free HZ spectrum was

$$\tilde{y}(\mathbf{x}) \equiv (-\nabla^2)^{1/4} \hat{y}(\mathbf{x}) \tag{4.11}$$

The study also considered the alternative where the operator playing the roll of the collapse operator $\hat{A}$ of Eq. (4.8) was taken to be linear in the field $\hat{\pi}(\eta, \vec{x})$ (the momentum conjugate to $\hat{y}$ and given by $\pi(x) = a(\eta)\frac{\partial\delta\phi}{\partial\eta}$) so that the evolution was given by the equation:

$$|\psi, \eta\rangle = \hat{\tau} e^{-i\int_{-\mathcal{T}}^{\eta} d\eta' \hat{H} - \frac{1}{4\lambda}\int_{-\mathcal{T}}^{\eta} d\eta' \int d\mathbf{x}'[w(\mathbf{x}',\eta') - 2\lambda \tilde{\pi}(\mathbf{x}')]^2} |\psi, -\mathcal{T}\rangle. \tag{4.12}$$

In this case, the collapse operator leading to adequate results turned out to be $\tilde{\pi}(\mathbf{x}) \equiv (-\nabla^2)^{-1/4}\hat{\pi}(\mathbf{x})$.

We do not know, at this point why are those particular choices of collapse operators the ones that work in this situation. Actually, it is clear that we need a general recipe for extending the spontaneous collapse theories that have been developed for the context of non-relativistic many particle quantum mechanics to more general settings

[31]The operator that drives the spontaneous collapse dynamics of the theory.

involving quantum fields, as well as special and general relativity. The general form of the collapse operator should be framed in such terms, and should, of course, reduce in the former context to the smeared position operators that have proven to successfully deal with the measurement problem, in the corresponding settings. This is clearly, at this stage of the research program, an open task. However, we might note that these seem to be rather natural choices, dimensionally speaking, in the sense that the constant $\tilde{\lambda}$, appearing in the above equations is of the correct dimensionality (i.e. s^{-1}). For further discussion on this point see [38].

Moreover, as shown in [38], the specific resulting prediction for the power spectrum is:

$$\mathcal{P}_S(k) \sim (1/k^3)(1/\epsilon)(V/M_{Pl}^4)\tilde{\lambda}\mathcal{T} \tag{4.13}$$

where $\mathcal{T}$ is the duration of the inflationary stage in conformal time taken for standard inflationary parameters as 10^8 MpC, V is the starting value of the inflationary potential, and ϵ is the slow roll parameter which is known to lead to a slight amplification of the spectrum even, in the context of spontaneous collapse theories [52]. When using standard estimates for the inflationary model, including the GUT scale for the inflation potential, and standard values for the slow-roll, the result of the detailed calculation leads to agreement with observation if one sets $\tilde{\lambda} \sim 10^{-5} MpC^{-1} \approx 10^{-19}$ s^{-1}. This is not very different from the GRW suggestion (a standard characteristic GRW value for this quantity in many particle non-relativistic applications of GRW and the corresponding one in CSL is often taken to be 10^{-17} s^{-1}). We find that result rather encouraging, in the sense that it provides hope for the existence of a general CSL like theory capable of simultaneously dealing with the present problem, and reducing to the standard versions of spontaneous collapse theories in the regimes appropriate to non relativistic many particle quantum mechanics, and thus adequate for the laboratory situations described elsewhere in this volume.

We should point out that treatments based on CSL adapted to the inflationary cosmology problem, but based on rather different specific implementations, have been carried out by other groups [39, 40].

4.1 Primordial Gravity Waves

We have seen that, while spontaneous collapse theories have, in principle, the features that allow them to resolve the very serious shortcoming of traditional inflationary accounts for the emergence of the seeds of cosmic structure out of quantum uncertainties in the early universe, a detailed qualitative and quantitatively successful treatment which is empirically adequate, requires rather specific features. In the preceding subsection we adjusted those so that the emergent "predictions" matched observations. This might be regarded as a search for clues of how a theory that was developed with the non-relativistic many particle quantum mechanics settings in mind could be extended to work in contexts involving quantum field theory in curved space-time, and, in particular, the cosmological contexts. It is clear we do not, at this time, have a general proposal for the universal form of such theory. However, once this is done, one can go beyond such adjustments and consider truly novel predictions. In particular, we will next concentrate on the generation, by the same mechanism, of the so called primordial gravity waves, also known as primordial tensor modes. Such waves are, indeed, also a generic prediction of the standard approaches. Extensive efforts are currently under way to detect the traces such gravity waves would leave in the CMB. The effects in question are expected to be observable in a certain type of anisotropy in the polarization patterns of the CMB radiation, known as the polarization B modes[32] [51]. The results of the search for such primordial B modes have, so far, failed to find any clear evidence of their existence at the levels that are expected from the simplest, and otherwise more compelling specific inflationary models. Empirical bounds on their amplitude are currently being employed to rule out many specific proposals [66]. We will show here that, regarding this specific issue, the results from the approach outlined in this review are rather different from those of the standard accounts. The former predict a much smaller amplitude for these primordial gravity waves than the usual approach [68], thus dramatically altering the conclusions regarding the viability of most inflationary models.

Our starting point for this calculation is again the corresponding component of the semi-classical version of Einstein's Eq. (3.1). This is the equation of motion for the tensor perturbations h_{ij}, which under the conditions we have previously set, and retaining only dominant terms, takes the form:

$$(\partial_0^2 - \nabla^2)h_{ij} + 2(\dot{a}/a)\dot{h}_{ij} = 16\pi G \langle (\partial_i \delta\phi)(\partial_j \delta\phi)\rangle_{Ren}^{tr-tr} \tag{4.14}$$

$tr - tr$ stands for the transverse trace-less part of the expression. The fact that the quantity has been subjected to a standard renormalization is indicated in the suffix (Ren) (for in-depth discussions of this see [59]).

We have retained in the right hand side just the largest non-vanishing term in the perturbation expansion of $\langle T_{\mu\nu}\rangle$, which, in this case, is quadratic in the collapsing quantities. The point is that the energy momentum tensor, for a field with simple quadratic potential, is simply quadratic in the field. As we have seen, the field might be written as $\phi(\eta, \vec{x}) = \phi_0(\eta) + \delta\phi(\eta, \vec{x})$. The part containing the relevant spatial dependences (leading to actual gravity waves) involves two kinds of terms (**a**) the term linear in ϕ_0 and linear in $\delta\phi$ and (**b**) the term quadratic in $\delta\phi$. When focusing on the component $\langle T_{00}\rangle$, the dominant contribution comes from terms of type (**a**), but those are absent when considering the component $\langle T_{ij}\rangle$ $(i \neq j)$ simply because ϕ_0, does not depend on $\vec{x}$ and thus $\frac{\partial}{\partial x^i}\phi_0 = 0$.

Passing to a Fourier decomposition, we need to solve the following equation,

$$\ddot{\tilde{h}}_{ij}(\vec{k}, \eta) + 2(\dot{a}/a)\dot{\tilde{h}}_{ij}(\vec{k}, \eta) + k^2\tilde{h}_{ij}(\vec{k}, \eta) = S_{ij}(\vec{k}, \eta), \tag{4.15}$$

[32]The other polarization modes, the so called E modes in the anisotropy in the polarization patterns, are expected to arise from well understood late time plasma physics effects, and have, in fact, been observed in complete accordance with expectations.

with vanishing initial data (i.e. $h_{ij} = \dot{h}_{ij} = 0$ on the initial hypersurface $\eta = -\mathcal{T}$), and source term given by:

$$S_{ij}(\vec{k}, \eta) = 16\pi G \int \frac{d^3x}{\sqrt{(2\pi)^3}} e^{i\vec{k}\vec{x}} \langle(\partial_i \delta\phi)(\partial_j \delta\phi)\rangle^{tr-tr}_{Ren}(\eta, \vec{x}). \tag{4.16}$$

As it is well known, general relativity has a well posed initial value formulation, thus once the gauge is fixed, the solution to the evolution equation is completely determined by the initial data (which are vanishing in this case), as well as the source terms.

Again, the quantity of interest is the following average over the ensemble of realizations of the stochastic processes, expressed as:

$$\overline{h_{ij}(\vec{k})h_{kl}(\vec{k}')} = \delta^{tr-tr}_{ikjl}(\vec{k})(2\pi)^3 \delta^3(\vec{k} - \vec{k}')\mathcal{P}_h(k) \tag{4.17}$$

where the symbol $\delta^{tr-tr}_{ikjl}(\vec{k})$ is 1 for the index structure compatible with the transverse traceless nature of the gravity waves, and the fact that distinct modes are uncorrelated and 0 otherwise. This expression can be taken as the adapted definition to that in Eq. (2.9) when considering the indexed quantity $h_{ij}(\vec{k})$, and thus defining the tensor mode power spectrum.

The calculation is rather involved, and the details can be found in [68]. The result turns out to be formally divergent, involving an integral over pairs of modes $\vec{q}, \vec{p}$ such that $\vec{p} + \vec{q} = \vec{k}$ arising from the term $\langle \partial_i \delta\phi(\eta, \vec{x})\rangle \langle \partial_j \delta\phi(\eta, \vec{x})\rangle$ (evaluated on the state resulting from the spontaneous collapse), which is the leading contribution left after the renormalization.[33] However, there are various clear physical reasons indicating that we must introduce an ultra-violet (or short wavelength) cut-off p^C_{UV} on the integral. In particular, we must consider the diffusion effects at late times, which would affect the gradients of density perturbations on very short scales, to the point of effectively damping their contribution to the generation of the gravity waves in question. The fact that other physical sources for a cut-off (for instance that which could be expected to result from GUT scale physics effects during the radiation domination) correspond to higher values of k suggests that we should take the cut-off at the scale of diffusion (Silk) dumping with $p^C_{UV} \approx 0.078$ MpC^{-1}.

After a long calculation, the resulting prediction for the power spectrum of the tensor perturbations is:

$$\mathcal{P}_h(k) \sim (1/k^3)(V/M^4_{Pl})^2(\tilde{\lambda}^2 \mathcal{T}^4 p^5_{UV}/k^3) \tag{4.18}$$

We note that the relation between the above power spectrum and that for the scalar perturbations given in Eq. (4.13) indicates that the latter is substantially smaller than

[33]This divergence is not the same as the standard divergences occurring in the evaluation of the expectation value of the energy momentum tensor. That is dealt with via a well established renormalization procedure, which in this setting corresponds to the so called "minimal subtraction" based on the Bunch Davies vacuum.

the former. Indeed, by comparing Eqs. (4.18) and (4.13), in terms of dimensionless quantities, we have that $\mathcal{P}_h(k)k^3 = \epsilon^2(\mathcal{P}_S(k)k^3)^2(\mathcal{T}^2 p_{UV}^5/k^3)$. Given that the scalar power spectrum is extremely small for the relevant values of k, it is clear that the tensor power spectrum is suppressed by a huge factor.

That is very different from the standardly obtained relation between them, which indicates that $\mathcal{P}_h(k) = r\mathcal{P}_S(k)$ with $r = 16\epsilon$ [70], (i.e. it is dictated by the slow roll parameter ϵ, which is, in turn, related to other observables, such as the scalar spectral tilt). Thus, in contrast with the expectations of the standard approach, we do not expect to see primordial tensor modes (and the corresponding polarization B modes) at the level they are currently being looked for. In this regard, our approach can be said to be empirically more adequate than the standard one, which, with the exception of few specific inflationary models, indicates that a detection should already have been made [67].

In [68], we have also considered a simpler spontaneous collapse model (the naive one designed in [7], with just this specific cosmological application in mind), and again obtained a **substantially reduced tensor mode amplitude**, but with a slightly, different shape. We take this, together with the general discussion before Eq. (4.14), as an indication of the robustness of the generic prediction of a substantial reduction for amplitude of the primordial gravity waves spectrum, in comparison with that of standard approaches. Further studies considering multiple specific inflationary models have been used to determine their viability within the present context [69].

5 Discussion

We have seen that despite its phenomenological success, the usual inflationary account for the emergence of the seeds of cosmic structure out of primordial quantum fluctuations of the vacuum suffers from a serious conceptual flaw: it is unable to account for the transition from a completely homogeneous and isotropic situation, as described both by the classical background and the quantum mechanical state of matter fields, to one that is not.

We have argued that spontaneous collapse theories contain, in principle, the elements needed to deal with that conceptual shortcoming. We have provided a theoretical framework whereby such spontaneous collapse theories might be incorporated within a semi-classical treatment of gravitation, and argued how that view might be reconciled, as a suitable effective treatment, with more traditional expectations regarding the nature of quantum gravity.

We have then proceeded to employ such approach in a simplified fashion within the context of inflationary cosmology, and obtained predictions for the primordial spectrum of both the scalar and tensor perturbations. We saw that by a suitable choice of collapse operators, using a simple inflationary model, with typical values of the inflationary parameters, the spontaneous collapse constant $\tilde{\lambda}$ of the theory can be adjusted so as to fit the scalar perturbation observations with values that are of a similar order of magnitude as those considered in the context of spontaneous collapse theories as applied to laboratory situations.

Furthermore, we have seen that, once such adjustments have been made, the approach leads to specific predictions for the tensor mode spectrum that differ substantially from the traditional expectations in their regard. Actually, according to this approach, the tensor perturbations occur only at a higher order in perturbation theory, and thus imply a substantially reduced amplitude, which naturally accounts for the lack of their experimental detection so far. Moreover, with the present approach, the predicted tensor spectrum has a mush steeper shape, indicating the the possibility of detection increases as one looks at longer wave-lengths. If such expectations were confirmed, this might represent the first case where spontaneous collapse theories lead to different predictions than those of the usual practice in quantum theory, which are such that the former are actually empirically preferred.

The above listed findings illustrate, not only the accuracy of J. Bell's observations concerning the need for cosmologists to become concerned about the conceptual problems surrounding quantum theory, but also the fact that present theoretical frameworks dealing with the early stages of our cosmological models, namely the quantum aspects of inflation, are actually inadequate without such a solid quantum theoretical foundation. We have seen that the incorporation of spontaneous collapse theories into the setting provides both a conceptual and an effective path for addressing such issues. We have also shown that some of the actual predictions naturally emerging from the approach differ from the standard ones in a manner that is favored by current observations.

A related development along this line of research is the analysis, within the context of quantum field theory of scalar fields, of the kind of collapse operators that have the property of maintaining the renormalizability of the expectation value of the energy momentum tensor. In other words, a characterization of spontaneous collapse theories for which the dynamics preserves the Hadamard properties of the quantum state [75]. Other works include an analysis showing how correlations arising from the spontaneous collapse dynamics could naturally account for an anomalous low power in the scalar CMB spectrum [71] at large angles [72]; the proposal of a speculative scenario based on spontaneous collapse dynamics that could dynamically account for the apparently spacial conditions characterized in Penrose's hypothesis[73] regarding the Weyl curvature of the initial state of the Universe [80]; the deveopment of a scheme whereby spontanous collapse theories could restore the viability of Higgs inflation [74]. Furthermore, we should mention a recent proposal that ties the violation of energy conservation characteristic of spontaneous collapse theories (as well as of other approches to deal with the "measurement problem" [81]) to the small current value of the cosmological constant [82], and a more recent refinement of that idea connecting the violation of energy momentum conservation to a space-time discreteness associated with quantum gravity which does a superb job in predicting the correct magnitude [83]. It must be said however that the latter is, at this time, not explicitly connected to spontaneous collapse dynamics. This last feature might decrease one's enthusiasm for some of the ideas discussed in this manuscript. However, this need not be so. The point is that it is not unreasonable to expect that the two might be connected, because, as argued in [41, 44], spontaneous collapse might be ultimately tied to quantum aspects of gravitation. Furthermore, the kind of connection that would be required in this specific context (i.e. one between some sort of space-time granularity and energy momentum non-conservation with spontaneous collapse theories), is, in my view, made plausible by the simultaneous consideration of a *flash type ontology*, the approach employed in [84] for the construction of a relativistic spontaneous collapse theory, and the proposal of [87] to incorporate intrinsic diffusion into the spontaneous collapse theories. Finally, it should be mentioned that we have applied spontaneous collapse theories to deal with a different problem involving the interface of quantum theory and gravitation, namely the so called "black hole information puzzle" [57, 58, 76–79] resulting in what we see as an overall self consistent and reasonable picture of the situation.

Needless is to say that there remain many issues requiring a deeper study and substantial development. Among those is the construction of a universal version of the spontaneous collapse theory that is applicable in general situations, including those pertaining to the laboratory conditions on which spontaneous collapse theories have been traditionally considered, as well as regimes such as cosmology, where they need to coexist with general relativity and quantum field theory. As far as the inflationary context is concerned, we made some adjustments in the theory particularly regarding the choice of the field operators that play the role of the collapse operators (or equivalently, the dependence of parameter on the mode's comoving wave number k) which we found to be dimensionally appropriate, but which we, at this point, could not otherwise justify in a clear manner. In this regard, the continuous search for versions of spontaneous collapse theories that are fully compatible, not just with special relativity (where there is already noteworthy progress [84–86]), but also with general relativity seems as an obligated path for future investigations.

More broadly speaking, the discussion presented in this manuscript offers support for my strong conviction that ignoring "the measurement problem", when discussing issues at the Gravity/Quantum interface, can be a serious source of confusion. The incorporation of proposals to seriously address it, such as spontaneous collapse theories, the subject to which G. C. Ghirardi contributed so dramatically to create, develop, and establish, will not just help in clarifying the overall physical picture, but, has the potential to contribute to the resolution of seemingly unconnected problems.

Acknowledgments I gladly acknowledge very useful discussions with A. Bassi, A. Perez, T. Maudlin and several other colleagues. I must specially thank G. León for carefully reading this manuscript and suggesting many important improvements. I must also be grateful to an anonymous referee who meticulously examined the manuscript, helping to remove multiple errors, and to clarify many delicate points. This work was supported by grant number FQXI-MGA-1920 from the Foundational Questions Institute and Fetzer Franklin Fund, a donor advised fund of Silicon Valley Community Foundation.

References

1. J. Bell, "Quantum mechanics for cosmologists", in *Speakable and unspeakable in quantum mechanics* (Cambridge U. Press 1987).
2. M. Gell-Mann, & J. B. Hartle, "Quantum Mechanics in the Light of Quantum Cosmology", e-Print: arXiv:1803.04605 [gr-qc]; J.B. Hartle "Generalizing quantum mechanics for quantum gravity", *Int. J. Theor. Phys.* **45**, 1390-6 (2006); J. B. Hartle, "Generalizing quantum mechanics for quantum space-time", Contributedto 23rd Solvay Conference in Physics: The Quantum Structure of Space and Time, Brussel, Belgium, 1–3 Dec. 2005, published in *The Quantum Structure of Space and Time*, pp. 21, (Brussels, 2005)
3. R.B. Griffiths, "Consistent histories and the interpretation of quantum mechanics", *J. Stat. Phys.* **36**, 219 (1984); "The consistency of consistent histories: A reply to d'Espagnat", *Found. Phys.* **23**, 1601(1993). R. Omnès, *J. Stat. Phys.* **53**, 893 (1988); R. Omnès, *The Interpretation of Quantum Mechanics*, Princeton University Press,Princeton (1994).M. Gell-Mann & J.B. Hartle in *Complexity, Entropy, and the Physics of Information*, SFI Studies in the Sciences of Complexity, Vol. VIII, (W. Zurek ed.), Addison Wesley, Reading (1990); "Classical equations for quantum systems", *Phys. Rev. D* **47**, 3345 (1993).
4. A. Kent, "Consistent sets contradict", *Phys. Rev. Lett.* **78**, 2874 (1997).
5. E. Okon & D. Sudarsky, "On the Consistency of the Consistent Histories Approach to Quantum Mechanics", *Found. Phys.* **44**, 19 (2014).
6. A. Guth, "Inflationary Universe: A Possible Solution to the Horizon and Flatness Problems", *Phys. Rev. D* **23**, 347 (1981).
7. A. Perez, H. Sahlmann & D. Sudarsky, "On the quantum origin of the seeds of cosmic structure", *Class. Quant. Grav.* **23**, 2317 (2006) [gr-qc/0508100].
8. P. Pearle, "Reduction of the state vector by a nonlinear Schrödinger equation", *Phys. Rev. D* **13**, 857 (1976).
9. P. Pearle, "Towards explaining why events occur", *Int. J. Theor. Phys.* **18**, 489 (1979).
10. G. Ghirardi, A. Rimini, T. Weber, "A model for a unified quantum description of macroscopic and microscopic systems," in *Quantum Probability and Applications*, pp. 223 A. L. Accardi (ed.), Springer, Heidelberg (1985).
11. G. Ghirardi, A. Rimini, T. Weber, "Unified dynamics for microscopic and macroscopic systems", *Phys. Rev. D* **34**, 470 (1986).
12. P. Pearle, "Combining stochastic dynamical state-vector reduction with spontaneous localization", *Phys. Rev. A* **39**, 2277 (1989).
13. G. Ghirardi, P. Pearle, A. Rimini, "Markov-processes in Hilbert-space and continuous spontaneous localization of systems of identical particles", *Phys. Rev. A* **42**, 7889 (1990).
14. P. Pearle, "Collapse models", arXiv: quant-ph/9901077.
15. A. Bassi & G. Ghirardi, "Dynamical reduction models", *Phys. Rep.* **379**, 257 (2003).
16. M. A. Amin, M. P. Hertzberg, David I. Kaiser, & J. Karouby "Nonperturbative dynamics of reheating after inflation: A review", *Int. J. Mod. Phys. D* **24**, 11530003 (2015).
17. S. Weinberg, "Cosmology", p. 476 (Oxford University Press, New York, 2008).
18. V. Muckhanov, "Physical Foundations of Cosmology", p. 348 (Cambridge University Press, Cambridge, 2005).
19. See Eq. 3.2.3 of R. M. Wald, *General Relativity* (University of Chicago Press, 1984).
20. J.M. Bardeen, "Gauge Invariant Cosmological Perturbations", *Phys. Rev. D* **22**, 1882 (1980); M.Bruni, P. K. S. Dunsby & G. F. R. Ellis "Cosmological Perturbations and the Physical meaning of Gauge Invariant Variables", *Astrophys. J.* **395**, 34 (1992).
21. Planck 2018 results. X. Constraints on inflation Planck Collaboration (Y. Akrami *et al.*), arXiv:1807.06211 [astro-ph.CO]
22. S. Landau, G. León & D. Sudarsky, "Quantum Origin of the Primordial Fluctuation Spectrum and its Statistics", *Phys. Rev. D* **88**(2), 023526 (2013), arXiv:1107.3054 [astro-ph.CO].
23. E. Harrison, " Fluctuations at the Threshold of Classical Cosmology", *Phys. Rev. D* **1**, 2726 (1970).

24. Y. B. Zeldovich "A Hypothesis Unifying Structure and Entropy of the Universe", *Mon. Not. R. Astr. Soc.* **160** 1P–3P (1972).
25. T. Maudlin, "Three measurement problems", *Topoi* **14** (1995).
26. C. Kiefer & D. Polarski, "Why do cosmological perturbations look classical to us?", arXiv:0810.0087 [astro-ph].
27. J. J. Halliwell, "Decoherence in Quantum Cosmology", *Phys. Rev. D* **39**, 2912 (1989); C. Kiefer, "Origin of classical structure from inflation", *Nucl. Phys. Proc. Suppl.* **88** 255 (2000). D. Polarski & A.A. Starobinsky, "Semiclassicality and decoherence of cosmological perturbations" *Class. Quant. Grav.* **13**, 377 (1996); W.H. Zurek, "Environment induced super selection in cosmology", in Moscow 1990, Proceedings, Quantum Gravity, pp. 456–472; R. Branderberger H. Feldman & V. Mukhanov, "Theory of cosmological perturbations. Part 1. Classical perturbations. Part 2. Quantum theory of perturbations. Part 3. Extensions", *Phys. Rep.* **215**, 203 (1992); R. Laflamme & A. Matacz, "Decoherence functional inhomogeneities in the early universe", *Int. J. Mod. Phys. D* **2**, 171 (1993); M. Castagnino & O. Lombardi, "The Selfinduced approach to decoherence in cosmology", *Int. J. Theor. Phys.* **42**, 1281 (2003); F. C. Lombardo & D. Lopez Nacir, "Decoherence during inflation: The Generation of classical inhomogeneities", Phys. Rev. D **72**, 063506 (2005); J. Martin, "Inflationary cosmological perturbations of quantum-mechanical origin", *Lect. Notes Phys.* **669**, 199 (2005); L.P. Grishchuk & J. Martin, "Best unbiased estimates for the microwave background anisotropies", *Phys. Rev. D* **56**, 1924 (1997); A.O. Barvinsky *et al.* "Decoherence in quantum cosmology at the onset of inflation", *Nucl. Phys. B* **551**, 374 (1999); J. Lesgourgues, D. Polarski & A. A. Starobinsky, "Quantum to classical tran-sition of cosmological perturbations for non vacuum initial states", [e-Print: gr-qc/961101904030].
28. D. Sudarsky, "Shortcomings in the Understanding of Why Cosmological Perturbations Look Classical," *Int. Jour. of Mod. Phys. D* **20**, 509 (2011); arXiv:0906.0315 [gr-qc].
29. D. N. Page & C. D. Geilker, "Indirect Evidence for Quantum Gravity", *Phys. Rev. Lett.* **47**, 979 (1981).
30. N. Huggett & C. Callender, "Why Quantize Gravity (Or Any Other Field for That Matter)?", *Phil. Sci.* **68**, No 3, S382 (2001).
31. J. Mattingly, "Is Quantum Gravity Necessary?", p. 325 in *The Universe of General Relativity*, (eds. Kox, A. J. & Eisenstaedt, J., Birkhäuser , 2005).
32. J. Mattingly, "Why Epply and Hannah's thought experiment fails?", *Phys. Rev. D*, **73** 064025 (2006).
33. S. Carlip "Is Quantum Gravity Necessary?", *Class. Quant. Grav.* **25**, 154010 (2008).
34. See section 1.1 of A. Perez, "The Spin Foam Approach to Quantum Gravity," *Living Rev. Rel.* **16**, 3 (2013); arXiv:1205.2019 [gr-qc].
35. See page 348 of R. M. Wald, *General Relativity* (University of Chicago Press, 1984).
36. N. Gouda & M. Sasaki, "Evolution of Gauge Invariant Cosmological Density Perturbations Through Decoupling Era", *Prog. Theor. Phys.* **76** 1036 (1986).
37. V.F. Mukhanov, "Quantum Theory of Gauge Invariant Cosmological Perturbations", *Sov. Phys. JETP67* , 1297 (1988); [Zh. Eksp. Teor. Fiz. 94 N7, 1 (1988 ZETFA,94,1-11.1988)].
38. P. Cañate, P. Pearl, & D. Sudarsky, "CSL Quantum Origin of the Primordial Fluctuation", *Phys. Rev. D*, **87**, 104024 (2013); arXiv:1211.3463 [gr-qc]
39. J. Martin, V. Vennin & P. Peter, "Cosmological Inflation and the Quantum Measurement Problem", *Phys. Rev. D* **86**, 103524 (2012).
40. S. Das, K. Lochan, S. Sahu & T. P. Singh, "Quantum to classical transition of inflationary perturbations: Continuous spontaneous localization as a possible mechanism", *Phys. Rev. D* **88**, 085020 (2013).
41. L. Diosi, "Gravitation and Quantum Mechanical Localization of Macro-Objects", *Phys. Lett. A* **105**, 199–202 (1984).
42. J. R. van Meter, "Schrodinger-Newton 'collapse' of the wave function", *Class. Quant. Grav.* **28**, 215013 (2011); arXiv:1105.1579.
43. D. Giulini, & A. Grobardt "The Schrodinger-Newton equation as a non-relativistic limit of self-gravitating Klein-Gordon and Dirac fields", *Class. Quantum Gravity* **29**, 215010 (2012).

44. R. Penrose, "On Gravity's Role in Quantum State Reduction", *Gen. Relat. Gravit.* **28**, 581–600 (1996).
45. B.S. De Witt, "Quantum theory of gravity. I. The canonical theory", *Phys. Rev.* **160**, 1113 (1967); J.A. Wheeler, in *Battelle Reencontres 1987*, eds. C. De Witt & J. A. Wheeler (Benjamin, New York, 1968) 94; C.J. Isham, "Canonical quantum gravity and the problem of time", GIFT Semminar 0157228 (1992).
46. C. Rovelli, "The century of the incomplete revolution: searching for general relativistic quantum field theory", *J. Math. Phys.* **41**, 3776 (2000), hep-th/9910131; "What is observable in classical and quantum gravity?", *Class. Quantum Gravity* **8**, 297 (1991).
47. R. Gambini, R.A. Porto & J. Pullin, "Realistic clocks, universal decoherence and the black hole information paradox" *Phys. Rev. Lett.* **93**, 240401 (2004); "Fundamental decoherence from relational time in discrete quantum gravity: Galilean covariance", *Phys. Rev. D* **70**, 124001 (2004).
48. T. Jacobson, "Thermodynamics of space-time: The Einstein equation of state", *Phys. Rev. Lett.* **75**, 1260 (1995); e-Print: gr-qc/9504004
49. D. Rideout & P. Wallden, Talk given at 13th Conference on Recent Developments in Gravity (NEB XIII), Thessaloniki, Greece, 4-6 Jun 2008, *J. Phys. Conf. Ser.* **189**, 012045 (2009).
50. N. Seiberg, in Brussels 2005, "The Quantum Structure of Space and Time", pp. 163–178.
51. See for instance S. Dodelson "Modern Cosmology", (Academic Press, 2003).
52. G. León & D. Sudarsky, "The slow roll condition and the amplitude of the primordial spectrum of cosmic fluctuations: Contrasts and similarities of standard account and the "collapse scheme"", *Class. Quant. Grav.* **27**, 225017 (2010).
53. "Planck 2015 results. XX. Constraints on inflation" Planck Collaboration (P.A.R. Ade *et. al.*), *Astonomy & Astrophysics* **594**, A20 (2016); arXiv:1502.02114 [astro-ph.CO].
54. "Planck 2018 results. VI. Cosmological parameters", Planck Collaboration (N. Aghanim (Orsay, IAS) *et al.*); arXiv:1807.06209 [astro-ph.CO].
55. M.P ía Piccirilli, G.León, S.J. Landau, M. Benetti, & D. Sudarsky, "Constraining quantum collapse inflationary models with current data: The semiclassical approach", *Int. J. Mod. Phys.* **D** **28**(02), 1950041 (2019); arXiv:1709.06237 [astro-ph.CO].
56. E. Okon & D. Sudarsky, "Benefits of Objective Collapse Models for Cosmology and Quantum Gravity", *Found. Phys.* **44**, 114–143 (2014); arXiv:1309.1730v1 [gr-qc].
57. E. Okon & D. Sudarsky, "The Black Hole Information Paradox and the Collapse of the Wave Function", *Found. Phys.* **45**,(4), 461–470 (2015).
58. E. Okon & D. Sudarsky, "Black Holes, Information Loss and the Measurement Problem", *Found. Phys.* **47**, 120 (2017); arXiv:1607.01255 [gr-qc].
59. R.M. Wald, *Quantum Field Theory In Curved Space-time and Black Hole Thermodynamics* (University of Chicago Press, 1994); L. Parker & D. Toms *Quantum Field Theory In Curved Space-time* (Cambridge University Press 2009).
60. B.A. Juárez-Aubry, B.S. Kay & D. Sudarsky, "Generally covariant dynamical reduction models and the Hadamard condition",*Phys. Rev. D*, **97**, 025010 (2018); arXiv:1708.09371 [gr-qc].
61. A. Diez-Tejedor & D. Sudarsky, "Towards a formal description of the collapse approach to the inflationary origin of the seeds of cosmic structure," *JCAP* **045**, 1207 (2012); arXiv:1108.4928 [gr-qc].
62. P. Cañate, E. Ramirez, & D. Sudarsky, "Semiclassical Self Consistent Treatment of the Emergence of Seeds of Cosmic Structure. The second order construction", *JCAP* **043**(08), 1808 (2018); arXiv:1802.02238 [gr-qc].
63. B. Juárez-Aubry, T. Miramontes & D. Sudarsky, "Semiclassical theories as initial value problems" in preparation.
64. M. Bahrami, A. Groardt, S. Donadi,& A. Bassi, "The Schrodinger-Newton equation and its foundations", arXiv:1407.4370.
65. G. León & D. Sudarsky, "Origin of structure: Statistical characterization of the primordial density fluctuations and the collapse of the wave function", *JCAP* **06**, 020 (2015).
66. See for instance section 4.2 of reference [21] above.

67. R. Adam et al. (Planck), Astron. Astrophys. **586**, A133 (2016); P. Ade et al. (BICEP2, Planck), *Phys. Rev. Lett.* **114**, 101301 (2015).
68. G. León, L. Kraiselburd, & S. J. Landau, "Primordial gravitational waves and the collapse of the wave function", *Phys. Rev. D* **92**(8), 083516 (2015); G. León, A. Majhi, E. Okon, & D. Sudarsky, "Reassessing the link between B-modes and inflation", *Phys. Rev. D* **96**, 101301(R) (2017); arXiv:1607.03523 [gr-qc]; "Expectation of primordial gravity waves generated during inflation", *Phys. Rev. D* **98**(2), 023512 (2018); arXiv:1712.02435 [gr-qc].
69. G. León, A. Pujol, S.J. Landau & M.P. Piccirilli, "Observational constraints on inflationary potentials within the quantum collapse framework", *Phys. Dark Univ.*, 100285 (2019); arXiv:1902.08696 [astro-ph.CO].
70. See for instance Luca Amendola *et al.* "Cosmology and fundamental physics with the Euclid satellite", *Living Rev. Rel.* **21**(1), 2 (2018).
71. C.J. Copi, D. Huterer, D.J. Schwarz, & G.D. Starkman, "Large angle anomalies in the CMB", *Adv. Astron.* 847541, (2010); arXiv:1004.5602.G.D; Starkman, C.J. Copi, D. Huterer, & D. Schwarz, "The oddly quiet universe: how the CMB challenges cosmologys standard model", arXiv:1201.2459.
72. G. León, & D. Sudarsky, "Novel possibility of nonstandard statistics in the inflationary spectrum of primordial inhomogeneities", *Sigma* **8**, 024 (2012).
73. R. Penrose, "Singularities and time-asymmetry General Relativity: An Einstein Centenary Survey" pp. 581 (ed. S. W. Hawking & W. Israel, Cambridge, Cambridge University Press, 1979).
74. S. Rodríguez & D. Sudarsky, "Revisiting Higgs inflation in the context of collapse theories," *JCAP* **1803**(03), 006 (2018); arXiv:1711.04912 [gr-qc].
75. B.A. Juárez-Aubry, B.S. Kay & D. Sudarsky, "Generally covariant dynamical reduction models and the Hadamard condition", *Phys. Rev. D* **97**(2), 025010 (2018); arXiv:1708.09371 [gr-qc].
76. E. Okon & D. Sudarsky, "Losing stuff down a black hole", *Found. Phys.* **48**, 411 (2018); arXiv:1710.01451 [gr-qc].
77. S.K. Modak, L. Ortíz, I. Peña & D. Sudarsky, "Non-Paradoxical Loss of Information in Black Hole Evaporation in a Quantum Collapse Model", *Phys. Rev. D* **91**(12), 124009 (2015); arXiv:1408.3062 [gr-qc].
78. S.K. Modak & D. Sudarsky, "Collapse of the wavefunction, the information paradox and backreaction", *Eur. Phys. J. C* **78**(7), 556 (2018); arXiv:1711.01509 [gr-qc].
79. D. Bedingham, S. K. Modak,& D. Sudarsky "Relativistic collapse dynamics and black hole information loss", *Phys. Rev. D* **94**(4), 045009 (2016); arXiv:1604.06537 [gr-qc].
80. E. Okon & D. Sudarsky, "A (not so?) novel explanation for the very special initial state of the universe", *Class. & Quant. Grav.* **33**(22), 225015 (2016); arXiv:1602.07006 [gr-qc].
81. T. Maudlin, E. Okon & D. Sudarsky, "The fate of conservation laws at the interface of quantum theory and gravitation", in press in *Studies in History and Philosophy of Modern Physics*; arXiv:1910.06473 [gr-qc].
82. T. Josset, A. Perez, & D. Sudarsky, "Dark energy as the weight of violating energy conservation", *Phys. Rev. Lett.* **118**, 021102 (2017); arXiv:1604.04183 [gr-qc].
83. A. Perez & D. Sudarsky, "Dark energy from quantum gravity discreteness", *Phys. Rev. Lett.* **122**(22), 221302 (2019); arXiv:1711.05183 [gr-qc]
84. D.J. Bedingham, "Relativistic state reduction dynamics", *Found. Phys.* **41**, 686 (2011); "Relativistic state reduction model", *J. Phys.: Conf. Ser.* **306**, 012034 (2011).
85. R. Tumulka, "A relativistic version of the Ghirardi-Rimini-Weber model", *J. Stat. Phys.* **125**, 821 (2006).
86. P. Pearle, "Relativistic dynamical collapse model", *Phys. Rev. D* **91**(10), 105012 (2015); arXiv:1412.6723 [quant-ph].
87. M. Bahrami, A. Smirne, & A. Bassi, "Gravity and the Collapse of the Wave Function: a Probe into Diósi-Penrose model", *Phys. Rev. A* **90**, 062135 (2014); "Dissipative Continuous Spontaneous Localization (CSL) model", A. Smirne & A. Bassi, *Sci. Rep.* **5**, 12518 (2015); A. Smirne & A. Bassi, "Toward an energy-conserving model of spontaneous wavefunction collapse", arXiv:1408.6446 [quant-ph].

A Relativistic GRW Flash Process with Interaction

Roderich Tumulka

Abstract In 2004, I described a relativistic version of the Ghirardi-Rimini-Weber (GRW) model of spontaneous wave function collapse for N non-interacting distinguishable particles. Here I present a generalized version for N interacting distinguishable particles. Presently, I do not know how to set up a similar model for indistinguishable particles or a variable number of particles. The present interacting model is constructed from a given interacting unitary Tomonaga-Schwinger type evolution between spacelike hypersurfaces, into which discrete collapses are inserted. I assume that this unitary evolution is interaction-local (i.e., no interaction at spacelike separation). The model is formulated in terms of Bell's flash ontology but is also compatible with Ghirardi's matter density ontology. It is non-local and satisfies microscopic parameter independence and no-signaling; it also works in curved space-time; in the non-relativistic limit, it reduces to the known non-relativistic GRW model.

1 Introduction

In this paper, I describe a relativistic model of spontaneous wave function collapse for N distinguishable particles with interaction, thereby generalizing my 2004 model without interaction [23, 24, 26]. The model involves, like the GRW model [7, 13] on which it is based, discrete jumps of the wave function with unitary evolution in between. I will describe the model in terms of Bell's flash ontology [2, 7, 27] by specifying the joint probability distribution of all flashes, but it could also be set

---

To the memory of GianCarlo Ghirardi (1935–2018).

R. Tumulka (✉)
Fachbereich Mathematik Eberhard-Karls-Universität, Auf der Morgenstelle 10, 72076 Tübingen, Germany
e-mail: roderich.tumulka@uni-tuebingen.de

V. Allori et al. (eds.), *Do Wave Functions Jump?*, Fundamental Theories of Physics 198,
https://doi.org/10.1007/978-3-030-46777-7_23

up using Ghirardi's matter density ontology [2, 9] along the lines described in [6]. Neither choice of ontology makes the problem easier or more difficult. In the model, one can, as already suggested by Aharonov and Albert in 1981 [1], associate a wave function ψ_Σ with every spacelike hypersurface Σ. For every flash in the past of Σ, a collapse operator gets applied to the wave function; thus, each collapse affects the wave function everywhere in the universe, although the model is fully relativistic. Since the flashes occur randomly, ψ_Σ is a random wave function and can thus be regarded as subject to a stochastic time evolution. In contrast to Bohmian mechanics but like the non-interacting 2004 model, the present model does not invoke a preferred foliation (i.e., slicing) of space-time into spacelike hypersurfaces.

Since, unlike Bohmian mechanics, the model does not involve trajectories in space-time, the word "particle" should not be taken literally. Rather, in this paper it means a space-time variable in the wave function (or in the configuration PVM on Hilbert space).

The interaction is incorporated by assuming the unitary part of the time evolution as given and including interaction. More precisely, I assume that a unitary Tomonaga-Schwinger type evolution between spacelike hypersurfaces is given and describe how to insert collapses in between unitary evolution operators. I assume that the unitary evolution is relativistic and interaction-local (i.e., involves no interaction terms between spacelike separated regions, see below). For example, such an evolution is rigorously known for N Dirac particles in $1 + 1$ dimensions with zero-range interaction [15–17]. For another example, the N particles could be taken to interact through a quantized field (which will neither be associated with local beables by itself nor with collapses).[1] If the unitary evolution is non-interacting, then the model reduces, up to small deviations, to the 2004 version. In particular, the model is non-local, i.e., two spacelike separated events a and b can influence each other, although there is no fact about the direction of the influence (whether a influenced b or b influenced a) [25]. Also like the 2004 version, the model obeys, up to small deviations, the condition that the distribution of the flashes up to a given spacelike hypersurface Σ does not depend on external fields in the future of Σ; this condition is a microscopic analog of the condition known as "parameter independence."

Much of the difficulty of devising a model with interaction arises from the fact that the collapse operators associated with different particles do not generally commute, whereas they do in the non-interacting case. This leads to a question of how to order the operators in the formula defining the joint probability density of the flashes, all the while with a need to ensure that the density integrates up to 1. The procedure proposed here is, roughly speaking, based on ordering the operator factors associated with two flashes in the temporal order when they are timelike separated, while the operators essentially still commute when the flashes are spacelike separated, except for certain details arising from the width σ of the collapses.

[1] However, quantized fields are usually mathematically ill defined due to ultraviolet divergence. It would be of interest to study carefully whether one of the few mathematically well defined evolutions for quantum fields (such as [14] in $1 + 1$ dimensions) can be put to work here.

Another difficulty arises precisely from the use of smeared-out collapse operators of width σ. Partly due to the use of different operator orderings depending on the space-time locations of the flashes, it turned out relevant to cut off the tails of the profile function, usually a Gaussian function of width σ, to ensure it vanishes exactly outside a certain admissible region. In fact, "cut off the tails" is a shorthand for a somewhat more involved procedure that will also change the shape of the profile function (away from a Gaussian shape) in the region where it does not vanish, as I will explain in Sect. 4.1. To make the model work, these several difficulties must jointly be dealt with.

Like the original GRW model, the present model has two parameters, the width σ of the collapse and the collapse rate λ per particle (or, equivalently, the expected waiting time $\tau = 1/\lambda$ for a collapse for a given particle). For our purposes, the waiting time is the relativistic timelike distance between two flashes associated with the same particle. We assume here the values suggested by GRW [13], $\sigma \approx 10^{-7}$ m and $\tau \approx 10^{16}$ s. The empirical predictions of the model are presumably, like those of the original GRW model, too close to those of standard quantum mechanics to allow for an experimental test with present technology; a careful study of its empirical predictions, its deviations from the original GRW model, and possible experimental tests would be of interest.

While the considerations of this paper also work in curved space-time, they will be formulated for Minkowski space-time $\mathbb{M}$.

Let me mention other proposals of relativistic collapse theories: Early attempts at a relativistic version of continuous spontaneous localization (CSL) [19, 20] are divergent and lead to infinite energy increase; see also [5]. A regularized relativistic version was developed by Bedingham and Pearle [3, 4, 21]. A model due to Dowker and Henson [10] lives on a discrete space-time and is relativistic in the appropriate lattice sense. A relativistic model due to Tilloy [22] is based on starting from a standard quantum field theory in a suitable regime, tracing out certain degrees of freedom, obtaining a master equation for the remaining ones, and finally using an unraveling of that master equation that should be empirically equivalent to the original quantum field theory in the regime considered.

The remainder of this paper is organized as follows. In Sect. 2, I review the non-interacting model. In Sect. 3, I describe the assumptions made on the unitary part of the time evolution. In Sect. 4, I define the interacting model. In Sect. 5, I discuss some of its properties.

2 Review of the Non-interacting Version

We begin with a brief summary of the 2004 model for N distinguishable non-interacting particles.

I will specify the joint distribution of the first n_i flashes for each particle number $i \in \{1, \ldots, N\}$. Let X_{ik} with $i \in \{1, \ldots, N\}$, $k \in \{1, \ldots, n_i\}$ be the random space-time points at which the flashes occur. Let $\underline{X}$ denote the collection of all X_{ik} with

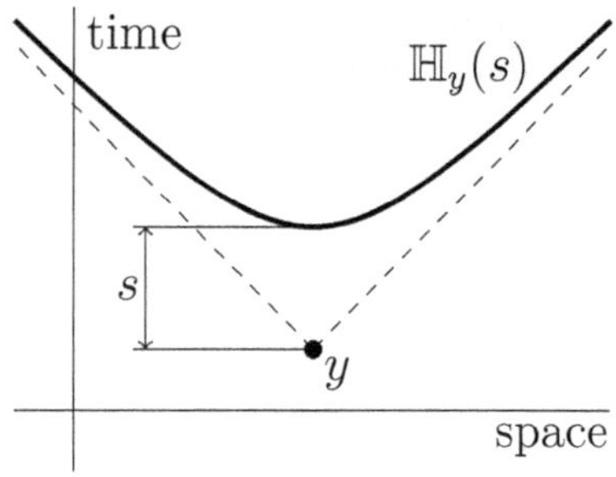

Fig. 1 In Minkowski space-time, the surface of constant timelike distance s from y in the future of y, $\mathbb{H}_y(s)$, has the shape of a hyperboloid that is asymptotic to the future light cone of y (dashed)

$1 \leq i \leq N$ and $1 \leq k \leq n_i$, likewise $\underline{x}$ the collection of the space-time points x_{ik}, and

$$d\underline{x} = \prod_{i=1}^{N} \prod_{k=1}^{n_i} d^4 x_{ik} \tag{1}$$

the volume element in 4ν dimensions (meaning either an infinitesimal set or its volume) with $\nu := n_1 + \cdots + n_N$. For each i, let x_{i0} be a given "seed" flash. The distribution of $\underline{X}$ is of the form

$$\mathbb{P}\big(\underline{X} \in d\underline{x}\big) = \langle \psi_0 | D(\underline{x}) | \psi_0 \rangle \, d\underline{x} \tag{2}$$

with operators D to be specified below and ψ_0 a wave function on a surface Σ_0 playing the role of an initial surface. In particular, the distribution of $\underline{X}$ is associated with a POVM $G(d\underline{x}) = D(\underline{x})\, d\underline{x}$ with density $D(\underline{x})$. Let $\mathscr{H}_{1\Sigma}$ be the 1-particle Hilbert space associated with the spacelike surface Σ, and $\mathscr{H}_{10} := \mathscr{H}_{1\Sigma_0}$, so $\psi_0 \in \mathscr{H}_0 := \mathscr{H}_{10}^{\otimes N}$. For each $i \in \{1, \ldots, N\}$, let $U_{i\Sigma}^{\Sigma'}$ be the unitary time evolution of particle i from the spacelike surface Σ to the spacelike surface Σ'. For all particles together, the unitary time evolution is $U_\Sigma^{\Sigma'} = U_{1\Sigma}^{\Sigma'} \otimes \cdots \otimes U_{N\Sigma}^{\Sigma'}$. We write $U_0^{\Sigma'}$ for $U_{\Sigma_0}^{\Sigma'}$.

Let $\mathbb{H}_y(s)$ be the surface of constant timelike distance s from $y \in \mathbb{M}$ in the future of y (henceforth called a hyperboloid, see Fig. 1); we also write $|\cdot|$ for the invariant (proper) length of a timelike 4-vector, $\mathbb{H}_y(x) := \mathbb{H}_y(|x - y|)$ for the hyperboloid containing $x \in \text{future}(y)$, and[2]

$$\mathbb{H}_{ik} := \mathbb{H}_{x_{ik-1}}(x_{ik}) \,. \tag{3}$$

Let $\tilde{g}_{yx}$ be the Gaussian function centered at x along the hyperboloid $\mathbb{H}_y(x)$,

$$\tilde{g}_{yx}(z) := \exp\left(-\frac{\text{s-dist}_{\mathbb{H}_y(x)}(x, z)^2}{4\sigma^2} \right), \tag{4}$$

[2]For definiteness, we take future(y) to be a closed set (i.e., the "causal future"), including the future light cone and y itself. Likewise for the past.

where s-dist$_\Sigma$ means the spacelike distance along Σ, and g_{yx} the version normalized in x,

$$g_{yx}(z) := \frac{1}{\|\tilde{g}_{yz}\|}\tilde{g}_{yx}(z)\,, \tag{5}$$

where, for a spacelike surface Σ and $f : \Sigma \to \mathbb{C}$,

$$\|f\| := \Big(\int_\Sigma d^3x\,|f(x)|^2\Big)^{1/2} \tag{6}$$

is the L^2 norm and d^3x means the invariant volume of a 3-surface element (defined by the 3-metric on Σ). For the multiplication operator by g_{yx} on $\Sigma = \mathbb{H}_y(x)$, we write $P(g_{yx})$. To the flash x_{ik} we associate the collapse operator

$$K(x_{ik}) := U^0_{i\mathbb{H}_{ik}}\,P(g_{x_{ik-1}x_{ik}})\,U^{\mathbb{H}_{ik}}_{i0}\,, \tag{7}$$

and to all flashes together the operator

$$L(\underline{x}) := \bigotimes_{i=1}^{N}\prod_{k=1}^{n_i} K(x_{ik}) \tag{8}$$

with the order in the product so that k increases from right to left. Then set

$$D(\underline{x}) := \Big(\frac{1}{\tau^\nu}\prod_{i=1}^{N}\prod_{k=1}^{n_i} 1_{x_{ik}\in\text{future}(x_{ik-1})}e^{-|x_{ik}-x_{ik-1}|/\tau}\Big)L(\underline{x})^\dagger\,L(\underline{x})\,. \tag{9}$$

It was shown in [23] (and it follows from the proofs below that apply to the more general interacting case) that

$$\int_{\mathbb{M}^\nu} d\underline{x}\;D(\underline{x}) = I \tag{10}$$

with I the unit operator; as a consequence, $\mathbb{P}$ is a probability distribution for every $\psi_0 \in \mathscr{H}_0$ with $\|\psi_0\| = 1$.

Equivalently, $G(d\underline{x}) = \otimes_{i=1}^N G_i(d^4x_{i1}\times\cdots\times d^4x_{in_i})$ and $D(\underline{x}) = \otimes_{i=1}^N D_i(x_{i1},\ldots,x_{in_i})$ with $G_i(d^4x_{i1}\times\cdots\times d^4x_{in_i}) = D_i(x_{i1},\ldots,x_{in_i})\,d^4x_{i1}\cdots d^4x_{in_i}$ and

$$\begin{aligned}D_i(x_{i1},\ldots,x_{in_i}) = \Big(\frac{1}{\tau^{n_i}}\prod_{k=1}^{n_i} 1_{x_{ik}\in\text{future}(x_{ik-1})}e^{-|x_{ik}-x_{ik-1}|/\tau}\Big)\times\\ K(x_{i1})^\dagger\cdots K(x_{in_i})^\dagger K(x_{in_i})\ldots K(x_{i1})\,.\end{aligned} \tag{11}$$

If ψ_0 factorizes into a tensor product $\otimes_{i=1}^N \psi_{0i}$, then $\mathbb{P}$ will factorize, and the flashes for different i will be independent of each other. But in general, even though G factorizes, $\mathbb{P}$ will not factorize, which leads to non-local correlations between the flashes associated with different i.

We now prepare for defining the novel interacting version of the model.

3 Assumptions

In this article, I put no emphasis on mathematical rigor. But the reasoning is actually rigorous if the assumption is satisfied that the unitary evolution is defined not only between Cauchy surfaces but also to hyperboloids or surfaces consisting of pieces of hyperboloids. For example, a sufficient class of surfaces would be the set $\mathscr{S}$ of those sets that are intersected exactly once by every timelike straight line; in the following, I will simply say "spacelike surface" for any $\Sigma \in \mathscr{S}$. For massive free Dirac particles, it is known [11] (see also [26]) that the unitary evolution is also defined from a Cauchy surface to a hyperboloid, so it seems plausible that it is also defined between any two surfaces belonging to $\mathscr{S}$.

So, we assume that the unitary part of the time evolution is given by a Tomonaga–Schwinger type evolution, more precisely, by a *unitary hypersurface evolution* [18] between spacelike surfaces. That is, we assume that with every spacelike surface $\Sigma \in \mathscr{S}$ there is associated a Hilbert space $\mathscr{H}_\Sigma$ (see [18] for examples), and that for any two spacelike surfaces Σ, Σ' we are given a unitary isomorphism $U_\Sigma^{\Sigma'} : \mathscr{H}_\Sigma \to \mathscr{H}_{\Sigma'}$ representing the time evolution without collapses, such that

$$U_\Sigma^\Sigma = I, \quad U_{\Sigma'}^{\Sigma''} U_\Sigma^{\Sigma'} = U_\Sigma^{\Sigma''} \tag{12}$$

for all Σ, Σ', and Σ''. Moreover, for each Σ we are given a position PVM P_Σ ("configuration observable") on Σ^N acting on $\mathscr{H}_\Sigma$. This completes the definition of "unitary hypersurface evolution." For a function $f : \Sigma^N \to \mathbb{R}$, we define the associated multiplication operator

$$P(f) := P_\Sigma(f) := \int_{\Sigma^N} P_\Sigma(d^3x_1 \times \cdots \times d^3x_N)\, f(x_1, \ldots, x_N)\,. \tag{13}$$

Of the unitary evolution we assume

Interaction locality (IL) [18]: *For any two spacelike hypersurfaces Σ, Σ', any set $A \subseteq \Sigma \cap \Sigma'$ in the overlap, and any $i \in \{1, \ldots, N\}$,*

$$P_{\Sigma'}\Big((\Sigma')^{i-1} \times A \times (\Sigma')^{N-i-1}\Big) = U_\Sigma^{\Sigma'}\, P_\Sigma\Big(\Sigma^{i-1} \times A \times \Sigma^{N-i-1}\Big)\, U_{\Sigma'}^\Sigma\,. \tag{14}$$

The condition expresses that the unitary evolution includes no interaction term between spacelike separated regions. Specifically, the unitary evolution from Σ to Σ' acts like the identity on $\Sigma \cap \Sigma'$. Here are some consequences of IL:

1. Fix a function f on $\Sigma \cap \Sigma'$ and a label $i \in \{1, \ldots, N\}$, and let $P(f)$ be the associated multiplication operator in the i-th variable; more precisely, let

$$P_\Sigma(f) := \int\limits_{x \in \Sigma \cap \Sigma'} P_\Sigma\Big(\Sigma^{i-1} \times d^3x \times \Sigma^{N-i-1}\Big)\, f(x)\,, \tag{15a}$$

$$P_{\Sigma'}(f) := \int\limits_{x \in \Sigma \cap \Sigma'} P_{\Sigma'}\Big(\Sigma'^{(i-1)} \times d^3x \times \Sigma'^{(N-i-1)}\Big)\, f(x)\,. \tag{15b}$$

Then

$$P_{\Sigma'}(f) = U_\Sigma^{\Sigma'}\, P_\Sigma(f)\, U_{\Sigma'}^\Sigma\,. \tag{16}$$

That is because, setting $A = d^3x$ in (14),

$$U_\Sigma^{\Sigma'}\, P_\Sigma(f)\, U_{\Sigma'}^\Sigma = \int_{\Sigma \cap \Sigma'} U_\Sigma^{\Sigma'}\, P_\Sigma(\Sigma^{i-1} \times d^3x \times \Sigma^{N-i-1})\, U_{\Sigma'}^\Sigma\, f(x) \tag{17a}$$

$$= \int_{\Sigma \cap \Sigma'} P_{\Sigma'}(\Sigma'^{(i-1)} \times d^3x \times \Sigma'^{(N-i-1)})\, f(x) \tag{17b}$$

$$= P_{\Sigma'}(f)\,. \tag{17c}$$

2. Let $A := \Sigma \cap \Sigma'$, $B := \Sigma \setminus A$, $B' := \Sigma' \setminus A$. Then

$$P_{\Sigma'}\Big((\Sigma')^{i-1} \times B' \times (\Sigma')^{N-i-1}\Big) = U_\Sigma^{\Sigma'}\, P_\Sigma\Big(\Sigma^{i-1} \times B \times \Sigma^{N-i-1}\Big)\, U_{\Sigma'}^\Sigma\,. \tag{18}$$

That is because, using the normalization $P_{\Sigma'}(\Sigma'^N) = I$,

$$P_{\Sigma'}\Big((\Sigma')^{i-1} \times B' \times (\Sigma')^{N-i-1}\Big) \tag{19a}$$

$$= I - P_{\Sigma'}\Big((\Sigma')^{i-1} \times A \times (\Sigma')^{N-i-1}\Big) \tag{19b}$$

$$\overset{(13)}{=} I - U_\Sigma^{\Sigma'}\, P_\Sigma\Big(\Sigma^{i-1} \times A \times \Sigma^{N-i-1}\Big)\, U_{\Sigma'}^\Sigma \tag{19c}$$

$$= U_\Sigma^{\Sigma'}\, P_\Sigma\Big(\Sigma^{i-1} \times B \times \Sigma^{N-i-1}\Big)\, U_{\Sigma'}^\Sigma\,. \tag{19d}$$

4 Interacting Model

4.1 A Simple Case

As a warm-up we consider the simple case of $N = 2$ particles and limit our attention to the first flash for each particle. That is, we define the joint distribution of two flashes, X_1 and X_2, each associated with a different particle. We take as given a seed flash for each particle, y_1 and y_2, and an initial wave function ψ_0. We postulate that the joint distribution is of the form

$$\mathbb{P}\big(X_1 \in d^4x_1, X_2 \in d^4x_2\big) = \langle\psi_0|D(x_1,x_2)|\psi_0\rangle\, d^4x_1\, d^4x_2 \tag{20}$$

with positive operator-valued density

$$D(x_1,x_2) = 1_{x_1\in\text{future}(y_1)}1_{x_2\in\text{future}(y_2)}\frac{1}{\tau^2}e^{-|x_1-y_1|/\tau}e^{-|x_2-y_2|/\tau}\,L(x_1,x_2)^\dagger\,L(x_1,x_2)\,, \tag{21}$$

where L will be defined below. As before, the distribution of (X_1,X_2) is determined by a POVM $G(dx_1\times dx_2) = D(x_1,x_2)\,dx_1\,dx_2$ with density $D(x_1,x_2)$. We will consider two relevant hyperboloids,

$$\mathbb{H}_i := \mathbb{H}_{y_i}(x_i) \quad \text{with } i=1,2, \tag{22}$$

and a profile function $g_{yx}(z)$ on $\mathbb{H}_y(x)$ that has a bump shape around x; the first thought would be to use the Gaussian function $\tilde{g}_{yx}$ centered at x given by (4), but we will refine this choice later. We write g_{yxi} for the function g_{yx} applied to the i-th variable; that is, g_{yx} is a function on a 3-surface Σ, and g_{yxi} a function on Σ^N (here with $N=2$).

A basic difference between the interacting and the non-interacting case is that in the non-interacting case, one can evolve particle 1 to Σ_1 and independently particle 2 to another surface Σ_2; in the interacting case, we can only evolve all particles *jointly* to a certain surface. Let us consider two candidates for $L(x_1,x_2)$,

$$L^{(21)}(x_1,x_2) := U^0_{\mathbb{H}_2}\,P_{\mathbb{H}_2}(g_{y_2x_2}2)U^{\mathbb{H}_2}_{\mathbb{H}_1}\,P_{\mathbb{H}_1}(g_{y_1x_1}1)\,U^{\mathbb{H}_1}_0\,, \tag{23a}$$

$$L^{(12)}(x_1,x_2) := U^0_{\mathbb{H}_1}\,P_{\mathbb{H}_1}(g_{y_1x_1}1)U^{\mathbb{H}_1}_{\mathbb{H}_2}\,P_{\mathbb{H}_2}(g_{y_2x_2}2)\,U^{\mathbb{H}_2}_0\,. \tag{23b}$$

We can think of each of (23a) and (23b) as a product of two multiplication operators, each Heisenberg-evolved to the initial surface Σ_0. Since the unitary evolution does not

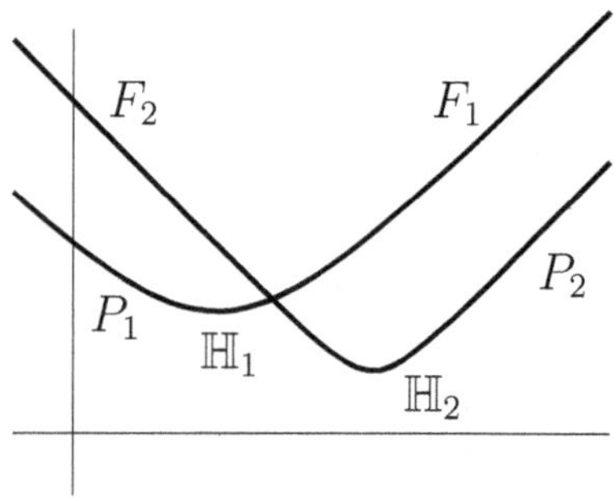

Fig. 2 Two hyperboloids $\mathbb{H}_1$, $\mathbb{H}_2$ are each subdivided according to (24) into two 3-regions F_i and P_i, above and below the other

commute with multiplication operators, the two multiplication operators on different surfaces do not commute with each other, so (23a) and (23b) are in general not equal, with two relevant exceptions: First, in the absence of interaction, the unitary evolution factorizes, and the two multiplication operators commute because they act on different factors. Second, if the supports of $g_{y_1x_1}$ and $g_{y_2x_2}$ are spacelike separated (i.e., if every point in the one set is spacelike from every point in the other), then they commute by virtue of IL. (It may appear pointless to talk about the support of g_{yx} if g_{yx} is a Gaussian because then its support is the entire surface $\mathbb{H}_y(x)$; but we will later cut off the Gaussian tails to create smaller supports.).

So, for the purpose of defining the operator $L(x_1, x_2)$, we are confronted with a problem of operator ordering. Roughly speaking, we choose the ordering according to the temporal ordering of x_1 and x_2: For x_1 in the past of x_2, we choose $L = L^{(21)}$ and vice versa. But the exact definition is a little more complicated, partly because we need to consider the support of $g_{y_1x_1}$, not just the point x_1.

To this end, we subdivide each $\mathbb{H}_i$ into two parts (see Fig. 2),

$$F_i := \mathbb{H}_i \cap \text{future}(\mathbb{H}_{3-i})\,, \quad P_i := \mathbb{H}_i \cap \text{past}(\mathbb{H}_{3-i})\,. \tag{24}$$

(Note that the interface $\mathbb{H}_1 \cap \mathbb{H}_2$ has measure 0 in $\mathbb{H}_1$ as well as in $\mathbb{H}_2$, except if $y_1 = y_2$ and $\tau_1 = \tau_2$, which happens with probability 0. Ignoring sets of measure 0, we can pretend that F_i and P_i form a partition of $\mathbb{H}_i$.)

For any set $A \subseteq \mathbb{H}_y(x)$, set

$$\|f\|_A = \Big(\int_A d^3z\,|f(z)|^2\Big)^{1/2} \tag{25}$$

and

$$g_{yAx}(z) := \frac{1}{\|\tilde{g}_{yz}\|_A} 1_{z\in A}\, 1_{x\in A}\, \tilde{g}_{yx}(z)\,. \tag{26}$$

Some functions of this type are depicted in Fig. 3.

Since

$$\tilde{g}_{yx}(z) = \tilde{g}_{yz}(x)\,, \tag{27}$$

it follows that for every $z \in \mathbb{H}_y(x)$,

$$\begin{aligned}
\int_A d^3x\, g_{yAx}(z)^2 &= 1_{z\in A} \int_A d^3x\, \frac{1}{\|\tilde{g}_{yz}\|_A^2} \tilde{g}_{yx}(z)^2 && \text{(28a)}\\
&= \frac{1_{z\in A}}{\|\tilde{g}_{yz}\|_A^2} \int_A d^3x\, \tilde{g}_{yx}(z)^2 && \text{(28b)}\\
&= \frac{1_{z\in A}}{\|\tilde{g}_{yz}\|_A^2} \int_A d^3x\, \tilde{g}_{yz}(x)^2 && \text{(28c)}\\
&= \frac{1_{z\in A}}{\|\tilde{g}_{yz}\|_A^2} \|\tilde{g}_{yz}\|^2 && \text{(28d)}
\end{aligned}$$

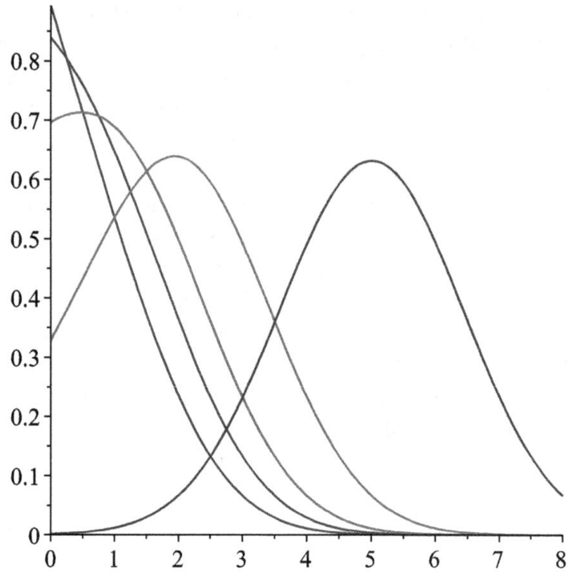

Fig. 3 Examples of functions of the type (26), but defined on the real line instead of a hyperboloid, here with $A = [0, \infty)$. The first factor on the right-hand side of (26) causes a deviation from the Gaussian shape which is small for x far from the boundary of A but visible for x close to it. The right axis shown is z, the functions plotted are $g_{Ax}(z) = \|g_z\|_A^{-1} \tilde{g}_x(z)$ with $\tilde{g}_x$ the Gaussian density with center x and width 1 for the values $x = 0, \frac{1}{2}, 1, 2, 5$ (in the order of centers from left to right, or of decreasing values of $g_{Ax}(0)$)

$$= 1_{z \in A} . \tag{28e}$$

Again, we write g_{yAxi} for the function g_{yAx} applied to the i-th variable. It follows from (28e) that, for any $i \in \{1, \ldots, N\}$ and $A \subseteq \mathbb{H}_y(s)$,

$$\int_A d^3x \, P_{\mathbb{H}_y(s)}(g_{yAxi})^2 = P_{\mathbb{H}_y(s)}\big(\mathbb{H}_y(s)^{i-1} \times A \times \mathbb{H}_y(s)^{N-i-1}\big) . \tag{29}$$

We define

$$L(x_1, x_2) := \begin{cases} U^0_{\mathbb{H}_2} \, P_{\mathbb{H}_2}(g_{y_2P_2x_22}) \, U^{\mathbb{H}_2}_{\mathbb{H}_1} \, P_{\mathbb{H}_1}(g_{y_1P_1x_11}) \, U^{\mathbb{H}_1}_0 & \text{if } x_1 \in P_1, x_2 \in P_2 \\ U^0_{\mathbb{H}_2} \, P_{\mathbb{H}_2}(g_{y_2F_2x_22}) \, U^{\mathbb{H}_2}_{\mathbb{H}_1} \, P_{\mathbb{H}_1}(g_{y_1P_1x_11}) \, U^{\mathbb{H}_1}_0 & \text{if } x_1 \in P_1, x_2 \in F_2 \\ U^0_{\mathbb{H}_1} \, P_{\mathbb{H}_1}(g_{y_1F_1x_11}) \, U^{\mathbb{H}_1}_{\mathbb{H}_2} \, P_{\mathbb{H}_2}(g_{y_2P_2x_22}) \, U^{\mathbb{H}_2}_0 & \text{if } x_1 \in F_1, x_2 \in P_2 \\ U^0_{\mathbb{H}_1} \, P_{\mathbb{H}_1}(g_{y_1F_1x_11}) \, U^{\mathbb{H}_1}_{\mathbb{H}_2} \, P_{\mathbb{H}_2}(g_{y_2F_2x_22}) \, U^{\mathbb{H}_2}_0 & \text{if } x_1 \in F_1, x_2 \in F_2 . \end{cases} \tag{30}$$

Proposition 1 *Interaction locality implies that*

$$\int d^4x_1 \int d^4x_2 \, D(x_1, x_2) = I . \tag{31}$$

As a consequence, $G(\cdot)$ is a POVM, and (20) *defines a probability distribution for every $\psi_0 \in \mathscr{H}_0$ with $\|\psi_0\| = 1$.*

Proof Since (coarea formula)

$$\int\limits_{\text{future}(y)} d^4x\, f(x,y) = \int\limits_0^\infty ds \int\limits_{\mathbb{H}_y(s)} d^3x\, f(x,y)\,, \tag{32}$$

and since

$$\int_0^\infty ds\, \tau^{-1} \exp(-s/\tau) = 1\,, \tag{33}$$

it suffices to show that for any two hyperboloids $\mathbb{H}_1, \mathbb{H}_2$ based at y_1 and y_2, respectively,

$$\int_{\mathbb{H}_1} d^3x_1 \int_{\mathbb{H}_2} d^3x_2\, L(x_1,x_2)^\dagger\, L(x_1,x_2) = I\,. \tag{34}$$

Let $\Sigma := P_1 \cup P_2$. Writing $\mathbb{H}_i = P_i \cup F_i$ yields four parts for $\mathbb{H}_1 \times \mathbb{H}_2$. We deal with each part separately, beginning with $P_1 \times P_2$: By the consequence (18) of IL,

$$U^\Sigma_{\mathbb{H}_2}\, P_{\mathbb{H}_2}(\mathbb{H}_2 \times P_2)\, U^{\mathbb{H}_2}_\Sigma = P_\Sigma(\Sigma \times P_2)\,. \tag{35}$$

By the consequence (16) of IL,

$$U^\Sigma_{\mathbb{H}_1}\, P_{\mathbb{H}_1}(g_{y_1 P_1 x_1} 1)\, U^{\mathbb{H}_1}_\Sigma = P_\Sigma(g_{y_1 P_1 x_1} 1) \tag{36}$$

for every $x_1 \in P_1$. Thus,

$$\int_{P_1} d^3x_1 \int_{P_2} d^3x_2\, L(x_1,x_2)^\dagger\, L(x_1,x_2) \tag{37a}$$

$$= \int_{P_1} d^3x_1 \int_{P_2} d^3x_2\, U^0_{\mathbb{H}_1}\, P_{\mathbb{H}_1}(g_{y_1 P_1 x_1} 1)\, U^{\mathbb{H}_1}_{\mathbb{H}_2}\, P_{\mathbb{H}_2}(g_{y_2 P_2 x_2} 2)^2\, U^{\mathbb{H}_2}_{\mathbb{H}_1}\, P_{\mathbb{H}_1}(g_{y_1 P_1 x_1} 1)\, U^{\mathbb{H}_1}_0 \tag{37b}$$

$$\overset{(29)}{=} \int_{P_1} d^3x_1\, U^0_{\mathbb{H}_1}\, P_{\mathbb{H}_1}(g_{y_1 P_1 x_1} 1)\, U^{\mathbb{H}_1}_{\mathbb{H}_2}\, P_{\mathbb{H}_2}(\mathbb{H}_2 \times P_2)\, U^{\mathbb{H}_2}_{\mathbb{H}_1}\, P_{\mathbb{H}_1}(g_{y_1 P_1 x_1} 1)\, U^{\mathbb{H}_1}_0 \tag{37c}$$

$$\overset{(35),(36)}{=} \int_{P_1} d^3x_1\, U^0_\Sigma\, P_\Sigma(g_{y_1 P_1 x_1} 1)\, P_\Sigma(\Sigma \times P_2)\, P_\Sigma(g_{y_1 P_1 x_1} 1)\, U^\Sigma_0 \tag{37d}$$

$$= \int_{P_1} d^3x_1\, U^0_\Sigma\, P_\Sigma(g_{y_1 P_1 x_1} 1)^2\, P_\Sigma(\Sigma \times P_2)\, U^\Sigma_0 \tag{37e}$$

$$\overset{(29)}{=} U^0_\Sigma\, P_\Sigma(P_1 \times \Sigma)\, P_\Sigma(\Sigma \times P_2)\, U^\Sigma_0 \tag{37f}$$

$$= U^0_\Sigma\, P_\Sigma(P_1 \times P_2)\, U^\Sigma_0\,. \tag{37g}$$

Here, we used in (37e) that the operators of a PVM commute, and in the last step that, for every PVM, $P(A)P(B) = P(A \cap B)$.

We now turn to the contribution from $P_1 \times F_2$. Here we exploit that, by the consequence (18) of IL applied to $\mathbb{H}_2$ and $\Sigma = P_1 \cup P_2$,

$$P_\Sigma(\Sigma \times P_1) = U^\Sigma_{\mathbb{H}_2}\, P_{\mathbb{H}_2}(\mathbb{H}_2 \times F_2)\, U^{\mathbb{H}_2}_\Sigma\,. \tag{38}$$

With the same strategy as in (37), we now obtain that

$$\int_{P_1} d^3x_1 \int_{F_2} d^3x_2 \, L(x_1, x_2)^\dagger \, L(x_1, x_2) = U_\Sigma^0 \, P_\Sigma(P_1 \times P_1) \, U_0^\Sigma \,. \tag{39}$$

For $F_1 \times P_2$, we interchange the order of integration so that the x_1 integration is carried out first (i.e., inside the x_2 integral). Exploiting that, by (18),

$$P_\Sigma(P_2 \times \Sigma) = U_{\mathbb{H}_1}^\Sigma \, P_{\mathbb{H}_1}(F_1 \times \mathbb{H}_1) \, U_\Sigma^{\mathbb{H}_1} \,, \tag{40}$$

we obtain through the same strategy as before that

$$\int_{P_2} d^3x_2 \int_{F_1} d^3x_1 \, L(x_1, x_2)^\dagger \, L(x_1, x_2) = U_\Sigma^0 \, P_\Sigma(P_2 \times P_2) \, U_0^\Sigma \,. \tag{41}$$

Likewise for $F_1 \times F_2$:

$$\int_{F_1} d^3x_1 \int_{F_2} d^3x_2 \, L(x_1, x_2)^\dagger \, L(x_1, x_2) = U_\Sigma^0 \, P_\Sigma(P_2 \times P_1) \, U_0^\Sigma \,. \tag{42}$$

Putting together (37g), (39), (41), and (42), we obtain that

$$\int_{\mathbb{H}_1} d^3x_1 \int_{\mathbb{H}_2} d^3x_2 \, L(x_1, x_2)^\dagger \, L(x_1, x_2) \tag{43a}$$

$$= U_\Sigma^0 \, P_\Sigma\Big((P_1 \times P_2) \cup (P_1 \times P_1) \cup (P_2 \times P_2) \cup (P_2 \times P_1)\Big) \, U_0^\Sigma \tag{43b}$$

$$= U_\Sigma^0 \, P_\Sigma(\Sigma^2) \, U_0^\Sigma = U_\Sigma^0 \, I \, U_0^\Sigma = I \,, \tag{43c}$$

as claimed in (34). □

4.2 General Case

Consider n_i flashes for particle i; they occur at the random points $X_{ik}, i \in \{1, \ldots, N\}$, $k \in \{1, \ldots, n_i\}$. Let $\underline{X}$ denote again the collection of all X_{ik} with $1 \le i \le N$ and $1 \le k \le n_i$, likewise $\underline{x}$ the collection of the space-time points x_{ik}, and $d\underline{x}$ as in (1). For each i, let x_{i0} be a given seed flash. The distribution of $\underline{X}$ is again of the form

$$\mathbb{P}\big(\underline{X} \in d\underline{x}\big) = \langle \psi_0 | D(\underline{x}) | \psi_0 \rangle \, d\underline{x} \tag{44}$$

with D again of the form

$$D(\underline{x}) = \left(\frac{1}{\tau^\nu} \prod_{i=1}^{N} \prod_{k=1}^{n_i} 1_{x_{ik} \in \text{future}(x_{ik-1})} e^{-|x_{ik}-x_{ik-1}|/\tau} \right) L(\underline{x})^\dagger \, L(\underline{x}) \,. \tag{45}$$

In particular, the distribution of $\underline{X}$ is again determined by a POVM $G(d\underline{x}) = D(\underline{x})\, d\underline{x}$ with density $D(\underline{x})$. We use the notation $\mathbb{H}_{ik}$ as in (3). The first, rough idea would be to take L to be something like

$$\text{“}\; L(\underline{x}) = \prod_{i=1}^{N} \prod_{k=1}^{n_i} U^{0}_{\mathbb{H}_{ik}} \, P_{\mathbb{H}_{ik}}(g_{x_{ik-1}x_{ik}i}) \, U_0^{\mathbb{H}_{ik}} \;\text{”} \tag{46}$$

with a problem of operator ordering. To address this problem, we need to construct the analogs of the 3-cells P_i and F_i of the previous section.

4.2.1 Division Into Cells

The connected components of $\mathbb{M} \setminus \cup_{ik} \mathbb{H}_{ik}$ (more precisely, their closures) we call *4-cells*. They can be labeled by $\underline{k} = (k_1, \ldots, k_N) \in \prod_{i=1}^{N} \{0, \ldots, n_i\}$: the 4-cell for $\underline{k}$ is defined as

$${}^4C_{\underline{k}} := \bigcap_{i=1}^{N} \Big(\text{future}(\mathbb{H}_{ik_i}) \cap \text{past}(\mathbb{H}_{ik_i+1}) \Big) \,, \tag{47}$$

where future$(\mathbb{H}_{i0})$ and past$(\mathbb{H}_{in_i+1})$ should be understood as $\mathbb{M}$; see Fig. 4 for an example. There are $\prod_i (n_i + 1)$ 4-cells. The 4-cells form a partition of space-time, except for overlap on the hyperboloids. For $\underline{k}$ with all $k_i = 0$ we write 0^N, and we write $\underline{n} = (n_1, \ldots, n_N)$, as well as ${}^4\mathscr{C}$ for the set of all 4-cells.

The faces of the 4-cells are pieces of hyperboloids henceforth called *3-cells*,

$${}^3C_{i\underline{k}} := \mathbb{H}_{ik_i} \cap \bigcap_{j \neq i} \Big(\text{future}(\mathbb{H}_{jk_j}) \cap \text{past}(\mathbb{H}_{jk_j+1}) \Big) \,, \tag{48}$$

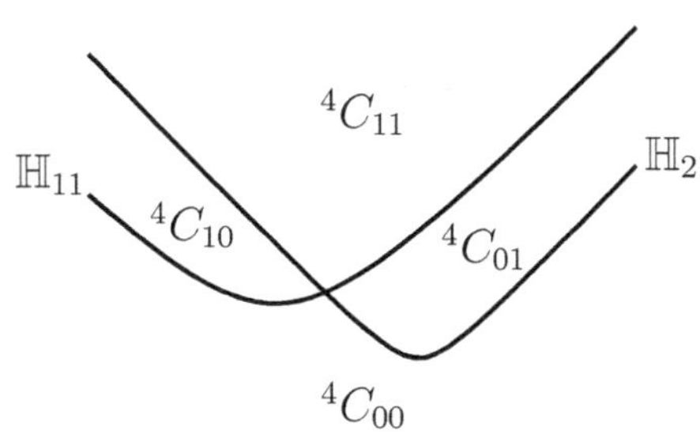

Fig. 4 Notation for 4-cells as in (47)

where $\underline{k}$ must be such that $k_i \geq 1$. In fact, ${}^3C_{i\underline{k}}$ is the common boundary of ${}^4C_{\underline{k}}$ and ${}^4C_{\underline{k}'}$, where $k'_i = k_i - 1$ and $k'_j = k_j$ for all $j \neq i$. For example, for two hyperboloids as in Figs. 2 and 4, $P_1 = {}^3C_{110}$, $P_2 = {}^3C_{201}$, and $F_i = {}^3C_{i11}$. The set of all 3-cells will be denoted by ${}^3\mathscr{C}$.

If two 4-cells border on each other along a 3-cell ${}^3C_{i\underline{k}}$, then the one in the future of $\mathbb{H}_{ik_i}$ will henceforth be said to be a *successor* of the one in the past of $\mathbb{H}_{ik_i}$, and conversely a *predecessor*. The predecessors of ${}^4C_{\underline{k}}$ are those for which one k_j in $\underline{k}$ has been replaced by $k_j - 1$.

We say that a set $S \subseteq \mathbb{M}$ is *past complete* if $\text{past}(S) \subseteq S$; correspondingly *future complete*. For example, $\emptyset$ and $\mathbb{M}$ are both past and future complete, the past of any set is past complete, an intersection of past complete sets is past complete, ${}^4C_{0^N}$ is past complete, ${}^4C_{\underline{n}}$ is future complete, and $\mathbb{M} \setminus {}^4C_{\underline{n}}$ is past complete. One easily verifies that the complement of a past complete set is future complete and vice versa.

Proposition 2 *Every (closed) past complete set S except $\emptyset$ and $\mathbb{M}$ is the past of its boundary, $S = \text{past}(\partial S)$, and ∂S is a spacelike-or-lightlike hypersurface.*

Proof For any $x \in S$, consider a timelike straight line (geodesic) γ through x; there must be a point on γ outside S, or else $S = \mathbb{M}$ by past completeness. Again by past completeness, γ must lie in S up to a point $\gamma(s_0)$ and outside from there onwards. So $\gamma(s_0)$ must lie on ∂S, and $x \in \text{past}(\gamma(s_0)) \subseteq \text{past}(\partial S)$. □

While the exact location and shape of ${}^4C_{\underline{k}}$ depends on the hyperboloids, many relations between the 4-cells, such as which one borders on which others along which 3-cells, can be read off from the index $\underline{k}$. That is why we also call $\underline{k} \in \prod_i \{0, \ldots, n_i\}$ an *abstract 4-cell* and a pair $(i, \underline{k})$ such that $k_i \geq 1$ an *abstract 3-cell*. The set of abstract 4-cells (respectively, 3-cells) is ${}^4\mathscr{A} := \prod_i \{0, \ldots, n_i\}$, respectively ${}^3\mathscr{A} := \{(i, \underline{k}) \in \{1 \ldots N\} \times {}^4\mathscr{A} : k_i \geq 1\}$. The *future faces* of the abstract 4-cell $\underline{k}$ are the abstract 3-cells $(i, \underline{k}')$ with $k'_i = k_i + 1$ and $k'_j = k_j$ for all $j \neq i$ (if they exist); the *past faces* of $\underline{k}$ are the $(i, \underline{k})$ with $k_i \geq 1$. The *predecessors* of $\underline{k}$ are those abstract 4-cells for which one k_j in $\underline{k}$ has been replaced by $k_j - 1$; correspondingly *successors*. A set V of abstract 4-cells will be called *predecessor complete* iff[3] it contains every predecessor of each of its elements; correspondingly *successor complete*. A set is successor complete iff its complement is predecessor complete.

Proposition 3 *If a set $V \subseteq {}^4\mathscr{A}$ is predecessor complete, then the corresponding space-time set $S(V) = \cup_{\underline{k} \in V} {}^4C_{\underline{k}}$ is past complete. Furthermore, if the hyperboloids are such that ${}^4C_{\underline{k}}$ has non-empty interior (or non-zero 4-volume) for every $\underline{k} \in {}^4\mathscr{A}$, then also the converse is true: $S(V)$ is past complete only if V is predecessor complete.*

Proof To see that $S(V)$ is past complete, consider $x \in S(V)$ and y in the past of x. The straight line (or any causal curve) from x to y, when crossing hyperboloids, enters a predecessor of the 4-cell, and thus remains in $S(V)$. We remark that if some ${}^4C_{\underline{k}}$ is empty (as would happen if $x_{ik_i-1} \in \text{future}(x_{jk_j+1})$ with $j \neq i$), then $S(\{\underline{k}\}) = {}^4C_{\underline{k}} = \emptyset$ is past complete although $V = \{\underline{k}\}$ is not predecessor complete.

[3] iff = if and only if.

Now assuming that the 4-cells have non-empty interior, if $S(V)$ were past complete but V not predecessor complete, then let $\underline{k}' \notin V$ be a predecessor of $\underline{k} \in V$. Since the interiors are non-empty, there are interior points $x \in {}^4C_{\underline{k}}$ and $y \in {}^4C_{\underline{k}'}$ such that $y \in \mathrm{past}(x)$, in contradiction to $y \in S(V)$. □

Let $\mathscr{N}$ be the set of predecessor complete sets of abstract 4-cells. It becomes a directed network by putting a directed edge from V_1 to V_2 whenever V_2 can be obtained from V_1 by adding one abstract 4-cell, $V_2 = V_1 \cup \{\underline{k}\}$. In particular, every edge is related to some abstract 4-cell, while the same abstract 4-cell can occur for several edges at different vertices. An *admissible sequence* $(V_1, \ldots, V_{r+1})$ is a path in $\mathscr{N}$ (using only edges in their direction) from the vertex $\emptyset$ to the vertex ${}^4\mathscr{A}$. We will show in Proposition 6 that admissible sequences exist.

We say that the admissible sequence $(V_1, \ldots, V_{r+1})$ *crosses the 4-cell* $\underline{k}$*in step* n iff $V_{n+1} = V_n \cup \{\underline{k}\}$. We say that it *crosses the 3-cell* $(i, \underline{k}) \in {}^3\mathscr{A}$ *in step* n iff $V_{n+1} = V_n \cup \{\underline{k}\}$.

Proposition 4 *Every admissible sequence crosses every 4-cell and every 3-cell exactly once.*

Proof Since each step in the path adds exactly one 4-cell, and since the last element of the sequence is the set of all 4-cells, each 4-cell must occur sooner or later, and cannot occur twice. The 3-cell $(i, \underline{k})$ gets crossed exactly when the 4-cell $\underline{k}$ gets crossed. □

In particular, r equals the number of 4-cells. Since the starting point is fixed, an admissible sequence can be characterized by specifying which edge to use in each step. Since the edges are labeled with abstract 4-cells, it can be specified by the sequence $(\underline{k}_1, \ldots, \underline{k}_r)$ of abstract 4-cells in the order in which they are crossed. Such a sequence is an ordering of the set of all abstract 4-cells. However, not every ordering of the 4-cells corresponds to an admissible sequence.

Proposition 5 *An ordering* $(\underline{k}_1, \ldots, \underline{k}_r)$ *of the 4-cells corresponds to an admissible sequence iff for every* $n \in \{1, \ldots, r\}$*, every predecessor of* $\underline{k}_n$ *occurred earlier.*

Proof "only if": Otherwise $V_{n+1} = V_n \cup \{\underline{k}_n\}$ is not predecessor complete, as $V_n = \{\underline{k}_1, \ldots, \underline{k}_{n-1}\}$.

"if": The sequence of 4-cells tells us in each step of the path in $\mathscr{N}$ which edge to take. In order to verify that such edges exist in $\mathscr{N}$, we need to check that, for each step from V_n to $V_{n+1} = V_n \cup \{\underline{k}_n\}$, V_{n+1} is predecessor complete. It is because each predecessor of $\underline{k}_n$ is contained in $V_n = \{\underline{k}_1, \ldots, \underline{k}_{n-1}\}$ by assumption, and because V_n is predecessor complete. Since every 4-cell occurs, the end point of the path in $\mathscr{N}$ is the set of all 4-cells. □

As a consequence, the sequence of 4-cells must begin with 0^N (the only one without predecessor) and end with $\underline{n}$ (the only one without successor). For $N = 2$ and $n_1 = 1 = n_2$ as in Figs. 4, 5, and 6, there are two orderings as described in Proposition 5: (00,01,10,11) and (00,10,01,11).

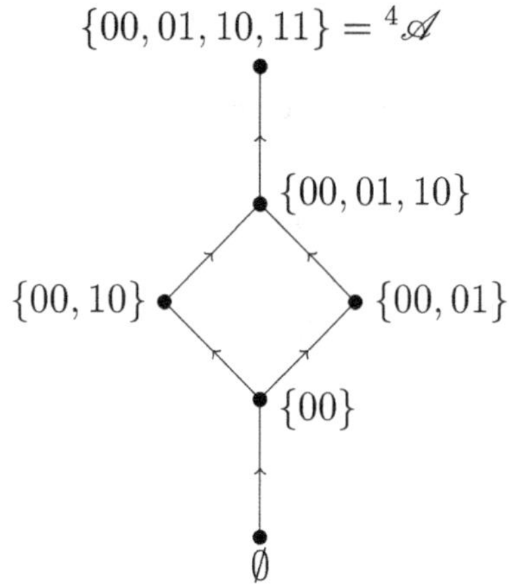

Fig. 5 The directed network $\mathscr{N}$ for two hyperboloids as in Fig. 4. There are two paths from $\emptyset$ to ${}^4\mathscr{A}$, both of which are admissible sequences

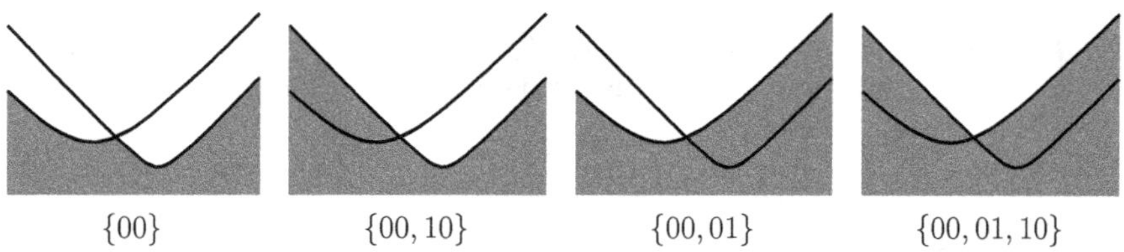

Fig. 6 The space-time sets (unions of 4-cells) $S(V)$ corresponding to some vertices V in $\mathscr{N}$ for two hyperboloids

Proposition 6 *For every choice of* $N, n_1, \ldots, n_N \in \mathbb{N}$*, there exists an admissible sequence.*

Proof For every $\underline{k}$, define $m(\underline{k}) = k_1 + \cdots + k_N$. We specify the ordering of 4-cells. Begin with $\underline{k} = 0^N$, the only 4-cell with $m(\underline{k}) = 0$. Then list, in arbitrary order, all 4-cells $\underline{k}$ with $m(\underline{k}) = 1$. Then, in arbitrary order, all 4-cells $\underline{k}$ with $m(\underline{k}) = 2$, and so on up to $m(\underline{k}) = \nu$, which occurs only for $\underline{k} = \underline{n}$. Then all 4-cells have occurred exactly once. Every predecessor $\underline{k}'$ of $\underline{k}$ occurred earlier than $\underline{k}$ because $m(\underline{k}') = m(\underline{k}) - 1$. (We remark that not every admissible sequence needs to have this structure.) □

Of two admissible sequences, we say that they differ by an *elementary deformation* if their associated orderings of 4-cells differ only by an exchange of two successive 4-cells, i.e., one is $(\underline{k}_1, \ldots, \underline{k}_r)$ and the other

$$\big(\underline{k}_1, \ldots, \underline{k}_{n-1}, \underline{k}_{n+1}, \underline{k}_n, \underline{k}_{n+2}, \ldots, \underline{k}_r\big) \tag{49}$$

for some $n \in \{1, \dots, r-1\}$. For example, in Fig. 5 this exchange corresponds to switching from the left path to the right one or vice versa. An exchange of 4-cells as in (49), applied to an admissible sequence, does not necessarily yield another admissible sequence, but here we need the converse fact:

Proposition 7 *Any two admissible sequences can be obtained from each other through finitely many elementary deformations.*

Proof Let $(\underline{k}_1, \dots, \underline{k}_r)$ be the ordering of 4-cells corresponding to one of the two admissible sequences and $(\underline{k}'_1, \dots, \underline{k}'_r)$ the other. Apply the following elementary deformations to the primed ordering. Find the place where $\underline{k}_1$ occurs and move $\underline{k}_1$ one place to the left in the primed ordering (by exchange with its left neighbor). The resulting ordering corresponds to an admissible sequence because $\underline{k}_1$ has no predecessor. Likewise, $\underline{k}_1$ can be moved again to the left, in fact repeatedly until it reaches the first position. Repeating the procedure, we can move $\underline{k}_2$ to the second position and so on until we have reached the unprimed ordering. In each intermediate ordering, predecessors always occur earlier, because they did in the two given orderings. □

4.2.2 Definition of L

We use an admissible sequence to define the operator ordering in $L(\underline{x})$, and then proceed to show that the operator $L(\underline{x})$ does not, in fact, depend on the choice of admissible sequence.

So fix an admissible sequence. Since each $\mathbb{H}_{ik}$ is partitioned into 3-cells, there is exactly one 3-cell ${}^3C(x_{ik})$ containing x_{ik} (except in the probability-0 case that x_{ik} lies on the boundary between two 3-cells on $\mathbb{H}_{ik}$, which we ignore). To the flash x_{ik} we associate the operator

$$K(x_{ik}) := U^0_{\mathbb{H}_{ik}}\, P_{\mathbb{H}_{ik}}\big(g_{x_{ik-1},{}^3C(x_{ik}),x_{ik},i}\big)\, U_0^{\mathbb{H}_{ik}}\,. \tag{50}$$

We define $L(\underline{x})$ as the product of the $K(x_{ik})$ in the order from right to left in which the 3-cells are crossed in the admissible sequence. Now in some steps of the sequence, several 3-cells are crossed in the same step. Among these, it does not matter which order we choose, as their operators commute:

Proposition 8 *Assume interaction locality and consider V and $V' = V \cup \{\underline{k}\}$ in $\mathscr{N}$. If ${}^3C(x_{ik})$ and ${}^3C(x_{j\ell})$ are two 3-cells in the common boundary of $S(V)$ and ${}^4C_{\underline{k}}$ (so $k = k_i$ and $\ell = k_j$), then $K(x_{ik})$ commutes with $K(x_{j\ell})$. As a consequence, every admissible sequence unambiguously defines a product $L(\underline{x})$.*

Proof Since ${}^3C(x_{ik}) \subseteq \mathbb{H}_{ik} \cap \partial S(V)$, and since $g_{x_{ik-1},{}^3C(x_{ik}),x_{ik}}$ vanishes outside of ${}^3C(x_{ik})$, the consequence (16) of interaction locality implies that multiplication by this g function can as well be carried out on $\partial S(V)$, i.e.,

$$K(x_{ik})\,K(x_{j\ell}) = U^0_{\mathbb{H}_{ik}}\,P_{\mathbb{H}_{ik}}\big(g_{x_{ik-1},{}^3C(x_{ik}),x_{ik},i}\big)\,U^{\mathbb{H}_{ik}}_{\mathbb{H}_{j\ell}}\,P_{\mathbb{H}_{j\ell}}\big(g_{x_{j\ell-1},{}^3C(x_{j\ell}),x_{j\ell},j}\big)\,U_0^{\mathbb{H}_{j\ell}} \tag{51a}$$

$$= U^0_{\partial S(V)}\,P_{\partial S(V)}\big(g_{x_{ik-1},{}^3C(x_{ik}),x_{ik},i}\big)\,P_{\partial S(V)}\big(g_{x_{j\ell-1},{}^3C(x_{j\ell}),x_{j\ell},j}\big)\,U_0^{\partial S(V)} \tag{51b}$$

$$= U^0_{\partial S(V)}\,P_{\partial S(V)}\Big(g_{x_{ik-1},{}^3C(x_{ik}),x_{ik},i}\;g_{x_{j\ell-1},{}^3C(x_{j\ell}),x_{j\ell},j}\Big)\,U_0^{\partial S(V)}\,. \tag{51c}$$

Since multiplication of the two g functions is commutative, $K(x_{j\ell})\,K(x_{ik})$ yields the same expression. □

Proposition 9 *Assuming interaction locality, any two admissible sequences lead to the same operator* $L(\underline{x})$.

Proof By Proposition 7, it suffices to consider two admissible sequences that differ by an elementary deformation as in (49). By Proposition 5, the two 4-cells $\underline{k}_n, \underline{k}_{n+1}$ that get exchanged must be such that neither is a predecessor of the other; that is, they do not have a 3-cell in common. Hence, for each of them the past boundary is a subset of $\partial S(V_n)$ with $V_n = \{\underline{k}_1, \ldots, \underline{k}_{n-1}\}$. For the same reasons as in the proof of Proposition 8, the K operators for any two flashes in 3-cells in the past boundaries of ${}^4C_{\underline{k}_n}$ and ${}^4C_{\underline{k}_{n+1}}$ commute (they are multiplication operators on a common spacelike surface). Thus, the different operator orderings associated with the two admissible sequences yield the same $L(\underline{x})$. □

This completes the definition of $L(\underline{x})$ and thus of $D(\underline{x})$ as in (45) and of the distribution of $\underline{X}$ as in (44). It remains to verify that $\mathbb{P}$ is a probability distribution.

4.2.3 Normalization

Proposition 10 *Interaction locality implies that*

$$\int_{\mathbb{M}^\nu} d\underline{x}\; D(\underline{x}) = I\,. \tag{52}$$

As a consequence, $G(\cdot)$ *is a POVM, and* (44) *defines a probability distribution for every* $\psi_0 \in \mathscr{H}_0$ *with* $\|\psi_0\| = 1$.

Proof Written out, (52) reads

$$\frac{1}{\tau^\nu}\Bigg(\prod_{ik}\;\int_{\mathrm{future}(x_{ik-1})} d^4x_{ik}\Bigg)\Bigg(\prod_{ik} e^{-|x_{ik}-x_{ik-1}|/\tau}\Bigg) L(\underline{x})^\dagger L(\underline{x}) = I \tag{53}$$

with the abuse of notation that $\prod_{ik}\int d^4x_{ik}$ means, not a product, but repeated integration over all x_{ik}, with each integral extending up to the equal sign.

By the coarea formula (32) and the normalization (33), it suffices to show that for all $s_{ik} > 0$,

$$\Bigg(\prod_{ik} \int\limits_{\mathbb{H}_{x_{ik-1}}(s_{ik})} d^3x_{ik}\Bigg) L(\underline{x})^\dagger L(\underline{x}) = I\,. \tag{54}$$

While Fubini's theorem allows us to exchange the order of integration, it must be noted here that the domain for x_{ik} depends on x_{ik-1}, so the x_{ik}-integral must occur to the right of the x_{ik-1}-integral. This limitation on the possible ordering of the integrals must be kept in mind; note also that the order of integrals is not a priori related to the order of factors in $L(\underline{x})$.

Fix the s_{ik} and let $\mathbb{H}_{ik} := \mathbb{H}_{x_{ik-1}}(s_{ik})$. We split the multiple integral, corresponding to the partition of each $\mathbb{H}_{ik}$ into the ${}^3C_{i\underline{k}}$'s with $k_i = k$, into a sum

$$\sum_{\substack{\underline{k}^{11}\ldots\underline{k}^{N n_N}\\ k_i^{ik}=k\,\forall ik}} \Bigg(\prod_{ik} \int\limits_{{}^3C_{i\underline{k}^{ik}}} d^3x_{ik}\Bigg) L(\underline{x})^\dagger L(\underline{x})\,. \tag{55}$$

Each summand is associated with a certain element of ${}^4\mathscr{A}^\nu$, and different summands with different elements of ${}^4\mathscr{A}^\nu$.

As a preparation for the general procedure, let us outline the first step of the induction. In each summand, consider the two innermost K factors of $L(\underline{x})^\dagger L(\underline{x})$, let them be $K(x_{j\ell})^\dagger K(x_{j\ell})$. We want to integrate them out using (29), resulting in a factor $P^{j\ell}_{{}^3C_{j\underline{k}^{j\ell}}}$ in the abbreviated notation

$$P^{j\ell}_{{}^3C_{i\underline{k}}} := U^0_{\mathbb{H}_{ik_i}}\, P_{\mathbb{H}_{ik_i}}\left(1_{x_{j\ell}\in {}^3C_{i\underline{k}}}\right) U_0^{\mathbb{H}_{ik_i}} \tag{56}$$

(which depends only on the 3-cell rather than on $\mathbb{H}_{ik_i}$ by interaction locality). But let us be slow and integrate out, at this step, $x_{j\ell}$ *only* if ${}^3C_{j\underline{k}^{j\ell}}$ lies on $\partial^4 C_{\underline{n}}$ (the futuremost surface formed by 3-cells). Since $k_j^{j\ell} = \ell$, and $\mathbb{H}_{j\ell}$ must border on ${}^4C_{\underline{n}}$, $\ell = n_j$; thus, there is no $\mathbb{H}_{j\ell+1}$, and therefore no obstacle to changing the order of integration so that the rightmost integral is over $x_{j\ell}$. That is, such an $x_{j\ell}$ can, in fact, be integrated out. Since factors corresponding to different 3-cells on $\partial^4 C_{\underline{n}}$ commute, we can integrate them all out. As a result, in each summand, there is no integration any more over any 3-cell on $\partial^4 C_{\underline{n}}$, but for each 3-cell on $\partial^4 C_{\underline{n}}$ involved in a summand, there is a factor of the form (56). The induction step will be about considering surfaces made up of 3-cells that lie further and further in the past, until we are done with the pastmost surface $\partial^4 C_{0^N}$ and all variables are integrated out. Now we give the details.

Fix an admissible sequence $(V_1, \ldots, V_{r+1})$. We will consider the sequence backwards and count down the index n of V_n from $r+1$ to 1. We write $V_n^c := {}^4\mathscr{A} \setminus V_n$ for the complement of V_n; since the corresponding space-time set $S(V_n^c) = S(V_n)^c$ is future complete, it is for every $n \neq r+1$ the future of some spacelike surface

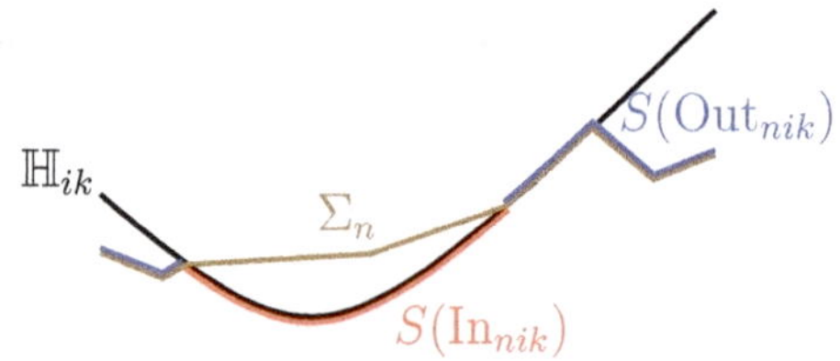

Fig. 7 The sets In and Out for a particular hyperboloid $\mathbb{H}_{ik}$ and Σ_n. Some pieces of hyperbola are drawn as straight lines. Some lines are drawn next to each other (rather than on top of each other) for better visibility

$$\Sigma_n := \partial S(V_n^c) = \partial S(V_n) = S(\partial V_n^c) = S(\partial V_n)\,. \tag{57}$$

At each stage of the process, each summand stemming from (55) is related to ν 3-cells. After integrating out one variable, we call the associated 3-cell an *out-cell*, while a 3-cell associated with a variable that has not yet been integrated out will be called an *in-cell*. In each step $n \to n-1$ of the induction, we will operate on the summands keeping their sum the same. Let $\mathscr{B}$ denote the set of abstract flashes, i.e., of all pairs (i,k):

$$\mathscr{B} := \big\{(i,k) \in \{1\dots N\} \times \mathbb{N} : 1 \le k \le n_i\big\}\,. \tag{58}$$

The abstract 3-cells on $\mathbb{H}_{ik}$ form the set

$${}^3\!\mathscr{A}_{ik} = \big\{(i,\underline{k}) \in {}^3\!\mathscr{A} : k_i = k\big\}\,. \tag{59}$$

Moreover, we define the sets that will turn out to be the sets of all in-cells (out-cells, respectively) by

$$\mathrm{In}_{nik} := \Big\{(i,\underline{k}) \in {}^3\!\mathscr{A}_{ik} : \text{earlier than } \partial V_n\Big\}\,, \tag{60a}$$

$$\mathrm{Out}_{nik} := \Big\{(j,\underline{\ell}) \in \partial V_n : \text{no later than } {}^3\!\mathscr{A}_{ik}\Big\}\,. \tag{60b}$$

They correspond to the space-time sets

$$S(\mathrm{In}_{nik}) = \mathbb{H}_{ik} \cap (\mathrm{past}(\Sigma_n) \setminus \Sigma_n)\,, \quad S(\mathrm{Out}_{nik}) = \Sigma_n \cap \mathrm{past}(\mathbb{H}_{ik})\,. \tag{61}$$

Together, they form the spacelike surface $\partial\big(\mathrm{future}(\mathbb{H}_{ik}) \cup \mathrm{future}(\Sigma_n)\big)$; see Fig. 7.

Induction hypothesis: *The summands are labeled by the elements of*

$$M_n := \Big\{\theta : \mathscr{B} \to {}^3\!\mathscr{A} \;:\; \forall ik \in \mathscr{B} : \theta(ik) \in \mathrm{In}_{nik} \cup \mathrm{Out}_{nik}\Big\} = \prod_{ik} (\mathrm{In}_{nik} \cup \mathrm{Out}_{nik}) \tag{62}$$

with $\prod$ *the Cartesian product, and the summand labeled* θ *reads*

$$\Big(\prod_{\substack{ik\in\mathscr{B}\\ \theta(ik)\in\mathrm{In}_{nik}}}\int_{^3C_{\theta(ik)}} d^3x_{ik}\Big)\Big(\prod_{\substack{ik\in\mathscr{B}\\ \theta(ik)\in\mathrm{In}_{nik}}} K(x_{ik})\Big)^{\dagger}\Big(\prod_{\substack{ik\in\mathscr{B}\\ \theta(ik)\in\mathrm{Out}_{nik}}} P^{ik}_{^3C_{\theta(ik)}}\Big)\Big(\prod_{\substack{ik\in\mathscr{B}\\ \theta(ik)\in\mathrm{In}_{nik}}} K(x_{ik})\Big), \tag{63}$$

where the product over $K(x_{ik})$ is understood in the order from right to left in which the 3-cells are crossed in the admissible sequence.

The form (63) of the summand labeled θ means, in particular, that the in-cells are integrated over, and the out-cells appear only in the projections in the middle. The order of the factors $P^{ik}_{^3C_{\theta(ik)}}$ need not be specified: they commute pairwise because all of the 3-cells $^3C_{\theta(ik)}$ lie on a common spacelike surface Σ_n.

The anchor of the induction is the case $n = r$, in which $\mathrm{In}_{rik} = \{(i, \underline{k}) \in {}^3\mathscr{A} : k_i = k\}$ and $\mathrm{Out}_{rik} = \emptyset$, so M_r as in (62) corresponds to those $(\underline{k}^{11} \ldots \underline{k}^{N n_N})$ with $k_i^{ik} = k$, and the summands agree with those of (55).

On the other end, for $n = 2$, we find that $V_2 = \{0^N\}$, $\Sigma_2 = \partial^4 C_{0^N}$, $\mathrm{In}_{2ik} = \emptyset$, and Out_{2ik} contains exactly the 3-cells on Σ_2. So the induction hypothesis, when proved, will imply that no summands involve integrals any more, and the sum reads

$$\sum_{\underline{k}^{11}\ldots\underline{k}^{N n_N}\in\partial\{0^N\}}\prod_{ik\in\mathscr{B}} P^{ik}_{^3C_{\underline{k}^{ik}}} = \prod_{ik\in\mathscr{B}}\underbrace{\Big(\sum_{\underline{k}\in\partial\{0^N\}} P^{ik}_{^3C_{\underline{k}}}\Big)}_{=I} = I\,, \tag{64}$$

as needed for (54).

So it remains to carry out the **induction step** $n \to n-1$, which consists of two parts. The first part deals with the projections in the middle of (63), the second with integrating out some of the variables.

First part: Interaction locality in the form (18) implies that the projection to the future boundary of a 4-cell can be "pulled across" the 4-cell, i.e., is equal to the projection to its past boundary,

$$P^{j\ell}_{\partial_+{}^4C} = P^{j\ell}_{\partial_-{}^4C}\,, \tag{65}$$

where $\partial_\pm$ denotes the future (past) boundary, which consists of one or more 3-cells, and $P^{j\ell}_{\partial_\pm{}^4C}$ equals the sum of the $P^{j\ell}_{^3C}$ over all 3-cells ^3C belonging to $\partial_\pm{}^4C$. The relevant 4-cell ^4C here is the one crossed by the admissible sequence between $n-1$ and n, $^4C = {}^4C_{\underline{k}}$ with $V_n = V_{n-1} \cup \{\underline{k}\}$. The future boundary of ^4C consists of 3-cells belonging to Σ_n, the past boundary of 3-cells belonging to Σ_{n-1}; in fact, the only difference between Σ_n and Σ_{n-1} is that the 3-cells belonging to the future boundary of ^4C are replaced by those belonging to the past boundary of ^4C.

For every $ik \in \mathscr{B}$,

$$\text{either all or none of the 3-cells in } \partial_+{}^4C \text{ belong to } \mathrm{Out}_{nik}. \tag{66}$$

Indeed, either ${}^4C \subseteq \text{future}(\mathbb{H}_{ik})$ or ${}^4C \subseteq \text{past}(\mathbb{H}_{ik})$. In the former case, $\partial_+{}^4C \subseteq \text{future}(\mathbb{H}_{ik}) \setminus \mathbb{H}_{ik}$; since Out_{nik} lies in the past of $\mathbb{H}_{ik}$, none of the 3-cells in $\partial_+{}^4C$ belong to Out_{nik}. In the latter case, $\partial_+{}^4C \subseteq \text{past}(\mathbb{H}_{ik})$; since all of the 3-cells in $\partial_+{}^4C$ belong to Σ_n, they all belong to Out_{nik}, which proves (66).

Now define

$$\widetilde{\text{Out}}_{nik} := \begin{cases} (\text{Out}_{nik} \setminus \partial_+{}^4C) \cup \partial_-{}^4C & \text{if } \partial_+{}^4C \subseteq \text{Out}_{nik} \\ \text{Out}_{nik} & \text{otherwise} \end{cases} \tag{67}$$

and $\widetilde{M}_n$ like M_n in (62) but with Out_{nik} replaced by $\widetilde{\text{Out}}_{nik}$.

Claim : The sum over $\theta \in M_n$ of (63) equals the sum over $\theta \in \widetilde{M}_n$ of (63) with Out_{nik} replaced by $\widetilde{\text{Out}}_{nik}$. (68)

To see this, think of M_n as the rightmost expression of (62). We take the following step successively for each $j\ell \in \mathscr{B}$ (in any ordering of $\mathscr{B}$): We replace $\text{Out}_{nj\ell}$ in (62) and (63) by $\widetilde{\text{Out}}_{nj\ell}$; that is, a P factor appears in each summand for each ik for which $\theta(ik) \in \text{Out}_{nik}$ respectively $\theta(ik) \in \widetilde{\text{Out}}_{nik}$, depending on whether the replacement step has already been done for ik. We check that each step leaves the sum unchanged; in fact, for every fixed choice of $\theta(ik)$ for all $ik \neq j\ell$, the sum of the summand remains unchanged. Indeed, this sum is a sum over all $\theta(j\ell) \in \text{In}_{nj\ell} \cup \text{Out}_{nj\ell}$. The summands with $\theta(j\ell) \in \text{In}_{nj\ell} \cup \text{Out}_{nj\ell} \setminus \partial_+{}^4C$ do not change. By (66), the summands with $\theta(j\ell) \in \text{Out}_{nj\ell} \cap \partial_+{}^4C$ together are either 0 or can be combined into one expression of the form (63) with $P^{j\ell}_{{}^3C_{\theta(j\ell)}}$ replaced by $P^{j\ell}_{\partial_+{}^4C}$. By (65), ∂_+ can be replaced by ∂_-, and by the same reasoning backwards, this equals the sum over $\theta(j\ell) \in \widetilde{\text{Out}}_{nj\ell} \cap \partial_-{}^4C$. Thus, each step leaves the sum unchanged, and after all steps (for all $j\ell$), we have proved the claim (68).

At this point, we have achieved in particular that all P factors refer to 3-cells on Σ_{n-1}.

Second part: We now wish to integrate out all variables that vary over 3-cells in Σ_{n-1}. We can do this for each summand individually, so focus on a particular $\tilde{\theta} \in \widetilde{M}_n$. The only 3-cells in Σ_{n-1} that were not included already in Σ_n are those in $\partial_-{}^4C$, and the only ones that any variable x_{ik} ever gets integrated over are those in In_{nik}. In the given summand $\tilde{\theta}$, there can be none or one or several variables x_{ik} for which $\partial_-{}^4C$ overlaps with In_{nik}. If none, we leave the summand unchanged. If one or more, we will treat them successively in an arbitrary order. So let $x_{j\ell}$ be one of them. The leftmost factors in the $\prod K(x_{ik})$ in (63) are those referring to 3-cells in $\partial_-{}^4C$; by Proposition 8, these factors commute with each other, so we can assume that $K(x_{j\ell})$ is the leftmost one.

Now we want to make sure that the integral over $x_{j\ell}$ is the rightmost integral. We can change the order of integration using Fubini's theorem, provided the domains of integration of the other integration variables do not depend on $x_{j\ell}$. The variables whose domain depends on $x_{j\ell}$ are $x_{j\ell+1}$ and higher ones for particle j. Since these domains all lie on $\mathbb{H}_{j\ell+1}$ or later, and thus in the future of $x_{j\ell}$, they lie on Σ_n or later, so by (60b) and the induction hypothesis, all of these variables have already been integrated out in previous induction steps ($\text{In}_{nj\ell+1} = \emptyset$), and we can assume that the $x_{j\ell}$ integral is the rightmost integral.

For carrying out the integral, we need that the space-time locations of the 3-cells ${}^3C_{\theta(\tilde{i}k)}$ in the factors $P^{ik}_{{}^3C_{\theta(\tilde{i}k)}}$ do not depend on $x_{j\ell}$. This follows if none of these 3-cells lies in the strict (open) future of $x_{j\ell}$. Now all of these 3-cells lie on Σ_{n-1}, a spacelike hypersurface containing $x_{j\ell}$, and thus not in the strict future of $x_{j\ell}$.

We also need that $K(x_{j\ell})$ commutes with the P's in the middle. That is the case because the P's are multiplication operators on their 3-cells and thus (by interaction locality) on Σ_{n-1}; likewise, $K(x_{j\ell})$ is by its definition (50) a multiplication operator on the 3-cell ${}^3C(x_{j\ell})$ containing $x_{j\ell}$ (which remains the same 3-cell ${}^3C := {}^3C_{\theta(\tilde{j}\ell)}$ during the integration over $x_{j\ell}$) and thus (by interaction locality) a multiplication operator on Σ_{n-1}. Since all multiplication operators on a common spacelike surface commute, we can pull $K(x_{j\ell})$ to the left of all P's, where it arrives next to $K(x_{j\ell})^\dagger$. Since none of the other factors (P's and K's) in the integrand depends on $x_{j\ell}$, they can be pulled out of the $x_{j\ell}$ integral. By (29), the integral can be carried out to yield

$$\int_{{}^3C} d^3x_{j\ell}\, K(x_{jk})^\dagger K(x_{jk}) = P^{j\ell}_{{}^3C}\,. \tag{69}$$

This factor joins the P factors, showing up in the correct position among all factors in the remaining integrand (63). In particular, still all P factors refer to 3-cells *on* Σ_{n-1}. We repeat this operation of carrying out the integral for all integrals over 3-cells on $\partial_-{}^4C$. Afterwards, in this summand $\tilde{\theta}$ the out-cells (with P factors) are those in $\widetilde{\text{Out}}_{nik}$ together with those in $\partial_-{}^4C$, and thus exactly those in $\text{Out}_{n-1,ik}$; the in-cells (with K factors) are those in In_{nik} except for those on Σ_{n-1} or later, and thus exactly those in $\text{In}_{n-1,ik}$. The summand has the form (63) with n replaced by $n-1$, and the index θ labeling the summands runs through M_{n-1}. We have thus proved the induction hypothesis for $n-1$, completed the induction step, and completed the proof of Proposition 10. □

4.2.4 Definition of the Theory

We have defined the model in (44) for chosen numbers n_i of flashes for each particle i. If we want to think of this model as a theory of the universe, and compare it to our empirical observations, we should take the limit $n_i \to \infty$ or choose n_i very large.

In contrast to the non-interacting 2004 model, in the present model the marginal distribution of the first $\tilde{n}_i$ flashes for each particle i (i.e., the distribution after integrating out the flashes after $\tilde{n}_i$) is *not* given by the same formula (44), although it is still given by *some* POVM. That is because the partition of the hyperboloids into 3-cells depends on the later flashes, and thus so does the procedure of cutting off the tails of the Gaussians. As a consequence, for the 2004 model we did not actually have to specify the numbers n_i, but now we have to; any choice of very large n_i should yield reasonable behavior of the theory, as well as the limit $n_i \to \infty$.

5 Properties of the Model

1. *Size of 3-cells.* Since the tails of the Gaussian profile function get cut off at the boundary of a 3-cell A, the width of the resulting profile function g_{yAx} could be smaller than σ if the diameter of A is, which could have undesirable consequences such as amplified empirical deviations of the model from standard quantum mechanics. I have made a crude estimate of the typical diameter of the 3-cells for condensed matter under everyday conditions and arrived at several millimeters or larger, which is much larger than GRW's suggested value of $\sigma = 10^{-7}$ m and thus suggests that the deviations are not amplified. Put differently, the tails are typically cut off at about 10^4 standard deviations, so the change is tiny. A more careful study of this question would be of interest.
   Matthias Lienert has made the interesting suggestion (personal communication) that since the Gaussians get cut off anyway, maybe they can be dispensed with altogether and replaced by a constant function (corresponding to the limit $\sigma \to \infty$); at each collapse, the wave function would then be localized to the size of a 3-cell. An investigation of whether such a theory is viable would be of interest.
2. *Stochastic evolution of the wave function.* In order to define a theory with flash ontology, it suffices to define the joint distribution of the flashes. But it is common to think of collapse model in terms of a stochastically evolving wave function. Such a wave function ψ_Σ can be defined for the present model for every spacelike surface Σ as follows. It should be related to the conditional probability distribution of $\underline{X}$, given the flashes up to Σ. To express this distribution, let $\mathscr{I} \subseteq \mathscr{B}$ be an arbitrary index set of ik's (with $\mathscr{B}$ as in (58) the set of all ik's), let $\mathscr{I}^c := \mathscr{B} \setminus \mathscr{I}$, and let $\underline{X}_{\mathscr{I}}$ be the collection of X_{ik} with $ik \in \mathscr{I}$; likewise $\underline{x}_{\mathscr{I}}$ etc., so we can write $\underline{x} = (\underline{x}_{\mathscr{I}}, \underline{x}_{\mathscr{I}^c})$. Then the conditional distribution of the flashes after Σ, given that those before Σ were at $\underline{x}_{\mathscr{I}} \in \text{past}(\Sigma)^{\mathscr{I}}$, is

$$\mathbb{P}\Big(\underline{X}_{\mathscr{I}^c} \in d\underline{x}_{\mathscr{I}^c} \Big| \underline{X}_{\mathscr{I}^c} \in \text{future}(\Sigma)^{\mathscr{I}^c} \text{ and } \underline{X}_{\mathscr{I}} = \underline{x}_{\mathscr{I}}\Big) = 1_{\underline{x}_{\mathscr{I}^c} \in \text{future}(\Sigma)^{\mathscr{I}^c}} \frac{\langle \psi_0 | D(\underline{x}) | \psi_0 \rangle}{\langle \psi_0 | W_\Sigma(\underline{x}_{\mathscr{I}})^2 | \psi_0 \rangle} \, d\underline{x}_{\mathscr{I}^c} \tag{70}$$

with positive operators

$$W_\Sigma(\underline{x}_{\mathscr{I}}) = \left(\frac{G(d\underline{x}_{\mathscr{I}} \times \text{future}(\Sigma)^{\mathscr{I}^c})}{d\underline{x}_{\mathscr{I}}} \right)^{1/2} = \left(\int\limits_{\text{future}(\Sigma)^{\mathscr{I}^c}} d\underline{x}_{\mathscr{I}^c} \, D(\underline{x}) \right)^{1/2}. \tag{71}$$

(The condition that $X_{ik+1} \in \text{future}(X_{ik})$ restricts the relevant index sets $\mathscr{I}$, but this fact does not change the validity of (70).) We therefore define, given that the flashes up to Σ were $\underline{x}_{\mathscr{I}}$,

$$\psi_\Sigma := \frac{U_0^\Sigma\, W_\Sigma(\underline{x}_{\mathscr{I}})\psi_0}{\|W_\Sigma(\underline{x}_{\mathscr{I}})\psi_0\|}\,, \tag{72}$$

in analogy to Eq. (25) of [24]. Considering a fixed pattern $\underline{x}$ of flashes and varying Σ, this wave function changes abruptly whenever Σ crosses one of the flashes (as $\mathscr{I}$ changes then). The conditional probability (70) can be expressed as

$$\mathbb{P} = 1_{\underline{x}_{\mathscr{I}^c}\in\text{future}(\Sigma)^{\mathscr{I}^c}} \Big\langle \psi_\Sigma \Big| U_0^\Sigma W_\Sigma(\underline{x}_{\mathscr{I}})^{-1} D(\underline{x}) W_\Sigma(\underline{x}_{\mathscr{I}})^{-1} U_\Sigma^0 \Big| \psi_\Sigma \Big\rangle \, d\underline{x}_{\mathscr{I}^c}\,. \tag{73}$$

3. *Non-interacting special case.* If the given unitary hypersurface evolution $U_\Sigma^{\Sigma'}$ is non-interacting, the situation simplifies as different particle variables x_j in the wave function can be evolved to different surfaces, and $K(x_{ik})$ commutes with $K(x_{j\ell})$ for $j \neq i$. If we could replace the cut-off Gaussians g_{yAx} of (26) in the definition (50) of the collapse by the original Gaussians $\tilde{g}_{yx}$ of (4), we would obtain exactly the 2004 model. Thus, whenever it is the case that the 3-cells A are typically much larger than the width σ of the Gaussians, then (with high probability) the cutting off does not make a big difference as it concerns only tiny tails of the Gaussian, and the 2004 model is a close approximation to the non-interacting case of the present model.
4. *Non-locality.* The collapse model presented here is non-local while being fully relativistic. In fact, it violates Bell's inequality. The non-locality corresponds to the fact that the joint distribution of two flashes is not a product even when the flashes are spacelike separated. Already the 2004 model was non-local, and further aspects of this property were discussed in [23–26].
5. *Microscopic parameter independence.* This is the property of a theory that the probability distribution of the local beables before any spacelike surface Σ does not depend on the external fields after Σ. For example, microscopic parameter independence is grossly violated in Bohmian mechanics (for Σ not belonging to the preferred foliation). The model presented here does not satisfy microscopic parameter independence exactly, but it does up to small deviations.
This is suggested by the following considerations. First, $U_0^{\Sigma'}$ does not depend on the external fields after Σ if both Σ_0 and Σ' lie in the past of Σ; by interaction locality, a collapse operator $K(x_{ik})$ does not depend on the external fields after Σ if ${}^3C(x_{ik})$ lies in the past of Σ. The space-time location of ${}^3C(x_{ik})$ (specifically, where its boundaries are) depends on other $x_{j\ell}$, but only on those before Σ. As a by-product of the proof of Proposition 10, the marginal distribution of the flashes in the past of ∂V_n for $V_n \in \mathscr{N}$ is given by the sum over $\theta \in M_n$ of the integrands in (63), so if $S(\partial V_n)$ lies in the past of Σ, this distribution will not depend on external fields after Σ. However, even if x_{ik} lies in the past of Σ, ${}^3C(x_{ik})$ need not lie in the past of Σ. Yet, it seems that the significant support of $g_{x_{ik-1},{}^3C(x_{ik}),x_{ik}}$ reaches no further than about $\sigma/c \approx 10^{-15}$ s into the future of Σ.
6. *No signaling.* This property means the impossibility for agents to transmit messages faster than light; it should follow from microscopic parameter independence,

as the message to be sent could be modeled as an external field and the message received would have to be some (coarse-grained) function of the local beables.

7. *Non-relativistic limit.* In the non-relativistic limit, the present model reduces to the non-relativistic GRW model, provided that the unitary evolution reduces to a non-relativistic unitary evolution. To see this, note that in the limit the hyperboloids become horizontal 3-planes, while the intersection between two hyperboloids escapes to infinity, so that every 3-cell becomes a full horizontal 3-plane and every 4-cell a layer between two such planes. Thus, cutting off the Gaussians becomes irrelevant, there is only one admissible sequence, K is just the Heisenberg-evolved multiplication by a Gaussian, and it becomes visible that the joint distribution of the flashes approaches that of the non-relativistic GRW model.

References

1. Y. Aharonov and D.Z. Albert: Can we make sense out of the measurement process in relativistic quantum mechanics? *Physical Review D* **24**: 359–371 (1981)
2. V. Allori, S. Goldstein, R. Tumulka, and N. Zanghì: On the Common Structure of Bohmian Mechanics and the Ghirardi-Rimini-Weber Theory. *British Journal for the Philosophy of Science* **59**: 353–389 (2008) http://arxiv.org/abs/quant-ph/0603027
3. D. Bedingham: Relativistic state reduction dynamics. *Foundations of Physics* **41**: 686–704 (2011) http://arxiv.org/abs/1003.2774
4. D. Bedingham: Relativistic state reduction model. *Journal of Physics: Conference Series* **306**: 012034 (2011) http://arxiv.org/abs/1103.3974
5. D. Bedingham and P. Pearle: On the CSL Scalar Field Relativistic Collapse Model. http://arxiv.org/abs/1906.11510
6. D. Bedingham, D. Dürr, G.C. Ghirardi, S. Goldstein, R. Tumulka, and N. Zanghì: Matter Density and Relativistic Models of Wave Function Collapse. *Journal of Statistical Physics* **154**: 623–631 (2014) http://arxiv.org/abs/1111.1425
7. J.S. Bell: Are there quantum jumps? In *Schrödinger. Centenary Celebration of a Polymath.* Cambridge University Press (1987). Reprinted as chapter 22 of [8].
8. J.S. Bell: *Speakable and unspeakable in quantum mechanics*. Cambridge University Press (1987)
9. F. Benatti, G.C. Ghirardi, and R. Grassi: Describing the macroscopic world: closing the circle within the dynamical reduction program. *Foundations of Physics* **25**: 5–38 (1995)
10. F. Dowker and J. Henson: Spontaneous Collapse Models on a Lattice. *Journal of Statistical Physics* **115**: 1327–1339 (2004) http://arxiv.org/abs/quant-ph/0209051
11. D. Dürr and P. Pickl: Flux-Across-Surfaces Theorem for a Dirac Particle. *Journal of Mathematical Physics* **44**: 423–456 (2003) http://arxiv.org/abs/math-ph/0207010
12. G.C. Ghirardi: Some Lessons From Relativistic Reduction Models. Pages 117–152 in H.-P. Breuer and F. Petruccione (editors), *Open Systems and Measurement in Relativistic Quantum Theory*, Heidelberg: Springer (1999)
13. G.C. Ghirardi, A. Rimini, and T. Weber: Unified dynamics for microscopic and macroscopic systems. *Physical Review D* **34**: 470–491 (1986)
14. J. Glimm and A. Jaffe: A $\lambda\phi^4$ Quantum Field Theory without Cutoffs. I. *Physical Review* **176**: 1945–1951 (1968)
15. M. Lienert: A relativistically interacting exactly solvable multi-time model for two mass-less Dirac particles in 1+1 dimensions. *Journal of Mathematical Physics* **56**: 042301 (2015) http://arxiv.org/abs/1411.2833

16. M. Lienert: *Lorentz invariant quantum dynamics in the multi-time formalism.* Ph.D. thesis, Mathematics Institute, Ludwig-Maximilians University, Munich, Germany (2015)
17. M. Lienert and L. Nickel: A simple explicitly solvable interacting relativistic N-particle model. *Journal of Physics A: Mathematical and Theoretical* **48**: 325301 (2015) http://arxiv.org/abs/1502.00917
18. M. Lienert and R. Tumulka: Born's Rule For Arbitrary Cauchy Surfaces. To appear in *Letters in Mathematical Physics* **110**: 753–804 (2020) http://arxiv.org/abs/1706.07074
19. P. Pearle: Toward a Relativistic Theory of Statevector Reduction. Pages 193–214 in A.I. Miller (editor), *Sixty-Two Years of Uncertainty: Historical, Philosophical, and Physical Inquiries into the Foundations of Quantum Physics*, NATO ASI Series B **226**, New York: Plenum Press (1990)
20. P. Pearle: Relativistic Collapse Model With Tachyonic Features. *Physical Review A* **59**: 80–101 (1999) http://arxiv.org/abs/quant-ph/9902046
21. P. Pearle: Relativistic dynamical collapse model. *Physical Review D* **91**: 105012 (2015) http://arxiv.org/abs/1412.6723
22. A. Tilloy: Interacting quantum field theories as relativistic statistical field theories of local beables. Preprint (2017) http://arxiv.org/abs/1702.06325
23. R. Tumulka: A relativistic version of the Ghirardi-Rimini-Weber model. *Journal of Statistical Physics* **125**: 821–840 (2006) http://arxiv.org/abs/quant-ph/0406094
24. R. Tumulka: Collapse and Relativity. Pages 340–352 in A. Bassi, D. Dürr, T. Weber, and N. Zanghì (editors), *Quantum Mechanics: Are there Quantum Jumps? and On the Present Status of Quantum Mechanics*, AIP Conference Proceedings **844**, American Institute of Physics (2006) http://arxiv.org/abs/quant-ph/0602208
25. R. Tumulka: Comment on "The Free Will Theorem." *Foundations of Physics* **37**: 186–197 (2007) http://arxiv.org/abs/quant-ph/0611283
26. R. Tumulka: The Point Processes of the GRW Theory of Wave Function Collapse. *Reviews in Mathematical Physics* **21**: 155–227 (2009) http://arxiv.org/abs/0711.0035
27. R. Tumulka: Paradoxes and Primitive Ontology in Collapse Theories of Quantum Mechanics. Pages 134–153 in S. Gao (editor), *Collapse of the Wave Function*, Cambridge University Press (2018) http://arxiv.org/abs/1102.5767

Non-Markov Processes in Quantum Theory

Bassano Vacchini

Abstract The study of quantum dynamics featuring memory effects has always been a topic of interest within the theory of open quantum system. The latter is concerned with providing useful conceptual and theoretical tools for the description of the reduced dynamics of a system interacting with an external environment. Definitions of non-Markovian processes have been introduced trying to capture the notion of memory effect by studying features of the quantum dynamical map providing the evolution of the system states, or changes in the distinguishability of the system states themselves. We introduce basic notions in the framework of open quantum systems. We stress in particular analogies and differences with models used for introducing modifications of quantum mechanics which should help in dealing with the measurement problem. We further discuss recent developments in the treatment of non-Markovian processes and their role in considering more general modifications of quantum mechanics.

1 Introduction

Quantum theory was born as a new mechanics, capable of providing the correct quantitative assessment of phenomena which could not find their explanation within the usual framework of classical mechanics. About a century after its introduction, many different facets and complementary presentations of the theory have been worked out. It has been put into evidence in particular that quantum theory indeed provides a new probabilistic framework for the prediction of outcomes of statistical experiments. It is therefore not only a "quantum" version of classical mechanics, it is indeed a "quantum" version of classical probability theory, containing into itself an often non trivial classical limit [1–3]. One of the most intriguing and delicate aspects of quantum theory is its irreducibly probabilistic structure, conflicting with the deterministic

B. Vacchini (✉)
Dipartimento di Fisica "Aldo Pontremoli", Università degli Studi di Milano,
Via Celoria 16, 20133 Milan, Italy
e-mail: bassano.vacchini@mi.infn.it

INFN Sezione di Milano, Via Celoria 16, 20133 Milan, Italy

V. Allori et al. (eds.), *Do Wave Functions Jump?*, Fundamental Theories of Physics 198,
https://doi.org/10.1007/978-3-030-46777-7_24

description we are accustomed to, as well as our everyday experience of the realization of definite events. From a classical viewpoint a probabilistic analysis is only necessary if not all degrees of freedom are under control or can be taken into account in detail. Not so for quantum theory. This state of affairs has led among others to the so-called "measurement problem". It refers to the difficulty in reconciling the classical description for macroscopic objects and the laws of quantum theory, predicting a statistical distribution rather than definite events [4]. On turn, this problem has led to consider alternatives to quantum theory, complying with its successes but leading to a different behavior for the prediction of events, effectively suppressing superposition of macroscopic objects. Among these theories one of the most renowned classes is given by collapse models, also known as dynamical reduction models [5, 6], arisen from the seminal paper [7]. Their distinctive trait is a stochastic non-linear modification of the Schrödinger equation, which on top of the standard evolution allows for the introduction of a collapse or localization mechanism. This mechanism, once accepted, avoids the measurement problem. Importantly, this mechanism has to be implemented at the level of the wavefunction, so as to allow for the suppression of superpositions. Nevertheless, at the level of experimental observations, it usually cannot be distinguished from other effects leading to a vanishing contribution of coherences.

The theory of open quantum systems is focused on the description of the reduced dynamics of a system interacting with other degrees of freedom, typically called environment, which are not described in detail [9, 10]. The environment therefore brings in an additional level of randomicity in the dynamics, on top of the unavoidable statistical aspect brought in by quantum theory. In this framework, the suppression of superposition states in a given basis is indeed predicted from a class of models known as decoherence models [8]. It thus appears that such models, bringing in another element of probabilistic description, typically provide the same average effect as dynamical reduction models, aimed at overcoming the inherent statistical structure of any quantum dynamics. In this respect, the two fields of dynamical reduction models and open quantum systems share some underlying mathematical structure. We will briefly address recent advancements in open quantum system having this perspective in mind. An important caveat to be mentioned is the fact that decoherence models do not provide a solution of the measurement problem in the sense addressed by collapse models: the suppression of macroscopic superpositions only takes place in the average and a whole statistical distribution of outcomes is predicted [11].

The contribution is organized as follows. In Sect. 2 we briefly outline the open quantum system viewpoint and address the term quantum process as used in the physical literature. The description of decoherence effects and their relationship to specific collapse models is worked out in Sect. 3. Finally Sect. 4 is devoted to introduce the notion of non-Markovian dynamics for an open system, and its influence on the elaboration of dynamical reduction models.

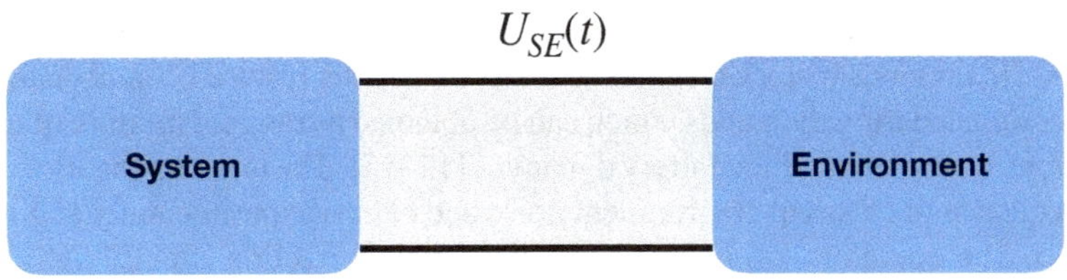

Fig. 1 Illustration of an open system interacting with an environment via a unitary coupling $U_{SE}(t)$

2 Open Systems and Quantum Processes

For the case in which a quantum system is not isolated from other quantum systems, the latter should be taken into account in the description of its dynamics. If the system and the other degrees of freedom, collectively named environment, do not share correlations at the initial time, one can describe the evolution of the system alone by introducing a collection of completely positive trace preserving maps $\{\Phi(t)\}_{t\in\mathbb{R}_+}$. They determine the statistics of any local observation once the initial state of the system $\rho_S(0)$ has been specified according to the formula

$$\langle A_S\rangle_t = \mathrm{Tr}\{A_S\Phi(t)[\rho_S(0)]\},$$

where A_S denotes a system observable. The collection of maps $\{\Phi(t)\}_{t\in\mathbb{R}_+}$ describes what is usually called a quantum process. The term process is here used in a loose sense, in analogy with the classical situation. It hints at the presence of an irreducible randomicity, here corresponding to the environmental degrees of freedom not accessible or described in detail, but affecting the system dynamics due to a unitary coupling with the environment $U_{SE}(t)$ as drawn in Fig. 1. If system and environment interaction can be neglected, and only in this case, $\Phi(t)$ is a unitary transformation, implying in particular a group composition law. In all other cases reversibility is lost, and the general mathematical structure of this collection of maps is not known. Some partial results are however available. A most famous and relevant class of reduced dynamics is obtained if we ask $\Phi(t)$ to obey a semigroup composition law forward in time. For this case we have $\Phi(t) = \exp(t\mathcal{L})$, with $\mathcal{L}$ in Lindblad form [9], that is

$$\mathcal{L}[\rho_S(t)] = -\frac{i}{\hbar}[H, \rho_S(t)] + \sum_k \lambda_k \left[A_k\rho_S(t)A_k^\dagger - \frac{1}{2}\{A_k^\dagger A_k, \rho_S(t)\}\right],$$

where $\{A_k\}$ and H denote system operators, with H an effective self-adjoint Hamiltonian, and λ_k are positive rates. A dynamics of this kind has always been called Markovian, since it arose as quantum counterpart of classical Markovian semigroups. The implicit idea is that the stochasticity in the dynamics arising due to interaction with the environment does not lead to effects that can be termed memory, making reference to previous history or states of the system. This feature is immediately lost even only considering dynamics which can be obtained as random mixture of unitary evolutions, so-called random unitary dynamics [12–15]. The latter might arise also as a consequence of classical environment noise and can be experimentally engineered [16, 17]. The operators $\{A_k\}$ describe microscopic interaction events, e.g. random localization or momentum transfer events for the case of decoherence as discussed in Sect. 3.

3 Events and Decoherence

Dynamical reduction models and open quantum system theory share a common root in the treatment of measurement in quantum mechanics. The description of measurement deals with a description of the outcomes of statistical experiments in which the interaction with the measurement apparatus is taken into account. Indeed, the first seminal contributions to open quantum systems were intimately connected with the description of measurement processes and its relevance for the foundations of quantum mechanics [18–20]. They put into evidence the relevance of the mathematical notion of complete positivity. Not by chance the original GRW paper, which introduced the first collapse model, was built upon work aimed at the quantum description of continuous measurement in time [21, 22], and started the treatment from a master equation describing decoherence in position [23].

To better work out this connection, let us consider in more detail how a collapse model can describe in the average a decoherence effect and how a microscopic description of decoherence can be related to a notion of event. In this spirit we briefly recall the formulation of the GRW model in the formulation via stochastic differential equations [5, 24]

$$\mathrm{d}|\psi(t)\rangle = -\frac{i}{\hbar}\hat{H}_0|\psi(t)\rangle \mathrm{d}t + \int_{\mathbb{R}} \mathrm{d}y \left(\frac{L(y,\hat{x})}{\| L(y,\hat{x})|\psi(t)\rangle \|} - \mathbb{1} \right) |\psi(t)\rangle \mathrm{d}N(y,t). \quad (1)$$

Here $\psi(t)$ is the system's wavefunction, $\hat{H}_0$ denotes the Hamiltonian appearing in the standard Schrödinger equation and the stochastic modification is determined by the collection of operators $\{L(y,\hat{x})\}_{y\in\mathbb{R}}$, with $\hat{x}$ the standard position operator, and the family of classical stochastic processes $\{N(y,t)\}_{y\in\mathbb{R}}$. Note in particular that this modification is non-linear in $\psi(t)$. In order to obtain suppression of spatial superposition of states, the L operators have to act as localization operators and to recover the original GRW model must be of the form

$$L(y,\hat{x}) = \frac{1}{\sqrt[4]{\pi} r_c} e^{-\frac{(y-\hat{x})^2}{2r_c^2}}. \quad (2)$$

The stochastic modification depends on the field of independent processes $\{N(y,t)\}_{y\in\mathbb{R}}$ such that $N(y,t)dy$ is the counting process giving the number of jumps taking place at time t in the space interval from y to $y+dy$. The collection of counting processes satisfies $\mathrm{d}N(x,t)\mathrm{d}N(y,t)=\delta(x-y)\mathrm{d}N(y,t)$, with rates given by

$$\mathbb{E}[\mathrm{d}N(y,t)]=\lambda\|L(y,\hat{x})|\psi(t)\rangle\|^2\mathrm{d}t,$$

where $\mathbb{E}[\cdot]$ denotes the stochastic average. The phenomenological parameters λ and r_c determine intensity and localization strength of the random jumps inducing a dynamical localization in position of the system. Averaging over the realization of the processes one obtains the state determining the statistics of observation on the system, namely

$$\rho(t)=\mathbb{E}[|\psi(t)\rangle\langle\psi(t)|],$$

which obeys the master equation

$$\frac{\mathrm{d}}{\mathrm{dt}}\rho(t)=-\lambda\left[\rho(t)-\int dyL(y,\hat{x})\rho(t)L(y,\hat{x})\right] \tag{3}$$

predicting a reduction of the off-diagonal matrix elements in the position representation according to

$$\langle x|\rho(t)|y\rangle=\exp\left(-\lambda t\left[1-\int dzL(z,x)L(z,y)\right]\right)\langle x|\rho(0)|y\rangle. \tag{4}$$

The obtained master Eq. (3) is in standard Lindblad form [9], describes decoherence in position according to Eq. (4), and in particular is characterised by translational invariance. Building on this aspect one realizes that it can be written in an explicit translationally covariant form [25–27] as follows

$$\frac{\mathrm{d}}{\mathrm{dt}}\rho(t)=-\lambda\left[\rho(t)-\int dq\tilde{L}(q)\mathrm{e}^{\frac{i}{\hbar}q\hat{x}}\rho(t)\mathrm{e}^{-\frac{i}{\hbar}q\hat{x}}\right] \tag{5}$$

with $\tilde{L}(q)$ Fourier transform of the function $L^2(y,0)$, that is again a Gaussian weight. It thus appears that the dynamics that can be observed as a consequence of the localization mechanism, described at the level of trajectories of the wavefunction in Hilbert space by the stochastic differential equation Eq. (1), is the same that would arise as a consequence of interaction of the system with an external environment whose effect can be described in terms of localisation events as in Eq. (3) or in terms of momentum transfers described by the collection of unitaries $\left\{\mathrm{e}^{\frac{i}{\hbar}q\hat{x}}\right\}_{q\in\mathbb{R}}$ as in Eq. (5). This viewpoint, connecting the open system based description of decoherence and the measurement based viewpoint of collapse models, implies in particular that the natural benchmark in the assessment of possible modifications of the quantum mechanical predictions due to a collapse mechanism is the estimate of possible decoherence effects affecting the considered dynamics. Indeed, this is one of the main difficulties in looking for experimental signatures of collapse mechanisms [6]. On the other hand, awareness of this relationship has opened the way to consider variants of dynamical reduction models. In particular, it has led to overcome an important intrinsic limitation of models such as Eq. (1), which predict an infinite growth of the system energy [24, 28]. A further natural extension of dynamical reduction models arising from analogy and differences shared with open quantum system models is the inclusion of memory effects [29–33], in view of a definition of non-Markovian dynamics as discussed in Sect. 4.

4 Non-Markovian Processes

In mentioning some of the basic tenets and results of the theory of open quantum systems, we have put into evidence the notion of quantum process as used and understood in the physical literature. In particular, the time evolutions arising as solutions of master equations in Lindblad form are typically termed quantum Markovian processes, since they provide the natural quantum counterpart of classical semigroup evolutions, arising in connection with homogeneous in time Markovian processes. A next natural step in this respect is considering time evolutions which can provide a quantum realization of a non-Markovian process. Given the looser definition of process considered in the quantum framework, as a collection of time dependent completely positive trace preserving maps describing a continuous quantum dynamics, one might consider a suitable definition of non-Markovian quantum process within this very same framework of dynamical maps. Indeed, providing a notion of non-Markovian quantum process in the same spirit as in the classical case, which gives an exact defintion of Markovian process in terms of conditions on the infinite hierarchy of conditional probability densities for the process, appears to be a very difficult task. Already from a conceptual point of view the situation does not appear to be neatly defined, since speaking about values of an observable at a given time calls for a measurement procedure which affects the subsequent values to be assumed by the quantity [34]. On the contrary, focusing on the collection of completely positive trace preserving maps giving the reduced dynamics has allowed to introduce clearcut definitions of Markovian, and in a complementary way non-Markovian, quantum process. Actually, there have been various proposals in this direction. We will here only focus on one of them, based on the behavior of the distinguishability of states in time, which is in direct relationship with a notion of divisibility of the time evolution maps. For more details and a complete treatment we refer the reader to recent reviews [35–38].

The basic insight can be summarized as follows. By interacting with the environmental degrees of freedom the system gets correlated with the environment and possibly leads to a change in time of the reduced state of the environment itself. As a consequence of the dynamics therefore, the capability of distinguishing two different initial system states, by performing measurements on the system degrees of freedom only, changes in time. Indeed, taking the partial trace necessary to define the reduced system state, which is all that is necessary in order to provide the statistics of measurements on the system, the whole information about correlations is no more available. To exploit this fact one can introduce a suitable quantifier of the distinguishability between states, such as the trace distance, given by the trace norm of the difference of the states

$$D(\rho_S^1(t), \rho_S^2(t)) = \frac{1}{2}\|\rho_S^1(t) - \rho_S^2(t)\|_1 \tag{6}$$

and consider its behavior in time. Being a contraction under the action of completely positive trace preserving transformations, the trace distance always diminishes with respect to its initial value, that is

$$D(\rho_S^1(t), \rho_S^2(t)) \leqslant D(\rho_S^1(0), \rho_S^2(0)).$$

In particular for the semigroup case, considered in Sect. 2 for the case of a quantum Markovian process, due to the composition law one has a monotonous reduction of the distance among states with time. In such a situation the distance between states, and therefore their distinguishability [39], gets smaller and smaller with elapsing time. The failure of this monotic decreasing behavior for at least a pair of possible initial states has been taken as indication of non-Markovian dynamics in the seminal paper [40]. Indeed, it amounts to a revival in the distinguishability between the states that can only arise as a consequence of previously established correlations with the environment or changes in the environmental state that affect the subsequent reduced system dynamics. This fact is schematically drawn in Fig. 2. The validity of this interpretation is substantiated by the inequality [41–43].

$$\begin{aligned} D(\rho_S^1(t), \rho_S^2(t)) - D(\rho_S^1(s), \rho_S^2(s)) \leqslant\ & D(\rho_{SE}^1(s), \rho_S^1(s) \otimes \rho_E^1(s)) \\ & + D(\rho_{SE}^2(s), \rho_S^2(s) \otimes \rho_E^2(s)) \\ & + D(\rho_E^1(s), \rho_E^2(s)), \end{aligned} \tag{7}$$

where it is assumed that $t \geqslant s$. The term at the lhs when positive provides a signature of non-Markovianity, so that the positivity of the rhs is a precondition for non-Markovianity, to be traced back to the effects mentioned above: correlations and influence of the system on the environment. While the notions of distinguishability, contractivity of the used distinguishability quantifier upon the action of a quantum transformation, and connection of the distinguishability revivals to the imprint of the system dynamics left in correlations or environment, provide the basic traits of this approach to the description of memory effects in quantum mechanics, many more subtle issues are involved in the definition of this framework. Importantly, there is a stringent mathematical connection between this viewpoint and divisibility properties of the time evolution, corresponding to the fact that the evolution over a finite time

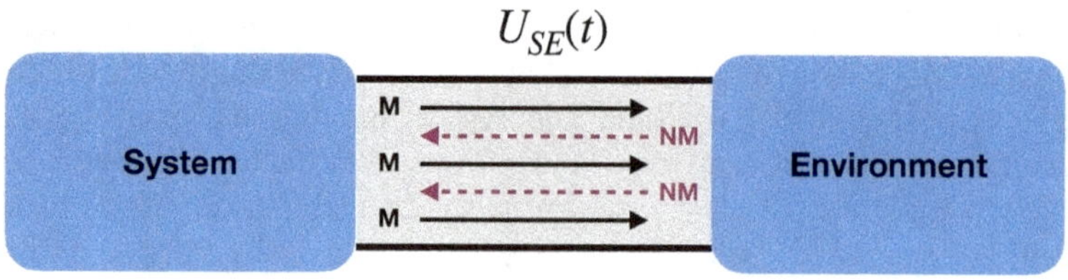

Fig. 2 Open system interacting with an environment via a unitary coupling $U_{SE}(t)$. Markovian effects (M) are depicted as an information flow from system to environment, while an information flow from environment to system (NM) is identified with memory effects

can always be split into evolutions over shorter times, each described by a proper quantum transformation [44–46].

Dynamics allowing for non-Markovian effects have also been considered in the above-mentioned framework of a decoherence dynamics driven by random events [47, 48], as well as in the introduction of more general dynamical reduction models [31, 49]. While in the context of decoherence allowing for non-Markovian dynamics is a way to consider more general and accurate description of the reduced dynamics, within the framework of dynamical reduction models non-Markovian models lead to possibly more stringent exclusion regions of the parameter values which characterise the model.

5 Conclusions and Outlook

In recent times a lot of work in the field of open quantum systems has been devoted to characterization and study of non-Markovian dynamics. This research has involved in the first instance the very definition and clarification of what can be meant as quantum dynamics featuring memory effects. It has further addressed the possible relevance of non-Markovian dynamics in the description of the reduced dynamics of non isolated quantum systems as well as related fields. In this contribution we have recalled in particular the relationship between the description of decoherence in open quantum systems and modifications of quantum mechanics such as dynamical reduction models introduced for the sake of better grasping the so-called quantum measurement problem. We have briefly discussed a natural physical interpretation of non-Markovian dynamics as related to information exchange between system and environment. We have further pointed to the use of the formalism of non-Markovian dynamics to consider more general collapse model which might help in improving the known bounds on the parameters characterizing the possible deviations from standard quantum mechanics. The relevance of the classification of non-Markovian dynamics itself as well as the role of memory effects in collapse mechanisms remain two open questions that will surely involve future research.

Acknowledgements The author acknowledges support from the Joint Project "Quantum Information Processing in Non-Markovian Quantum Complex Systems" funded by FRIAS, University of Freiburg and IAR, Nagoya University, from the FFABR project of MIUR and from the Unimi Transition Grant H2020.

References

1. R. F. Streater, J. Math. Phys. **41**, 3556 (2000)
2. F. Strocchi, *An introduction to the mathematical structure of quantum mechanics* (World Scientific, 2005)
3. B. Vacchini, *Covariant Mappings for the Description of Measurement, Dissipation and Decoherence in Quantum Mechanics*, in *Theoretical Foundations of Quantum Information Processing and Communication*, edited by E. Bruening and F. Petruccione (Springer, Berlin, 2010), Lecture Notes in Physics 787, pp. 39–77
4. M. Schlosshauer, Rev. Mod. Phys. **76**, 1267 (2004)
5. A. Bassi and G. Ghirardi, Phys. Rep. **379**, 257 (2003)
6. A. Bassi, K. Lochan, S. Satin, T. P. Singh, and H. Ulbricht, Rev. Mod. Phys. **85**, 471 (2013)
7. G. C. Ghirardi, A. Rimini, and T. Weber, Phys. Rev. D **34**, 470 (1986)
8. K. Hornberger, *Introduction to Decoherence Theory*, in *Entanglement and Decoherence*, edited by Andreas Buchleitner and Carlos Viviescas and Markus Tiersch (Springer, Berlin, 2009), Lecture Notes in Physics 768, pp. 221–276
9. H.-P. Breuer and F. Petruccione, *The Theory of Open Quantum Systems* (Oxford University Press, Oxford, 2002)
10. A. Rivas and S. F. Huelga, *Open Quantum Systems: An Introduction* (Springer, 2012)
11. E. Joos, H. D. Zeh, C. Kiefer, D. Giulini, J. Kupsch, and I.-O. Stamatescu, *Decoherence and the Appearance of a Classical World in Quantum Theory*, 2nd edn. (Springer, Berlin, 2003)
12. K. M. R. Audenaert and S. Scheel, New Journal of Physics **10**, 023011 (2008)
13. A. Pernice, J. Helm, and W. T. Strunz, Journal of Physics B: Atomic, Molecular and Optical Physics **45**, 154005 (2012)
14. B. Vacchini, J. Phys. B **45**, 154007 (2012)
15. D. Chruściński and F. A. Wudarski, Phys. Lett. A **377**, 1425 (2013)
16. S. Cialdi, M. A. C. Rossi, C. Benedetti, B. Vacchini, D. Tamascelli, S. Olivares, and M. G. A. Paris, Applied Physics Letters **110**, 081107 (2017)
17. M. A. C. Rossi, C. Benedetti, D. Tamascelli, S. Cialdi, S. Olivares, B. Vacchini, and M. G. A. Paris, International Journal of Quantum Information **15**(08), 1740009 (2017)
18. E. B. Davies, *Quantum Theory of Open Systems* (Academic Press, London, 1976)
19. G. Ludwig, *Foundations of quantum mechanics.* (Springer-Verlag, New York, 1983)
20. K. Kraus, *States, Effects, and Operations*, Vol. 190 of *Lecture Notes in Physics* (Springer, Berlin, 1983)
21. A. Barchielli, L. Lanz, and G. M. Prosperi, Nuovo Cimento B **72**, 79 (1982)
22. A. Barchielli, L. Lanz, and G. M. Prosperi, Found. Phys. **13**, 779 (1983)
23. B. Vacchini, J. Phys. A: Math. Theor. **40**, 2463 (2007)
24. A. Smirne, B. Vacchini, and A. Bassi, Phys. Rev. A **90**, 062135 (2014)
25. A. S. Holevo, Rep. Math. Phys. **32**, 211 (1993)
26. B. Vacchini, J. Math. Phys. **42**, 4291 (2001)
27. B. Vacchini, Phys. Rev. Lett. **95**, 230402 (2005)
28. A. Bassi, E. Ippoliti, and B. Vacchini, J. Phys. A: Math. Gen. **38**, 8017 (2005)
29. A. Bassi and L. Ferialdi, Phys. Rev. Lett. **103**, 050403 (2009)
30. A. Bassi and L. Ferialdi, Phys. Rev. A **80**, 012116 (2009)
31. L. Ferialdi and A. Bassi, Phys. Rev. A **86**, 022108 (2012)
32. L. Ferialdi and A. Bassi, Phys. Rev. Lett. **108**, 170404 (2012)

33. L. Ferialdi and A. Smirne, Phys. Rev. A **96**, 012109 (2017)
34. B. Vacchini, A. Smirne, E.-M. Laine, J. Piilo, and H.-P. Breuer, New J. Phys. **13**, 093004 (2011)
35. H.-P. Breuer, J. Phys. B **45**, 154001 (2012)
36. A. Rivas, S. F. Huelga, and M. B. Plenio, Rep. Prog. Phys. **77**, 094001 (2014)
37. H.-P. Breuer, E.-M. Laine, J. Piilo, and B. Vacchini, Rev. Mod. Phys. **88**, 021002 (2016)
38. I. de Vega and D. Alonso, Rev. Mod. Phys. **89**, 015001 (2017)
39. C. A. Fuchs and J. van de Graaf, IEEE Trans. Inf. Th. **45**, 1216 (1999)
40. H.-P. Breuer, E.-M. Laine, and J. Piilo, Phys. Rev. Lett. **103**, 210401 (2009)
41. E.-M. Laine, J. Piilo, and H.-P. Breuer, EPL **92**, 60010 (2010)
42. H.-P. Breuer, G. Amato, and B. Vacchini, New Journal of Physics **20**, 043007 (2018)
43. S. Campbell, M. Popovic, D. Tamascelli, and B. Vacchini, New Journal of Physics **21**, 053036 (2019)
44. A. Rivas, S. F. Huelga, and M. B. Plenio, Phys. Rev. Lett. **105**, 050403 (2010)
45. D. Chruscinski, A. Kossakowski, and A. Rivas, Phys. Rev. A **83**, 052128 (2011)
46. S. Wißmann, H.-P. Breuer, and B. Vacchini, Phys. Rev. A **92**, 042108 (2015)
47. B. Vacchini, Phys. Rev. A **78**, 022112 (2008)
48. A. Smirne and B. Vacchini, Phys. Rev. A **82**, 042111 (2010)
49. S. L. Adler and A. Bassi, J. Phys. A: Math. Theor. **40**, 15083 (2007)

Experimental Physics

Eight Oxford Questions: Quantum Mechanics Under a New Light

N. Ares, A. N. Pearson, and G. A. D. Briggs

Abstract Conceptual and experimental advances are opening up possibilities for addressing new questions in quantum theory. What is changing is the potential for relating conceptual and theoretical developments to foreseeable experimental tests. It is becoming feasible to rule out certain interpretations, maybe even to look for new ones, as well as addressing the various open questions in quantum mechanics, such as the role of gravity. We set out eight questions as a manifesto for future study and research.

1 Background

Apart from violations of inequalities of the type of Bell's, which have been implemented to stunning precision [1–9], until recently it was, for the most part, doubted that experiments could ever discriminate between different variants and interpretations of quantum theory (QT). However, we now believe that avenues for such experimental tests are opening up. The steady improvement of experimental techniques [10, 11] for manipulating quantum systems might even allow us now to explore the post-quantum territory.

Fundamentally different theories of quantum reality, such as Everettian QT [12, 13], collapse-variants of QT [14–18], the pilot-wave theory [19–21] and Quantum Bayesianism [22], disagree on crucial issues like locality, reversibility, universality, completeness and determinism. For each variant there are different interpretations for entities appearing in the theory—for instance, concerning the reality of the wave function (which motivated the so-called ontological models framework [23]). It might be argued that since the empirically accessible part of quantum theory is the same for different interpretations/variants of QT, we should not worry about the above differences, but the key point is that differences may become testable in the near future.

N. Ares · A. N. Pearson · G. A. D. Briggs (✉)
Department of Materials, University of Oxford, Oxford OX1 3PH, UK
e-mail: andrew.briggs@materials.ox.ac.uk

V. Allori et al. (eds.), *Do Wave Functions Jump?*, Fundamental Theories of Physics 198,
https://doi.org/10.1007/978-3-030-46777-7_25

Looking for post quantum theories, a good place to start would be where gravitational effects are important. The problem of integrating general relativity (GR) with QT remains open. Loop quantum gravity [24] and string theory [25] are the two pre-eminent approaches that have been taken to studying such issues. GR and QT—the best available fundamental theories in their own domains of applicability—clash with one another at the fundamental level. Different quantum physical realities, even if empirically equivalent, may suggest different ways of going about reconciling QT with GR, of searching for the successor of QT, and of designing new experiments for QT. For example, a local, deterministic theory might be easier to reconcile with GR than a non-local, stochastic one.

To help make progress with such deep, long-unsolved problems, it is useful to find some specific key questions which could be addressed in the shorter term, rather than simply revisiting a familiar cycle of arguments which have now been around for decades. Fresh pieces of evidence are of the essence. We therefore ask where one should look for that evidence, and what kinds of new experimental (and theoretical) tools are needed to look for it? Although the realisation of experiments that would shine light on some of the long-standing problems will be challenging, the mere exercise of trying to pose the appropriate questions stimulates the intellectual environment and may hopefully create a fertile territory for scientific breakthroughs. This was the fundamental motivation for the questions which follow, which we have collected during a conference on Experimental Tests of Quantum Reality. This conference followed a conference on Quantum Physics and the Nature of Reality that led to *The Oxford Questions on the foundations of quantum physics* [26].

These new questions aim for a productive interplay between theory and experiment. Our questions reinforce the conviction that experimental tests are of the essence in discerning which foundational theories to accept and which to reject. We expect that the interplay between experimental and theoretical tools will in turn lead to the development of new ideas. The questions are charted in Fig. 1. The questions pertain to two territories: one, within QT and the other, post-QT, and some overlap between them.

2 The Questions

1. **Is it possible to devise an experiment that might allow one to falsify one of the interpretations or variants of QT?**

The quantum realities described by the variants of QT show profound differences from one another. For example, **Everettian quantum theory (EQT)** is universal [13], *deterministic* and *reversible*, and non-probabilistic, in that it does not have the Born Rule axiom. Moreover, the theory is deemed to be *local* by some [27], and non-local by others [28].

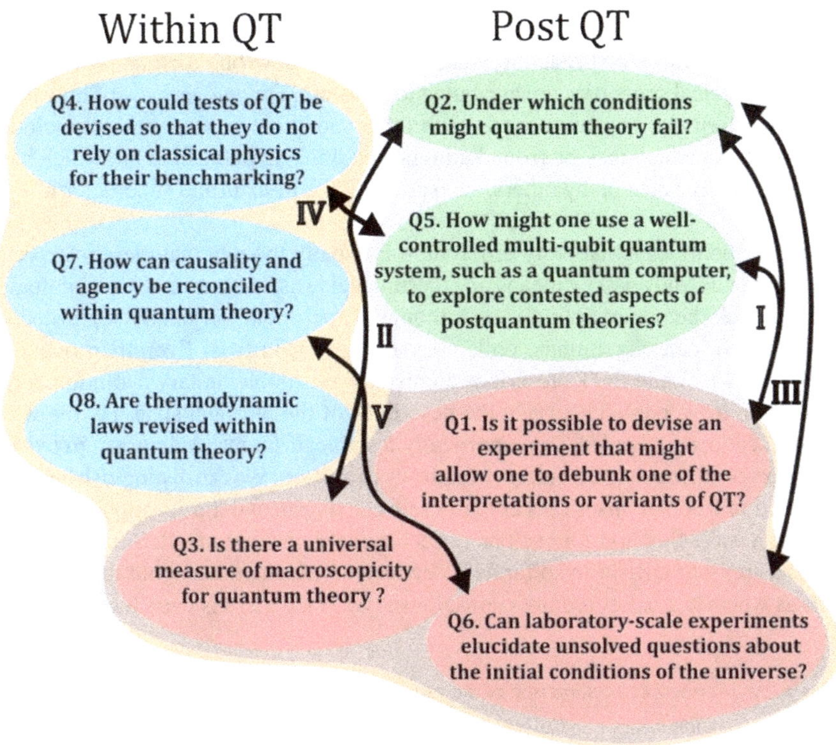

Fig. 1 The questions relate to two territories: within QT and post QT, with questions 1, 3 and 6 having bearings both within and post QT. Arrows indicate links between questions. (I) highlights the link between certain interpretations of QT being disproved and a more radical failure of quantum theory. It also links these questions with the possibility of advancement of our understanding enabled by the use of quantum technologies to look beyond our current formulation of QT. (II) brings out the question whether at a certain level of macroscopicity QT will fail and give way to classical physics. (III) shows that experiments on the early conditions of the universe might point to a breakdown of QT. (IV) links questions 4 and 5: the development of a quantum computer will allow for new tests of QT looking beyond the benchmark of classical physics. (V) draws on questions that are raised about the initial conditions of the universe, specifically with regards to measurement, when considering causality and agency

Collapse variants of QT (CQT) are stochastic and prescribe that there be an *irreversible evolution of the wave function* (the *collapse*), whenever a measurement is completed. EQT implies the existence of the Multiverse, where the probabilistic predictions of the Born Rule are recovered via the decision-theoretic approach to probability [13, 29] (there is an ongoing debate in regard to that approach, see e.g. Ref. [30] for a critique). According to CQT, there is instead a single, irreversibly (although this is challenged in Ref. [31]) and stochastically evolving universe.

Bohmian mechanics is deterministic, non-local and avails itself of an additional equation of motion [32] but it is limited in that it has not fully been extended to relativistic quantum field theories of the Standard Model.

Other relational interpretations of QT such as **QBism** [33] and **Contexts, Systems and Modalities** [34, 35] redefine what is meant by/how one assigns probabilities and how physical properties are attributed to a system. For an overview of how these different interpretations compare to each other see Table 1. We have not included interpretations which may be from distinguished thinkers in the field but which have not been widely taken up by others, such as cellular automata and superdeterminism [36, 37].

Despite being so remarkably different, it is usually thought that these interpretations cannot be told apart by means of experimental tests. However, new technologies might unlock the potential to do so. A thought experiment, originally suggested by Deutsch [38], can discriminate collapse variants (CQT) from Everettian quantum theory (EQT). While EQT prescribes a universal reversible unitary evolution, CQT requires an irreversible change in the descriptor of the physical state, i.e. the wave function collapse. This leads to empirically testable different predictions, **provided an observer can undergo a coherent unitary evolution**. We can distinguish between the question whether this is conceptually self-consistent, and the question whether in practice a wave function for such an observer could be determined. Similar considerations apply to QBism in so far as it depends on assumptions about the observer, such as whether a belief requires consciousness. Since there is even less consensus

Table 1 An overview and comparison of the variants of QT mentioned in the text

Everettian quantum theory	Universal, deterministic and reversible. Debateable whether local [27] or non-local [41]. The wave function is ontic	Everett [12] and Wallace [13]
Measurement collapse variants of quantum theory	Non-universal, stochastic, irreversible, non-local and violates energy conservation. The wave function is epistemic	Ghirardi et al. [14–16] and Bassi et al. [17, 18]
Dynamical collapse variants of quantum theory	Universal, stochastic, irreversible, non-local and violates energy conservation. The wave function is ontic	Ghirardi et al. [14–16] and Bassi et al. [17, 18]
Pilot Wave Mechanics	Universal, deterministic, reversible and non-local. It is proposed that the wave function is ontic, either (quasi) material or nomological	Dürr et al. [19], Bricmont [20], Norsen [21], Dürr et al. [42]
QBism	Piece-wise universal (there's nothing, from one's own perspective, with the exception of one's own experiences which QT does not apply to) and local. The wave function is doxastic	Fuchs et al. [33]

about consciousness than about interpretations of quantum theory [39, 40], we shall not follow that path.

Deutsch's thought experiment can be described schematically as follows: a spin ½ particle is prepared in an eigenstate $|\uparrow\rangle$ of, say, the z-component of the spin. This is an equally weighted superposition of the eigenstates $|0\rangle$ and $|1\rangle$ of the x-component of the spin, X. An automaton is then coupled with the particle **so that in the EQT interpretation it would undergo a unitary evolution that corresponds to its measuring the observable X**. The automaton is programmed so that once the measurement is complete, it writes on a piece of paper that the measurement is indeed complete and that it sees a definite outcome, *without writing down which value it sees*. According to CQT, the particle and the automaton's register by this point have undergone an irreversible collapse, ending up in either the pure state $|00\rangle$ or the pure state $|11\rangle$, with probability ½ each. In the case of EQT, the composite system of the automaton and the spin is now in an equally weighted superposition of $|00\rangle$ and $|11\rangle$: the spin and the automaton are entangled. Then one applies the time-reversal of the unitary transformation that implemented the reversible measurement of the automaton on the spin—acting on the spin and the automaton's register only, but *not* on the piece of paper; finally, the z-component of the spin is measured. In the case of CQT, the prediction is that the outcome 'up' or 'down' is observed, each one with probability ½. According to EQT, since the above was an interference experiment, the outcome will be invariably $|\uparrow\rangle$. The fundamental irreversibility of CQT therefore makes the difference, and this difference can be empirically tested in this scenario.

This experiment touches on several of the problems mentioned below—chiefly, **macroscopicity** (see Q3) and **the role of the observer**. Implementing such an experiment might require technology beyond current capabilities; but understanding what technologies would be needed, and which kinds of approximations to that experiment could be currently conceived is a productive line of enquiry. Tests of CQT are discussed in more detail in Q2.

Another interesting line of experimental tests are **non-local hidden variable 'super-deterministic' theories**, as proposed by Hossenfelder [43]. She proposes searching for evidence for correlations generated by non-local hidden variables via a time-resolved single-photon double-slit experiment and a Stern-Gerlach type experiment in a spatial loop, at low temperature and with minimal sources of noise. This should probe regimes where the corrections to quantum theory predicted by those hidden-variable models become relevant.

Concerning **the reality of the wave function**, the problem is whether the quantum wave function encodes our knowledge of a quantum system (the so-called 'psi-epistemic' view), or whether it describes something objective about reality, irrespective of us (the 'psi-ontic' view). New, general no-go theorems provide the theoretical inspiration for experimentally testing the difference between ontic and epistemic views of the wave function: the Pusey-Barrett-Rudolph (PBR) [44] theorem, further advanced in the Barrett-Cavalcanti-Lal-Maroney (BCLM) [45], and Branciard theorems [46]. **These theorems have started to be tested experimentally with the results all favouring the psi-ontic view** [47–50]. In addition, recent experimental work by S. Simmons and co-authors (private communication) ruled out 'maximally psi-epistemic' models using a single electron-nuclear two-spin system in isotopically purified silicon, achieving the low degree of errors required by the BCLM test. For an overview of testable theorems and their experimental status see Table 2.

Table 2 Theorems used to test aspects of QT, with their underlying postulates and experiments performed to date

Theorem	Postulates	Experimental status
Bell	Correlations are **locally explicable** and no causal influence can travel faster than light [51][a]	The inequality has been violated [1–5]
Leggett-Garg (Bell type inequality for time)	**Macrorealism per se**—A macroscopic object, which has available to it two or more macroscopically distinct states, is at any given time in a definite one of those states. **Noninvasive measurability**—It is possible in principle to determine which of these states the system is in without any effect on the state itself, or on the subsequent system dynamics [52]	The inequality has been violated in microscopic systems leading to the conclusion that "All accurate descriptions of systems of this type must include a concept similar to that of quantum superposition, and/or an exotic notion of measurement similar to that of wavefunction collapse" [7–9] **but tests using a macroscopic system have yet to be carried out**. It should be noted that this conclusion has been debated philosophically [53]
Bell-Kochen-Specker	No non-contextual hidden variable theorem (i.e. one in which the values of the physical observables are the same whatever the experimental context in which they appear) can reproduce the predictions of quantum theory [54, 55]	The inequality has been violated in microscopic systems showing that the observed phenomena cannot be described by non-contextual models [56, 57]. In addition it has been shown experimentally that there is a monogamy relation between the violation of either a Bell inequality or a Bell-Kochen-Specker inequality [58]
PBR/BCLM/Branciard	The quantum state is not purely epistemic (informational) [44–46]	Bounds have been put on maximally psi-epistemic models but further tests are needed to rule out partially psi-epistemic models [47–50]

[a]In Bell's original 1964 treatment [51] the existence and determinism of underlying hidden variables was derived from the conjunction of locality with the existence of perfect (anti-)correlations for parallel measurements on a singlet state. In later discussions (e.g. 1976, 1981, 1990 [175]) he relaxed the perfect correlations assumption, whilst retaining the idea that correlations ought to be explicable, even if only probabilistically

2. **Under which conditions might quantum theory fail?**

The **continuous spontaneous localisation (CSL)** model [14–16] is one of many theoretical efforts to explain wave function collapse [17, 18]. In this model, the wave function collapses spontaneously, and *the collapse rate is proportional to the mass*, hence certain superposition states of macroscopic objects (e.g. involving a localised mass being in a superposition of significantly different positions) are very difficult to observe.

In order to distinguish the effects of CSL from decoherence stimulated by interactions with the environment, a system in which noise induced by the environment is minimised is required. Potential phenomena due to CSL include:

1. ***the decoherence of a superposition state*** [59–61],
2. ***the linewidth broadening*** [62] ***and heating of a mechanical oscillator (i.e. a violation of energy conservation due to the collapse of the wave function)*** [63–65],
3. ***diffusion in free space*** [66, 67].

It might be that the most practical way to test for CSL is to look for thermally induced delocalisation due to the collapse process. A detailed analysis has been evaluated for an experiment to detect the heating due to CSL of a trapped nanosphere [68] and also of a charged macroscopic object in an ion trap [69]. Of all the possible causes of unwanted decoherence, the dominant ones are likely to be mechanical and electrical noise and molecular collisions. The calculations suggest that although the practical demands exceed what has already been achieved, the experiment should be within reach. Using a high quality factor cantilever, a nonthermal force noise of unknown origin which could be due to the CSL heating rate predicted by Adler has been recently detected [65].

Trying to create macroscopic quantum superpositions is another way of testing the ground where quantum mechanics might fail. See Q3 for a discussion on macroscopicity.

To perform *laboratory-scale* experiments of QT where gravity would be important, the challenge is to engineer quantum states of mechanical systems in which gravitational effects must be taken into account to describe the dynamics [70]. In such scenarios, QM may need to be modified in a yet unknown way in order to account for gravitational effects such as decoherence and gravitational self-interaction [71–74], or on the other hand the gravitational force may be quantum coherent [75, 76].

In a quantum theory of gravity, quantum fluctuations in the underlying field that mediates the gravitational interaction between matter degrees of freedom may appear as an additional source of noise [77]. Such effects might be thought to be restricted to the Planck scale and thus seem unlikely to arise in table-top experiments. Surprisingly, proposals of Penrose [72] and Diosi [73], and later by Kafri et al. [78], amongst others [79–82], would indicate that this is not the case and that, given sufficient quantum control over macroscopic mechanical degrees of freedom, gravitational decoherence might be revealed. Like CSL, the most accessible way to test for gravitational decoherence might be through sensitive detection of heating effects. Optomechanical systems might be a good platform for this goal, enabling us to measure minuscule heating rates [83].

3. Is there a universal measure of macroscopicity for quantum theory?

Leggett used the term ‘macroscopic’ to—amongst other things—articulate how, in our ‘us-sized’ lives, we experience events and outcomes that are definite and predictable in a way that seems quite different from the mystery of quantum superposition [84]. ***He questioned whether (whereas Bohr assumed) there was some different kind of reality at the macroscopic level from that which is found at the quantum level***, and he sought to devise a rigorous test of this proposition in the form of experimentally measureable inequalities. What his inequalities actually put under test is, however, still under debate [53].

There could be many dimensions of *macroscopicity*. Does it lie in a greater number of atoms or photons, in a greater mass or spatial size, in greater complexity (if so how should this be quantified?), or in a greater number of dimensions in Hilbert space? Does it perhaps lie at the threshold where life begins [85]? Another quantifiable possibility may be that it depends on limits to the linearity of the system. ***If so, can we quantify the degree of non-linearity of dynamical evolution that would be required to prevent macroscopic superpositions or entanglement occurring, and can we characterise the kinds of contexts in which this limit would arise*** [86]? Are they for example related to the issue of thermalisation and heat baths that interact with quantum systems [87]? Each of these dimensions of macroscopicity needs to be explored in order to extend the tests of macroscopic realism.

The task of defining macroscopicity measures within quantum theory is confounded by a fundamental problem of an *ad hoc* selection of distinguishable observables. As Nimmrichter and Hornberger put it, ‘the more macroscopic [something is] the better its experimental demonstration allows one to rule out even a minimal modification of quantum mechanics, which would predict a failure of the superposition principle on the macroscale’ [88]. As that paper showed, the question of macroscopicity applies as much to collapse theories as to Leggett-type inequalities.

The issue of whether a *universal* measure of macroscopicity exists is central to our understanding of quantum reality. On the one hand, wave function collapse variants of quantum mechanics ***require there to be a limit to the domain of applicability of reversible unitary quantum theory, whence the necessity of specifying, quantitatively, where this limit exactly is***. On the other hand, if there is not such a fundamental limit to quantum coherence, a universal measure of macroscopicity would still be highly desirable in order to monitor technological progress, and compare results of different experiments. Indeed, the question of how far can we demonstrate quantum behaviour it is not only at the heart of foundations of QT but it is crucial for the development of new technologies. It has also big implications on how we understand complex systems and even life; for example, can macroscopic living entities make use of quantum coherence? The importance of quantum effects in biological processes has been highlighted for olfaction [89], magneto-reception in the avian compass [90] and photosynthesis [91, 92].

Finally, even if there is no universal measure of macroscopicity (see Ref. [93] for an outline of 14 different measures of macroscopicity), it is fruitful to ***search for measures of how hard it is to maintain a physical system in a given superposition or to implement a unitary gate to arbitrarily high degree of accuracy***—whence the modified question of what particular measures of macroscopicity might arise from operational/experimental considerations.

Theoretical proposals for creating superpositions of macroscopically distinct states include capacitively coupling a resonator to a superconducting qubit [94], flux coupling a nanotube to a superconducting qubit [95] and using an interferometer to optomechanically couple a mirror to a photon in a superposition [61, 96, 97]. On the experimental front, although yet to be fully realised, much progress has been made toward the goal of creating a superposition state in a micromechanical resonator coupled to a superconducting qubit [98, 99]. In millimetre sized [100, 101] resonators, the ground state has been reached, which is the first step towards creating a quantum superposition. Interference experiments have been carried out with molecules of up to 1×10^4 atomic mass units [102] and entanglement has been demonstrated between a single photon and a single collective atomic excitation in a 1 cm long crystal [103], as well as between two mechanical resonators [104, 105].

4. **How could tests of QT be devised so that they do not rely on classical physics for their benchmarking?**

Most tests involving QT over the past decades have been designed to corroborate the idea that QT largely violates our classical expectations. Indeed, QT's predictions have been tested against rival theories **sharing the common feature of *keeping one or the other basic principle of the classical physics intact***:

1. hidden-variable models involved in Bell-type experiments assign definite values to outcomes of unperformed measurements;
2. non-linear Schrödinger equations allow solutions with localised wave-packets to resemble classical trajectories;
3. collapse-type models restore macrorealism by suppressing superpositions between macroscopically distinct states.

While of great importance in the problem-situation of demonstrating fundamental differences between quantum mechanics and the classical world-view, such approaches to testing quantum reality may not be very fruitful when considering *different problems* that are now coming to the fore. A particularly prominent example of problems calling for experiments with a different benchmark than classical physics is the search for the successor of quantum theory: for a 'post-quantum' theory may be expected to break not only principles of classical but also of quantum physics. This is a case where interplay between experiment and theory promises to be particularly fruitful. There have indeed been a number of proposals for theoretical frameworks for thinking of viable post-quantum theories, against which QT could then be tested. These frameworks could thus be the source of such new tests.

One logic, suggested by Dakic and Bruckner [106], is to reconstruct QT from a set of axioms (see, e.g., Hardy [107]; Clifton et al. [108]; Chiribella et al. [109]), and then weaken or drop some of the axioms to get broader theoretical structures, whereby we can conceive of QT's generalisations. This has led to **generalised probabilistic theories** [110]—generalisations of quantum theory, which permit phenomena such as interference, randomness of individual results or violation of Bell's inequalities, but in more extreme ways than quantum theory does.

Another approach is to define a set of theoretical possibilities designed so that they share with quantum theory (some of) its main features (and reproduce its testable predictions). For example, when the features are chosen to be quantum theory's information-theoretic properties, a *local, non-probabilistic framework* for generalisations of quantum theory can be accommodated in the recently-proposed **constructor-theory of information** [111, 112]. Other general frameworks that could provide tools to devise rivals against which to test QT are ontological models frameworks [23].

Quantum simulators, annealers and computers provide a playground for testing QT without relying in classical physics (see Q5).

5. **How might one use a well-controlled multi-qubit quantum system, such as a quantum computer, to explore contested aspects of postquantum theories?**

In his Nobel lecture, Robert B. Laughlin starts by saying that to deduce phenomena such as superfluidity from first principles is an impossible task; superfluidity, he says, is an emergent phenomenon, a low energy collective effect of huge number of particles that cannot be deduced from the microscopic equations of motion in a rigorous way and that disappears completely when the system is taken apart [113].

Could it be that new questions and new answers arise from **emergent phenomena in multi-qubit quantum systems** like quantum annealers, simulators, computers or different types of quantum networks? Can these systems help us to explore and formulate postquantum theories [114–116]? In the same way we use classical computers to calculate quantum (post-classical) predictions, could we use a quantum computer to calculate postquantum predictions?

Large quantum networks provide an appealing route to a scalable universal quantum computer, which is built by networking together several simple processor nodes (as opposed to a monolithic structure) [117]. Other applications are made possible via the network directly. One is so-called **blind quantum computation** [118], where a remote person can control a quantum computer which is run and maintained by another person, who nevertheless cannot know what particular computational task the computer is performing. This is crucial to preserve the privacy of a computation.

Other applications arise for quantum networks with some amount of computation at each node (as implementable by, e.g., the ion trap architecture being pursued at Oxford [119]). For instance, cryptographic applications beyond quantum key distribution; verifiable quantum computation (which allows a user to verify the results of a quantum computation with certainty) [120]; quantum homomorphic encryption (a form of encryption which allows operations to be performed on the encrypted data without access to the secret key) [121]; a quantum internet [122] which can distribute quantum software [123]; and even long baseline astronomy [124].

Large scale quantum computers—not necessarily universal ones—open up a new frontier because they are systems with large amounts of entanglement. In particular, when the internal entanglement of a system becomes sufficiently high, our ability to simulate the system with anything other than a quantum computer falls away, due to the apparent difference between the computational complexity classes P and BQP [125]. Experimentally demonstrating quantum algorithms which show a large advantage over the best classical algorithms (particularly in terms of provable advantages such as in the case of query complexity for search and collision algorithms) would provide further insights into a proof of clear separation between computation allowed by quantum physics and computations achievable by a classical computation model. Another notable phenomenon is the breakdown of thermalisation for open systems. Although we expect systems in a confined volume in contact with a heat bath eventually to reach a Gibbs distribution, it is known that the ground states of certain classes of local Hamiltonian are QMA hard to compute [126]; they cannot efficiently be reached by physical systems. Thus, existing conjectures about computational complexity [127] would imply the possibility of constructing systems which cannot thermalize in less than exponential time. This may bring about an important transition in our understanding of chemistry and condensed matter physics, implying that absolute energy structure is less important for understanding the behaviour of large quantum systems and materials.

Quantum machine learning might be another promising tool. In addition to quantum machine learning showing quadratic improvements in learning efficiency and exponential improvements in performance over limited time periods over classical machine learning [128] it is hoped that a quantum artificial intelligence may be able to recognize patterns that are difficult to recognize classically [129]. This could be a powerful tool for research into post-quantum theories.

6. **Can laboratory-scale experiments elucidate open questions in the evolution of the universe?**

About 400,000 years after the Big Bang, the *last scattering* occurred; photons decoupled from matter and travelled freely through the universe, constituting what we observe today as cosmic microwave background radiation. There is no good theory for how the quantum fluctuations arising during inflation get changed to classical fluctuations by the time of last scattering, when they become seeds for large scale structure formation. Any experiments that elucidate the classical to quantum transition as a real or effective physical process have the potential to help throw light on this. It has been proposed on the one hand that this may be due to decoherence [130] and on the other that this is related to variants of CSL [131, 132].

In order to discriminate between such proposals, one can speculate on whether these proposals might make a difference to the expected classical fluctuations at the end of inflation. For example, might they be scale-dependent and hence cause a breakdown in the prediction of an almost scale-invariant spectrum of perturbations, affecting Cosmic Microwave Background and large scale structure observations at the present time [133]? Would they affect the usual assumption of Gaussianity of fluctuations at the end of inflation? If either were to be true these would be observational tests of such quantum theory variants [134]. Test of quantum theory variants can be devised as further bench-top experiments, enabling us to answer questions about the growth of structure in the Universe in the laboratory, as well as using data from cosmological phenomena to probe QT in extreme conditions [135, 136].

Finally, when considering the problem of the initial conditions of the universe one gets into the domain of quantum gravity (see also Q2). We are unable to access the required energies to test quantum gravity theories in colliders, and in the cosmological context inflation smooths out any pre-inflationary structures there might be, so we cannot see their cosmological outcomes either, although one avenue for exploring quantum gravity is Black Hole evaporation [137]. *String cosmology* suggests some inflationary potential on the basis of string theory, but this cannot be applied to the inflationary regime where the relevant energies are quite different.

The issues here are twofold. First, do the same principles of ordinary quantum theory apply to quantum gravity or do we need new foundational principles for the nature of space-time, or for some kind of (probably discrete) pre-spacetime structure, and hence for quantum theory? Loop quantum gravity [24] and string theory [25] are the two pre-eminent approaches that have been taken to studying such issues, but others such as causal set theory [138] provide more radical departures points, because they assume space-time structure is discrete. What justifies use of the same principles as those of ordinary quantum theory in these circumstances?

The more direct relation to the questions posed here arise as regards the second point: does the start of the universe in some sense correspond to a measurement event? How does the idea of measurement work out in quantum gravity theories, whatever they are?

7. How can causality and agency be reconciled within quantum theory?

Puzzling situations can arise where the causal order of events (in a fixed spacetime background) is not necessarily fixed, but is subject to quantum uncertainty. Could there be indefiniteness with respect to the question of whether an interval between two events is time-like or space-like, or even whether event A is prior to or after event B? Might, this correspond to the "superpositions of situations where, 'A is in the past of B' and 'B is in the past of A' jointly" [139]? This problem has bearings on quantum gravity studies: a theory unifying collapse-based variants of QT with GR, causal structure might plausibly be both be dynamic, as in general relativity, as well as indefinite, due to quantum features. A framework for the dynamics of quantum causal structures of something of this kind is given in Ref. [140].

Oreshkov, Costa and Brukner have put forward theoretical ***models where there is no fixed causal order and the dynamics is specified in terms of linear operators*** [141]. In 2017 an experiment was carried out in Vienna which implemented a measurement which was described as a superposition of causal orders [142], and subsequently a demonstration described as an entanglement of temporal orders [143] violating a Bell inequality for temporal order [144] (although it is not yet a loophole free test). The development of this inequality for temporal order allows for a quantitative method for investigating quantum aspects of space-time and gravity, and the demonstration of its violation could lead to the conclusion that nature is incompatible with a local definite temporal order. An improved experiment to render the causal order between operations indistinguishable by their spacetime location has yielded a causal witness 18 standard deviations beyond the definite-order bound [145].

Other approaches to investigating modified causal orders within quantum theory are based on the framework of closed time-like curves (see Ref. [139] for a discussion), and have recently inspired experimental simulations—see e.g. Ref. [146].

A special kind of causal order is that pertaining to events caused by the action of *agents*: things which react to environmental stimuli in flexible yet sensible ways, and to whose active powers we attribute many of the happenings around us. In this regard, there is an ongoing rich debate about how *attributable agency* can be incorporated within physics, and in particularly reconciled with quantum theory. In the case of collapse-endowed variants of QT, the problem, according to some (see Ref. [147] for a discussion of the problem of agency), is that there appears to be no room for attributable agency in the context of a stochastic theory; it is therefore a challenge to accommodate a prominent feature of physical reality, i.e., the existence of agents, within those variants of quantum theory. Progress has been made via a theoretical model of *projective simulation*, where the concrete outcome of a random process can be consistently attributed to an agent [147]. This has recently inspired a model for an implementation via measurement-based quantum computation [148]. Could the source of asymmetry between cause and effect be simply the act of intervention itself? Milburn and Shrapnel put forward the view that it is the temporally symmetric laws of physics that underwrite the agent-based interventions through which asymmetric causal relations are discovered [149].

An additional aspect to consider when discussing causality is the possibility of retrocausality; of events in the future being able to influence events in the past. Retrocausality would be one way of allowing for a Lorentz-invariant explanation of Bell correlations without action at a distance and it has been proposed that retrocausality follows directly from the quantization of light, provided that fundamental physics is time-symmetric and that one does not take an ontic interpretation of the quantum state [150, 151]. Of course, some models of QT are not time symmetric—the introduction of a collapse event for the wave function is said by some to introduce time-asymmetry, in which case retrocausality would not be introduced. The transactional interpretation introduces a form of retrocausality, although in this interpretation the future does not influence the past [152]. In this case the predictions of QT are interpreted to be due to an exchange of advanced and retarded waves, with the predictions of the theory being the same as standard quantum mechanics.

8. **Are thermodynamic laws revised in quantum theory?**

Standard thermodynamics offers a bird's eye view of a system consisting of vast numbers of particles by describing it using a few parameters such as temperature, volume and pressure. Although this simplicity allows for an elegant approach when dealing with systems with large numbers of particles, the downside is that ***as the system size decreases the thermodynamic approach starts to lose accuracy as fluctuations of the parameters become relevant***. Stochastic thermodynamics can be used to describe fluctuations in the thermodynamic quantities due to thermal effects and to describe non equilibrium systems however, once quantum effects come into play we need a theory of quantum thermodynamics [153].

The famous Maxwell demon case, with regards to the second law, highlights the link between work and information. Using knowledge of the system the demon can extract work from the system without increasing its entropy, seemingly violating the second law. We achieve a neat resolution to this puzzle upon realising that the ***information (stored in the demon's memory, which is used to enable it to extract work) must also be accounted for thermodynamically, and achieving a thermodynamic cycle requires that this memory be erased, incurring a waste of energy as heat*** [154].

Early experiments to study non-equilibrium phenomena in nanoscale systems have been realized with molecules and soft matter at ambient temperatures [155, 156]. Heat-to-work conversion has been demonstrated with a dimeric polystyrene bead suspended in a fluid by measuring if the particle has moved up and, depending on the result, modifying an external potential to ensure the particle continues climbing. A micrometre-sized stochastic heat engine and a Carnot engine were realized with a single optically trapped particle as the working substance [157, 158]. A direct measurement of the entropy change along symmetry-breaking transitions for a Brownian particle in a bistable potential has also been achieved [159].

These experiments, however, do not provide an easy path towards incorporating quantum effects. A single-atom heat engine has been demonstrated in a single calcium ion in a tapered ion trap [159]. The study of quantum fluctuation relations with spin-1/2 system [160] and a trapped ion [161], as well as a demonstration of the Landauer principle in the quantum regime with a three-nuclear-spin molecule [162] and ultracold ions [163], were achieved. Very recently, spin heat engines were realised in an ion trap and a spin ½ system [164, 165]. The limitation in these cases is that either they are restricted to closed systems or the reservoir is not much larger than the system; reservoirs in the conventional sense are those of open systems and it is the study of open quantum systems that will answer the most pressing questions about energy harvesting, dissipation and thermalisation in quantum circuits.

A glimpse of the potential of solid-state circuits became evident when the Jarzynski equality, a Szilard engine and an autonomous Maxwell's demon were demonstrated with a single electron box [166–168]. A superconducting qubit has been used to demonstrate a quantum Maxwell demon [169] and an ensemble of nitrogen-vacancy centres in diamond has been used as a quantum heat engine [170]. A quantum heat valve [171], a quantum-dot heat engine [172] and a quantum-dot energy harvester [173] have also been realised in the solid-state.

The development of these experimental techniques opens the way for testing disagreements which are beginning to emerge about aspects of nonequilibrium thermodynamics in nanoscale quantum systems. To some extent these different viewpoints arise because of the different communities in which they originate, such as statistical physics, mesoscopic physics, quantum information theory, and many-body theory. Open questions include the definition of work, how quantum systems thermalize, and the efficiency and power of quantum engines [174]. We confidently hope and anticipate that experimental testing will serve to evaluate the validity of different approaches in different contexts, and will elucidate those concepts which are presently obscure.

3 Discussion and Conclusions

Thinking back to the first set of Oxford questions [26], these new questions, although sharing the same theme, are less focused on reconciling quantum physics with classical physics and ideas *and are more focused on experiments to test the boundaries of QT with post quantum theories.* With the development of new technologies it is important to continue to think of experiments to expand the boundaries of our understanding of the quantum realm.

With regards to collecting fresh evidence, good progress has been made even in the short period of time since the first set of Oxford questions. There has been further theoretical development of the *PBR theorem on the reality of the wave function* to the BCLM and Branciard theorems, with experimental tests pointing towards a psi-ontic interpretation [47–50]. *A framework for indefinite causal order within quantum theory* has been developed including a Bell inequality for temporal order with the first experimental verifications being recently published [142, 143]. The most recent experiment of a CSL type heating effect has measured *a nonthermal force noise of unknown origin,* down to the level of the CSL heating predicted by Adler [65], although the authors are not willing to claim CSL heating until every other possible source has been ruled out. We are still some way off making sense of many of the aspects of QT, but as these experiments are improved and theoretical proposals are brought into realisation we can expect the murky waters of the foundations of QT to become clearer.

Acknowledgments The authors are grateful for input from C. Marletto. C.G. Timpson contributed throughout and especially to Q1, G.F.R. Ellis to Q6 and C. Brukner to Q7. We are grateful for comments and observations from S.C. Benjamin, F. Bussières, J. Fitzsimons, E.M. Gauger, G.C. Knee, E.A. Laird, T. Oosterkamp, J. Schmiedmayer, A.M. Steane, H. Ulbricht, T. van der Reep and J. van Wezel. We thank Maria and David Willetts for sustained support.

Funding We gratefully acknowledge grants from Templeton World Charity Foundation. The opinions expressed in this publication are those of the authors and do not necessarily reflect the views of Templeton World Charity Foundation. N. Ares thanks the Royal Society for a University Research Fellowship.

Competing interests We have no competing interests.

Authors' contributions N. Ares, A.N. Pearson and G.A.D. Briggs drafted the manuscript. All authors gave final approval for publication.

References

1. Hensen B *et al.* 2015 Loophole-free Bell inequality violation using electron spins separated by 1.3 kilometres. *Nature* **526**, 682–686. (https://doi.org/10.1038/nature15759)
2. Hensen B *et al.* 2016 Loophole-free Bell test using electron spins in diamond: second experiment and additional analysis. *Sci. Rep.* **6**, 30289. (https://doi.org/10.1038/srep30289)
3. Dehollain JP *et al.* 2016 Bell's inequality violation with spins in silicon. *Nat. Nanotechnol.* **11**, 242–246. (https://doi.org/10.1038/nnano.2015.262)
4. Rosenfeld W, Burchardt D, Garthoff R, Redeker K, Ortegel N, Rau M, Weinfurter H. 2017 Event-Ready Bell Test Using Entangled Atoms Simultaneously Closing Detection and Locality Loopholes. *Phys. Rev. Lett.* **119**, 010402. (https://doi.org/10.1103/physrevlett.119.010402)
5. Handsteiner J *et al.* 2017 Cosmic Bell Test: Measurement Settings from Milky Way Stars. *Phys. Rev. Lett.* **118**, 060401. (https://doi.org/10.1103/physrevlett.118.060401)
6. Riedinger R, Hong S, Norte RA, Slater JA, Shang J, Krause AG, Anant V, Aspelmeyer M, Gröblacher S. 2016 Nonclassical correlations between single photons and phonons from a mechanical oscillator. *Nature* **530**, 313–316. (https://doi.org/10.1038/nature16536)
7. George RE *et al.* 2013 Opening up three quantum boxes causes classically undetectable wavefunction collapse. *Proc. Natl. Acad. Sci. U. S. A.* **110**, 3777–81. (https://doi.org/10.1073/pnas.1208374110)
8. Knee GC *et al.* 2012 Violation of a Leggett-Garg inequality with ideal non-invasive measurements. *Nat. Commun.* **3**, 606. (https://doi.org/10.1038/ncomms1614)
9. Katiyar H, Brodutch A, Lu D, Laflamme R. 2017 Experimental violation of the Leggett – Garg inequality in a three-level system. *New J. Phys* **19**. (https://doi.org/10.1088/1367-2630/aa5c51)
10. Veldhorst M *et al.* 2015 A two-qubit logic gate in silicon. *Nature* **526**, 410–414. (https://doi.org/10.1038/nature15263)
11. Kovachy T, Asenbaum P, Overstreet C, Donnelly CA, Dickerson SM, Sugarbaker A, Hogan JM, Kasevich MA. 2015 Quantum superposition at the half-metre scale. *Nature* **528**, 530–533. (https://doi.org/10.1038/nature16155)
12. Everett H. 1957 On the Foundations of Quantum Mechanics (Phd Thesis). Princeton University.
13. Wallace D. 2012 *The emergent multiverse: Quantum theory according to the Everett interpretation.* Oxford University Press.
14. Ghirardi GC, Rimini A, Weber T. 1986 Unified dynamics for microscopic and macroscopic systems. *Phys. Rev. D* **34**, 470–491. (https://doi.org/10.1103/physrevd.34.470)
15. Ghirardi GC, Pearle P, Rimini A. 1990 Markov processes in Hilbert space and continuous spontaneous localization of systems of identical particles. *Phys. Rev. A* **42**, 78–89. (https://doi.org/10.1103/physreva.42.78)
16. Ghirardi GC, Grassi R, Benatti F. 1995 Describing the macroscopic world: Closing the circle within the dynamical reduction program. *Found. Phys.* **25**, 5–38. (https://doi.org/10.1007/bf02054655)
17. Bassi A, Ghirardi GC. 2003 Dynamical Reduction Models. *Phys. Rep.* **379**, 257–426. (https://doi.org/10.1016/s0370-1573(03)00103-0)

18. Bassi A, Lochan K, Satin S, Singh TP, Ulbricht H. 2013 Models of wave-function collapse, underlying theories, and experimental tests. *Rev. Mod. Phys.* **85**, 471–527. (https://doi.org/10.1103/revmodphys.85.471)
19. Dürr D, Teufel S. 2009 *Bohmian Mechancis*. Springer-Verlag Berlin Heidelberg.
20. Bricmont J. 2016 The de Broglie-Bohm Theory. In *Making Sense of Quantum Mechanics*, Springer International Publishing.
21. Norsen T. 2017 The Pilot-Wave Theory. In *Foundations of Quantum Mechanics*, Springer International Publishing.
22. Fuchs CA. 2010 QBism, the Perimeter of Quantum Bayesianism. *arxiv.1003.5209*
23. Harrigan N, Spekkens RW. 2010 Einstein, Incompleteness, and the Epistemic View of Quantum States. *Found. Phys.* **40**, 125–157. (https://doi.org/10.1007/s10701-009-9347-0)
24. Rovelli C. 2008 Loop quantum gravity. *Living Rev. Relativ.* **11**, 5–69. (https://doi.org/10.12942/lrr-2008-5)
25. Polchinski JG. 2005 *String Theory, Volumes 1 and 2*. Cambridge University Press.
26. Briggs GAD, Butterfield JN, Zeilinger A. 2013 The Oxford Questions on the foundations of quantum physics. *Proc. R. Soc. A.* **469**, 20130299. (https://doi.org/10.1098/rspa.2013.0299)
27. Deutsch D, Hayden P. 2000 Information flow in entangled quantum systems. *Proc. R. Soc. A* **456**, 1759–1774. (https://doi.org/10.1098/rspa.2000.0585)
28. Norsen T. 2016 Quantum Solipsism and Non-Locality. In *Quantum Nonlocality and Reality: 50 Years of Bell's Theorem*, Cambridge University Press.
29. Deutsch D. 1999 Quantum theory of probability and decisions. *Proc. R. Soc. A* **455**, 3129–3137. (https://doi.org/10.1098/rspa.1999.0443)
30. Kent A. 2010 One World Versus Many: The Inadequacy of Everettian Accounts of Evolution, Probability and Scientific Confirmation. In *Many Worlds?: Everett, Quantum Theory, & Reality.* Oxford University Press.
31. Bedingham DJ, Maroney OJE. 2017 Time symmetry in wave-function collapse. *Phys. Rev. A* **95**, 042103. (https://doi.org/10.1103/physreva.95.042103)
32. Bohm D. 1952 A Suggested Interpretation of the Quantum Theory in Terms of 'Hidden' Variables. I. *Phys. Rev.* **85**, 166–179. (https://doi.org/10.1103/physrev.85.166)
33. Fuchs CA, Mermin ND, Schack R. 2014 An Introduction to QBism with an Application to the Locality of Quantum Mechanics. *Am. J. Phys.* **82**, 749. (https://doi.org/10.1119/1.4874855)
34. Auffèves A, Grangier P. 2016 Contexts, Systems and Modalities: A New Ontology for Quantum Mechanics. *Found. Phys.* **46**, 121–137. (https://doi.org/10.1007/s10701-015-9952-z)
35. Auffèves A, Grangier P. 2017 Recovering the quantum formalism from physically realist axioms. *Sci. Rep.* **7**, 43365. (https://doi.org/10.1038/srep43365)
36. t Hooft G. 2016 *The Cellular Automaton Interpretation of Quantum Mechanics.* Springer Nature
37. t Hooft G. 2017 Free Will in the Theory of Everything.*arXiv:1709.02874*
38. Deutsch D. 1986 Three connections between Everett's interpretation and experiment. In *Quantum Concepts of Space and Time* (eds R Penrose, C Isham), pp. 215–225. Clarendon Press Oxford.
39. Ball P. 2019 Neuroscience Readies for a Showdown Over Consciousness Ideas. See https://www.quantamagazine.org/neuroscience-readies-for-a-showdown-over-consciousness-ideas-20190306/ (accessed on 31 October 2019).
40. Reardon S. 2019 'Outlandish' competition seeks the brain's source of consciousness. See https://www.sciencemag.org/news/2019/10/outlandish-competition-seeks-brain-s-source-consciousness (accessed on 31 October 2019).
41. Wallace D, Timpson CG. 2007 Non-locality and Gauge Freedom in Deutsch and Hayden's Formulation of Quantum Mechanics. *Found. Phys.* **37**, 951–955. (https://doi.org/10.1007/s10701-007-9135-7)
42. Dürr D, Goldstein S, Zanghi N. 2013 Reality and the Role of the Wave Function in Quantum Theory. In *Quantum Physics Without Quantum Philosophy*, pp. 263–278. Springer Berlin Heidelberg.

43. Hossenfelder S. 2011 Testing Super-Deterministic Hidden Variables Theories. *Found. Phys.* **41**, 1521–1531. (https://doi.org/10.1007/s10701-011-9565-0)
44. Pusey MF, Barrett J, Rudolph T. 2012 On the reality of the quantum state. *Nat. Phys.* **8**, 475–478. (https://doi.org/10.1038/nphys2309)
45. Barrett J, Cavalcanti EG, Lal R, Maroney OJE. 2014 No ψ -epistemic model can fully explain the indistinguishability of quantum states. *Phys. Rev. Lett.* **112**, 250403. (https://doi.org/10.1103/physrevlett.112.250403)
46. Branciard C. 2014 How ψ-epistemic models fail at explaining the indistinguishability of quantum states. *Phys. Rev. Lett.* **113**, 020409. (https://doi.org/10.1103/physrevlett.113.020409)
47. Patra MK, Olislager L, Duport F, Safioui J, Pironio S, Massar S. 2013 Experimental refutation of a class of ψ-epistemic models. *Phys. Rev. A* **88**, 032112. (https://doi.org/10.1103/physreva.88.032112)
48. Ringbauer M, Duffus B, Branciard C, Cavalcanti EG, White AG, Fedrizzi A. 2015 Measurements on the reality of the wavefunction. *Nat. Phys.* **11**, 249–254. (https://doi.org/10.1038/nphys3233)
49. Liao KY, Zhang XD, Guo GZ, Ai BQ, Yan H, Zhu SL. 2016 Experimental test of the no-go theorem for continuous ψ-epistemic models. *Sci. Rep.* **6,** 26519. (https://doi.org/10.1038/srep26519)
50. Nigg D, Monz T, Schindler P, Martinez EA, Hennrich M, Blatt R, Pusey MF, Rudolph T, Barrett J. 2016 Can different quantum state vectors correspond to the same physical state? An experimental test. *New J. Phys.* **18,** 013007. (https://doi.org/10.1088/1367-2630/18/1/013007)
51. Bell JS. 1964 On the Einstein Podolsky Rosen Paradox. *Phys. Phys. Fiz.* **1**, 195–200.
52. Leggett AJ, Garg A. 1985 Quantum mechanics versus macroscopic realism: Is the flux there when nobody looks? *Phys. Rev. Lett.* **54**, 857–860. (https://doi.org/10.1103/physrevlett.54.857)
53. Maroney OJE, Timpson CG. 2014 Quantum- vs. Macro-Realism: What does the Leggett-Garg Inequality actually test? *arXiv:1412.6139*
54. Kochen S, Specker E. 1967 The Problem of Hidden Variables in Quanutm Mechanics. *J. Math. Mech.* **17**, 59–87. (https://doi.org/10.1512/iumj.1968.17.17004)
55. Bell JS. 1966 On the Problem of Hidden Variables in Quantum Mechanics. *Rev. Mod. Phys.* **38,** 447. (https://doi.org/10.1103/revmodphys.38.447)
56. Kirchmair G, Zähringer F, Gerritsma R, Kleinmann M, Gühne O, Cabello A, Blatt R, Roos CF. 2009 State-independent experimental test of quantum contextuality. *Nature* **460**, 494–497. (https://doi.org/10.1038/nature08172)
57. Ahrens J, Amselem E, Cabello A, Bourennane M. 2013 Two Fundamental Experimental Tests of Nonclassicality with Qutrits. *Sci. Rep.* **3**, 2170. (https://doi.org/10.1038/srep02170)
58. Zhan X, Zhang X, Li J, Zhang Y, Sanders BC, Xue P. 2016 Realization of the Contextuality-Nonlocality Tradeoff with a Qubit-Qutrit Photon Pair. *Phys. Rev. Lett.* **116,** 090401. (https://doi.org/10.1103/physrevlett.116.090401)
59. Romero-Isart O, Pflanzer AC, Blaser F, Kaltenbaek R, Kiesel N, Aspelmeyer M, Cirac JI. 2011 Large quantum superpositions and interference of massive nanometer-sized objects. *Phys. Rev. Lett.* **107**, 020405. (https://doi.org/10.1103/physrevlett.107.020405)
60. Romero-Isart O. 2011 Quantum superposition of massive objects and collapse models. *Phys. Rev. A* **84**, 052121. (https://doi.org/10.1103/physreva.84.052121)
61. Pepper B, Ghobadi R, Jeffrey E, Simon C, Bouwmeester D. 2012 Optomechanical superpositions via nested interferometry. *Phys. Rev. Lett.* **109**, 023601. (https://doi.org/10.1103/physrevlett.109.023601)
62. Bahrami M, Paternostro M, Bassi A, Ulbricht H. 2014 Proposal for a Noninterferometric Test of Collapse Models in Optomechanical Systems. *Phys. Rev. Lett.* **112**, 210404. (https://doi.org/10.1103/physrevlett.112.210404)
63. Adler SL. 2005 Stochastic collapse and decoherence of a non-dissipative forced harmonic oscillator. *J. Phys. A* **38**, 2729–2745. (https://doi.org/10.1088/0305-4470/38/12/014)

64. Vinante A, Bahrami M, Bassi A, Usenko O, Wijts G, Oosterkamp TH. 2016 Upper Bounds on Spontaneous Wave-Function Collapse Models Using Millikelvin-Cooled Nanocantilevers. *Phys. Rev. Lett.* **116**, 090402. (https://doi.org/10.1103/physrevlett.116.090402)
65. Vinante A, Mezzena R, Falferi P, Carlesso M, Bassi A. 2017 Improved Noninterferometric Test of Collapse Models Using Ultracold Cantilevers. *Phys. Rev. Lett.* **119**, 110401. (https://doi.org/10.1103/physrevlett.119.110401)
66. Collett B, Pearle P. 2003 Wavefunction collapse and random walk. *Found. Phys.* **33**, 1495–1541. (https://doi.org/10.1023/a:1026048530567)
67. Bera S, Motwani B, Singh TP, Ulbricht H. 2015 A proposal for the experimental detection of CSL induced random walk. *Sci. Rep.* **5**, 7664. (https://doi.org/10.1038/srep07664)
68. Goldwater D, Paternostro M, Barker PF. 2016 Testing wave-function-collapse models using parametric heating of a trapped nanosphere. *Phys. Rev. A* **94**, 010104. (https://doi.org/10.1103/physreva.94.010104)
69. Li Y, Steane AM, Bedingham D, Briggs GAD. 2017 Detecting continuous spontaneous localization with charged bodies in a Paul trap. *Phys. Rev. A* **95**, 32112. (https://doi.org/10.1103/physreva.95.032112)
70. Schmole J, Dragosits M, Hepach H, Aspelmeyer M. 2016 A micromechanical proof-of-principle experiment for measuring the gravitational force of milligram masses. *Class. Quantum Gravity* **33**, 125031. (https://doi.org/10.1088/0264-9381/33/12/125031)
71. Karolyhazy F. 1966 Gravitation and quantum mechanics of macroscopic objects. *Nuovo Cim. A* **42**, 390–402. (https://doi.org/10.1007/bf02717926)
72. Penrose R. 1996 On Gravity's role in Quantum State Reduction. *Gen. Relativ. Gravit.* **28**, 581–600. (https://doi.org/10.1007/bf02105068)
73. Diósi L, Elze HT, Fronzoni L, Halliwell J, Prati E, Vitiello G, Yearsley J. 2011 5th International Workshop DICE2010: Space-Time-Matter - Current Issues in Quantum Mechanics and beyond. *J. Phys. Conf. Ser.* **306**, 011001. (https://doi.org/10.1088/1742-6596/306/1/011001)
74. Colin S, Durt T, Willox R. 2014 Can quantum systems succumb to their own (gravitational) attraction? *Class. Quantum Gravity* **31**, 245003–54. (https://doi.org/10.1088/0264-9381/31/24/245003)
75. Marletto C, Vedral V. 2017 Witness gravity's quantum side in the lab. *Nature.* **547**, 156–158. (https://doi.org/10.1038/547156a)
76. Bose S *et al.* 2017 Spin Entanglement Witness for Quantum Gravity. *Phys. Rev. Lett.* **119**, 240401. (https://doi.org/10.1103/physrevlett.119.240401)
77. Rovelli C. 2004 *Quantum Gravity*. Cambridge University Press.
78. Kafri D, Taylor JM, Milburn GJ. 2014 A classical channel model for gravitational decoherence. *New J. Phys.* **16**, 065020. (https://doi.org/10.1088/1367-2630/16/6/065020)
79. Pfister C, Kaniewski J, Tomamichel M, Mantri A, Schmucker R, McMahon N, Milburn G, Wehner S. 2015 Understanding nature from experimental observations: a theory independent test for gravitational decoherence. *arXiv:1503.00577*
80. Großardt A, Bateman J, Ulbricht H, Bassi A. 2016 Optomechanical test of the Schrödinger-Newton equation. *Phys. Rev. D* **93**, 096003. (https://doi.org/10.1103/physrevd.93.096003)
81. Gan CC, Savage CM, Scully SZ. 2016 Optomechanical tests of a Schrödinger-Newton equation for gravitational quantum mechanics. *Phys. Rev. D* **93**, 124049. (https://doi.org/10.1103/physrevd.93.124049)
82. Pikovski I, Vanner MR, Aspelmeyer M, Kim MS, Brukner Č. 2012 Probing planck-scale physics with quantum optics. *Nat. Phys.* **8**, 393–397. (https://doi.org/10.1038/nphys2262)
83. Bowen WP, Milburn GJ. 2016 *Quantum Optomechanics*. CRC Press.
84. Leggett a J. 2002 Testing the limits of quantum mechanics: motivation, state of play, prospects. *J. Phys. Condens. Matter* **14**, R415–R451. (https://doi.org/10.1088/0953-8984/14/15/201)
85. Davies P. 2019 *The Demon in the Machine*. Penguin.
86. Ellis GFR. 2012 On the limits of quantum theory: Contextuality and the quantum-classical cut. *Ann. Phys. (N. Y).* **327**, 1890–1932. (https://doi.org/10.1016/j.aop.2012.05.002)
87. Drossel B. 2017 Ten reasons why a thermalized system cannot be described by a many-particle wave function. *Stud. Hist. Philos. Sci. B.* **58**, 12–21. (https://doi.org/10.1016/j.shpsb.2017.04.001)

88. Nimmrichter S, Hornberger K. 2013 Macroscopicity of mechanical quantum superposition states. *Phys. Rev. Lett.* **110**, 160403. (https://doi.org/10.1103/physrevlett.110.160403)
89. Turin L. 1996 A Spectroscopic Mechanism for Primary Olfactory Reception. *Chem. Senses* **21**, 773–791. (https://doi.org/10.1093/chemse/21.6.773)
90. Gauger EM, Rieper E, Morton JJL, Benjamin SC, Vedral V. 2011 Sustained quantum coherence and entanglement in the avian compass. *Phys. Rev. Lett.* **106**, 040503. (https://doi.org/10.1103/physrevlett.106.040503)
91. Mohseni M, Rebentrost P, Lloyd S, Aspuru-Guzik A. 2008 Environment-assisted quantum walks in photosynthetic energy transfer. *J. Chem. Phys.* **129**, 174106. (https://doi.org/10.1063/1.3002335)
92. Sarovar M, Ishizaki A, Fleming GR, Whaley KB. 2010 Quantum entanglement in photosynthetic light-harvesting complexes. *Nat. Phys.* **6**, 462–467. (https://doi.org/10.1038/nphys1652)
93. Fröwis F, Sekatski P, Dür W, Gisin N, Sangouard N. 2018 Macroscopic quantum states: Measures, fragility, and implementations. *Rev. Mod. Phys.* **90,** 025004. (https://doi.org/10.1103/revmodphys.90.025004)
94. Armour AD, Blencowe MP, Schwab KC. 2002 Entanglement and Decoherence of a Micromechanical Resonator via Coupling to a Cooper-Pair Box. *Phys. Rev. Lett.* **88**, 148301. (https://doi.org/10.1103/physrevlett.88.148301)
95. Khosla KE, Vanner MR, Ares N, Laird EA. 2018 Displacemon Electromechanics: How to Detect Quantum Interference in a Nanomechanical Resonator. *Phys. Rev. X* 8, 021052. (https://doi.org/10.1103/physrevx.8.021052)
96. Marshall W, Simon C, Penrose R, Bouwmeester D. 2003 Towards Quantum Superpositions of a Mirror. *Phys. Rev. Lett.* **91**, 130401. (https://doi.org/10.1103/physrevlett.91.130401)
97. Akram U, Bowen WP, Milburn GJ. 2013 Entangled mechanical cat states via conditional single photon optomechanics. *New J. Phys.* **15,** 093007. (https://doi.org/10.1088/1367-2630/15/9/093007)
98. O'Connell AD *et al.* 2010 Quantum ground state and single-phonon control of a mechanical resonator. *Nature* **464**, 697–703. (https://doi.org/10.1038/nature08967)
99. Pirkkalainen J-M, Cho SU, Li J, Paraoanu GS, Hakonen PJ, Sillanpää MA. 2013 Hybrid circuit cavity quantum electrodynamics with a micromechanical resonator. *Nature* **494**, 211–5. (https://doi.org/10.1038/nature11821)
100. Yuan M, Singh V, Blanter YM, Steele GA. 2015 Large cooperativity and microkelvin cooling with a three-dimensional optomechanical cavity. *Nat. Commun.* **6**, 8491. (https://doi.org/10.1038/ncomms9491)
101. Noguchi A, Yamazaki R, Ataka M, Fujita H, Tabuchi Y, Ishikawa T, Usami K, Nakamura Y. 2016 Ground state cooling of a quantum electromechanical system with a silicon nitride membrane in a 3D loop-gap cavity. *New J. Phys.* **18**, 103036. (https://doi.org/10.1088/1367-2630/18/10/103036)
102. Eibenberger S, Gerlich S, Arndt M, Mayor M, Tüxen J. 2013 Matter-wave interference of particles selected from a molecular library with masses exceeding 10,000 amu. *Phys. Chem. Chem. Phys.* **15**, 14696–700. (https://doi.org/10.1039/c3cp51500a)
103. Clausen C, Usmani I, Bussiéres F, Sangouard N, Afzelius M, De Riedmatten H, Gisin N. 2011 Quantum storage of photonic entanglement in a crystal. *Nature* **469**, 508–511. (https://doi.org/10.1038/nature09662)
104. Riedinger R, Wallucks A, Marinković I, Löschnauer C, Aspelmeyer M, Hong S, Gröblacher S. 2018 Remote quantum entanglement between two micromechanical oscillators. *Nature* **556**, 473–477. (https://doi.org/10.1038/s41586-018-0036-z)
105. Ockeloen-Korppi CF, Damskägg E, Pirkkalainen J-M, Asjad M, Clerk AA, Massel F, Woolley MJ, Sillanpää MA. 2018 Stabilized entanglement of massive mechanical oscillators. *Nature* **556**, 478–482. (https://doi.org/10.1038/s41586-018-0038-x)
106. Dakic B, Brukner Č. 2011 Quantum Theory and Beyond: Is Entanglement Special? In *Deep Beauty: Understanding the Quantum World through Mathematical Innovation* (ed H Halvorson), pp. 365–392. Cambridge University Press.

107. Hardy L. 2001 Quantum Theory From Five Reasonable Axioms. *arXiv:quant-ph/0101012*
108. Clifton R, Bub J, Halvorson H. 2003 Characterizing quantum theory in terms of information-theoretic constraints. *Found. Phys.* **33**, 1561–1591. (https://doi.org/10.1023/a:1026056716397)
109. Chiribella G, D'Ariano GM, Perinotti P. 2011 Informational derivation of quantum theory. *Phys. Rev. A* **84**, 012311. (https://doi.org/10.1103/physreva.84.012311)
110. Barrett J. 2007 Information processing in generalized probabilistic theories. *Phys. Rev. A.* **75**, 032304. (https://doi.org/10.1103/physreva.75.032304)
111. Deutsch D, Marletto C. 2015 Constructor theory of information. *Proc. R. Soc. A* **471**, 20140540. (https://doi.org/10.1098/rspa.2014.0540)
112. Marletto C. 2016 Constructor theory of probability. *Proc. R. Soc. A* **472**, 20150883. (https://doi.org/10.1098/rspa.2015.0883)
113. Laughlin RB. 1999 Nobel Lecture: Fractional quantization. *Rev. Mod. Phys.* **71**, 863–874. (https://doi.org/10.1103/revmodphys.71.863)
114. Paterek T, Dakić B, Brukner Č. 2010 Theories of systems with limited information content. *New J. Phys.* **12**, 053037. (https://doi.org/10.1088/1367-2630/12/5/053037)
115. Sinha U, Couteau C, Jennewein T, Laflamme R, Weihs G. 2010 Ruling out multi-order interference in quantum mechanics. *Science.* **329**, 418–421. (https://doi.org/10.1126/science.1190545)
116. Dahlsten OCO, Garner AJP, Vedral V. 2014 The uncertainty principle enables non-classical dynamics in an interferometer. *Nat. Commun.* **5**, 4592. (https://doi.org/10.1038/ncomms5592)
117. Nickerson NH, Li Y, Benjamin SC. 2013 Topological quantum computing with a very noisy network and local error rates approaching one percent. *Nat. Commun.* **4**, 1756. (https://doi.org/10.1038/ncomms2773)
118. Broadbent A, Fitzsimons J, Kashefi E. 2009 Universal Blind Quantum Computation. In *2009 50th Annual IEEE Symposium on Foundations of Computer Science*, pp. 517–526. IEEE. (https://doi.org/10.1109/focs.2009.36)
119. Schäfer VM, Ballance CJ, Thirumalai K, Stephenson LJ, Ballance TG, Steane AM, Lucas DM. 2018 Fast quantum logic gates with trapped-ion qubits. *Nature.* **555**, 75–78. (https://doi.org/10.1038/nature25737)
120. Hayashi M, Morimae T. 2015 Verifiable Measurement-Only Blind Quantum Computing with Stabilizer Testing. *Phys. Rev. Lett.* **115**, 220502. (https://doi.org/10.1103/physrevlett.115.220502)
121. Broadbent A, Schaffner C. 2015 Quantum cryptography beyond quantum key distribution. *Des. Codes Cryptogr.* **78**, 351–382. (https://doi.org/10.1007/s10623-015-0157-4)
122. Kimble HJ. 2008 The quantum internet. *Nature.* **453**, 1023–1030. (https://doi.org/10.1038/nature07127)
123. Preskill J. 1999 Plug-in quantum software. *Nature.* **402**, 357–358. (https://doi.org/10.1038/46434)
124. Gottesman D, Jennewein T, Croke S. 2012 Longer-baseline telescopes using quantum repeaters. *Phys. Rev. Lett.* **109**, 070503. (https://doi.org/10.1103/physrevlett.109.070503)
125. Bennett CH, Bernstein E, Brassard G, Vazirani U. 1997 Strengths and Weaknesses of Quantum Computing. *SIAM J. Comput.* **26**, 1510–1523. (https://doi.org/10.1137/s0097539796300933)
126. Kempe J, Kitaev A, Regev O. 2006 The Complexity of the Local Hamiltonian Problem. *SIAM J. Comput.* **35**, 1070–1097. (https://doi.org/10.1137/s0097539704445226)
127. Impagliazzo R, Paturi R. 2001 On the Complexity of k-SAT. *J. Comput. Syst. Sci.* **62**, 367–375. (https://doi.org/10.1006/jcss.2000.1727)
128. Dunjko V, Taylor JM, Briegel HJ. 2016 Quantum-Enhanced Machine Learning. *Phys. Rev. Lett.* **117**, 130501. (https://doi.org/10.1103/physrevlett.117.130501)
129. Biamonte J, Wittek P, Pancotti N, Rebentrost P, Wiebe N, Lloyd S. 2017 Quantum machine learning. *Nature.* **549**, 195–202. (https://doi.org/10.1038/nature23474)
130. Kiefer C, Polarski D. 2009 Why do cosmological perturbations look classical to us? *Adv. Sci. Lett.* **2**, 164–173. (https://doi.org/10.1166/asl.2009.1023)

131. Bengochea GR, Cañate P, Sudarsky D. 2015 Inhomogeneities from quantum collapse scheme without inflation. *Phys. Lett. B* **743**, 484–491. (https://doi.org/10.1016/j.physletb.2015.03.016)
132. Okon E, Sudarsky D. 2016 A (not so?) novel explanation for the very special initial state of the universe. *Class. Quantum Gravity* **33**, 225015. (https://doi.org/10.1088/0264-9381/33/22/225015)
133. León G, Sudarsky D. 2015 Origin of structure: Statistical characterization of the primordial density fluctuations and the collapse of the wave function. *J. Cosmol. Astropart. Phys.* **2015,** 020. (https://doi.org/10.1088/1475-7516/2015/06/020)
134. Castagnino M, Fortin S, Laura R, Sudarsky D. 2017 Interpretations of Quantum Theory in the Light of Modern Cosmology. *Found. Phys.* **47**, 1387–1422. (https://doi.org/10.1007/s10701-017-0100-9)
135. Valentini A. 2007 Astrophysical and cosmological tests of quantum theory. *J. Phys. A Math. Theor* **40**, 3285. (https://doi.org/10.1088/1751-8113/40/12/s24)
136. Valentini A. 2010 Inflationary cosmology as a probe of primordial quantum mechanics. *Phys. Rev. D.* **82,** 063513. (https://doi.org/10.1103/physrevd.82.063513)
137. Braunstein SL, Patra MK. 2011 Black hole evaporation rates without spacetime. *Phys. Rev. Lett.* **107**, 071302. (https://doi.org/10.1103/physrevlett.107.071302)
138. Henson J. 2009 The causal set approach to Quantum Gravity. In *Approaches to Quantum Gravity: Toward a New Understanding of Space, Time and Matter*, pp. 393–414. Cambridge University Press.
139. Brukner Č. 2014 Quantum causality. *Nat. Phys.* **10**, 259–263. (https://doi.org/10.1038/nphys2930)
140. Castro-Ruiz E, Giacomini F, Brukner Č. 2018 Dynamics of Quantum Causal Structures. *Phys. Rev. X* **8**, 011047. (https://doi.org/10.1103/physrevx.8.011047)
141. Oreshkov O, Costa F, Brukner Č. 2012 Quantum correlations with no causal order. *Nat. Commun.* **3,** 1092. (https://doi.org/10.1038/ncomms2076)
142. Rubino G, Rozema LA, Feix A, Araújo M, Zeuner JM, Procopio LM, Brukner Č, Walther P. 2017 Experimental verification of an indefinite causal order. *Sci. Adv.* **3**, 1602589. (https://doi.org/10.1126/sciadv.1602589)
143. Rubino G, Rozema LA, Massa F, Araújo M, Zych M, Brukner Č, Walther P. 2018 Experimental Entanglement of Temporal Orders. *arXiv:1712.06884*
144. Zych M, Costa F, Pikovski I, Brukner Č. 2019 Bell's theorem for temporal order. *Nat. Commun.* **10,** 3772. (https://doi.org/10.1038/s41467-019-11579-x)
145. Goswami K, Giarmatzi C, Kewming M, Costa F, Branciard C, Romero J, White AG. 2018 Indefinite Causal Order in a Quantum Switch. *Phys. Rev. Lett.* **121,** 090503. (https://doi.org/10.1103/physrevlett.121.090503)
146. Ringbauer M, Broome MA, Myers CR, White AG, Ralph TC. 2014 Experimental simulation of closed timelike curves. *Nat. Commun.* **5**, 4145. (https://doi.org/10.1038/ncomms5145)
147. Briegel HJ, Müller T. 2015 A Chance for Attributable Agency. *Minds Mach.* **25**, 261–279. (https://doi.org/10.1007/s11023-015-9381-y)
148. Tiersch M, Ganahl EJ, Briegel HJ. 2015 Adaptive quantum computation in changing environments using projective simulation. *Sci. Rep.* **5**, 12874. (https://doi.org/10.1038/srep12874)
149. Milburn G, Shrapnel S. 2018 Classical and quantum causal interventions. *Entropy* **20,** 687. (https://doi.org/10.3390/e20090687)
150. Price H. 2012 Does time-symmetry imply retrocausality? How the quantum world says 'Maybe'? *Stud. Hist. Philos. Sci. B* **43**, 75–83. (https://doi.org/10.1016/j.shpsb.2011.12.003)
151. Price H, Wharton KB. 2017 Dispelling the Quantum Spooks: A Clue that Einstein Missed? In *Time of Nature and the Nature of Time.*, pp. 123–137. Springer International Publishing.
152. Cramer JG. 1986 The transactional interpretation of quantum mechanics. *Rev. Mod. Phys.* **58**, 647–687. (https://doi.org/10.1103/revmodphys.58.647)
153. Millen J, Xuereb A. 2016 Perspective on quantum thermodynamics. *New J. Phys.* **18**, 11002. (https://doi.org/10.1088/1367-2630/18/1/011002)

154. Landauer R. 1961 Irreversibility and Heat Generation in the Computing Process. *IBM J. Res. Dev.* **5**, 183–191. (https://doi.org/10.1147/rd.53.0183)
155. Alemany, A. & Ritort F. 2010 Fluctuation theorems in small systems: Extending thermodynamics to the nanoscale. *Eur. News* **41**, 27–30. (https://doi.org/10.1051/epn/2010205)
156. Alemany A, Ribezzi-Crivellari M, Ritort F. 2013 Recent Progress in Fluctuation Theorems and Free Energy Recovery. In *Non-equilibrium Statistical Physics of Small Systems: Fluctuation Relations and Beyond*. Wiley-VCH.
157. Blickle V, Bechinger C. 2012 Realization of a micrometre-sized stochastic heat engine. *Nat. Phys.* **8**, 143–146. (https://doi.org/10.1038/nphys2163)
158. Martínez IA, Roldán É, Dinis L, Petrov D, Parrondo JMR, Rica RA. 2015 Brownian Carnot engine. *Nat. Phys.* **12**, 67-70. (https://doi.org/10.1038/nphys3518)
159. Roldan E, Martinez IA, Parrondo JMR, Petrov D. 2014 Universal features in the energetics of symmetry breaking. *Nat. Phys.* **10**, 457–461. (https://doi.org/10.1038/nphys2940)
160. Batalhão TB *et al.* 2014 Experimental Reconstruction of Work Distribution and Study of Fluctuation Relations in a Closed Quantum System. *Phys. Rev. Lett.* **113**, 140601. (https://doi.org/10.1103/PhysRevLett.113.140601)
161. An S, Zhang J-N, Um M, Lv D, Lu Y, Zhang J, Yin Z-Q, Quan HT, Kim K. 2015 Experimental test of the quantum Jarzynski equality with a trapped-ion system. *Nat. Phys.* **11**, 193–199. (https://doi.org/10.1038/nphys3197)
162. Peterson JPS, Sarthour RS, Souza AM, Oliveira IS, Goold J, Modi K, Soares-Pinto DO, Céleri LC. 2016 Experimental demonstration of information to energy conversion in a quantum system at the Landauer limit. *Proc. R. Soc. A* **472**, 20150813. (https://doi.org/10.1098/rspa.2015.0813)
163. Yan LL *et al.* 2018 Single-Atom Demonstration of the Quantum Landauer Principle. *Phys. Rev. Lett.* **120**, 210601. (https://doi.org/10.1103/physrevlett.120.210601)
164. von Lindenfels D, Gräb O, Schmiegelow CT, Kaushal V, Schulz J, Mitchison MT, Goold J, Schmidt-Kaler F, Poschinger UG. 2019 Spin Heat Engine Coupled to a Harmonic-Oscillator Flywheel. *Phys. Rev. Lett.* **123**, 80602. (https://doi.org/10.1103/physrevlett.123.080602)
165. Peterson JPS, Batalhão TB, Herrera M, Souza AM, Sarthour RS, Oliveira IS, Serra RM. 2019 Experimental characterization of a spin quantum heat engine. *Phys. Rev. Lett.* **123,** 240601. (https://doi.org/10.1103/PhysRevLett.123.240601)
166. Koski J V, Maisi VF, Sagawa T, Pekola JP. 2014 Experimental observation of the role of mutual information in the nonequilibrium dynamics of a Maxwell demon. *Phys. Rev. Lett.* **113,** 030601. (https://doi.org/10.1103/physrevlett.113.030601)
167. Koski JV, Maisi VF, Pekola JP, Averin DV. 2014 Experimental realization of a Szilard engine with a single electron. *Proc. Natl. Acad. Sci. U. S. A.* **111**, 13786–13789. (https://doi.org/10.1073/pnas.1406966111)
168. Koski JV, Kutvonen A, Khaymovich IM, Ala-Nissila T, Pekola JP. 2015 On-Chip Maxwell's Demon as an Information-Powered Refrigerator. *Phys. Rev. Lett.* **115**, 260602. (https://doi.org/10.1103/PhysRevLett.115.260602)
169. Cottet N *et al.* 2017 Observing a quantum Maxwell demon at work. *Proc. Natl. Acad. Sci.* **114**, 7561–7564. (https://doi.org/10.1073/pnas.1704827114)
170. Klatzow J *et al.* 2019 Experimental Demonstration of Quantum Effects in the Operation of Microscopic Heat Engines. *Phys. Rev. Lett.* **122**, 110601. (https://doi.org/10.1103/physrevlett.122.110601)
171. Ronzani A, Karimi B, Senior J, Chang Y-C, Peltonen JT, Chen C, Pekola JP. 2018 Tunable photonic heat transport in a quantum heat valve. *Nat. Phys.* **14**, 991–995. (https://doi.org/10.1038/s41567-018-0199-4)
172. Josefsson M, Svilans A, Burke AM, Hoffmann EA, Fahlvik S, Thelander C, Leijnse M, Linke H. 2018 A quantum-dot heat engine operating close to the thermodynamic efficiency limits. *Nat. Nanotechnol.* **13**, 920–924. (https://doi.org/10.1038/s41565-018-0200-5)
173. Jaliel G, Puddy RK, Sánchez R, Jordan AN, Sothmann B, Farrer I, Griffiths JP, Ritchie DA, Smith CG. 2019 Experimental Realization of a Quantum Dot Energy Harvester. *Phys. Rev. Lett.* **123**, 117701. (https://doi.org/10.1103/physrevlett.123.117701)

174. Vinjanampathy S, Anders J. 2016 Quantum thermodynamics. *Contemp. Phys.* **57**, 545–579. (https://doi.org/10.1080/00107514.2016.1201896)
175. Bell JS. 2004 *Speakable and unspeakable in quantum mechanics*. Cambridge University Press.

Interferometric Tests of Wave-Function Collapse

Stefan Gerlich, Yaakov Y. Fein, and Markus Arndt

Abstract Among the various commonly proposed interpretations of quantum mechanics, models of wave function collapse are unique in being empirically falsifiable. Here we review experiments which place bounds on the parameter space of these models, with a particular focus on matter-wave interferometry. Proving the persistence of superposition states is the most direct way to test collapse models, and by performing interferometry with macromolecules we can exclude large regions of parameter space due to the quadratic scaling of the collapse rate with mass.

1 How the Puzzle Started

When Louis de Broglie proposed the wave nature of matter in 1923, he did this with the conviction that "*by means of these new ideas, it will probably be possible to ... solve almost all the problems brought up by quanta*" [1]. De Broglie was certainly right in many respects: Already in 1926, his idea inspired Erwin Schrödinger's wave mechanics [2], which became the most successful and best-verified non-relativistic theory of modern physics. Born's rule, which states that observable probabilities are predicted by the squared modulus of the quantum wave function, has also been verified in many experiments [3–5]. Quantum mechanics seems to be here to stay.

And yet we challenge de Broglie's statement with the counterclaim that these new ideas *created* almost all (philosophical) problems brought up by quanta. By this we refer to the longstanding debate about the meaning of the wave function and the measurement process. The persistence of a surprisingly large number of interpretations of quantum mechanics is evidence of this discomfort, with one of the key questions being whether the wave function is something that exists as an ontological entity, or rather describes what we can know, an epistemological concept.

The success of quantum theory should not be understated: It underlies countless modern technologies, and has been corroborated by every experimental test so far.

S. Gerlich · Y. Y. Fein · M. Arndt (✉)
Faculty of Physics, University of Vienna, Boltzmanngasse 5, 1090 Vienna, Austria
e-mail: markus.arndt@univie.ac.at

V. Allori et al. (eds.), *Do Wave Functions Jump?*, Fundamental Theories of Physics 198,
https://doi.org/10.1007/978-3-030-46777-7_26

And yet there persist fundamental questions, such as the apparent transition to a classical regime and the fuzzy concepts of measurement and observers. There are many approaches toward a satisfactory resolution of these questions, which we will not attempt to list here, but there is one approach that stands out, because of its experimental falsifiability: objective collapse, and in particular continuous spontaneous localization (CSL) models.

Here we describe ongoing experiments to test CSL models, as originally proposed by Ghirardi, Rimini and Weber [6] and Pearle [7], as well as by Diosi [8] and Penrose [9] with a focus on the gravitational origin of objective wavefunction collapse. We discuss the current state of the art and the possibility to falsify such models, in particular with matter-wave interferometry of massive objects.

2 Experimental Approaches to Testing Objective Collapse

Several versions of collapse models have been described in the literature and also in this book. They all assume the wave function to shrink to a certain localization length r_c with a certain localization rate λ. All models are phenomenological and extend the Schrödinger equation by a stochastic non-linear term, with the purpose to effectively destroy mesoscopic or macroscopic spatial superposition states. The collapse-inducing agent is typically unspecified. On the experimental side, there are three main approaches for testing wave function collapse [10, 11]:

Absence of anomalous heating: If collapse confines the wave function this would lead to heating that should be observable in ultra-cold systems. This can be studied with a wide range of even classical systems, such as levitated nanoparticles [12], cantilevers [13], neutron stars [14, 15], the gravitational wave detector LIGO [16–18] or the satellites of LISA pathfinder. This approach is described in other chapters of this book and provides strong bounds on the CSL parameter space. However, the concept also has one apparent loophole: colored and dissipative extensions of the postulated noise field may not lead to any heating at all [20]. Even spontaneous cooling is not excluded. The absence of anomalous heating does not prove by itself the non-existence of the hypothetical background field, but rather constrains its possible properties.

Absence of spontaneous radiation: If light charged particles such as electrons undergo spontaneous wave function collapse, the random recoil associated with that process should lead to X-ray emission. This unexpected radiation has been searched for so far without success, which provides some of the most constraining bounds on CSL models [19].

Persistence of quantum superposition: CSL theory was introduced to effectively destroy quantum superpositions beyond a certain scale, so the most direct way to test it is to demonstrate quantum interference with increasingly massive particles over longer times. Unlike the two previous tests, the bounds produced by an interferometric test are not altered by the characteristics of the collapsing field [20]. In the following we will therefore focus on an account of matter-wave interference experiments which were in part motivated by the work of Giancarlo Ghirardi.

Matter-wave interferometry is an established and highly active field of research, nearly a century after the publication of Louis de Broglie's wave hypothesis [1]. The quantum wave nature of electrons has become a key element of modern surface science, used in electron microscopy [21], diffraction [22], holography [23] and multidimensional electron scattering [24] even on the femtosecond timescale [25]. Coherent neutron scattering has become a routine tool of modern condensed matter research, elucidating the 3D order and structure of metals [26], proteins [27] and even atoms [28]. Atom interferometers have become sensitive tools for fundamental physics [29], for testing general relativity [30–32] and measuring fundamental constants [33, 34], and are also sensitive devices for inertial sensing [35], geodesy and navigation [36].

But how macroscopic can such superpositions become? Here we review in particular our own work on macromolecule, cluster and nanoparticle interferometry [37–39] and put this into perspective with references to neutrons and atoms split over large distances, masses entangled over long distances and the prospects for entangled condensates or macroscopic bodies.

In doing this, we will refer to a measure of quantum macroscopicity which was introduced by Nimmrichter and Hornberger [40] and recently refined and extended [41]. The quantum macroscopicity μ was introduced to provide a way to compare how well experiments exclude modifications to the Schrödinger equation. The models and hypotheses underlying the derivation of macroscopicity are similar to those of continuous spontaneous localization. However, while CSL models focus primarily on spontaneous momentum kicks, quantum macroscopicity considers both spontaneous kicks and displacements which lead to a stochastic nonlinear term in the Schrödinger equation.

As discussed above, CSL models define a localization length r_c and rate λ, whereas the macroscopicity is independent of the length scale [40]. The measure is defined as

$$\mu = \log\left[\frac{1}{|f|}\frac{\tau}{1s}\left(\frac{m}{m_e}\right)^2\right] \tag{1}$$

which depends on the time τ the matter-wave remains coherent and the mass m of the delocalized particles compared to that of a reference particle, here chosen to be the electron with mass m_e. The fidelity f of the observed signal also enters, which in matter-wave interferometry can be estimated as the ratio of the experimental to the theoretically expected fringe contrast.

In the following we outline, roughly in order of increasing macroscopicity, the various categories of matter-wave interferometry experiments potentially relevant for placing bounds on CSL.

Electron interferometry
Electron matter-waves can cover macroscopic areas with billions or trillions of electrons in a superconducting current of a SQUID [42]. However, even in the presence of such large numbers of particles, only a small subset of electrons rotating in the clockwise or anti-clockwise direction are distinguishable in momentum space. Because of that, SQUIDS have low macroscopicity [40]. CSL bounds are expected to be similarly weak because of the quadratic mass scaling of the collapse rate.

Neutron interferometry
Mass increases by a factor of 1836 when we go from electrons to neutrons. This is a sizable increase and spatial separations can be as large as 10 cm with an enclosed area of 80 cm, as demonstrated in a perfect-crystal neutron interferometer in [43]. Such experiments are an impressive achievement, but the bounds on CSL models are still moderate because the neutron is light compared to atoms or molecules.

Bose Einstein condensates
One might rightfully argue that Bose Einstein condensates are huge macroscopic matter-waves with anywhere from a few hundred to a few hundred million atoms contributing to the condensate. They can extend over millimeters when freely expanded and the participation of many atoms in the effect seems to imply a large mass and macroscopicity. However, while BECs represent large coherent ensembles of atoms, the atoms are not entangled with one another. During the beam splitting process the system is better described by the single particle product state $\psi_{\text{tot}} \propto (|L\rangle + |R\rangle)^{\otimes \text{N}}$ rather than by the n-particle NOON-state $\psi_{\text{tot}} \propto |L\rangle^{\otimes \text{N}} + |R\rangle^{\otimes \text{N}}$. Without entanglement, the large number of atoms does not add to the mass term in Eq. (1). This will change if it becomes feasible to prepare high-NOON states, i.e. true Schrödinger cat states.

Atom interferometry
Atoms can be much better controlled than neutrons. High phase space densities and ultra-low temperatures are accessible with laser cooling. The mass of a heavy alkali exceeds that of a single neutron by more than two orders of magnitude. Separations of a few millimeters were achieved early in the history of atom interferometry. The recent record of up to half a meter separation in a 10 m high atomic fountain [44] is a stunning realization and represents a new bound on the parameter space of collapse models, particularly for large values of the localization radius r_c, with values up to several tens of centimeters. Obtaining a larger splitting over longer times in such experiments may become possible in drop towers or fountains with up to 300 m height [45]. This can stretch the superposition state and increase the coherence time by a factor of 5.5 over the 10 m fountain, as the speed rapidly increases in free fall.

Atom interferometry using coherent storage of trapped wave packets
While atomic fountains seem to be the most obvious approach to achieving long decoherence-free evolution times, a recent experiment in Berkeley demonstrated that trapped atoms could maintain their coherence for up to 20 s after the beam splitting process in an off-resonant optical lattice before being recombined [46]. A fountain would need to be 1000 m high to achieve a comparable storage time. In present-day experiments the optical trap still reduces the matter-wave coherence by about 30% within 5 s. One could envision, however, that future lattice-interferometry experiments with lower atom temperatures, improved beam splitter efficiency and optimized trapping conditions may tighten CSL bounds, in particular for r_c on the micron or millimeter scale.

Macromolecule and cluster interferometry
We focus now on experiments at the University of Vienna, which exploit the regime of short beam path separation and high mass. Because the CSL rate increases with the mass of the particle squared, an experiment with 10^4 amu particles in a quantum superposition [47] is thus a 10^{14} times better test of CSL models than an equivalent experiment with electrons [40].

If we are searching for localization lengths $r_C = 0.1 - 1\ \mu\text{m}$ [48], there is no big gain in splitting the wave function of electrons or atoms much more than this amount. Mass then becomes the most relevant parameter in interferometric tests of CSL once the coherence extends beyond this scale. This is the approach and parameter range that our group has pursued in recent years.

Starting with the first diffraction of hot fullerenes [49], numerous devices were built and tested in Vienna to push the mass and complexity of particles in matter-wave interference experiments. In the following we focus first on far-field diffraction, due to its conceptual simplicity, and then on the near-field interferometer LUMI, as the current mass record holder in matter-wave interferometry.

2.1 *Far-Field Diffraction*

The diffraction of massive particles at a double slit or grating constitutes perhaps the most intuitive demonstration of the wave-particle duality of matter. Following the seminal double slit or grating diffraction experiments with electrons [50], neutrons [51], atoms [52] and dimers [53], quantum interference was also observed with complex molecules [49].

To observe molecule interference on the single-molecule level in real time [54], phthalocyanine molecules (m = 514 amu) were chosen for the experiment since they are fluorescent and can thus be imaged with established microscopy techniques. They were evaporated by a micron-focused laser beam into the vacuum chamber with velocities in the range of 160 m/s. The beam was collimated to about 10 μrad and then diffracted at a 10 nm thick SiN_x grating with period 100 nm and a slit opening of 50 nm. The arrival of each single molecule onto a quartz slide behind

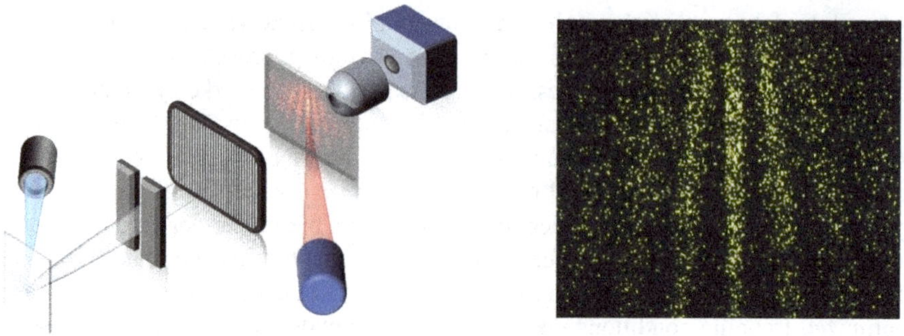

Fig. 1 Far-field diffraction. This is the most intuitive scheme to illustrate the wave particle duality of quantum mechanics. Molecules are laser desorbed, collimated and sent through a nanomechanical grating. The resulting diffraction pattern is visible on a quartz slide further downstream. The arriving molecules are imaged as they arrive one-by-one via fluorescence microscopy [54]

the grating was imaged via fluorescence microscopy and localized to within 10 nm accuracy [54, 55]. The molecular interference pattern observed on the quartz slide is the expected probability amplitude distribution to find the molecules in certain places on the detector plane. A schematic of the far-field experiment and sample diffraction fringes are shown in Fig. 1. The interference pattern reveals a number of non-classical aspects of quantum mechanics.

Quantum indistinguishability

Matter-wave interference requires coherence, i.e. a sufficiently well-defined and constant phase across the beam when it arrives at the diffraction slit or grating. Spatial coherence is the condition for a massive particle to have many possible paths through the diffractive element that are fundamentally indistinguishable. This is what is sometimes colloquially referred to as a molecule being in two or more places at once: the wave function covers several slits in the grating and there is no a priori way to know which path it will take.

The probabilistic nature of reality

While classical mechanics contains deterministic chaos and ignorance, quantum mechanics adds an element of true randomness, in which a measurement acts as a stochastic projection of the quantum system into one of its observable eigenstates. Deterministic laws describe the probabilities, but randomness seems to determine the selection of any individual observation. While it is fundamentally impossible to predict where on the quartz slide any particular molecule will land, quantum mechanics still allows us to describe the observed distribution with incredible accuracy.

It is such properties of the realm of quantum physics that are so foreign to our everyday experience in the classical world which inspired Giancarlo Ghirardi to consider stochastic non-linear extensions of the Schrödinger equation. The bounds placed by this far-field diffraction experiment on collapse models have recently been analyzed [56]. However, because of the low mass of the molecules involved and the moderate signal to noise ratio, this experiment is predominantly a conceptual introduction to molecule interference rather than one to restrict CSL models.

2.2 Near Field Interference with a Long Baseline Interferometer

The CSL rate increases with the square of the mass of the object put in quantum superposition, suggesting interference experiments with particles of increasing mass as the natural and most effective way to probe the largest area of CSL parameter space. However, quantum experiments with massive objects much more complex than elementary particles or single atoms involve multiple challenges. In particular, high mass is associated with extremely small de Broglie wavelengths. The phthalocyanine molecules investigated in the far-field diffraction experiments above, for example, correspond to a de Broglie wavelength of only 4 pm, while a molecular beam experiment with particles in the mass range of 100,000 amu would require the handling of de Broglie wavelengths below 50 fm. In far-field diffraction, as shown in Fig. 1, matter with $\lambda_{dB} = 50$ fm diffracted at a grating of d = 100 nm would be diffracted at an angle of $\vartheta = \lambda_{dB}/d = 200$ nrad. This would require impractically high beam collimation of order 100 nrad and long beam lines.

Clauser suggested that a near-field Talbot-Lau interferometer (TLI) is well adapted to the task [57], because it is shorter and with a higher particle throughput than any far-field experiment. The Talbot-Lau concept also makes rather moderate demands on coherence and scales very favorably with respect to the particle wavelength, making it ideal for high-mass interference experiments.

A TLI with mechanical gratings [58] was proven to work with complex molecules [59] but the van der Waals interactions with mechanical gratings suggested that an optical grating would be better suited as the central diffraction element when working with highly polarizable particles [60]. This was confirmed in the realization of a Kapitza-Dirac-Talbot-Lau Interferometer (KDTLI) [61], which employs an optical phase grating (KDTLI) as the second grating. The optical grating induces an electric dipole moment proportional to the molecule's polarizability which interacts with the laser field to induce a phase shift of the molecules in the anti-nodes of the grating. The Optical Time-domain Ionizing Matter-wave Interferometer (OTIMA) [62] also employs optical gratings, but relies on photo-depletion rather than the Kapitza-Dirac effect. The most recent addition, the Long-baseline Universal Matter-wave Interferometer (LUMI) [47], combines the capabilities of both an all-mechanical TLI and a mixed optical-mechanical KDTLI scheme with a ten-fold increased interferometer length compared to the original KDTLI experiment.

At the heart of LUMI are three gratings of period 266 nm which are spaced equidistantly by 0.98 m. The outer gratings are nanofabricated masks etched into a silicon-nitride wafer, while for the experiments presented here, the central grating is

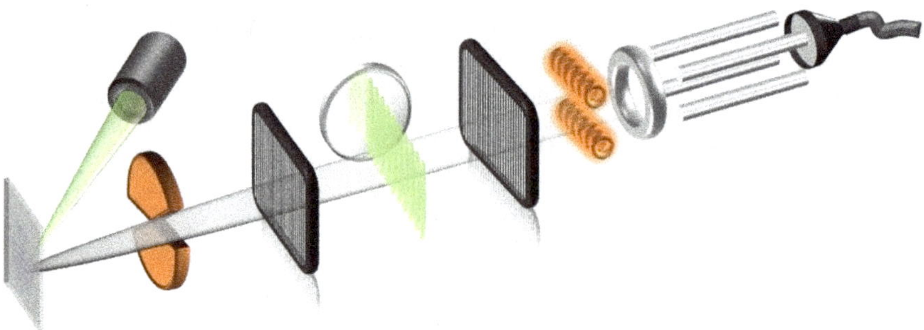

Fig. 2 Schematic of the LUMI experiment. The molecules are desorbed from a glass slide by nanosecond laser pulses (532 nm, 10^8 W/cm^2). They pass a set of three gratings of identical periods of 266 nm, spaced by 0.98 m. While silicon-nitride nanostructures serve as the first and the last grating, the central grating is realized as a pure phase grating generated by retro-reflection of a continuous high-power laser beam (532 nm). The molecules are then ionized by electron impact and sent through a quadrupole for mass selection. The time-of-flight can be measured by encoding a pseudo-random sequence on the beam with a chopper

realized as an optical phase grating by retro-reflection of a 532 nm laser beam. The interferometer scheme is illustrated in Fig. 2.

LUMI can accept spatially incoherent illumination of the first grating and can be operated with broad velocity distributions. The alignment of the three gratings with respect to each other and to gravity requires an accuracy on the order of 100 μrad. A set of 18 slip-stick piezo actuators allows the positioning of the gratings along all axes with the necessary precision. The interferometer is also highly sensitive to vibrations and drifts, as well as the rotation of the Earth. The vibrational noise has been damped to better than 10 nm across the relevant frequency band (>2 Hz) by means of a vibration isolation system that combines a pendulum suspension, mechanical springs and eddy current brakes.

The molecules used in the experiment are perfluoroalkyl-functionalized porphyrin-tetramer derivatives, which come as a family of molecules with up to 60 attached perfluoroalkyl chains. The molecules were designed and synthesized by Marcel Mayor and his team at the University of Basel to ensure high mass as well as moderate polarizability and intermolecular bond strengths, even though these bonds are still too strong for sublimation in a Knudsen cell. Previous experiments have shown that nanosecond desorption can lead to intact volatilization of such large molecules [63]. The beam consists of a molecular library, where each molecule is composed of different combinations and numbers of the perfluoralkyl chains, connected to different combinations of nodes. One molecule from this library is shown in Fig. 3.

The diffraction at the second grating leads to a modulation in the density of the beam that is roughly sinusoidal with the same period as the gratings. Depending on the transverse position of the third grating, molecules can either be transmitted

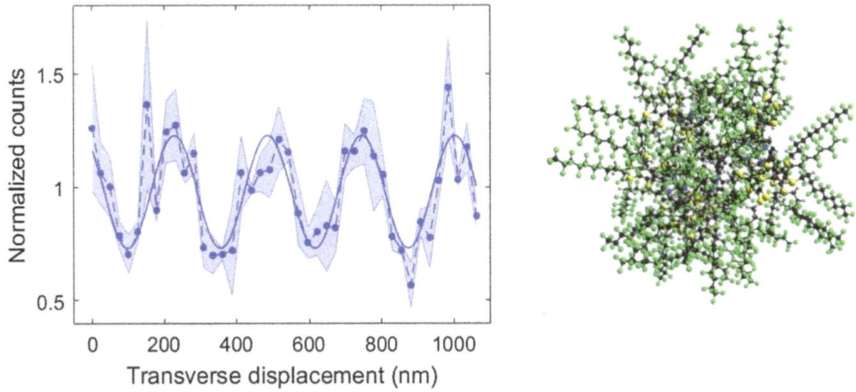

Fig. 3 **Left**: Molecular interference pattern with a phase grating power of 0.8 W, as recorded by scanning the third grating across the molecular density pattern. The counts have been referenced to a non-interfering signal to compensate for fluctuations of the signal intensity. The displayed points are an average of three scans and the shaded area is the standard deviation of the mean. **right:** The beam consists of a molecular library of functionalized oligoporphyrins with the largest compounds containing up to 2000 atoms. The mass of this library is centered around m = 27,000 amu and has a mass spread of about 15% [47]

or blocked, leading to a sinusoidal variation in flux as the third grating is moved. The transmitted molecules are ionized by electron impact and mass filtered in a quadrupole. The mass-filtered molecules are accelerated onto a dynode and release electrons which are counted by a secondary electron multiplier. A sample of the observed interference signal is shown in Fig. 3. A key parameter in determining the agreement of the observed signal with quantum theory is the visibility, $V = (S_{\max} - S_{\min})/\big((S_{\max} + S_{\min}\big)$. The visibility can also be measured as a function of the phase grating laser power for a direct comparison with theory. The resulting bounds on the CSL parameter space are plotted in Fig. 4.

3 A Look into the Future of Matter-Wave Tests of CSL Models

Quantum interference experiments with beams of even heavier particles depend on the further development of efficient sources for neutral massive particles and potentially new interferometer schemes. Metal clusters launched from a magnetron sputter source have been identified as a feasible candidate to access the mass regime beyond 10^5 amu in LUMI interferometry [67], as plotted in Fig. 5. The high density of metal clusters and the possibility to ionize and neutralize them in free-flight with available lasers is an asset of this material class [68]. UV laser gratings will allow a further reduction of the grating periods in an all-optical interferometer that relies on the neutralization, ionization [62] or fragmentation [69] in the standing light waves.

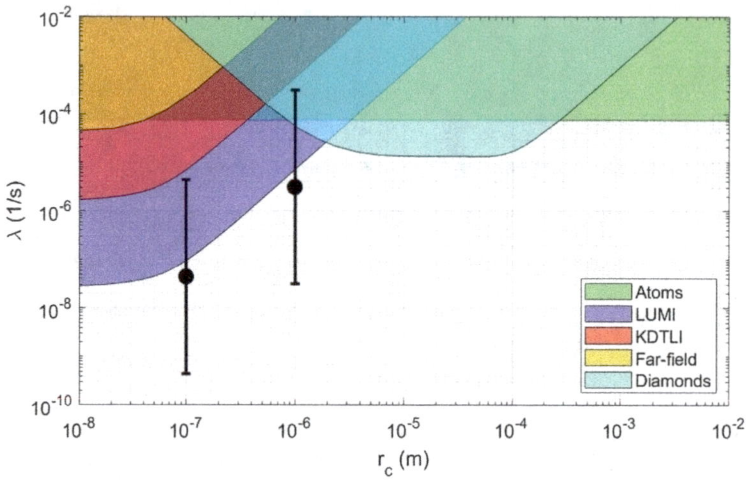

Fig. 4 CSL exclusion plot indicating the current limits imposed by superposition tests: long-baseline atom interferometry [44], long-baseline molecule interference (LUMI) [47], KDTLI molecule interference [64], far-field diffraction of molecules [54] and entangled NV centers in nanodiamonds [65]. The plot shows the combinations of localization radius r_c and event rate λ which are excluded by superposition experiments. Non-interferometric experiments currently exclude larger areas, but are a less direct and robust test [66]

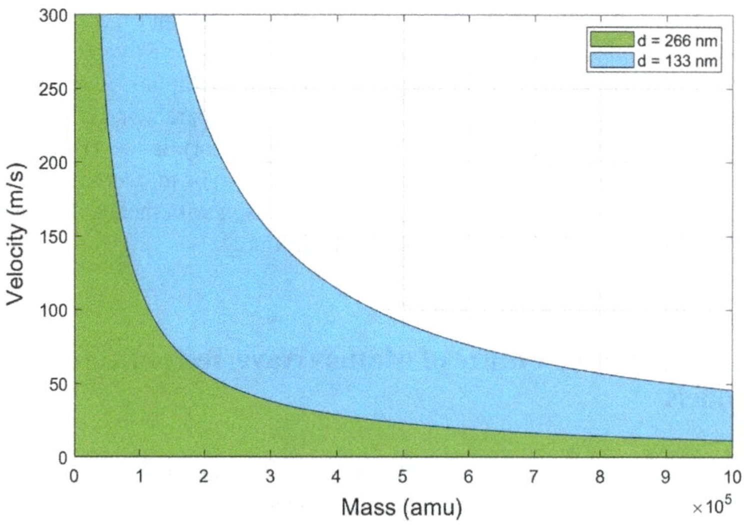

Fig. 5 Mass-velocity combinations accessible with LUMI and with the next generation of long-baseline universal matter-wave interferometers. De Broglie wavelengths as tiny as 35 fm can be handled in LUMI already. A future enhanced version of LUMI will operate with grating separations of L = 1 m but reduce the grating period by a factor of two, thus reducing the minimum accessible de Broglie wavelength by a factor of up to four

Sputter sources can generate nanoparticles with an internal temperature of 80 K which suppresses thermal decoherence even with masses up to 10^7 amu. Cooling to 10 K is conceivable and may allow scaling to even higher masses.

Cooling to even lower translational temperatures is in sight for individual dielectric particles. The fields of parametric feedback cooling [70–73], cavity assisted self-induced feedback cooling [74–76] and cavity assisted scattering cooling are progressing at a fast pace [77, 78]. However, while cooling works best for large systems with large polarizability and therefore large coupling to the electric field, quantum delocalization over large distances is attained more easily with lower masses: The width of the harmonic oscillator ground state $\Delta x = \sqrt{\hbar/2m\omega}$ scales with $1/\sqrt{m}$, and the diffraction angle grows directly with $\lambda_{dB} = h/mv$, i.e. with $1/m$.

The idea of extending the concept of Talbot and Talbot-Lau interferometry to high masses has been proposed [57] and refined a number of times [39, 79, 80] including the proposal of using cavity cooled nanoparticles for that purpose [81].

While these schemes assumed optical gratings as beam splitters, recent proposals also focused on using magnetic beam splitters coupled to an NV center in nanodiamond [82], or magnetic splitting of magnetically levitated superconducting spheres [83].

Orientational interference and localization by collapse
A new branch of macroscopic quantum superposition experiments has recently been opened with the idea of utilizing the closed space of rotating nanorods [84]. Advances in preparing and manipulating nanorods in optical tweezers [85–87] have triggered proposals for how to prepare coherent superpositions of rotational states and observe their revivals [88, 89] and how to use this to search for quantum wave function collapse [90].

4 Summary: On the Distinction of Collapse and Decoherence

As of today, all tests of spontaneous wave function collapse have provided negative results. Standard quantum mechanics is still safe. This is positive news as more and more teams are pushing emergent quantum technologies and an objective limit in mass or number of qubits involved in a quantum computer, simulator or sensor would be a road stop for important emergent technologies. At present, all indications are that these systems are still in the clear.

Interestingly, a wide range of models, including versions of collapse models with particle sizes smaller than r_c, gravity-induced collapse models and solutions of the Schrödinger-Newton equation [91], would predict a scaling with m^2. Even models exploring the diffusion of matter-waves on spontaneous conformal fluctuations of space-time would predict the same scaling [92, 93].

But if we ever observe an absence of interference at a certain mass scale, what claim could we make regarding collapse models? The naïve answer is: Absence of interference shows absence of interference. Identifying the cause for an unexpected loss of fringe visibility is always hard and particularly challenging since many potentially decohering effects also scale with particle mass. A key challenge would be to distinguish spontaneous collapse from decoherence. While we know how to test for collisional [94, 95] or thermal decoherence [96] there may still be unexpected new contributions as we scale up the particle mass. Successful high contrast interference with delocalization over more than 100 nm of masses larger than 10^9 amu would allow us to exclude the parameters originally proposed by Giancarlo Ghirardi. Non-interferometric tests are guiding the way, but genuine quantum superposition tests will be required to demonstrate that quantum physics ultimately holds in this mass regime.

References

1. L. De Broglie, Nature **112**, 540 (1923).
2. E. Schrödinger, Annalen der Physik **79**, 361 (1926).
3. J. P. Cotter, C. Brand, C. Knobloch, Y. Lilach, O. Cheshnovsky, and M. Arndt, Science Advances **3**, e1602478 (2017).
4. U. Sinha, C. Couteau, T. Jennewein, R. Laflamme, and G. Weihs, Science **329**, 418 (2010).
5. A. Sinha, A. H. Vijay, and U. Sinha, Scientific Reports **5**, 10304 (2015).
6. G. C. Ghirardi, A. Rimini, and T. Weber, Phys. Rev. D **34**, 470 (1986).
7. P. Pearle, Phys. Rev. A **39**, 2277 (1989).
8. L. Diosi, Phys. Rev. A **40**, 1165 (1989).
9. R. Penrose, Gen. Rel. Grav. **28**, 581 (1996).
10. A. Bassi and G. Ghirardi, Physics Reports **379**, 257 (2003).
11. A. Bassi and H. Ulbricht, Journal of Physics: Conference Series **504**, 012023 (10 pp.) (2014).
12. A. Vinante, A. Pontin, M. Rashid, M. Toroš, P. F. Barker, and H. Ulbricht, Phys. Rev. A **100** (2019).
13. A. Vinante, R. Mezzena, P. Falferi, M. Carlesso, and A. Bassi, Phys. Rev. Lett. **119** (2017).
14. A. Tilloy and T. M. Stace, arXiv:1901.05477v1 [quant-ph] 16 Jan 2019 (2019).
15. S. L. Adler, A. Bassi, M. Carlesso, and A. Vinante, Physical Review D **99** (2019).
16. M. Carlesso, A. Bassi, P. Falferi, and A. Vinante, Physical Review D **94** (2016).
17. B. Helou, B. J. J. Slagmolen, D. E. McClelland, and Y. Chen, Physical Review D **95** (2017).
18. M. Carlesso, M. Paternostro, H. Ulbricht, A. Vinante, and A. Bassi, New J. Phys. **20** (2018).
19. K. Piscicchia, A. Bassi, C. Curceanu, R. Grande, S. Donadi, B. Hiesmayr, and A. Pichler, Entropy **19** (2017).
20. M. Toroš, G. Gasbarri, and A. Bassi, Phys. Lett. A **381**, 3921 (2017).
21. G. F. Missiroli, G. Pozzi, and U. Valdre, J. Phys. E **14**, 649 (1982).
22. M. A. V. Hove and S. Y. Tong, *Surface crystallography by LEED* (Springer-Verlag New York, 1979).
23. A. Tonomura, *Electron Holography* (Springer, 1999).
24. P. Baum and A. H. Zewail, Chem. Phys. **366**, 2 (2009).
25. J. R. Dwyer, C. T. Hebeisen, R. Ernstorfer, M. Harb, V. B. Deyirmenjian, R. E. Jordan, and R. J. Dwayne Miller, Philosophical Transactions of the Royal Society A: Mathematical, Physical and Engineering Sciences **364**, 741 (2006).
26. G. E. Bacon, *Neutron diffraction* (Clarendon Press; Oxford, 1975), 3rd ed. edn.

27. J. Fitter, T. Gutberlet, and J. Katsaras, *Neutron scattering in biology: techniques and applications* (Springer Science & Business Media, 2006).
28. L. Cser, G. Toeroek, G. K. an D I. Sharkov, and B. Farago, Phys. Rev. Lett. **89 No. 17**, 175504 (2002).
29. G. Tino and M. Kasevich, *Atom Interferometry* (IOS, Varenna, 2014), Vol. 188, Proceedings of the International School of Physics "Enrico Fermi".
30. P. Asenbaum, C. Overstreet, T. Kovachy, D. D. Brown, J. M. Hogan, and M. A. Kasevich, Phys. Rev. Lett. **118**, 183602 (2017).
31. S. Dimopoulos, P. W. Graham, J. M. Hogan, and M. A. Kasevich, Phys. Rev. Lett. **98** (2007).
32. P. Febvre *et al.*, E3S Web of Conferences **4** (2014).
33. R. Bouchendira, P. Cladé, S. Guellati-Khélifa, F. Nez, and F. Biraben, Phys. Rev. Lett. **106**, 080801 (2011).
34. R. H. Parker, C. Yu, W. Zhong, B. Estey, and H. Müller, Science **360**, 191 (2018).
35. R. Geiger *et al.*, Nat Commun **2**, 474 (2011).
36. B. Barrett, L. Antoni-Micollier, L. Chichet, B. Battelier, T. Leveque, A. Landragin, and P. Bouyer, Nat Commun **7**, 13786 (2016).
37. M. Arndt, Phys. Today **67**, 30 (2014).
38. M. Arndt, N. Dörre, S. Eibenberger, P. Haslinger, J. Rodewald, K. Hornberger, S. Nimmrichter, and M. Mayor, in *Atom Interferometry*, edited by G. M. Tino, and M. A. Kasevich (IOS Press, Varenna, 2014).
39. M. Arndt and K. Hornberger, Nat. Phys. **10**, 271 (2014).
40. S. Nimmrichter and K. Hornberger, Phys. Rev. Lett. **110**, 160403 (2013).
41. B. Schrinski, S. Nimmrichter, B. A. Stickler, and K. Hornberger, Phys. Rev. A **100** (2019).
42. J. R. Friedman, V. Patel, W. Chen, S. K. Tolpygo, and J. E. Lukens, Nature **406**, 43 (2000).
43. M. Zawisky, M. Baron, R. Loidl, and H. Rauch, Nucl. Instr. and Meth. in Phys. Res. A **481** 406 (2002).
44. T. Kovachy, P. Asenbaum, C. Overstreet, C. A. Donnelly, S. M. Dickerson, A. Sugarbaker, J. M. Hogan, and M. A. Kasevich, Nature **528**, 530 (2015).
45. M.-S. Zhan *et al.*, International Journal of Modern Physics D, 1940005 (2019).
46. V. Xu, M. Jaffe, C. D. Panda, S. L. Kristensen, L. W. Clark, and H. Müller, arXiv:1907.030 54v1 (2019).
47. Y. Y. Fein, P. Geyer, P. Zwick, F. Kiałka, S. Pedalino, M. Mayor, S. Gerlich, and M. Arndt, Nat. Phys. (2019).
48. S. L. Adler, Journal of Physics A: Mathematical and Theoretical **40**, 2935 (2007).
49. M. Arndt, O. Nairz, J. Voss-Andreae, C. Keller, G. van der Zouw, and A. Zeilinger, Nature **401**, 680 (1999).
50. C. Jönsson, Z. Phys. **161**, 454 (1961).
51. A. Zeilinger, R. Gähler, C. G. Shull, W. Treimer, and W. Mampe, Rev. Mod. Phys. **60**, 1067 (1988).
52. D. W. Keith, M. L. Schattenburg, H. I. Smith, and D. E. Pritchard, Phys. Rev. Lett. **61**, 1580 (1988).
53. W. Schöllkopf and J. P. Toennies, Science **266**, 1345 (1994).
54. T. Juffmann, A. Milic, M. Müllneritsch, P. Asenbaum, A. Tsukernik, J. Tüxen, M. Mayor, O. Cheshnovsky, and M. Arndt, Nature Nanotechn. **7**, 297 (2012).
55. C. Brand *et al.*, Nature Nanotechnology **10**, 845 (2015).
56. Marko Toros and A. Bassi, J. Phys. A **51**, 115302 (2018).
57. J. F. Clauser, in *Experimental Metaphysics*, edited by R. S. Cohen, M. Horne, and J. Stachel (Kluwer Academic, 1997), pp. 1.
58. B. Brezger, L. Hackermüller, S. Uttenthaler, J. Petschinka, M. Arndt, and A. Zeilinger, Phys. Rev. Lett. **88**, 100404 (2002).
59. L. Hackermüller, S. Uttenthaler, K. Hornberger, E. Reiger, B. Brezger, A. Zeilinger, and M. Arndt, Phys. Rev. Lett. **91**, 090408 (2003).
60. B. Brezger, M. Arndt, and A. Zeilinger, J. Opt. B. **5**, 82 (2003).
61. S. Gerlich *et al.*, Nat. Phys. **3**, 711 (2007).

62. P. Haslinger, N. Dörre, P. Geyer, J. Rodewald, S. Nimmrichter, and M. Arndt, Nat. Phys. **9**, 144 (2013).
63. U. Sezer, P. Schmid, L. Felix, M. Mayor, and M. Arndt, J. Mass Spectrom. **50**, 235 (2015).
64. S. Eibenberger, S. Gerlich, M. Arndt, M. Mayor, and J. Tüxen, Phys. Chem. Chem. Phys. **15**, 14696 (2013).
65. S. Belli, R. Bonsignori, G. D'Auria, L. Fant, M. Martini, S. Peirone, S. Donadi, and A. Bassi, Phys. Rev. A **94** (2016).
66. M. Bahrami, M. Paternostro, A. Bassi, and H. Ulbricht, Phys. Rev. Lett. **112**, 210404 (2014).
67. F. Kiałka, B. Stickler, K. Hornberger, Y. Y. Fein, P. Geyer, L. Mairhofer, S. Gerlich, and M. Arndt, Phys. Scr. **94**, 034001 (2018).
68. E. Reiger, L. Hackermüller, M. Berninger, and M. Arndt, Opt. Commun. **264**, 326 (2006).
69. N. Dörre, J. Rodewald, P. Geyer, B. von Issendorff, P. Haslinger, and M. Arndt, Phys. Rev. Lett. **113**, 233001 (2014).
70. T. Li, S. Kheifets, and M. G. Raizen, Nat. Phys. **7**, 527 (2011).
71. P. Mestres, J. Berthelot, M. Spasenović, J. Gieseler, L. Novotny, and R. Quidant, Appl. Phys. Lett. **107**, 151102 (2015).
72. J. Gieseler, B. Deutsch, R. Quidant, and L. Novotny, Phys. Rev. Lett. **109**, 103603 (2012).
73. J. Vovrosh, M. Rashid, D. Hempston, J. Bateman, M. Paternostro, and H. Ulbricht, J. Opt. Soc. Am. B **34** (2017).
74. N. Kiesel, F. Blaser, U. Delic, D. Grass, R. Kaltenbaek, and M. Aspelmeyer, Proc. Natl. Acad. Sci. USA **110**, 14180 (2013).
75. P. Asenbaum, S. Kuhn, S. Nimmrichter, U. Sezer, and M. Arndt, Nature Communications **4**, 2743 (2013).
76. J. Millen, P. Z. G. Fonseca, T. Mavrogordatos, T. S. Monteiro, and P. F. Barker, Phys. Rev. Lett. **114** (2015).
77. U. Delic, M. Reisenbauer, D. Grass, N. Kiesel, V. Vuletic, and M. Aspelmeyer, arXiv:1812.09358v1 [quant-ph] 21 Dec 2018 (2018).
78. D. Windey, C. Gonzalez-Ballestero, P. Maurer, L. Novotny, O. Romero-Isart, and R. Reimann, Phys. Rev. Lett. **122**, 123601 (2019).
79. M. Arndt, K. Hornberger, and A. Zeilinger, Phys. World **3**, 35 (2005).
80. M. Arndt, M. Aspelmeyer, and A. Zeilinger, Fortschr. Phys. **57**, 1153 (2009).
81. J. Bateman, S. Nimmrichter, K. Hornberger, and H. Ulbricht, Nature Communications **5**, 4788 (2014).
82. M. Scala, M. Kim, G. Morley, P. Barker, and S. Bose, Phys. Rev. Lett. **111**, 180403 (2013).
83. O. Romero-Isart, L. Clemente, C. Navau, A. Sanchez, and J. I. Cirac, Phys. Rev. Lett. **109**, 147205 (2012).
84. B. A. Stickler, B. Papendell, S. Kuhn, B. Schrinski, J. Millen, M. Arndt, and K. Hornberger, New J. Phys. **20**, 122001 (2018).
85. S. Kuhn, B. A. Stickler, A. Kosloff, F. Patolsky, K. Hornberger, M. Arndt, and J. Millen, Nature Communications **8**, 1670 (2017).
86. S. Kuhn, A. Kosloff, B. A. Stickler, F. Patolsky, K. Hornberger, M. Arndt, and J. Millen, Optica **4**, 356 (2017).
87. S. Kuhn *et al.*, Nano Lett. **15**, 5604 – 5608 (2015).
88. B. A. Stickler and K. Hornberger, Phys. Rev. A **92** (2015).
89. B. A. Stickler, F. T. Ghahramani, and K. Hornberger, Phys. Rev. Lett. **121**, 243402 (2018).
90. B. Schrinski, B. A. Stickler, and K. Hornberger, J. Opt. Soc. Am. B **34** (2017).
91. D. Giulini and A. Großardt, Class. Quantum Grav. **30**, 155018 (2013).
92. C. T. Wang, R. Bingham, and J. T. Mendonca, Class. Quant. Grav. **23**, L59 (2006).
93. P. M. Bonifacio, C. H. T. Wang, J. T. Mendonça, and R. Bingham, Class. Quantum Grav. **26**, 145013 (2009).

94. K. Hornberger, S. Uttenthaler, B. Brezger, L. Hackermüller, M. Arndt, and A. Zeilinger, Phys. Rev. Lett. **90**, 160401 (2003).
95. K. Hornberger and J. E. Sipe, Phys. Rev. A **68**, 12105 (2003).
96. L. Hackermüller, K. Hornberger, B. Brezger, A. Zeilinger, and M. Arndt, Nature **427**, 711 (2004).

Tests in Space

Rainer Kaltenbaek

Abstract The technological progress on ground and in space have rendered space an increasingly attractive platform for highprecision fundamental tests of physics. Here, we will discuss a specific example: the mission proposal MAQRO for a medium-sized space mission with the goal of testing quantum physics with high-mass dielectric particles. In particular, we will provide an overview of the origins of MAQRO, the on-going development of the mission and payload design and the remaining challenges that need to be faced before a possible launch. The origins of the MAQRO proposal have benefited greatly from the work of G. C. Ghirardi. His work provided the framework for estabilishing experimental benchmarks to meet when testing for potential deviations from the pedictions of quantum physics.

Over the last decades, there has been impressive progress in space technology thanks to missions like LISA [1] and LISA Pathfinder [2]. The rapid development of space technology in combination with the possibility of harnessing the unique environment of space provided fertile ground for our mission proposal MAQRO [3]: the vision of realizing a space platform for testing the foundations of quantum physics. In developing the concept of MAQRO, collapse models have played a central role by providing challenging bench marks for the mission design. The main motivation for MAQRO has been to develop a quantum physics platform in space to test quantum physics with macroscopic test masses. Gravitational collapse models like the ones of Diósi [4, 5] and Penrose [6] were of particular interest because they indicated one could expect deviations from quantum physics due to gravitational effects for test masses on the order of 10^9 atomic mass units (amu) or more.

Although I never had the pleasure of meeting G. C. Ghirardi, his work on collapse models and the influence of his research on others have proved crucial in developing MAQRO. He and others developed the mathematical framework for the Ghirardi-Rimini-Weber (GRW) [7] and continuous spontaneous localization (CSL) [8] models

R. Kaltenbaek (✉)
Faculty of Mathematics and Physics, University of Ljubljana, Ljubljana, Slovenia
e-mail: rainer.kaltenbaek@fmf.uni-lj.si

IQOQI Vienna, Austrian Academy of Sciences, Vienna, Austria

V. Allori et al. (eds.), *Do Wave Functions Jump?*, Fundamental Theories of Physics 198,
https://doi.org/10.1007/978-3-030-46777-7_27

and for parameterizing and describing potential deviations from the coherent evolution of quantum superpositions. In the following, we will describe the motivation behind performing experiments in space. We will give an overview of the development of the mission concept of MAQRO and of recent results of a study performed together with the European Space Agency (ESA) at their Concurrent Design Facility (CDF) about a potential future realization of a quantum physics platform (QPPF) in space [9].

1 Beyond State of the Art

Recently, there have been suggestions to use not center-of-mass superpositions to test quantum physics and collapse models in order to circumvent some of the challenges one needs to face to realize and to maintain macroscopic center-of-mass superpositions. For example, it was suggested to use rotational degrees of freedom instead [10], which could be less sensitive to decoherence in position space. The most favorable argument for this approach would be that one does not require the implementation of an interferometer and therefore does not need to keep a large experimental setup interferometrically stable. Especially in the low-frequency regime, this could prove a significant advantage. Nevertheless, it can be argued that testing center-of-mass superpositions is the most direct way of testing the quantum superposition principle in the spirit of G. C. Ghirardi and others. In addition, the technological readiness of interferometric setups is well established in comparison to using rotational degrees of freedom. A central advantage of interferometric tests is that they provide unequivocal bounds on collapse models like the CSL model. For comparison, indirect experimental tests look, e.g., for spurious heating of the center-of-mass motion due to collapse processes [11, 12]. Bounds resulting from such tests can depend on the specific noise model for the collapse.

One can confirm the successful preparation of a center-of-mass superposition by observing matter-wave interferometry. The preceding chapters showed the state of the art of such experiments with the mass record being held by the group of M. Arndt with masses on the order of several 10^4 amu [13, 14]. Matter-wave experiments with high-mass test particles have not yet achieved combative restrictions on the GRW or CSL collapse models. Instead, some of the strongest bounds result from the evaluation of data from LISA Pathfinder and from tests not observing spurious heating of the motion of cantilevers [11, 15]. Significant progress in observing matter-wave interferometry with increasing test masses promises novel, stronger bounds on the CSL model. To cover significant parts of the remaining untested CSL parameter range, one has to increase by several orders of magnitude (a) the size of the superposition and/or (b) the test masses. Both approaches face specific challenges:

(a) increasing the superposition size in Talbot-type interferometers is difficult because the required coherence time scales with the square of the superposition size. As an alternative, one can consider non-classical state preparation like pulsed quantum optomechanics [16] or Bordé-type matter-wave interferometry harnessing spin to center-of-mass coupling [17]. It has also been suggested to dynamically modify the size of quantum states of levitated objects in a potential landscape [18]. For now, these methods are far from realization and from the technical readiness of state-of-the-art Talbot-type matter-wave interferometry.

(b) the mass of the test particles used in Talbot-type matter-wave interferometry has continuously increased over the last decades. Apart from technical challenges, the required coherence time in such experiments scales linearly with the test mass. To cover all or at least most of the remaining untested parameter regime of the CSL model, one would need to increase the test mass from below 10^5 amu at the moment to about 10^9 amu or 10^{10} amu. This will require coherence times on the order of tens or even hundreds of seconds. If the test masses are not suspended or trapped in some way, this also means that *the test particles need to be in free fall for* tens or hundreds *of seconds*. As we will see, such long free-fall times are a very good argument to perform such experiments in space. Novel proposals using charged [19] or magnetically suspended [18] test particles might overcome this limitation in the future, but these proposals may well suffer from other practical limitations not yet fully understood. For comparison: when it was first suggested to perform quantum experiments with optically trapped particles, the central toolbox for dealing with optically trapped dielectric particles was already well established due to efforts of the atom-trapping community (see, e.g., Ref. [20]). Still it took about 10 years to achieve sufficient control over optically trapped particles to cool them close to the ground-state of motion [21], but significant challenges still remain in order to use optomechanically prepared dielectric particles for high-mass matter-wave interferometry. For example, while there has been progress [22, 23], there is still no reliable way to load single, massive and neutral dielectric particles into an optical trap in ultra-high vacuum. Also, there are significant challenges in creating non-classical states for matter-wave interferometry. For example, as the test particles become more massive, their radius becomes comparable to or even larger than the wavelength of the light used for the standing-wave gratings in a Talbot-type interferometer [24, 25], and gas collisions may pose an ultimate limitation to such experiments [9]. For comparison, the theoretical proposals for using charged particles or magnetically suspended particles are still very recent, so it is difficult to estimate which technical or fundamental challenges these approaches may face in the future. For example, low-frequency vibrations could pose a significant challenge, and the vacuum requirements for particles with a mass of $\sim 10^{13}$ amu may prove challenging even at very low temperatures.

2 Why Space?

As we will argue below, a deep-space environment could provide significant advantages compared to ground-based experiments, but one has to be keenly aware that space-based experiments are also extremely expensive, time consuming and technologically challenging. For these reasons, one should not only have a good argument *why* one wants to do an experiment in space, but one also has to show that it is *possible* and that *no-one* could do a similar experiment on ground within the foreseeable future. That means, within the next *20 years*, which is usually about the time it takes to get a space mission going from the moment one already has a reasonable concept.

Needing a long free-fall time would not be a sufficient reason to go to space. For example, there are also parabolic flights, where one can have tens of seconds of free-fall time, and there are sounding rockets with a few minutes of free-fall time (see e.g. Ref. [26]). The crucial point is that in order to test quantum physics using high-mass matter-wave interferometry one not only has to have free-fall times on the order of tens or hundreds of seconds but one also must prevent vibrations from washing out the interference pattern. In particular, one requires a rather high-quality microgravity environment between 10^{-6} and 10^{-9} g, depending on the test masses [27]. At present, such conditions are potentially achievable on Earth for times on the order of one second, for example, by using a free-fall capsule within a free-fall capsule in the drop tower in Bremen [28]. Longer free-fall times are achievable in parabolic flights, but the microgravity environment there is by far not good enough. In sounding rockets, it may be conceivable to isolate the experiment well enough, but in this case one faces the problem that one would have to collect a sufficient number of data points in one shot. For the high-mass test particles we are interested in, typically one would only get one data point or maybe a few during one flight of a sounding rocket. This is far from the thousands or tens of thousands of data points one would need to resolve *a single* interference pattern [29].

But how sure are we that no-one will come up with a clever idea within the near future to perform comparable tests on ground? As we mentioned earlier, there already have been suggestions of using alternatives to Talbot-type interferometry to test quantum physics with high test masses. However, these are just ideas so far, and it is unclear if and when it will be possible to realize them. One should also note that quantum physics is one of the most successful physical theories we have. If we do an experiment with the potential to see deviations from the predictions of quantum physics, we would want to be *very* certain indeed that we can trust the results. In addition, we have the question of scaling: state-of-the-art experiments have confirmed quantum physics with test masses up to a few 10^4 amu. Ground-based tests in the lab or in a drop tower could successfully test quantum physics up to masses of possibly 10^6 amu. MAQRO would aim for tests between 10^8 amu and 10^{10} amu [29]. Using non-interferometric tests, MAQRO could potentially cover a wider mass range. The proposal of O. Romero-Isart's group would aim for test particles with a mass beyond ~10^{12} amu [18]. In order to test the predictions of quantum physics as well as the predictions for standard decoherence mechanisms and for physical models predicting deviations from quantum physics, it will be crucial to have a continuous coverage of the range of masses. In this sense, ground-based experiments and experiments in microgravity and in space will perfectly complement each other, and they will provide sanity checks in parameter regimes covered by multiple platforms.

3 MAQRO—Macroscopic Quantum Resonators

In 2010, me and a few colleagues pondered on the question whether we could use a space environment to perform tests of quantum physics or to realize applications of quantum physics using novel space technology originally developed for LISA and LISA pathfinder as well as GAIA and the James-Webb Space Telescope in combination with the novel idea of using optically trapped particles as a quantum optomechanical system.

After investigating a large variety of ideas in terms of their feasibility, we came to the conclusion that it should indeed be possible to perform matter-wave interferometry with high-mass dielectric particles in space using optomechanical state preparation in combination with a post-selection procedure for preparing non-classical quantum superposition states. We developed this idea into a mission concept and submitted it as a mission proposal called "MAQRO" in response to ESA's call for proposals for a medium-sized mission in 2010 [3].

MAQRO contained many unique ideas, but let us concentrate on those of imminent interest in the context of collapse models and tests of quantum physics:

- MAQRO was the first dedicated mission proposal to test the foundations of quantum physics. While there had been proposals of using atom interferometry in space, the science goals of these mission proposals were to use atom interferometry as a tool to test general relativity or to test the applicability of general relativity to quantum objects [30].
- Not taking into account environmental decoherence, the post-selection based method of preparing quantum superpositions in this first version of the proposal would have allowed to prepare (nearly) arbitrarily large superpositions. The central idea was to focus a short-wavelength pulse of light much narrower than the width of the wave function at the very center of the wavefunction. The width of the superposition would then depend on the size of the laser waist and the laser intensity. Within the laser beam and close to it, the quantum state would decohere due to light scattering, but parts of the wavefunction sufficiently far from the beam center would remain coherent [3, 27]. For practical reasons, we later replaced this method with a Talbot-type approach [31]. Some of the problems were: (1) the focused laser beam would decohere most of the quantum state and therefore lead to only a limited amount of interference visibility on top of the background of fully localized particles, (2) lasers at the required short wavelength (~30 nm) are far from being space-qualified, (3) the short wavelength could induce color centers in the test particles used, and (4) the required free-fall times would have been several 100 s [3].
- Our spacecraft design was optimized to reduce any environmental decoherence effects. That means, we aimed to achieve extremely high vacuum and cryogenic temperatures while at the same time ascertaining a very high quality of microgravity comparable to LISA Pathfinder.
- The environmental conditions were to be achieved by directly harnessing a deep-space environment without the need for active cooling. This allowed for extremely

good microgravity, low weight, (comparatively) low cost and extended lifetimes in order to collect high amounts of data and achieve high statistical significance [3, 32, 33].

Our proposal was well received and it was followed by a lot of activity over the following years in order to further develop the mission concept, the mission design as well as some of the core technology. We will provide a short overview of these developments in the following.

From 2011 to 2012, I led an ESA study (MQES) with members of the Aspelmeyer group, and in collaboration with G. Hechenblaikner and U. Johann (Airbus Defence & Space) with the goal of defining experiments using quantum optomechanics in space, to devise possible designs and to derive the corresponding scientific and technological requirements [27].

The study proposed and focused on two experiments:

- **DECIDE**: this was a more detailed development of the central experiment proposed in MAQRO. That means, quantum optomechanics is used to prepare an optically trapped dielectric test particle in a low-entropy state. After this state preparation, matter-wave interferometry is used to confirm the preparation of the macroscopic superposition, and the test results are compared with the predictions of quantum physics and with deviations predicted by collapse models or other alternative theoretical models.
- **WAX**: the layout for this experiment is simpler than for DECIDE. Instead of performing matter-wave interferometry, one simply lets the wavefunction expand over time, and the width of the wavepacket determined in the experiment is compared with the predictions of quantum theory and with the predictions of alternative theoretical models. This experiment has less demanding technical requirements than DECIDE and could operate even in the presence of stronger sources of decoherence.

Since 2012, we have investigated specific core technologies of MAQRO in more detail: (a) methods for loading dielectric nanoparticles into an optical trap in ultra-high vacuum (UHV) [34] and (b) optimized designs of the thermal shield design of MAQRO, which is necessary to achieve the required vacuum and cryogenic temperature conditions [32, 33].

In 2015, ESA published another call for proposals for a medium sized mission. This call required a very high technological readiness level (TRL) for the missions proposed in order to be launched on a short time scale. While MAQRO did not satisfy these conditions, we used this opportunity to further improve the mission design. The most noteworthy changes were the following:

- we adapted the payload to perform Talbot-type near-field interferometry instead of the far-field interferometry approach used in the original MAQRO proposal.
- we proposed to use multiple cavity modes in order to achieve 3D cooling of the center-of-mass motion of the test particle.
- we took into account the results of the MQES study in order to better define the scientific and technical requirements of MAQRO.

- we replaced the original loading mechanism for test particles (launching them via surface-acoustic waves from a piezoelectric substrate) with a more sophisticated approach, where the particles are initially charged and loaded into hollow-core fibers inside the spacecraft. They are then transported outside the spacecraft using linear Paul traps to confine them and buffer gas to cool their internal temperature. Outside the spacecraft, the particles are then discharged and loaded into an optical trap for the experiment.
- we adopted recent results of two finite-element studies on passive radiative cooling using the heat shield design of MAQRO.

4 QPPF—A Quantum Physics Platform in Space

In 2017, testing quantum physics with high test masses and testing a possible transition between the predictions of quantum physics and classical physics has been chosen as one of the New Science Ideas of ESA. As a consequence, ESA performed a detailed study at their Concurrent Design Facility (CDF). The study investigated the engineering details of a "Quantum Physics Platform" (QPPF) in space based on the MAQRO mission design [9].

The study provided an updated design, where the payload was to be inside a protective cover outside the space craft. The conical heat shields of MAQRO were replaced with a V-grove structure to accommodate a larger payload volume. In addition, the payload was to be also actively cooled by a hydrogen sorption cooler in order to achieve the necessary cryogenic temperatures with certainty.

The results of the study showed that the mission is, in principle, feasible and could be realized in the mid 2030 s. In the CDF study, we identified several critical issues that need to be addressed:

- **Vacuum**: because the payload is not open to space, the achievable vacuum will not be as good as required for MAQRO (10^{-15} mbar). Either we find ways to achieve the necessary vacuum, or we have to restrict ourselves to smaller test particles and/or to shorter free-fall times.
- **Loading mechanism**: ESA suggested a different loading mechanism, but in any case the loading mechanism [9], the transport of the test particles and the state preparation and the discharging of the test particles remain to be demonstrated.
- **Light scattering**: as the test particle size is comparable to the grating wavelength in a Talbot-type approach, the scattering of light may decohere the quantum state [25]. Either we show that this approach can still work, or we need a different way to prepare our non-classical states.

5 The Experiments to Be Performed

We already described the two central experiments very briefly in the context of the MQES study above, where we denoted these experiments as DECIDE and WAX. We later dropped these acronyms as the respective ideas developed further. There are still (at least) two experiments to be performed: wave-packet expansion and matter-wave interferometry. In either case, the first step of the experiment is always to optically trap a dielectric test particle and to cool its center-of-mass motion. How much we have to cool the motion depends on the free fall time, the size of the optical modes used for measuring the particle position and on the sensitivity we want to achieve.

In order to achieve the necessary optomechanical interaction while reducing potential surface interactions, we always assumed to use a long, high-finesse cavity in MAQRO. This requires very narrow modes such that the test particle will typically move out of the cavity mode during long free-fall times. In order to later detect the particle's position via optical means, we need an additional, wider optical mode perpendicular to the cavity mode.

To observe wavepacket expansion, the experimental steps are:

1. Load and then optomechanically cool a test particle.
2. Release the particle and let it evolve freely for some time t.
3. Optically detect the position of the test particle.

These steps are repeated many times, and the standard deviation of the positions measured will give the width of the wavepacket. This can be done for different times t and for different particle sizes and materials. It should be noted that the test particle remains close to its original position defined by the optical modes because in space the whole setup will be in free fall.

To observe matter-wave interferometry, we have to add an additional step: in the case of Talbot-type interferometry, we spatially overlap the transverse infrared beam perpendicular to the cavity mode with a short-wavelength pulse of light to realize a phase grating for the dielectric test particle. The steps of the experiment then are the following:

1. Load and then optomechanically cool a test particle
2. Release the particle and let it evolve freely for some time t_1.
3. Use a short-wavelength pulse to realize a phase grating (pulse duration much shorter than the mechanical oscillation period) [35].
4. Let the particle evolve freely again for a time t_2.
5. Optically detect the position of the test particle.

6 Using the Results to Test Quantum Physics

If we want to test quantum physics, we need to have some measure for any deviations we might see. Any realistic experiments will of course not give the exact values predicted by quantum theory. For example, we will never see 100% interference visibility. The question then is: when do we consider a deviation to be significant?

Typically, the most significant deviations from free quantum evolution will result from environmental decoherence. Examples are: (1) collisions with gas particles, (2) emission, absorption, scattering of blackbody radiation, (3) scattering of (cosmic) radiation, (4) stray electromagnetic fields. Of course, one could extend this list. For example, one can also include higher-order effects of some of these decoherence mechanisms. The question is: when is our measurement precise enough?

One of the big benefits of collapse models in the context of MAQRO is that they provide us with specific numeric predictions of deviations from free evolution. This then gives us a benchmark for how well we have to isolate our quantum system from environmental decoherence and, for example, how good our micro-gravity environment needs to be such that the interference pattern is not washed out too much.

Collapse models typically also predict a different parameter dependence of decoherence compared to environmental decoherence. The same may hold true for deviations from quantum physics due to quantum gravitational effects or similar.

7 Recent Results and Current Efforts

After the first proposals in 2009 and 2010 to use optically trapped particles to do quantum optomechanics [36–38], researchers were optimistic that it would not take long to bring these systems into the quantum regime by cooling the center-of-mass motion close to the quantum ground state. Unfortunately, achieving this goal has proved surprisingly difficult, often due to very mundane reasons. For example, getting clean dielectric particles trapped in ultra-high vacuum has long proved a challenge. It was achieved in the Novotny group in 2012 by using feedback cooling to keep particles stably trapped even at low pressures [39]. In 2013, the first demonstration of side-band cooling of optically trapped dielectric particles was reported [40], and very recently the Aspelmeyer group demonstrated the use of coherent scattering for optomechanical cooling [41]. The same approach was then also implemented by the Novotny group [42]. A very important benefit of this method is that it requires less laser power and therefore reduces the effects of laser shot noise and optical heating on the trapped particles. Using this technique, ground-state cooling is now achievable [21].

Using hollow-core fibers for particle transport has been demonstrated by the Aspelmeyer group in 2016 [23], and the Northup group and the Barker group have reported a range of interesting experiments using charged dielectric particles (e.g., Refs. [19, 43]). Controlling the charges of dielectric particles has been reported by the groups of Geraci [44, 45] and Novotny [46].

8 The Next Steps Towards Experiments in Space

Now that optically trapped dielectric particles are entering the quantum regime, we can soon expect a series of proof-of-principle experiments and potentially also first applications. At the same time, efforts are on-going to make the loading of the test particles more reliable and to achieve optical trapping at extremely high vacuum levels in order to limit collisional decoherence.

The loading and the delivery of test particles have to be combined with reliable techniques of charge control. In the context of future space experiments, some of the methods used/proposed in the laboratory will not be applicable, and we will have to adapt more space-suitable techniques in some cases. For example, particle loading via laser desorption could be challenging in space due to mass and power limitations as well as the lack of space-proof lasers necessary for this technique. Alternatives may be piezoelectric or MEMS devices [3, 9, 34]. Suggested methods for charge control using plasma close to the dielectric particles [46] is only feasible in the presence of some residual gas—not in the case of near perfect vacuum required for MAQRO.

9 Conclusions

Efforts towards future space-based tests of quantum physics have come a long way. Ground-based experiments have provided proof-of-principle demonstrations of many of the techniques involved, and the success of the LISA Pathfinder mission has shown it is possible to realize an extremely good microgravity environment in space [47]. The selection of QPPF as a New Science Idea by ESA shows that testing quantum physics has attracted significant interest from space agencies, and the study showed that such tests are, in principle, feasible within the foreseeable future [9].

Without the efforts of G. C. Ghirardi and others this would not have been possible. They have shown us the way towards testing for possible deviations from the predictions of quantum physics. Who knows where this road will lead? An interesting analogy might be that the quest for detecting gravitational waves fueled efforts towards increasingly sensitive measurements. Apart from leading to the seminal success of detecting gravitational waves, these efforts resulted in high-sensitivity measurements even beyond the quantum limit and they led to the completely new field of quantum optomechanics. In a similar way, the quest to isolate increasingly macroscopic systems to study novel sources of decoherence may lead to novel discoveries we cannot yet point the finger to, but we can be excited about the opportunities arising.

References

1. P. Amaro-Seoane et al., arXiv:1702.00786 (2017).
2. M. Armano et al., Class. Quantum Gravity **26**, 094001 (2009).
3. R. Kaltenbaek et al., Exp. Astron. **34**, 123 (2012).
4. L. Diósi, Phys. Lett. A **105**, 199 (1984).
5. L. Diósi, J. Phys. A Math. Theor. **40**, 2989 (2007).
6. R. Penrose, Gen. Relativ. Gravit. **28**, 581 (1996).
7. G. C. Ghirardi, A. Rimini, and T. Weber, Phys. Rev. D **34**, 470 (1986).
8. G. C. Ghirardi, P. Pearle, and A. Rimini, Phys. Rev. A **42**, 78 (1990).
9. A. Pickering, T. Voirin, P. Falkner, CDF Study Report: QPPF - Assessment of a Quantum Physics Payload Platform (2018).
10. B. A. Stickler et al., New J. Phys. **20**, 122001 (2018).
11. A. Vinante et al., Phys. Rev. Lett. **119**, 110401 (2017).
12. M. Carlesso et al., New J. Phys. **20**, 083022 (2018).
13. S. Eibenberger et al., Phys. Chem. Chem. Phys. **15**, 14696 (2013).
14. J. Schätti et al., Commun. Chem. **1**, 93 (2018).
15. M. Carlesso, A. Bassi, P. Falferi, and A. Vinante, Phys. Rev. D **94**, 124036 (2016).
16. O. Romero-Isart et al., Phys. Rev. Lett. **107**, 20405 (2011).
17. S. Bose et al., Phys. Rev. Lett. **119**, 240401 (2017).
18. H. Pino et al., Quantum Sci. Technol. **3**, 025001 (2018).
19. D. Goldwater et al., Quantum Sci. Technol. (2018).
20. G. Hechenblaikner, M. Gangl, P. Horak, and H. Ritsch, Phys. Rev. A **58**, 3030 (1998).
21. U. Delić, M. Reisenbauer, K. Dare, D. Grass, V. Vuletić, N. Kiesel, M. Aspelmeyer, Science **367**, 892–895 (2020).
22. P. Mestres et al., Appl. Phys. Lett. **107**, 151102 (2015).
23. D. Grass et al., Appl. Phys. Lett. **108**, 221103 (2016).
24. S. Nimmrichter, *Macroscopic Matter Wave Interferometry* (Springer International Publishing, Cham, 2014).
25. A. Belenchia et al., Phys. Rev. A **100**, 033813 (2019).
26. D. Becker et al., Nature **562**, 391 (2018).
27. R. Kaltenbaek et al., Macroscopic Quantum Experiments in Space Using Massive Mechanical Resonators, Study under Contract with ESA, Po P5401000400 (2012).
28. H. Selig, H. Dittus, and C. Lämmerzahl, Microgravity Sci. Technol. **22**, 539 (2010).
29. R. Kaltenbaek et al., EPJ Quantum Technol. **3**, 5 (2016).
30. D. N. Aguilera et al., Class. Quantum Gravity **31**, 115010 (2014).
31. R. Ursin et al., Nature **430**, 849 (2004).
32. G. Hechenblaikner et al., New J. Phys. **16**, 013058 (2014).
33. A. Pilan Zanoni et al., Appl. Therm. Eng. **107**, 689 (2016).
34. P. Schmid et al., Study under Contract with ESA, AO/1-6889/11/NL/CBi (2014).
35. J. Bateman, S. Nimmrichter, K. Hornberger, and H. Ulbricht, Nat. Commun. **5**, 4788 (2014).
36. D. E. Chang et al., Proc. Natl. Acad. Sci. U. S. A. **107**, 1005 (2010).
37. O. Romero-Isart, M. L. Juan, R. Quidant, and J. I. Cirac, New J. Phys. **12**, 33015 (2010).
38. P. F. Barker and M. N. Shneider, Phys. Rev. A **81**, 23826 (2010).
39. J. Gieseler, B. Deutsch, R. Quidant, and L. Novotny, Phys. Rev. Lett. **109**, 103603 (2012).
40. N. Kiesel et al., Proc. Natl. Acad. Sci. U. S. A. **110**, 14180 (2013).
41. U. Delić et al., Phys. Rev. Lett. **122**, 123602 (2019).
42. D. Windey et al., Phys. Rev. Lett. **122**, 123601 (2019).
43. J. Millen et al., Phys. Rev. Lett. **114**, 123602 (2015).
44. J. H. Stutz, Master thesis, University of Nevada, 2014.
45. G. Ranjit et al., Phys. Rev. A **91**, 051805 (2015).
46. M. Frimmer et al., Phys. Rev. A **95**, 061801 (2017).
47. M. Armano et al., Phys. Rev. Lett. **116**, 231101 (2016).

Sneaking a Look at Ghirardi's Cards: Collapse Models Mapped with the Spontaneous Radiation

K. Piscicchia, R. Del Grande, M. Laubenstein, and C. Curceanu

Abstract The collapse models were proposed more than 30 years ago to save the poor Schrödinger's cat from its zombie-fate. In these models the standard Schrödinger equation is modified with the introduction of non-linear and stochastic terms, naturally collapsing the wave function in space. Ghirardi was pioneering these models, which predict deviations from the standard Quantum Mechanics. One of these predictions is the emission of a "spontaneous radiation", which we explored to set the most stringent limits on the collapse models parameters in a broad range. This allowed us paraphrasing the title of his famous book, to sneak a look at Ghirardi's cards.

1 Introduction

Quantum Mechanics has been experimentally confirmed with outstanding precision and represents the most complete and successful theory of the microscopic world. However, since its formulation, the tension among the linear and unitary character of the Schrödinger equation and the wave packet reduction principle—needed to account for the measurement process—demands for a deeper understanding. Moreover the superposition principle does not seems to apply to the macroscopic objects, the scale which marks the transition among quantum and classical worlds being unknown.

In this context GianCarlo Ghirardi proposed collapse models as phenomenological solutions to overcome what is usually referred as the *measurement problem* (see

K. Piscicchia (✉) · R. Del Grande
Centro Fermi - Museo Storico della Fisica e Centro Studi e Ricerche "Enrico Fermi", 00184 Rome, Italy
e-mail: kristian.piscicchia@centrofermi.it

K. Piscicchia · R. Del Grande · C. Curceanu
Istituto Nazionale di Fisica Nucleare (INFN), Laboratori Nazionali di Frascati, 00044 Frascati, Italy

M. Laubenstein
Istituto Nazionale di Fisica Nucleare (INFN), Laboratori Nazionali ded Gran Sasso, 67100 L'Aquila, Italy

V. Allori et al. (eds.), *Do Wave Functions Jump?*, Fundamental Theories of Physics 198,
https://doi.org/10.1007/978-3-030-46777-7_28

e.g. Ref. [1]); they consist in a dynamical reformulation of the standard Quantum Mechanics, which is realised by modifying the linear and unitary evolution given by the Schrödinger equation by adding non-linear terms interacting with a stochastic noise field. Collapse models account for the wave function collapse in space, which is characterised by an amplification mechanism, the biggest the mass the faster the reduction of the wave packet. The quantum to classical transition is then realised by ensuring that macroscopic objects always have well defined positions. On the other hand the interaction with the noise field is very small at the microscopic level, where the standard Schrödinger evolution dominates.

Phenomenological collapse models embed slight deviations from the standard quantum mechanics predictions, in particular an unavoidable effect of the non-linear interaction with the noise field represented by emission of a *spontaneous radiation* [2]. When collapse occurs the centre of mass is shifted towards the localized wave function position, and since the process is random this results in a diffusion motion which determines, for charged particles, the emission of photons.

We will review in this work the experimental constrains on the so-called Ghirardi-Rimini-Weber (GRW) and Continuous Spontaneous Localization (CSL) models, obtained by analysing X-ray spectra measured in high precision low-background experiments, exploiting the predicted spontaneous radiation emission process.

The paper is organized as follows: in Sect. 2, we briefly describe the GRW and CSL models and introduce their characteristic physical parameters λ and r_C. In the same section we also introduce the gravity related collapse model introduced by L. Diósi and R. Penrose (DP). In Sect. 3 we review recent constrains on the GRW and CSL models (see also Ref. [3, 4]) based on the spontaneous radiation emission search. In Sect. 4 we present the ongoing experimental efforts and future perspectives for further improving the limits on λ and r_C and for setting stringent constraints on the DP model.

2 Collapse Models and Characteristic Physical Parameters

In the GRW [5] model particles experience spontaneous localizations in space around appropriate positions, at random times according to a Poisson distribution with mean rate λ. The mechanism is designed such that the Born rule is recovered. The wave function localization at the position $\mathbf{a}$ is realised by the operator $\mathbf{L_a} = (\pi r_C^2)^{-3/4} e^{-\frac{(\mathbf{q}-\mathbf{a})^2}{2r_C^2}}$ (where $\mathbf{q}$ is the position operator), among subsequent localizations the evolution is determined by the Schrödinger equation. Considered the superposition of two one dimensional wave packets characterised by a width $l \ll r_C$ and a separation $2a \gg r_C$, then the application of $\mathbf{L_a}$ (with appropriate normalization) determines the localization around $\mathbf{a}$ with a probability which recovers the Born rule.

In the GRW model the localization process does not preserve the symmetry of the wave function, hence it can not be applied to a system of identical particles. The problem was solved with the introduction of the CSL model [6, 7], developed in the second quantization formulation. In the CSL the standard Hamiltonian is modified with the introduction of non-linear and stochastic terms, characterized by the interaction with a continuous set of independent noises $w(\mathbf{x}, t)$ (one for each point of the space, which is why this set is often referred to as "noise field") having zero average and white correlation in time, i.e., $E[w(\mathbf{x}, t)] = 0$ and $E[w(\mathbf{x}, t)w(\mathbf{y}, s)] = \delta(\mathbf{x} - \mathbf{y})\delta(t - s)$ where $E[...]$ denotes the average over the noises. Two phenomenological parameters (λ and r_C) are introduced in the model. The parameter λ has the dimensions of a rate and sets the strength of the collapse, while r_C is a correlation length which determines the spatial resolution of the collapse: the collapse is weaker for a superposition with size much smaller than r_C, compared to the case in which the delocalization is much larger than r_C. The values originally proposed by Ghirardi for λ and r_C are [5] $\lambda = 10^{-16}$ s^{-1}, $r_C = 10^{-7}$ m. Higher values for λ were also suggested by Adler [8], up to $\lambda = 10^{-8\pm2}$ s^{-1}.

The energy distribution of the spontaneous radiation, emitted as a consequence of the interaction of free electrons with the collapsing stochastic field, was first calculated by Fu [2] and later on studied in more detail in [9–12], in the framework of the non-relativistic CSL model. If the stochastic field is assumed to be a white noise, coupled to the particle mass density (mass proportional CSL model), the spontaneous emission rate is given by:

$$\frac{d\Gamma(E)}{dE} = \frac{e^2\lambda}{4\pi^2 r_C^2 m_N^2 E}, \tag{1}$$

where e is the charge of the proton, m_N represents the nucleon mass and E is the energy of the emitted photon. In the non-mass proportional case, the rate takes the expression:

$$\frac{d\Gamma(E)}{dE} = \frac{e^2\lambda}{4\pi^2 r_C^2 m_e^2 E}, \tag{2}$$

with m_e the electron mass.

Using the measured radiation emitted in an isolated slab of Germanium [13] corresponding to an energy of 11 keV, and comparing it with the predicted rate in Eqs. (1) and (2), Fu extracted the following upper limits on λ for the two cases:

$$\lambda \leq 2.20 \cdot 10^{-10}\ \text{s}^{-1} \quad \text{mass prop.}, \tag{3}$$

$$\lambda \leq 0.55 \cdot 10^{-16}\ \text{s}^{-1} \quad \text{non-mass prop.}, \tag{4}$$

assuming that the correlation length value is $r_C = 10^{-7}$ m. In his estimate, Fu considered the contribution to the spontaneous X-ray emission of the four valence electrons in the Germanium atoms. Such electrons can be considered as *quasi-free*, since their binding energy (of the order of $\sim$10 eV) is much less than the emitted photons' energy. In Ref. [8], the author argues that an erroneous value for the fine structure constant is used in Ref. [2]. This correction is taken into account in the analysis described in Sect. 3. Further, the preliminary TWIN data set [13] used by Fu to estimate the upper limit on λ turned out to be underestimated by a factor of about 50 at 10 keV.

A new analysis was performed in Ref. [14]. Based on the improved data presented in Ref. [15], the limits corresponding to the footnote [7] in Ref. [14], for the cases of mass proportional and non-mass proportional CSL models, were:

$$\lambda \leq 8 \cdot 10^{-10}\,\mathrm{s}^{-1} \quad \text{mass prop.,} \tag{5}$$

$$\lambda \leq 2 \cdot 10^{-16}\,\mathrm{s}^{-1} \quad \text{non-mass prop..} \tag{6}$$

To conclude this section, among the most intriguing collapse models, of great charm, is the gravity related collapse model named Diósi-Penrose (DP) after the authors [16] (first proposed by Diósi, and later by Penrose, based on independent arguments on the behavior of spacetime in presence of quantum superpositions). DP proposes that gravity might be the ultimate explanation for the wave function collapse. Gravity is universal, and its magnitude increases with increasing mass, matching the amplification requirements of the collapse. Since no experimental evidence is available so far for gravity quantization, this could give the non-linear coupling which is necessary for the quantum linearity breakdown.

More in detail DP sets the correlation function of the noise equal to the Newtonian gravitational potential. The DP introduces only one cut-off length phenomenological parameter R_0, which cures the ultraviolet divergence of the gravitational interaction. The effective collapse rate, analogous to λ, is given by $(Gm_N^2)/(\sqrt{\pi}\hbar R_0)$, while R_0 describes how well an object is localized, in analogy to r_C.

3 Recent Upper Limits on λ from Spontaneous Radiation Search

Recently the most stringent limits on the collapse rate parameter λ were set, in a broad range of the $\lambda - r_C$ parameters space, and in particular for $r_C = 10^{-7}$ m, looking for signature of spontaneous radiation emission in extremely low background X-ray spectra. In Ref. [3, 4] the data collected by the IGEX experiment [17] is analysed based on various techniques. IGEX is a low-background experiment originally conceived for the neutrinoless double beta decay ($\beta\beta0\nu$) search. The analyses described in Ref. [3, 4] refer to the data set published in Ref. [18], the measurement exploited one High Purity Germanium detector (HPGE) (active mass of about 2 kg) with an 80 kg day exposure (the total mass which is considered to be possible signal source multiplied by the total acquisition time). The shielding and the cryostat were produced following ultra-low background techniques, since the main contamination is represented by the radionuclides emission. The experiment had an overburden of

2450 m.w.e. (metres of water equivalent) , corresponding to a muon flux of $2 \cdot 10^{-7}$ $\text{cm}^{-2}\ \text{s}^{-1}$, moreover a cosmic muon veto covered the top and the sides of the shield. In Ref. [18] a Monte Carlo (MC) simulation of the background from known emission processes is not given, nor a simulation of the detection efficiency. The main inefficiency sources are due to the muon veto anti-coincidence and the adopted pulse shape analysis. The probability of rejecting non-coincident events with the muon veto was found to be less than 0.01. The pulse shape analysis contributes negligibly to the efficiency loss for energies above 4 keV.

In Ref. [3] the X-ray experimental spectrum published in [18] is fitted in the range $\Delta E = (4.5 \div 48.5)$ keV by minimising a χ^2 function, assuming the expected number of counts in each 1 keV bin to be given by the theoretical prediction Eqs. (1) and (2), i.e. the fitting function is:

$$\frac{d\Gamma(E)}{dE} = \frac{\alpha(\lambda)}{E}. \tag{7}$$

ΔE is compatible with the non-relativistic assumption (for electrons) used in the calculation of the predicted rate. Moreover Eqs. (1) and (2) are valid for free electrons; this assumption is accurate if the spontaneous emission of the 22 external electrons of each Ge atom is considered, down to the 3s orbit. The binding energy of the 3 s orbit is 180.1 eV—much less then the lower energetic measured photon—the 22 outermost electrons can then be considered as *quasi-free*. From the fit the value $\alpha(\lambda) = 110 \pm 7$ is obtained, corresponding to a reduced $\chi^2/(n.d.f. - n.p.) = 1.1$ ($n.d.f.$ represents the number of degrees of freedom, $n.p.$ is the number of free parameters of the fit). The upper limits on λ are then extracted using Eqs. (1) and (2):

$$\frac{d\Gamma(E)}{dE} = c\,\frac{e^2\lambda}{4\pi^2 r_C^2 m^2 E} \leq \frac{121.48}{E}, \tag{8}$$

where the factor c is given by:

$$c = \left(8.29 \times 10^{24}\,\frac{\text{atoms}}{\text{kg}}\right) \cdot (80\,\text{kg} \times \text{day}) \cdot \left(8.64 \times 10^4\,\frac{\text{n.of seconds}}{\text{day}}\right) \cdot (22), \tag{9}$$

the first bracket accounts for the particle density of Germanium, the second represents the amount of emitting material expressed in kg × day, the third term is the number of seconds in one day and 22 represents the number of spontaneously emitting electrons for each Ge atom. Applying Eq. (8), the following upper limits for the reduction rate parameter are obtained:

$$\lambda \leq 9.4 \cdot 10^{-12}\ \text{s}^{-1} \quad \text{mass prop.}, \tag{10}$$

$$\lambda \leq 2.8 \cdot 10^{-18}\ \text{s}^{-1} \quad \text{non-mass prop.}, \tag{11}$$

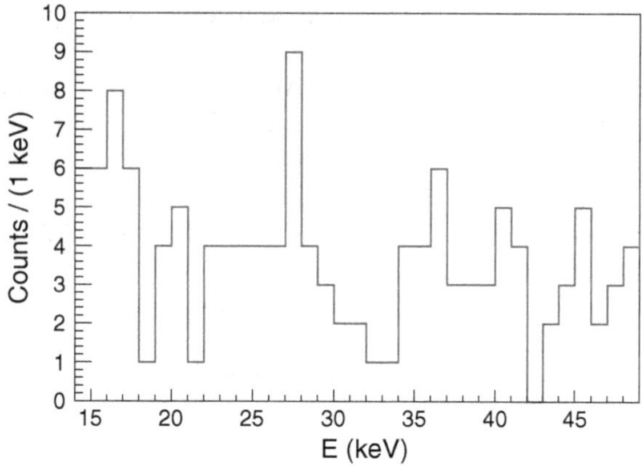

Fig. 1 X-ray emission spectrum measured by the IGEX experiment [17, 18]

corresponding to a probability of 0.95.

This procedure is implicitly based on two assumptions: first, the measured spectrum is assumed to be background free, that is to say that the upper limits on λ correspond to the case in which all the measured X-ray emission would be originated in spontaneous emission processes. This ansatz is conservative, and is necessary since in Ref. [18] a MC description of the background is missing. The second assumption, consistent with the analysis presented in Ref. [18], is that the detector efficiency, in the range ΔE, is one, and that the inefficiencies which are introduced by the muon veto anticoincidence and the pulse shape analysis, performed to extract the experimental spectrum in Ref. [18], are very small for energies greater than 4 keV.

Based on these considerations and with the aim to account for the spontaneous emission of the 30 outermost electrons of the Ge atoms (considered as quasi-free) a new analysis of the same data set is performed in Ref. [4]. The measured spectrum is analysed in the range $\Delta E' = (14.5 \div 48.5)$ keV, the binding energy of the 2 s orbit in Ge is still one order of magnitude smaller than 14.5 keV, justifying the quasi-free hypothesis. The experimental spectrum is shown in blue in Fig. 1.

The spectrum is scarcely populated in the range $\Delta E'$, the bin contents y_i are then considered to fluctuate around the mean values Λ_i according to Poisson distributions. By applying the Bayes theorem the probability distribution function for the collapse rate parameter λ is then obtained:

$$G'(\lambda) \propto \left(\sum_{i=1}^{n} \frac{\alpha(\lambda)}{E_i} + 1 \right)^{y} e^{-\left(\sum_{i=1}^{n} \frac{\alpha(\lambda)}{E_i} + 1\right)}. \tag{12}$$

In Eq. (12) the expected number of signal counts is calculated from Eq. (7), n is the number of bins each corresponding to an energy E_i, the total number of measured counts is $y = 130$. The upper limits on the λ parameter are obtained by solving the

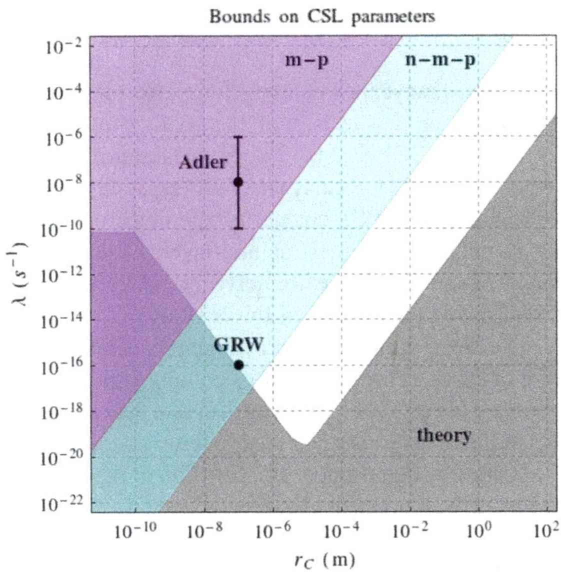

Fig. 2 Mapping of the $\lambda - r_C$ Continuous Spontaneous Localization (CSL) parameters: the originally proposed theoretical values (GRW, Adler) are shown as black points; the region excluded by theory (theory) is represented in gray. The excluded region according to our analysis is shown in cyan for the non-mass proportional case (n-m-p) and in magenta for the mass proportional case (m-p)

following integral equation, i.e. by equating the cumulative distribution function to 0.95:

$$\int_0^{\lambda_0} G'(\lambda)\, d\lambda = 0.95, \tag{13}$$

from which the upper limits on λ are

$$\lambda \leq 6.8 \cdot 10^{-12}\ \text{s}^{-1} \quad \text{mass prop.,} \tag{14}$$

$$\lambda \leq 2.0 \cdot 10^{-18}\ \text{s}^{-1} \quad \text{non-mass prop.,} \tag{15}$$

with probability 0.95 having set $r_C = 10^{-7}$ m.

Figure 2 represents the exclusion plot in the $\lambda - r_C$ plane letting the r_C parameter to vary. The originally proposed values are shown, together with the results of the Bayesian analysis, in cyan the region excluded for the non-mass proportional case and in magenta for the mass proportional case. The gray band is excluded by theory (see Ref. [19]).

Figure 2 can be compared with Fig. 2 in Ref. [20], where the mapping is obtained using other measurements. It is interesting to note that, for a collapse induced by a white noise, the allowed parameter space is confined to a drastically reduced region.

4 Ongoing Experimental Efforts and Future Perspectives

In the last few years we devoted big effort to design experimental setups dedicated to collapse models tests, with the goal to improve the experimental limits obtained by using the IGEX data. Two data taking runs were performed corresponding to exposures of 124 kg × day and 141.4 kg × day respectively. The experimental setups were based on HPGe coaxial p-type detectors (about 2 kg active area) surrounded by a complex shielding structure (layers of radio-pure electrolytic copper and roman lead), operated in the extremely low-background environment of the underground Gran Sasso Laboratories of INFN (Italy).

The most significant improvement is represented by a complete MC characterization of the detector and all of its components, the simulation based on the GEANT4 software library. This allows an accurate determination of the background from residual radionuclides contained in the setup materials. Moreover the detection efficiency was carefully determined, as a function of the energy, for spontaneous photons emitted in each component close to the crystal.

The accurate knowledge of the background, as well as the determination of the spontaneous emission contributions from all the apparatus components will improve the bound on λ by at least one order of magnitude, and will allow to set the first limit ever on the characteristic length R_0 of the DP model.

Acknowledgments We acknowledge the support of the Centro Fermi—Museo Storico della Fisica e Centro Studi e Ricerche "Enrico Fermi" (*Open Problems in Quantum Mechanics project*), EU COST Action CA 15220 and EU FET project TEQ (grant agreement 766900). Furthermore, these studies were made possible by the support of the Foundational Questions Institute, FOXi.

References

1. For reviews and references, see Bassi, A.; Ghirardi, G. C. Dynamical reduction models. *Phys. Rep.* **2003**, *379*, 257; Pearle, P. Collapse models Open Systems and Measurements in Relativistic Quantum Field Theory. *Lecture Notes in Physics* **1999**, vol 526 ed H-P. Breuer and F. Petruccione (Berlin: Springer); Diósi, L. Models For Universal Reduction Of Macroscopic Quantum Fluctuations. *Phys. Rev. A* **1989**, *40*, 1165; Bassi, A. Collapse models: analysis of the free particle dynamics. Available online: https://arxiv.org/abs/quant-ph/0410222.pdf (accessed on 25 March 2009); Adler, S. L. Quantum Theory as an Emergent Phenomenon. (Cambridge: Cambridge University Press) **2004** ch 6. Alternative choices of the correlation function are discussed in Weber, T. Quantum mechanics with spontaneous localization revisited. *Nuovo Cimento B* **1991***106*, 1111.
2. Fu, Q. Spontaneous radiation of free electrons in a nonrelativistic collapse model. *Phys. Rev. A* **1997**, *56*, 1806.
3. Curceanu, C.; Hiesmayr, B.C.; Piscicchia, K. X-rays help to unfuzzy the concept of measurement. *J. Adv. Phys.* **2015**, *4*, 263–266.
4. K. Piscicchia et al., Entropy 2017, 19(7), 319.
5. Ghirardi, G.; Rimini, A.; Weber, T. Unified dynamics for microscopic and macroscopic systems. *Phys. Rev. D* **1986**, *34*, 470.

6. Pearle, P. Combining stochastic dynamical state-vector reduction with spontaneous localization. *Phys. Rev. A* **1989**, *39*, 2277.
7. Ghirardi, G.C.; Pearle, P.; Rimini, A. Markov processes in Hilbert space and continuous spontaneous localization of systems of identical particles. *Phys. Rev. A* **1990**, *42*, 78.
8. Adler, S.L. Lower and Upper Bounds on CSL Parameters from Latent Image Formation and IGM Heating. *J. Phys. A* **2007**, *40*, 2935–2958.
9. Adler, S.L.; Ramazanoglu, F.M. Photon emission rate from atomic systems in the CSL model. *J. Phys. A* **2007**, *40*, 13395–13406.
10. Adler, S.L.; Bassi, A.; Donadi, S. On spontaneous photon emission in collapse models. *J. Phys. A* **2013**, *46*, 245304.
11. Bassi A.; Donadi S. Spontaneous photon emission from a non-relativistic free charged particle in collapse models: A case study. *Phys. Lett. A* **2014**, *378*, 761.
12. Donadi, S.; Bassi, A.; Deckert, D.-A. On the spontaneous emission of electromagnetic radiation in the CSL model. *Ann. Phys.* **2014**, *340*, 70–86.
13. Miley, H.S.; Avignone, F.T.; Brodzinski, R.L., III.; Collar, J.I.; Reeves, J.H. Suggestive evidence for the two neutrino double beta decay of Ge-76. *Phys. Rev. Lett.* **1990**, *65*, 3092.
14. Laloë, F.; Mullin, W.J.; Pearle, P. Heating of trapped ultracold atoms by collapse dynamics. *Phys. Rev. A* **2014**, *90*, 52119.
15. Collett, B.; Pearle, P.; Avignone, F.; Nussinov, S. Constraint on collapse models by limit on spontaneous x-ray emission in Ge. *Found. Phys.* **1995**, *25*, 1399–1412.
16. Diósi, L. Models for universal reduction of macroscopic quantum fluctuations. *Phys. Rev. A* **1989**, *40*, 11651174.
17. Aalseth, C.E.; Avignone, F.T., III; Brodzinski, R.L.; Collar, J.I.; Garcia, E.; González, D.; Hasenbalg, F.; Hensley, W.K.; Kirpichnikov, I.V.; Klimenko, A.A.; et al. Neutrinoless double-beta decay of Ge-76: First results from the International Germanium Experiment (IGEX) with six isotopically enriched detectors. *IGEX collab. Phys. Rev. C* **1999**, *59*, 2108.
18. Morales, A.; Aalseth, C. E.; Avignone, F. T.; Brodzinski, R. L., III; Cebrian, S.; Garcia, E.; Irastorza, I. G.; Kirpichnikov, I. V.; Klimenko, A. A.; Miley, H. S.; et al. Improved constraints on WIMPs from the international Germanium experiment IGEX. *IGEX collab. Phys. Lett. B* **2002**, *532*, 8-14.
19. Toroš, M.; Bassi, A. Available online: https://arxiv.org/pdf/1601.03672.pdf (accessed on 31 May 2017).
20. Carlesso, M.; Bassi, A.; Falferi, P.; Vinante, A. Experimental bounds on collapse models from gravitational wave detectors. *Phys. Rev. D* **2016**, *94*, 124036.

New Avenues for Testing Collapse Models

Andrea Vinante and Hendrik Ulbricht

1 Introduction

There is an increasing interest in developing experiments aimed at testing collapse models, in particular the Continuous Localization Model (CSL), the natural evolution of the GRW model initially proposed by Ghirardi et al. [1–4]. Current experiments and related bounds on collapse parameters are partially discussed in other contributions in this review. Our aim here is to discuss some of the most promising directions towards future improvements. The paper is organized as follows. In Sects. 2, 3, 4 we will discuss noninterferometric techniques. In detail, in Sect. 2 we will discuss mechanical experiments, both with conventional and levitated mechanical resonators, in Sect. 3 we will consider proposed experiment looking at bulk thermal heating of solid bodies and in Sect. 4 we will briefly discuss the use of cold atoms or macroscopic condensates. In Sect. 5 we will outline proposals of matter-wave interference with massive nano/microparticles. We will end in Sect. 6 with some ideas on how precision experiments can be used for testing collapse models.

2 Noninterferometric Mechanical Tests of Collapse Models

2.1 Key Concepts

Noninterferometric spontaneous heating experiments have emerged in recent years as a powerful and effective way to test collapse models. Some examples which

---

A. Vinante (✉) · H. Ulbricht
School of Physics and Astronomy University of Southampton, Southampton SO17 1BJ, UK
e-mail: a.vinante@soton.ac.uk

H. Ulbricht
e-mail: h.ulbricht@soton.ac.uk

V. Allori et al. (eds.), *Do Wave Functions Jump?*, Fundamental Theories of Physics 198,
https://doi.org/10.1007/978-3-030-46777-7_29

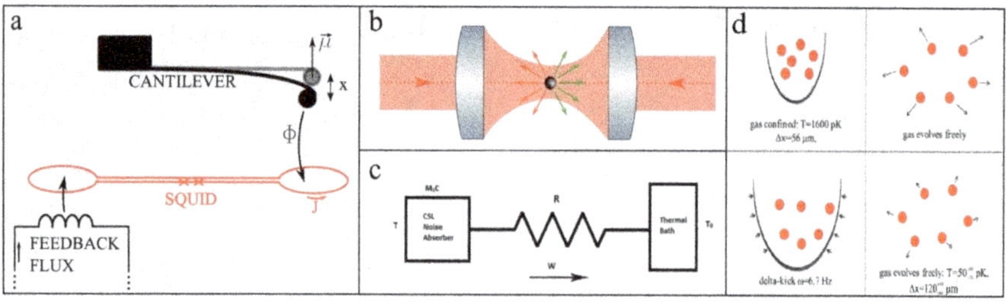

Fig. 1 Simplified sketch of some of the noninterferometric methods to test collapse models discussed in this contribution. **a** Measuring the mechanical noise induced by CSL using an ultracold cantilever detected by a SQUID (adapted from Ref. [15]); **b** Measuring the mechanical noise induced by CSL using a levitated nanoparticle detected optically (adapted from Ref. [24]); **c** Measuring the heating induced by CSL in a solid matter object cooled to very low temperature (adapted from Ref. [42]); **d** Measuring the increase of kinetic energy induced by CSL in a ultracold atoms cloud (adapted from Ref. [44])

will be discussed in this contribution are shown in Fig. 1. We start by discussing purely mechanical experiments. The underlying idea [5–7] is that a mechanism which continuously localizes the wavefunction of a mechanical system, which can be either a free mass or a mechanical resonator, must be accompanied by a random force noise acting on its center-of-mass. This leads in turn to a random diffusion which can be possibly detected by ultrasensitive mechanical experiments.

In a real mechanical system such diffusion will be masked by standard thermal diffusion arising from the coupling to the environment, i.e. from the same effects which lead to decoherence in quantum interference experiments [8]. In practice there will be additional nonthermal effects, due to external nonequilibrium vibrational noise (seismic/acoustic/gravity gradient). Moreover, one has to ensure that the back-action from the measuring device is negligible.

Under the assumption that thermal noise is the only significant effect, the (one-sided) power spectral density of the force noise acting on the mechanical system is given by:

$$S_{ff} = \frac{4k_B T m\omega}{Q} + 2\hbar^2\eta. \tag{1}$$

where k_B is the Boltzmann constant, T is the temperature, m is the mass, ω the angular frequency, Q is the mechanical quality factor. η is a diffusion constant associated to spontaneous localization, and can be calculated explicitly for the most known models. For CSL, it is given by the following expression

$$\eta = \frac{2\lambda}{m_0^2} \iint \mathrm{d}^3\mathbf{r}\,\mathrm{d}^3\mathbf{r}'\, \exp\left(-\frac{|\mathbf{r}-\mathbf{r}'|^2}{4r_C^2}\right) \frac{\partial\varrho(\mathbf{r})}{\partial z}\frac{\partial\varrho(\mathbf{r}')}{\partial z'} \tag{2}$$

$$= \frac{(4\pi)^{\frac{3}{2}}\,\lambda r_C^3}{m_0^2} \int \frac{\mathrm{d}^3\mathbf{k}}{(2\pi)^3}\, k_z^2\, e^{-\mathbf{k}^2 r_C^2}\, |\tilde{\varrho}(\mathbf{k})|^2 \tag{3}$$

with $\mathbf{k} = (k_x, k_y, k_z)$, $\tilde{\varrho}(\mathbf{k}) = \int d^3\mathbf{x}\, e^{i\mathbf{k}\cdot\mathbf{r}}\, \varrho(\mathbf{r})$ and $\varrho(\mathbf{r})$ the mass density distribution of the system. In the expressions above m_0 is the nucleon mass and r_C and λ are the free parameters of CSL. The typical values proposed in CSL literature are $r_C = 10^{-7}$ m and 10^{-6} m, while for λ a wide range of possible values has been proposed, which spans from the GRW value $\lambda \approx 10^{-16}$ Hz [1, 2] to the Adler value $\lambda \approx 10^{-8\pm 2}$ Hz at $r_C = 10^{-7}$ m [7]. The possibility for such non-interferometric tests, which aim to directly test the non-thermal noise predicted by collapse models has been pointed out first by Bahrami et al. [9] and the ideas has been picked-up rapidly by many others [10–14].

An experiment looking for CSL-induced noise has to be designed in order to maximize the signal to noise ratio between the CSL term and the thermal noise. From Eq. (1) it follows that this is achieved by lowest possible temperature T, lowest possible damping time or linewidth, $1/\tau = \omega/Q$, and highest possible η/m ratio. The first two conditions express the requirement of lowest possible power exchange with the thermal bath, the third condition is inherently related to the details of the specific model.

For CSL we can distinguish two relevant limits. When the characteristic size L of the system is small, $L \ll r_C$, then the CSL field cannot resolve the internal structure of the system, and one finds $\eta/m \propto m$. When the characteristic length of the system in the direction of motion L is large, $L \gg r_C$, then $\eta/m \propto \rho/L$, where ρ is the mass density [10, 12, 14]. The expressions in the two limits imply that, for a well defined characteristic length r_C, the optimal system is a plate or disk with thickness $L \sim r_C$ and the largest possible density ρ.

Among other models proposed in literature, we mention the gravitational Diosi-Penrose (DP) model, which leads to localization and diffusion similarly to CSL. The diffusion constant η_{DP} is given by [12]:

$$\eta_{DP} = \frac{G\rho m}{6\sqrt{\pi}\hbar}\left(\frac{a}{r_{DP}}\right)^3, \tag{4}$$

where a is the lattice constant and G is the gravitational constant, so that he ratio η_{DP}/m depends only on the mass density. Unlike CSL, there is no explicit dependence on the shape or size of the mechanical system.

2.2 Cantilevers and Other Clamped Resonators

Experiments based on ultrasensitive cryogenic cantilevers have been historically the first serious attempt to bound collapse models using diffusive mechanical experiments. The micro/nanocantilever employed in these experiments are optimized devices developed in the context of atomic force microscopy. They are characterized by low stiffness, relatively low frequency $f_0 \sim$ kHz and high Q factors in the range $10^5 - 10^7$. Operation at millikelvin temperature has been enabled by the use of SQUIDs for detection.

Current cantilever-based experiments bound the CSL collapse rate to be lower than 10^{-8} Hz at $r_C = 10^{-7}$ m, and 10^{-10} Hz at $r_C = 10^{-6}$ m, values which are already partially excluding the Adler parameters [14, 15].

It is not easy to push much further the current limits. Operation at lower temperature appears challenging due to increasing thermalization problems, while mechanical Q can be hardly improved over current values due to clamping losses. It can be noticed that, unlike cantilevers, micromembranes with much higher quality factor up to 10^9 have been demonstrated using an optimized design to suppress clamping losses [16]. However, these outstanding values are obtained only in high stress membrane at relatively high frequency ~0.1–1 MHz. The ratio Q/f_0 is not improved by this trick. Finally, pushing micromechanical systems to lower frequencies is possible but this approach has not been much investigated so far.

A novel route towards a significant improvement, specifically valid for the CSL model, has been recently proposed. The idea is to optimize the shape of a test mass to be attached on the cantilever, in order to maximize the effect at a given value of r_C [17]. The proposed optimized shape is a multilayer structure, where many different layers of two alternate materials with large difference in mass density are stacked together. This configuration is predicted to enhance the effect of CSL for $r_C \lesssim 3d$, where d is the layer thickness, at the expense of reducing the effect at larger r_C. First experiments in this direction have been able to bound the CSL collapse rate well below 10^{-9} Hz at $r_C = 10^{-7}$ m and are thus close to exclude completely the parameter range proposed by Adler [18].

2.3 Levitated Particles

One of the most promising approaches towards a significant leap forward in the achievable sensitivity to spontaneous collapse effects is by levitation of nanoparticles or microparticles. The main benefits of levitation are the absence of clamping mechanical losses and wider tunability of mechanical parameters. In addition, several degrees of freedom can be exploited, either translational or rotational [13, 19, 20]. This comes at the price of higher complexity, poor dynamic range and large nonlinearities, which usually require active feedback stabilization over multiple degrees of freedom. However, levitated systems hold the promise of much better isolation from the environment, therefore higher quality factor. One relevant example, in the macroscopic domain, is the space mission LISA Pathfinder (LPF), currently setting the strongest bound on collapse models over a wide parameter range [21]. LPF test masses were nominally in free-fall, but a complex electrostatic system was used to control the spacecraft and the laser readout position with respect to the test masses. This makes LPF substantially similar to an electrostatically levitated system, with the advantage that the near free-fall condition achievable in space allows to operate with very low electrostatic coupling, thus minimizing the effect of external disturbances.

Several levitation methods for micro/nanoparticles are currently being investigated. The most developed is optical levitation using force gradients induced by laser fields, the so called optical tweezer approach [22]. While this is a very effective and flexible approach to trap nanoparticles, in this context it is inherently limited by two factors: the relatively high trap frequency, in the order of 100 kHz, and the high internal temperature of the particles, induced by laser power absorption, which leads ultimately to strong thermal decoherence. Alternative approaches have to be found, featuring lower trap frequency and low or possible null power dissipated in the levitated particle. The two possible classes of techniques are electrical levitation and magnetic levitation.

Electrical levitation has been deeply developed in the context of ion traps. The standard tool is the Paul trap, which allows to trap an ion, or equivalently a charged nanoparticle, using a combination of ac and dc bias electric fields applied through a set of electrodes [23]. The power dissipation is much lower than in the optical case, and the technology is relatively well-established. However, the detection of a nanoparticle in a Paul trap still poses some technological challenge.

This issue has been extensively investigated in a recent paper [24], specifically considering a nanoparticle in a cryogenic Paul trap in the context of collapse model testing. Three detection schemes have been considered: an optical cavity, an optical tweezer, and a all-electric readout based on SQUID. It was found that to detect the nanoparticle motion with good sensitivity, optical detection has to be employed. Unfortunately, optical detection is not easily integrated in a cryogenic environment, and leads to a nonnegligible internal heating and excess force noise. On the other hand, an all-electrical readout would potentially allow for a better ultimate test of collapse models, but at the price of a very poor detection sensitivity, which could make the experiment hardly feasible. The authors argue that a Paul-trapped nanoparticle, with an oscillating frequency of 1 kHz, cooled in a cryostat at 300 mK with an optical readout may be able to probe the CSL collapse rate down to 10^{-12} Hz at $r_C = 10^{-7}$ m. A SQUID-based readout, if viable, could theoretically allow to reach 10^{-14} Hz.

A recent experiment employing a nanoparticle in a Paul trap with very low secular frequencies at $\sim$100 Hz and low pressure has demonstrated ultranarrow linewidth $\gamma/2\pi = 82\ \mu$Hz [25]. This result has been used to set new bounds on the dissipative extension of CSL. This experiment may be able to probe the current limits on the CSL model in the near future, once it will be performed at cryogenic temperature and the main sources of excess noise, in particular bias voltage noise, will be removed.

Magnetic levitation, while less developed, has the crucial advantage of being completely passive. Furthermore the trap frequencies can be quite low, in the Hz range. Three possible schemes can be devised: levitation of a diamagnetic insulating nanoparticle with strong external field gradients [26, 27], levitation of a superconducting particle using external currents [28, 29], and levitation of a ferromagnetic particle above a superconductor [30].

The first approach has been recently considered in the context of collapse models [27]. The experiment was based on a polyethylene glycol microparticle levitated in the static field generated by neodymium magnets and optical detection. The experiment has been able to set an upper bound on the CSL collapse rate $\lambda < 10^{-6.2}$ Hz at

$r_C = 10^{-7}$ m, despite being performed at room temperature. A cryogenic version of this experiment should be able to approach the current experimental limits on CSL.

The second and third approach based on levitating superconducting or ferromagnetic particles are currently investigated by a handful of groups [18, 28–31], but no experiment has so far reached the experimental requirements needed to probe collapse models. However, a significant progress has been recently achieved: a ferromagnetic microparticle levitated above a type I superconductor (lead) and detected using a SQUID, has demonstrated mechanical quality factors for the rotational and translational rigid body mechanical modes exceeding 10^7, corresponding to a ring-down time larger than 10^4 s [18]. The noise in this experiment is still dominated by external vibrations. However, as the levitation is completely passive and therefore compatible with cryogenic temperatures, this appears as an excellent candidate towards near future improved tests of collapse models.

3 Bulk Heating Experiments

A different strategy towards testing the violation of energy conservation caused by collapse models is to search for anomalous heating of specific systems. The difference compared to mechanical experiments is that in the latter case one looks for an energy increase in an individual degree of freedom, which is a purely mechanical effect, while in the general case one looks for the total increase of internal energy (i.e. of temperature) of a macroscopic body, which is a thermal effect.

According to Adler [7], the heating of a macroscopic body due to CSL can be generally written as:

$$\frac{dE}{dt} = \frac{3}{4}\frac{\lambda\hbar^2 M}{r_C^2 m_N^2} \tag{5}$$

where M is the total mass of the system. This expression, initially derived for a gas of noninteracting particles, has been shown to be very general. For instance, it holds for standard solid state systems [33–35] and for nonstandard matter such as Fermi liquids [35].

An important caveat has to be pointed out in the case of condensed matter systems, in particular of solids. The λ factor has to be regarded as an effective value λ_{eff}, averaged over the frequencies of the internal (phononic) modes of the system. This average value is shown to be $\lambda_{\mathrm{eff}} \approx \lambda(\omega_0)$, where ω_0 is the frequency corresponding to phonons with wavelength $\sim r_C$ [33–35]. For typical solid matter, this corresponds to $\omega_0 \approx 10^{11}\ s^{-1}$. A consequence of this fact is that any bound from heating of solid matter would be evaded by a nonwhite CSL noise with a low-pass cutoff at frequency lower than ω_0 [36–38].

As the bulk heating scales with the mass, one possible experimental approach is to estimate this effect in astronomical objects. For instance by analyzing the intergalactic medium, mainly composed of cold hydrogen, one can infer a bound on the CSL collapse rate $\lambda < 10^{-8}$ Hz [7] at $r_C = 10^{-7}$ m (in the following of this chapter we will always assume $r_C = 10^{-7}$ m). More recently, it has been suggested that neutron stars can set much stronger bounds [39]. The actual bounds inferred from current observational data are however not yet competitive, at level $\lambda_{\text{eff}} < 10^{-7}$ Hz [35, 39], but speculative bounds based on the capabilities of future astronomical surveys suggest that much stronger bounds can be obtained in the future. Stronger bounds, at level $\lambda_{\text{eff}} < 10^{-10}$ Hz, can be inferred from the astronomical data on planets of the solar system, in particular Neptune [35]. This is essentially due to the very low temperature of these planets. An even better bound, at level $\lambda_{\text{eff}} < 10^{-11}$ Hz, can be inferred from the earth thermal balance, once primordial and radiogenic sources of Earth heat are very carefully taken into account [33]. A more speculative prospect is to test collapse models by evaluating their effect at cosmological level, for instance in the Cosmic Microwave Background.

Here, we focus instead on the possibility of detecting very small heating in controlled laboratory experiments. Bulk massive objects can be routinely cooled down to very low temperatures. Dilution refrigerators can be used to cooldown relatively massive objects down to $\sim$10 mK. The most massive object ever cooled in this way is probably the CUORE detector looking at neutrinoless beta decay [40], with a mass of $\sim$1 ton cooled to 10 mK. Much lower temperatures, even below 100 μK, can be reached by adiabatic nuclear demagnetization cryostats [41]. Here, the typical mass which can be cooled is of the order of several kilograms.

As the thermalization of any object becomes increasingly difficult at lower and lower temperature, a crucial requirement of these experiments is to suppress as much as possible any heat leak. The dominant residual heat sources are (i) vibrations, (ii) relaxation of internal stress and two-level systems and (iii) the background of radioactivity and muons from cosmic rays [41]. The first two sources can be efficiently suppressed by proper mechanical isolation, proper choice of materials and by waiting for long relaxation times. Overall, the best residual heat leak estimated in current experiments is of the order of 10^{-11} W/kg, limited by background muons, which corresponds to $\lambda_{\text{eff}} < 3 \times 10^{-11}$ Hz [33].

It is important to note that, due to the high penetration of cosmic muons, their background heating scales with the experimental mass and is therefore a fundamental barrier, unless the experiment is performed heavily underground. This idea has been considered recently by Mishra et al. [42], who have estimated the achievable upper limit on λ_{eff} which could be detected by an ideal cosmic-background-limited detector placed underground. It has been found that the shielding provided by the deepest existing underground laboratory (the China Jinping Underground Laboratory, placed at 6.7 km of "water-equivalent" depth) would be sufficient to test the CSL model down to the GRW parameters, i.e $\lambda_{\text{eff}} \approx 10^{-16}$ Hz. Slightly worse performance is expected by operating in alternative sites, such as the Gran Sasso Laboratory in Italy.

Of course such an experiment would require a systematic and very efficient suppression of any parasitic heating source, such as vibrations or internal relaxation, by several orders of magnitude. This appears a tough challenge, which is however purely technical and not related to fundamental limits. The technology developed for existing underground cryogenic experiments looking for Dark Matter or neutrinoless double beta decay, such as CUORE [40], is probably already good enough to push the current bounds by 1–2 orders of magnitude. However, a specific experimental design is needed, specifically optimized to detect a constant heating source such as CSL. In particular, thermometry with very good absolute accuracy is needed, in contrast with existing detectors which are optimized for very high responsivity in order to resolve individual high energy events.

4 Cold Atoms and Condensates

Cold atoms represent another possible system to detect effects related with collapse models, due to (i) the very low kinetic temperatures that can be achieved, which can be as low as a few pK and (ii) the flexibility of these systems which allow the realization of a variety of quantum states involving a relatively large number of atoms, in particular Bose-Einstein Condensates (BEC). An obvious drawback, which partially compensates these advantages, is the much lower density, i.e. a comparatively lower total number of atoms compared to solid state systems.

Some first investigations of using cold atoms in order to probe collapse models, in particular CSL, have been reported in literature [43–45] and include an outlook on future improvements.

The first approach considered in Ref. [43] consists in analyzing the lifetime of BEC condensates. The spontaneous heating induced by CSL results into an exponential decay of the ground state population, which can be bounded by experiments. Analysis of current experiments yielded a bound on the CSL collapse rate $\lambda < 10^{-7}$ Hz at the standard $r_C = 10^{-7}$ m. The authors remark that the analysis is provisional, and that new experiments specifically tailored to estimate the rate of energy increase will significantly improve the bounds. Specific improvements will be the use of heavy atomic mass (like cesium), lower background of foreign atoms, suppression of three-body recombination, and a sufficiently high barrier to eliminate evaporative cooling.

A second approach was investigated by Bilardello et al. [44], by analyzing an experiment performed by Kasevich et al. [45] with a diluted cloud of ultracold rubidium atoms. The last stage of this experiment consisted in a delta-kick optical-lensing cooling, which enabled free evolution of the cloud on a time scale of seconds at an extremely low temperature below 100 pK. This bounds the CSL collapse rate to $\lambda < 5 \times 10^{-8}$ Hz at $r_C = 10^{-7}$ m, slightly improving BEC limits. The authors of the experiment estimate that this technique can be improved by 2–3 orders of magnitude before reaching fundamental limits imposed by diffraction limited collimation temperature [45]. The analysis performed in [44] also shows that cold atoms experiments are particularly efficient in testing non Markovian [46–49] and dissipative [50–52] extensions of the CSL model.

A third option is to use cold atoms to perform interferometric experiments. This will be discussed in the next section.

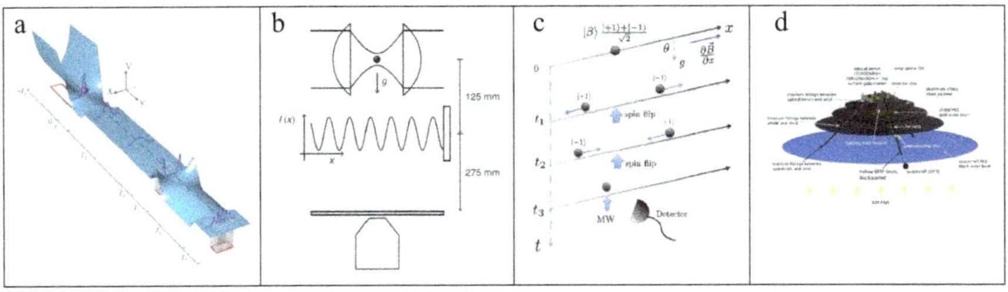

Fig. 2 Illustration of some of the proposed schemes for matterwave interferometry with nano- and micro-particles to test the quantum superposition principle directly, and therefore also collapse models. **a** The cryogenic skatepark for a single superconducting micro-particle (adapted from Ref. [53]); **b** The nanoparticle Talbot interferometer (adapted from Ref. [54]); **c** The Ramsey scheme addressing the electron Spin of a NV-centre diamond coupled to an external magnetic field gradient $(\partial B/\partial x)$ (adapted from Ref. [55]); **d** The adaptation of an interferometer at a free falling satellite platform in space to allow form longer free evolution times (adapted from Ref. [56])

5 Matter-Wave Interferometry

Matterwave interferometry is directly testing the quantum superposition principle. Relevant for mass-scaling collapse models, such as CSL, are matterwave interferometers testing the maximal macroscopic extend in terms of mass, size and time of spatial superpositions of single large-mass particles. Such beautiful, but highly challenging experiments have been pushed by Markus Arndt's group in Vienna to impressive particle masses of 10^4 atomic mass units (amu), which is still not significantly challenging CSL. More details can be found in chapter **XXX** of this volume. Therefore the motivation remains to push matterwave interferometers to more macroscopic systems. Here we will discuss some possible ideas, some of which are summarized in Fig. 2 [60–63]. Predicted bounds on collapse models set by large-mass matterwave interferometers are worked out in detail in [51, 57].

As usual in open quantum system dynamics treatments, non-linear stochastic extensions of the Schrödinger equation on the level of the wavefunction [58] correspond to a non-uniquely defined master equation on the level of the density matrix ρ to describe the time evolution of the quantum system, say the spatial superposition across distance $|x-y|$, where the conserving von Neumann term $\partial\rho_t(x,y)/\partial t = -(i/\hbar)[H,\rho]$, is now extended by a Lindblad operator L term:

$$\frac{\partial\rho_t(x,y)}{\partial t} = -\frac{i}{\hbar}[H,\rho_t(x,y)] + L\rho_t(x,y), \tag{6}$$

where H is the Hamilton operator of the quantum system and different realisations of a Lindblad operator are used to describe both standard decoherence (triggered by the immediate environment of the quantum system) [59] as well as spontaneous collapse of the wavefunction triggered by the universal classical noise field as predicted by collapse models.

Now the dynamics of the system is very different with and without the Lindbladian, where with the Lindbladian the unitary evolution breaks down and the system dynamics undergoes a quantum-to-classical transition witnessed by a vanishing of the fringe visibility of the matterwave interferometer. In the state represented by the density matrix the off-diagonal terms vanish as the system evolves according to the open system dynamics, the coherence/superposition of that state is lost. The principal goal of interference experiments with massive particles is then to explore and quantify the relevance of the $(L\rho_t(x, y))$-term—as collapse models predict a break down of the quantum superposition principle for a sufficient macroscopic system. An intrinsic problem is the competition with known and unknown environmental decoherence mechanisms, if a visibility loss is observed. However solutions seem possible.

In order to further increase the macroscopic limits in interference some ambitious proposals have been made utilizing nano- and micro-particles, c.f. Fig.(2). The main challenge is to allow for a long enough free evolution time of the prepared quantum superposition state in order to be sensitive to the collapsing effects. The free evolution—the spatial spreading of the wavefunction $\Psi(r, t)$ with time—according to the time-dependent Schrödinger equation with the potential $V(r) = 0$,

$$\frac{\partial}{\partial t}\Psi(r, t) = -i\frac{\hbar}{2m}\nabla^2\Psi(r, t), \tag{7}$$

describes a diffusive process for probability amplitudes similar to a typical diffusion equation with the imaginary diffusion coefficient $(-i\hbar/2m)$. Therefore the spreading of $\Psi(r, t)$ scales inversely with particle mass m. For instance for a 10^7 amu particle it already takes so long to show the interference pattern in a matterwave experiment that the particle would significantly drop in Earth's gravitational field, in fact it would drop on the order of 100 m. This requires a dramatic change in the way large-mass matterwave interferometry experiments have to be performed beyond the mass of 10^6 amu [54].

Different solutions are thinkable. One could of course envisage building a 100 m fountain, but that seems very unfeasible also given that no sufficient particle beam preparation techniques exist (and don't seem to be likely to be developed in the foreseeable future) to enable the launch and detection of particles in the mass range in question over a distance of 100 m. One can consider to levitate the particle by a force field to compensate for the drop in gravity, but here we face a high demand on the fluctuations of that levitating field, which have to be small compared to the amplitudes of the quantum evolution. This requirement does not appear to be feasible with current technology. A maybe possible option is to coherently boost/accelerate the evolution of the wavefunction spread by a beam-splitter operation. The proposals in Refs. [53, 55] are such solutions, which are still awaiting their technical realisation for large masses. A more realistic alternative, given current technical capabilities, is to allow for long enough free evolution by freely fall the whole interferometer apparatus in a co-moving reference frame with the particle. This is the idea of the MAQRO proposal, a dedicated satellite mission in space to perform large-mass matterwave interference experiments with micro- and nano-particles [56].

Another interesting approach is to consider the use of cold or ultra-cold ensembles of atoms such as cloud in a magneto optical trap (MOT) or an atomic Bose-Einstein Condensate (BEC) as also there we find up to 10^8 atoms of alkali species such as rubidium or caesium. On closer look it turns out that such weakly interacting atomic ensembles are not of immediate use for the purpose to test macroscopic quantum superpositions in the context of collapse model test. For instance, a crucial property for testing the CSL model is the mass-proportional (number of particles N, more precisely the number of nucleons: protons and neutrons in the nuclei of the atoms) amplification which in principle can even go with N^2. This effect can be seen in Eq. (2), and can be thought as arising from the classical collapse noise (treated as a wave with correlation length r_c) coherently scattering off the particle in the quantum superposition state. Naively, if the CSL noise is collapsing the wavefunction of only one of the constituent nucleons, then the total wavefunction of the whole composite object collapses. While this holds for a solid nanoparticle consisting of many atoms (and therefore nucleons), it is not the case for a weakly interacting atomic ensemble. If one atom is collapsing then the total atomic wavefunction remains intact and the one atom is lost from the ensemble.

This may change if the atoms in the cold or ultra-cold ensemble can be made strongly interacting, without running into the complications of chemistry which may forbid condensation of the atomic—then molecular—cloud at all. However there is hope to circumvent this problem by means of quantum optical state preparation techniques applied after a BEC has been formed. For instance, collective NOON or squeezed states, featuring macroscopic entanglement between individual atoms, would enable N and even N^2 scaling in the fashion fit for testing wavefunction collapse. This approach is extremely challenging, and is discussed in detail in Ref. [60].

A different scenario might arise if the physical mechanism responsible for the collapse of the wavefunction, which remains highly speculative at present, is in any way related to gravity [61], then there might be hope that atomic ensembles even in the weakly interacting case can be used to test CSL-type models. The condition to fulfil is that the atomic ensemble is interacting gravitationally strong enough so that it acts collectively under collapse, even if just a single constituent atom (nucleon) is affected by the collapsing effect. That hope is possibly very weak.

6 Some Concluding Remarks

We have discussed avenues for non-interferometric and interferometric tests of the linear superposition principle of quantum mechanics in direct comparison to predictions from collapse models which break the linear/unitary evolution of the wavefunction. As matters stand both non-interferometric and interferometric set already bounds on the CSL collapse model, while those from non-interferometric tests are stronger by orders of magnitude. The simple reason lies in the immense difficulty to experimentally generate macroscopic superposition states, however a number of proposals have been made and experimentalists are set to approach the challenge.

We want to close by mentioning that there are possibly other experimental platforms which could set experimental bounds on collapse models and it would be of interest to study those in detail. Collapse models predict a universal classical noise field to fill the Universe and in principle couple to any physical system. In the simplest approach the experimental test particle can be regarded as a two-level system, as typically described in quantum optics. Then the collapse noise perturbs the two-level system and emissive broadening and spectral shifts can be expected, unfortunately out of experimental reach at the moment [62]. The minuscule collapse effect on a single particle (nucleon) needs some sort of amplification mechanism which usually comes with an increase of the number of constituent particles. However, ultra-high precision experiments have improved a lot in recent years. For instance much improved ultra-stable Penning ion traps are used to measure the mass of single nuclear particles, such as the electron, proton, and neutron, with an ultra-high precision to test quantum electrodynamics predictions [63]. In principle also here the effect of collapse models should become apparent. Any theoretical predictions are difficult as relativistic versions of collapse models still represent a serious formal challenge [64–66]. Other high potentials for testing collapse are ever more precise spectroscopies of simple atomic species with analytic solutions such as transitions in hydrogen [67] and needless to say atomic clocks [68].

As tests move on to set stronger and stronger bounds, we have to remain open to actually find something new. It is so easy to disregard tiny observed effects as unknown technical noise. In the case of direct testing collapse noise it is a formidable theoretical challenge to think about possible physics responsible for collapse, satisfying the constraints given by the structure of the collapse equation: the noise has to be classical, stochastic and nonlinear. Such concrete physics models will predict a clear frequency fingerprint, should we ever observe the collapse noise field.

Acknowledgements We acknowledge support from the EU project TEQ (grant agreement 766900). We would like to thank our collaborators, discussion and debating partners on the topic of experimental testing collapse models for many enlightening discussions over the years: Angelo Bassi, Steve Adler, Mauro Paternostro, TP Singh, Marko Toroš, Matteo Carlesso, Giulio Gasbarri, Sandro Donadi, Luca Ferialdi, Marco Bilardello, Catalina Curceanu, Andrew Briggs, Caslav Bruckner, Markus Arndt, Daniel Bedingham, Ward Stuyve, Kinjalk Lochan, Michael Drewsen, Peter F. Barker, Antonio Pontin, Anis Rahman, James Bateman, Sougato Bose, Myungshik Kim, Mohammad Bahrami, Stefan Nimmrichter, Klaus Hornberger, Andrew Steane, Oriol Romero-Isart and Tjerk Oosterkamp. The paper is dedicated to the late Gian Carlo Ghirardi, who had a great idea and fought for it all his life.

References

1. G.C. Ghirardi, A. Rimini, and T. Weber, Phys. Rev. D **34**, 470 (1986).
2. G.C. Ghirardi, P. Pearle, and A. Rimini, Phys. Rev. A **42**, 78 (1990).
3. A. Bassi, and G. C. Ghirardi, Phys. Rep. **379**, 257 (2003).

4. A. Bassi, K. Lochan, S. Satin, T. P. Singh, and H. Ulbricht, Rev. Mod. Phys. **85**, 471 (2013).
5. B. Collett and P. Pearle, Found. Phys. **33**, 1495 (2003).
6. S.L. Adler, J. Phys. A **38**, 2729 (2005).
7. S.L. Adler, J. Phys. A **40**, 2935 (2007).
8. K. Hornberger, S. Gerlich, P. Haslinger, S. Nimmrichter and M. Arndt, Rev. Mod. Phys. **84**, 157 (2012).
9. M. Bahrami, M. Paternostro, A. Bassi, and H. Ulbricht, Phys. Rev. Lett. **112**, 210404 (2014).
10. S. Nimmrichter, K. Hornberger, and K. Hammerer, Phys. Rev. Lett. **113**, 020405 (2014).
11. S. Bera, B. Motwani, T.P. Singh, and H. Ulbricht, Sci. Rep. **5**, 7664 (2015).
12. L. Diosi, Phys. Rev. Lett. **114**, 050403 (2015).
13. D. Goldwater, M. Paternostro, and P.F. Barker, Phys. Rev. A **94**, 010104 (2016).
14. A. Vinante, M. Bahrami, A. Bassi, O. Usenko, G. Wijts, and T.H. Oosterkamp, Phys. Rev. Lett. **116**, 090402 (2016).
15. A. Vinante, R. Mezzena, P. Falferi, M. Carlesso, and A. Bassi, Phys. Rev. Lett. **119**, 110401 (2017).
16. D. Mason, J. Chen, M. Rossi, Y. Tsaturyan, and Albert Schliesser, Nature Phys. **15**, 745 (2019).
17. M. Carlesso, A. Vinante, and A. Bassi, Phys. Rev. A **98**, 022122 (2018).
18. A. Vinante, M. Carlesso, A. Bassi, A. Chiasera, S. Varas, P. Falferi, B. Margesin, R. Mezzena, H. Ulbricht, arXiv:2002.09782
19. B. Schrinski, B. A. Stickler, and K. Hornberger, J. Opt. Soc. Am. B **34**, C1 (2017).
20. M. Carlesso, M. Paternostro, H. Ulbricht, A. Vinante, A. Bassi, New J. Phys. **20**, 083022 (2018).
21. M. Carlesso, A. Bassi, P. Falferi, and A. Vinante, Phys. Rev. D **94**, 124036 (2016).
22. A. Ashkin, J.M. Dziedzic, J.E. Bjorkholm, S. Chu, Opt. Lett. **11**, 288(1986).
23. W. Paul, Rev. Mod. Phys. **62**, 531 (1990).
24. A. Vinante, A. Pontin, M. Rashid, M. Toroš, P.R. Barker, H. Ulbricht, Phys. Rev. A **100**, 012119(2019).
25. A. Pontin, N.P. Bullier, M. Toroš, and P. Barker, Phys. Rev. Research **2**, 023349 (2020).
26. B.R. Slezak, C.W. Lewandowski, J-F. Hsu and B. D'Urso, New J. Phys. **20**, 063028 (2018).
27. Di Zheng et al., Phys. Rev. Research **2**, 013057 (2020).
28. O. Romero-Isart, L. Clemente, C. Navau, A. Sanchez, and J.I. Cirac, Phys. Rev. Lett. **109**, 147205 (2012).
29. B. van Waarde, *The lead zeppelin: a force sensor without a handle*, PhD Thesis, Leiden University (2016).
30. J. Prat-Camps, C. Teo, C.C. Rusconi, W. Wieczorek, and O. Romero-Isart, Phys. Rev. Appl. **8**, 034002 (2017).
31. C. Timberlake, G. Gasbarri, A. Vinante, A. Setter, H. Ulbricht, Appl. Phys. Lett. **115**, 224101 (2019).
32. A. Vinante, P. Falferi, G. Gasbarri, A. Setter, C. Timberlake, H. Ulbricht, Phys. Rev. Applied **13**, 064027 (2020)
33. S.L. Adler and A. Vinante, Phys. Rev. A **97**, 052119 (2018).
34. M. Bahrami, Phys. Rev. A **97**, 052118 (2018).
35. S.L. Adler, A. Bassi, M. Carlesso, and A. Vinante, Phys. Rev. A **99**, 103001 (2019).
36. S.L. Adler, A. Bassi, J. Phys. A **40**, 15083 (2007).
37. S.L. Adler, A. Bassi, S. Donadi, J. Phys. A **46**, 245304 (2013).
38. M. Carlesso, L. Ferialdi, A. Bassi, Eur. Phys. J. D **72**, 159 (2018).
39. A. Tilloy and T.M. Stace, Phys. Rev. Lett. **123**, 080402 (2019).
40. See URL https://cuore.lngs.infn.it/
41. F. Pobell, *Matter and Methods at Low Temperatures*, 3rd ed.(Springer, Berlin, 2007).
42. R. Mishra, A. Vinante, T.P. Singh, Phys. Rev. D **98** 052121 (2018).
43. F. Lalöe, W.J. Mullin, and P. Pearle, Phys. Rev A **90**, 052119 (2014).
44. M. Bilardello, S. Donadi, A. Vinante, and A. Bassi, Physica A **462**, 764 (2016).
45. T. Kovachy, J.M. Hogan, A. Sugarbaker, S.M. Dickerson, C.A. Donnelly, C. Overstreet, and M.A. Kasevich, Phys. Rev. Lett. **114**, 143004 (2015).
46. S.L. Adler and A. Bassi, J. Phys. A **40**, 15083 (2007)

47. A. Bassi and L. Ferialdi, Phys. Rev. A **80**, 012116 (2009)
48. M. Toroš, G. Gasbarri, and A. Bassi, Phys. Lett. A **381**, 3921 (2017)
49. M. Carlesso, L. Ferialdi, A. Bassi, Eur. Phys. J. D **72**, 159 (2018)
50. A. Smirne and A. Bassi, Sci. Rep. **5**, 12518 (2015).
51. M. Toroš, G. Gasbarri, and A. Bassi, Phys. Lett. A **381**, 3921 (2017).
52. J. Nobakht, M. Carlesso, S. Donadi, M. Paternostro, and A. Bassi, Phys. Rev. A **98**, 042109 (2018).
53. H. Pino, J. Prat-Camps, K. Sinha, B. P. Venkatesh, and O. Romero-Isart, Quantum Sci. Technol. **3**, 025001 (2018).
54. J. Bateman, S. Nimmrichter, K. Hornberger, and H. Ulbricht, Nat. Commun. **5**, 4788 (2014).
55. C. Wan, M. Scala, G. W. Morley, ATM A. Rahman, H. Ulbricht, J. Bateman, P. F. Barker, S. Bose, and M. S. Kim, Phys. Rev. Lett. **117**, 143003 (2016).
56. R. Kaltenbaek, et al. EPJ Quantum Technol. **3**, 5 (2016).
57. M. Toroš, and A. Bassi, J. Phys. **A 51**, 115302 (2018).
58. C. Gardiner, and P. Zoller. *Quantum noise: a handbook of Markovian and non-Markovian quantum stochastic methods with applications to quantum optics.* Vol. 56. Springer Science and Business Media (2004).
59. H. P. Breuer, and F. Petruccione. *The theory of open quantum systems.* Oxford University Press (2002).
60. M. Bilardello, A. Trombettoni, and A. Bassi, Phys. Rev. A **95**, 032134 (2017).
61. M. Bahrami, A. Smirne, and A. Bassi, Phys. Rev. **A 90**, 062105 (2014).
62. M. Bahrami, A. Bassi, and H. Ulbricht, Phys. Rev, **A 89**, 032127 (2014).
63. S. Sturm, F. Köhler, J. Zatorski, A. Wagner, Z. Harman, G. Werth, W. Quint, C. H. Keitel, and K. Blaum, Nature **506**, 467 (2014).
64. R. Tumulka, J. Stat. Phys **125**, 821 (2006).
65. D. Bedingham, Detlef Dürr, G.C. Ghirardi, S. Goldstein, R. Tumulka, and N. Zanghí, J. Stat. Phys **154**, 623 (2014).
66. C. Jones, T. Guaita, and A. Bassi, arXiv:1907.02370 (2019).
67. M. Weitz, A. Huber, F. Schmidt-Kaler, D. Leibfried, and T. W. Hänsch, Phys. Rev. Lett. **72**, 328 (1994).
68. E. Oelker, R. B. Hutson, C. J. Kennedy, L. Sonderhouse, T. Bothwell, A. Goban, D. Kedar et al., preprint Nature Photonics 13, 714–719 (2019).